Magnetic Resonance Technology
Hardware and System Component Design

New Developments in NMR

Editor-in-Chief:
Professor William S. Price, *University of Western Sydney, Australia*

Series Editors:
Professor Bruce Balcom, *University of New Brunswick, Canada*
Professor István Furó, *Industrial NMR Centre at KTH, Sweden*
Professor Masatsune Kainosho, *Tokyo Metropolitan University, Japan*
Professor Maili Liu, *Chinese Academy of Sciences, Wuhan, China*

Titles in the Series:
1: Contemporary Computer-Assisted Approaches to Molecular Structure Elucidation
2: New Applications of NMR in Drug Discovery and Development
3: Advances in Biological Solid-State NMR
4: Hyperpolarized Xenon-129 Magnetic Resonance: Concepts, Production, Techniques and Applications
5: Mobile NMR and MRI: Developments and Applications
6: Gas Phase NMR
7: Magnetic Resonance Technology: Hardware and System Component Design

How to obtain future titles on publication:
A standing order plan is available for this series. A standing order will bring delivery of each new volume immediately on publication.

For further information please contact:
Book Sales Department, Royal Society of Chemistry, Thomas Graham House, Science Park, Milton Road, Cambridge, CB4 0WF, UK
Telephone: +44 (0)1223 420066, Fax: +44 (0)1223 420247
Email: booksales@rsc.org
Visit our website at www.rsc.org/books

Magnetic Resonance Technology
Hardware and System Component Design

Edited by

Andrew G Webb
Leiden University Medical Center, Leiden, The Netherlands
Email: a.webb@lumc.nl

ROYAL SOCIETY OF CHEMISTRY

THE QUEEN'S AWARDS FOR ENTERPRISE: INTERNATIONAL TRADE 2013

New Developments in NMR No. 7

Print ISBN: 978-1-78262-359-5
PDF eISBN: 978-1-78262-387-8
EPUB eISBN: 978-1-78262-817-0
ISSN: 2044-253X

A catalogue record for this book is available from the British Library

Published by The Royal Society of Chemistry,
Thomas Graham House, Science Park, Milton Road,
Cambridge CB4 0WF, UK

Registered Charity Number 207890

For further information see our web site at www.rsc.org

Printed in the United Kingdom by CPI Group (UK) Ltd, Croydon, CR0 4YY, UK

Preface

This book is intended for the NMR or MRI operator who sits down at the console for the first, second or even one hundredth time and wonders "why is this component designed this way, what does this button REALLY do, why didn't they do it this way?" Historically, many of the pioneers of magnetic resonance (Bloch, Gutowsky, Schlichter, Lauterbur and Mansfield, to name a few) also constructed their own equipment. The seed of a new experiment required a deep knowledge of the quantum mechanical behaviour of the spin system, but to actually perform an experiment it was necessary to build the equipment oneself, and not to get on the cell phone to complain to the relevant vendor. While acknowledging that the days of being able to fix a system with a sheath of circuit diagrams and a well aimed soldering iron are regrettably over, we believe that understanding how a system works expands the opportunities for new science and engineering to flourish.

This book was inspired by the seminal books of authors such as Eiichi Fukushima and David Hoult: volumes that provide a deep level of understanding coupled with unbridled enthusiasm and excitement for the subject matter. In a world of ever-increasing specialization, it is perhaps worth noting that these eminent scientists are also renowned in the mountain climbing and operatic worlds, respectively. To this end, a mixture of academic and industrial scientists were invited to contribute chapters, the latter to provide insights that too commonly are not amply represented in the academic literature. Regrettably, the list is male dominated: as one who has personally and professionally been inspired by the contributions of female engineers in many related fields, this is a sign of a too-slowly changing society, at all levels of education. Should further editions evolve, we hope that the author contributions become naturally more balanced.

New Developments in NMR No. 7
Magnetic Resonance Technology: Hardware and System Component Design
Edited by Andrew G Webb

Published by the Royal Society of Chemistry, www.rsc.org

Chapter 1 provides a summary of the phenomenon of magnetic resonance, linked to the relevant hardware components, described more fully in the later chapters of this book. The appendices provide an outline of mathematical constructions and approaches, including the Biot–Savart law and spherical harmonics, which are widely used in the design of many of the different hardware components covered in the specific chapters.

Chapter 2 covers the principles of designing superconducting magnets for magnetic resonance. In high resolution NMR, field strengths have broken through the 23.5 tesla ceiling required for 1 GHz operation, and new hybrid magnets with low temperature superconductors supplemented by high temperature superconducting inserts will soon result in field strengths above 30 tesla. Similarly, human-sized MRI magnets of 11.7 tesla are now available, coupled with new designs for making high field magnets of similar size and footprint to conventional lower field systems.

Chapter 3 describes the wide variety of radiofrequency coils that have been developed for high resolution NMR, and human and animal MRI. Basic electromagnetic principles behind the geometries used, methods of impedance matching, multiple-frequency tuning, active detuning and other concepts are all explained from a basic level. The ubiquitous use of multi-element receive arrays and increasing use of transmit arrays are also covered, with a final section devoted to new types of RF coil used for very high field human MRI.

Chapter 4 is concerned with the design of shim coils, which are used to maximize the static magnetic field homogeneity within the volume of interest for the particular magnetic resonance experiment. Shim coil design based on spherical harmonics is described: these coils are used for high resolution MR systems as well as human MRI systems. The chapter also considers alternative designs that are particularly applicable to human MRI at high field.

Chapter 5 describes the design of magnetic field gradients that enable spatial information to be encoded for MRI, and also form the basis of coherence selection in high resolution NMR, and molecular diffusion measurements in both solution state NMR and human MRI. Specific examples of gradient coil design are outlined including strong small diameter gradients for animal imaging, as well as the strongest yet designed for human use, the so-called connectome gradients.

Chapter 6 concentrates on the basis of designing radiofrequency power amplifiers, which are used to provide power to the RF coils. The basic operation of a MOSFET amplifier is used to provide detailed analysis of amplifier behaviour and design. Different types of amplifier are considered, including new developments in current source and low output impedance amplifiers, and the pros and cons of each design discussed.

Chapter 7 provides an outline of the receive chain of the magnetic resonance system. The system is analyzed in terms of minimizing the noise figure of the chain. Specific designs of preamplifiers and quadrature hybrids are outlined. Different forms of data sampling, including the use of undersampling, are discussed as well as the future use of optical and wireless techniques for massively parallel receive systems.

Chapter 8 describes methods and applications of electromagnetic simulations for magnetic resonance. Interactions of the human body with the main magnetic field, magnetic field gradients and electric fields produced by the RF coil are discussed in detail. The combination of electromagnetic simulations with the Bloch equations provides a platform for simulating reconstructed images produced by different imaging sequences.

Andrew Webb

Contents

Chapter 1 The Principles of Magnetic Resonance, and Associated Hardware 1
Andrew Webb

1.1 Introduction 1
1.2 The Superconducting Magnet and Nuclear Polarization 5
1.3 The Transmitter Coil to Generate Radiofrequency Pulses 7
1.4 Precession 10
1.4.1 Chemical Shift 10
1.4.2 Scalar Coupling 12
1.4.3 Relaxation Processes 13
1.5 The Receiver Coil for Detecting the MR Signal 15
1.6 The Receiver: Signal Demodulation, Digitization and Fourier Transformation 16
1.6.1 Receiver Electronics 16
1.6.2 Signal Processing 17
1.7 Shim Coils 18
1.8 Gradient Coils 19
1.8.1 Crusher Gradients to Dephase Transverse Magnetization 21
1.8.2 Gradients for Coherence Selection in High-Resolution NMR 22
1.8.3 Measurements of Apparent Diffusion Coefficients Using Gradients 23
1.8.4 Gradient-Based Shimming 24
1.8.5 Gradients in MRI 25

New Developments in NMR No. 7
Magnetic Resonance Technology: Hardware and System Component Design
Edited by Andrew G Webb

Published by the Royal Society of Chemistry, www.rsc.org

1.9 The Deuterium Lock Channel and Field Monitoring 26
1.10 Magic Angle Spinning Solid-State NMR: Principles and Instrumental Requirements 28
1.11 Magnetic Resonance Imaging: Principles and Instrumental Requirements 31
Appendices 35
References 46

Chapter 2 Magnets **48**
Rory Warner and Simon Pittard

2.1 Introduction 48
2.2 Magnet Types 49
2.2.1 Air-Cored Resistive Magnets 49
2.2.2 Permanent Magnets 50
2.2.3 Iron-Cored Resistive Magnets 50
2.2.4 Iron-Cored Superconducting Magnets 50
2.2.5 Superconducting Cylindrical Magnets 51
2.3 Magnetic Field Generation 51
2.3.1 Basic Physics 51
2.3.2 Field Homogeneity 53
2.3.3 Magnetic Shielding 56
2.3.4 System Shielding from External Interference 56
2.3.5 Magnetic Field Shimming 58
2.4 Superconductivity 60
2.4.1 Superconducting Materials 61
2.4.2 Energising a Superconducting Magnet 64
2.4.3 Superconducting Switch 64
2.4.4 Superconducting Joints 65
2.4.5 Quenching 66
2.4.6 Quench Protection 66
2.4.7 Stress Limits 67
2.5 Heat Transfer and Cryostat Design 70
2.5.1 Cryo-Refrigerators 71
2.5.2 Sub-Atmospheric Operation 72
2.5.3 Gradient-Induced Heating 73
2.6 Practical Considerations 74
2.6.1 Safety 74
2.6.2 Installation Issues 75
2.7 Future Developments 76
2.7.1 High and Ultra-High Field Magnets 76
2.7.2 Helium-Free Technology 78
References 79

Chapter 3 Radiofrequency Coils 81
Andrew Webb

3.1 Introduction 81
3.2 General Electromagnetic Principles for RF Coil Design 82
3.2.1 Maxwell's Equations and the Biot–Savart law 84
3.2.2 Transmit (B_1^+) and Receive (B_1^-) Magnetic Fields 85
3.2.3 Linear and Circular Polarization 87
3.2.4 Conservative and Non-Conservative Electric Fields 89
3.2.5 Electromagnetic Simulations 91
3.3 Electrical Circuit Analysis 91
3.3.1 RF Coil Impedance 92
3.3.2 Resonant Circuits 94
3.3.3 Capacitive Impedance Matching 95
3.3.4 Inductive Impedance Matching 96
3.3.5 Impedance Matching Using Transmission Line Elements 98
3.3.6 Baluns and Cable Traps 99
3.3.7 RF Coil Loading—The Effect of the Sample 101
3.4 RF Coils Producing a Homogeneous Magnetic Field (Volume Coils) 104
3.4.1 Birdcage Coils 106
3.4.2 Transverse Electromagnetic Mode (TEM) Resonators 107
3.4.3 Partial-Volume Coils 108
3.4.4 Solenoids and Loop Gap Resonators 109
3.5 Surface Coils 111
3.5.1 Transmit/Receive Surface Coils 111
3.5.2 Quadrature Surface Coils 112
3.6 Detuning Circuits for Transmit-Only Volume Coils and Receive-Only Surface Coils 115
3.7 Receive Arrays 116
3.7.1 Array Optimization 117
3.7.2 Preamplifier Decoupling in Receive Arrays 118
3.8 Multiple-Frequency Circuits 120
3.8.1 Multiple-Pole Circuits 120
3.8.2 Transformer Coupled Circuits 121
3.8.3 Multiple-Tuned Volume Coils 122
3.8.4 Multiple-Tuned Surface Coils 124
3.9 RF coils for NMR Spectroscopy 124

3.9.1 Probes for High Resolution Liquid-State NMR 124
3.9.2 Microprobes for High Resolution NMR and Hyphenated Microseparations 127
3.9.3 Probes for Solid-State NMR 130
3.9.4 Cryoprobes 132
3.10 RF Coils for Small Animal Imaging and MR Microscopy 135
3.10.1 Small Animal Imaging Coils 136
3.10.2 RF coils for MR Microscopy and Combined MR/Optical Histology 138
3.11 RF Coils for Clinical Imaging Systems 142
3.11.1 Single-Channel and Dual-Channel Transmit Coils for Clinical Systems 142
3.11.2 Receive Arrays for Clinical Systems 144
3.12 RF Coils for Very High Field Human Imaging 145
3.12.1 Multi-Channel Transmit Arrays for High Field Imaging 146
3.13 Dielectric Resonators 146
3.13.1 HEM_{11} Mode Resonators 150
3.13.2 TE_{01} Mode Resonators 150
3.14 Antennae for Travelling Wave MRI 150
Appendix A 155
References 158

Chapter 4 B_0 Shimming Technology 166
Robin A. de Graaf and Christoph Juchem

4.1 Introduction 166
4.2 The Origins of Magnetic Field Inhomogeneity 167
4.3 Static Spherical Harmonic Shimming 172
4.3.1 Theory 172
4.3.2 Magnetic Field Mapping 178
4.3.3 Calibration of Shim Coil Efficiency 182
4.3.4 Static Spherical Harmonic Shimming of the Human Brain 185
4.4 Dynamic Spherical Harmonic Shimming 189
4.4.1 Principle of Dynamic Shimming 189
4.4.2 Practical Considerations for Dynamic Shimming 190
4.4.3 Dynamic Spherical Harmonic Shimming of the Human Brain 193
4.5 Alternative Shimming Methods 196
4.5.1 Passive Approaches 196
4.5.2 Active Approaches 198
References 202

Chapter 5 Magnetic Field Gradients **208**
Ralph Kimmlingen

5.1 Introduction 208
5.1.1 Linear Magnetic Field Gradients 208
5.1.2 Spatial Encoding and Geometric Distortion 209
5.1.3 Classification of Design Methods 211
5.1.4 Biot–Savart Methods 211
5.1.5 Current Density Methods 213
5.1.6 Methods Using Spherical Harmonics 219
5.1.7 Definition of Gradient Performance Parameters 219
5.1.8 Developments in "Conventional" Gradient Designs 221
5.1.9 Integrated Gradient and RF Designs 225
5.1.10 Increased Bore-Size Systems 225
5.1.11 Non-Cylindrical Designs 229
5.2 Gradient System 230
5.2.1 Overview 230
5.2.2 Gradient Coil System 231
5.2.3 Gradient Power Amplifier (GPA) and Connectors 234
5.2.4 Gradient Cooling System and Temperature Supervision 236
5.2.5 Gradient Control System and "Safety Watchdog" 241
5.3 Examples of Specific Gradient Coil Designs 244
5.3.1 High Strength Gradients for a 7 T Horizontal Bore Animal Magnet 244
5.3.2 A Whole-Body Modular Gradient Set with Continuously Variable Field Characteristics 250
5.3.3 A Head Gradient Coil Insert 255
5.3.4 Ultra-Strong Whole-Body Gradients 258
References 262

Chapter 6 Radiofrequency Amplifiers for NMR/MRI **264**
Neal A. Hollingsworth, Krishna Kurpad, and Steven M. Wright

6.1 Introduction 264
6.2 Principles of RF Amplification 266
6.2.1 The RF Power MOSFET 266
6.2.2 DC Characteristics 267
6.2.3 RF Characteristics 270
6.2.4 Amplifier Classes 272

6.2.5 Switch-Mode Amplifiers 275
6.2.6 Mechanism of RF Power Amplification 275
6.3 Matching Networks for Amplifiers 279
6.3.1 Basics of Matching Networks 279
6.3.2 Narrowband Matching 280
6.3.3 Broadband Matching 282
6.4 Amplifier Performance Considerations 282
6.4.1 Linearity 283
6.4.2 Noise Gating 284
6.4.3 Dynamic Range 284
6.4.4 Efficiency 284
6.4.5 Stability 285
6.4.6 Technical Specifications for Commercial RFPAs for MRI 285
6.5 Amplifiers for Multi-Channel Transmission 285
6.5.1 Mutual Coupling in Transmit Arrays 287
6.6 Current Source Amplifiers 290
6.6.1 Matching Networks for Current Source Amplifiers 291
6.6.2 Analysis of the Current Source Network 292
6.7 Low Output Impedance Amplifiers 294
6.7.1 Coil Matching Network for an LOI amplifier 297
6.8 Testing and Comparison of Amplifiers Architectures 301
6.8.1 Amplifier Bench Measurements 301
6.8.2 Amplifier Testing Using MRI 303
6.9 Selection of Amplifier Architecture 304
References 305

Chapter 7 The MR Receiver Chain **308**
Dennis W. J. Klomp and Andrew Webb

7.1 Introduction 308
7.2 Signal Levels and Dynamic Ranges of MR Data 310
7.3 Overall Noise Figure of the Receive Chain 312
7.4 Design of Transmit/Receive Switches 313
7.5 Low-Noise Preamplifiers 316
7.6 Data Sampling 319
7.6.1 Frequency Demodulation 320
7.6.2 Direct Detection Using Undersampling 321
7.7 Analogue-to-Digital Converters 322
7.8 Optical and Wireless Data Transmission 328
References 329

Chapter 8 Electromagnetic Modelling **331**
Christopher M. Collins, Andrew G. Webb, and Jan Paška

8.1 Introduction 331
8.2 Simulating Electromagnetic Fields for Magnetic Resonance 333
8.2.1 Static Magnetic (B_0) Fields 334
8.2.2 Switched Gradient Fields (G_x, G_y, G_z) 336
8.2.3 Radiofrequency Magnetic (B_1) Fields 337
8.3 The Role of Simulations in Assessing MR Safety and Bioeffects 355
8.3.1 Static Field Effects 356
8.3.2 Gradient-Induced Peripheral Nerve Stimulation (PNS) 357
8.3.3 RF-Induced Heating in the Human Body 360
8.3.4 Safety of Devices and Implants 364
8.3.5 MR Safety in Practice 365
8.4 Calculating the Effects of Electromagnetic Fields on MR Images 366
8.4.1 Calculation of the Intrinsic Signal-to-Noise Ratio (ISNR) 366
8.4.2 Simulating MR Images 369
8.5 Methods for Validating Simulations 372
Acknowledgements 374
References 374

Subject Index **378**

CHAPTER 1

The Principles of Magnetic Resonance, and Associated Hardware

ANDREW WEBB*[a]

[a]C.J.Gorter Center for High Field MRI, Department of Radiology, Leiden University Medical Center, Leiden, The Netherlands
*E-mail: a.webb@lumc.nl

1.1 Introduction

The diversity of magnetic resonance (MR) experiments is enormous, ranging from simple one-dimensional proton nuclear magnetic resonance (NMR) spectroscopy through multi-dimensional multi-nuclear spectra to full three-dimensional magnetic resonance imaging (MRI) of morphology and function in animals and humans. Some examples of the types of data produced from different MR experiments are shown in Figure 1.1.

Despite the widely different information content of these data, the fundamental hardware systems for NMR spectroscopy (in both the liquid and solid states) and MRI (human and animal) are very similar. The basic components include the following:

New Developments in NMR No. 7
Magnetic Resonance Technology: Hardware and System Component Design
Edited by Andrew G Webb

Published by the Royal Society of Chemistry, www.rsc.org

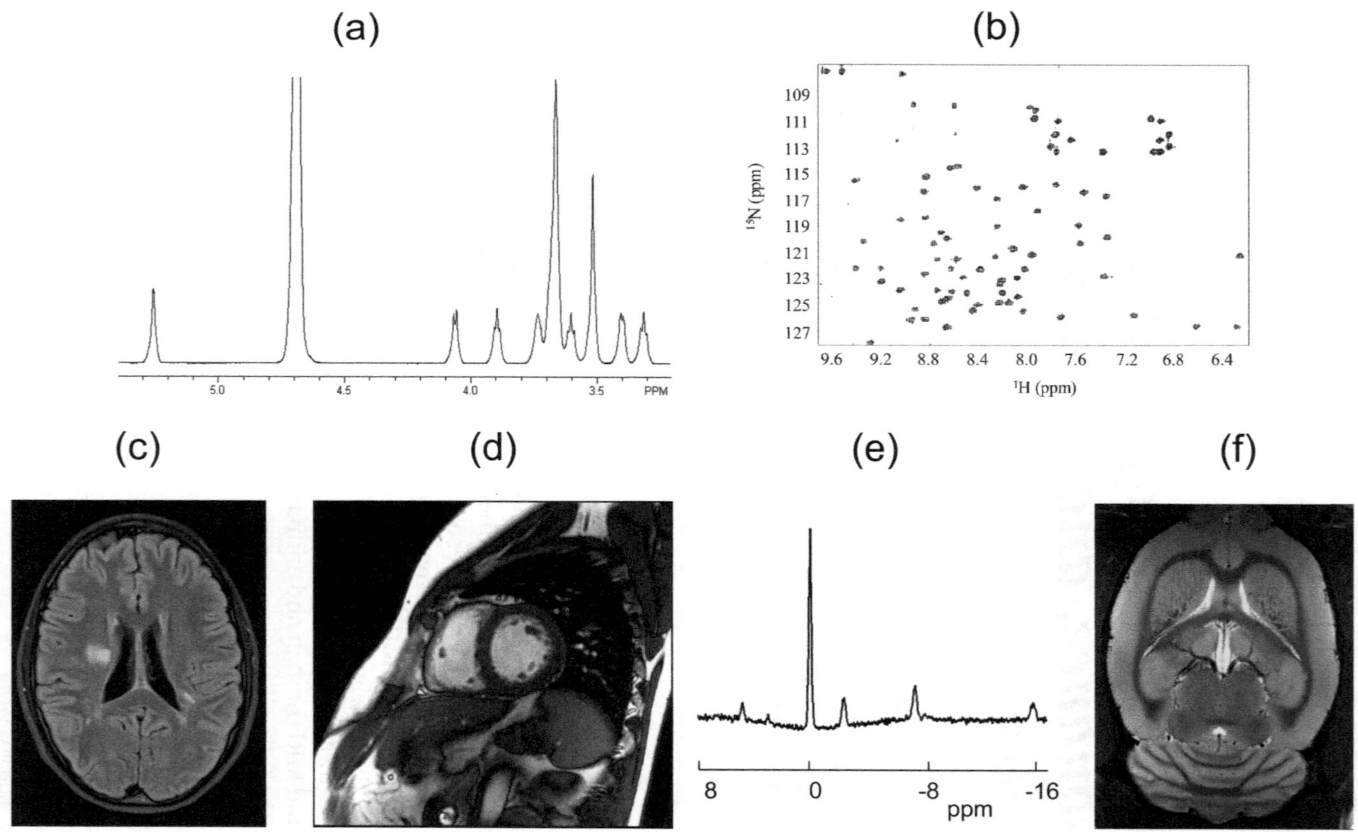

Figure 1.1 Examples of data produced from different magnetic resonance experiments. (a) One-dimensional proton NMR spectrum, (b) two-dimensional proton-nitrogen NMR spectrum, (c) proton image of the brain, (d) electrocardiogram-triggered proton image of the human heart, (e) phosphorus spectrum from the human calf muscle, and (f) proton image of a rat brain.

(i) The magnet, which polarizes the nuclei and produces a net magnetization within the sample.
(ii) The radiofrequency coil(s), which transmit pulses of electromagnetic (EM) energy into the sample and detect the precessing magnetization, which constitutes the MR signal.
(iii) Magnetic field gradient coils, which induce a spatial dependence of the nuclear precession frequency and can be used for coherence order selection, measurements of diffusion, and MRI.
(iv) Shim coils, which are used to produce as homogeneous a magnetic field as possible throughout the sample.
(v) The receiver electronics and circuitry, which amplify, filter and digitize the MR signal for data storage and post-processing.

The physical arrangement of components (i) to (iv) is shown in Figure 1.2 for a vertical bore magnet.

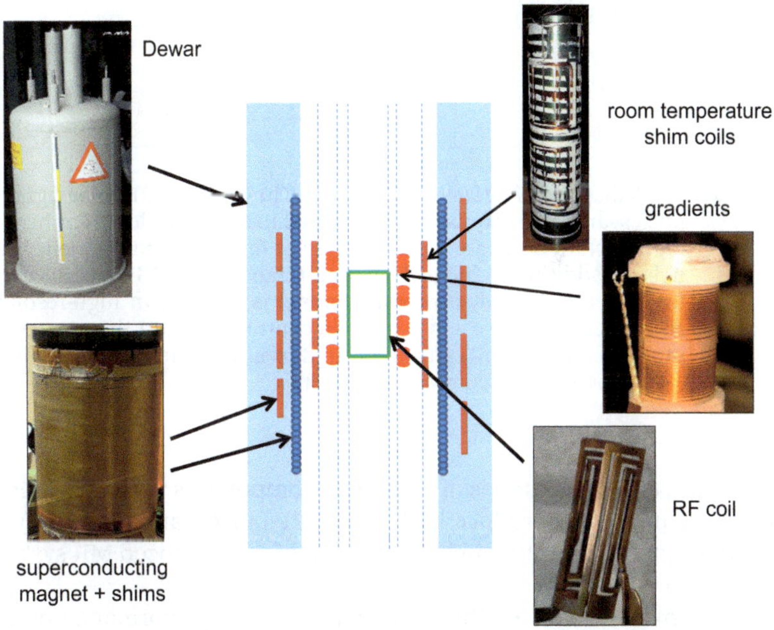

Figure 1.2 Schematic of the setup for high-resolution liquid-state NMR experiments. The vertical-bore superconducting magnet has a clear bore of either 89 mm (wide-bore) or 51 mm (narrow-bore). Sets of superconducting shim coils are embedded within the cryostat. Additional room temperature shims are located inside the magnet bore. The next level inwards comprises the magnetic field gradient set, which may consist of one- or three-axis gradients. The innermost structure is the RF coil containing the NMR sample.

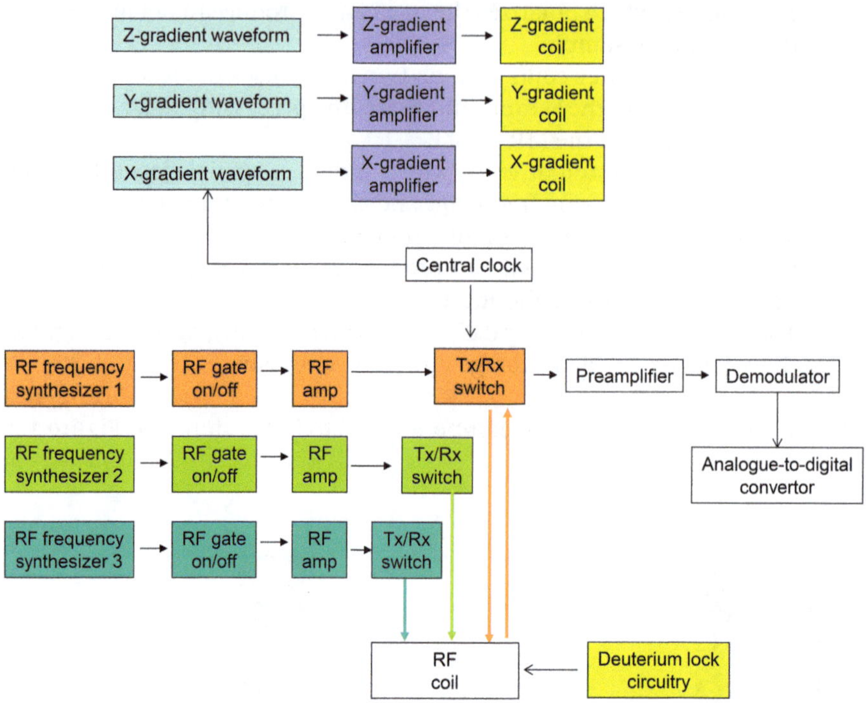

Figure 1.3 Block diagram of a generic MR system. The timing of RF pulses, gradient pulses, and data acquisition are all tied to a central clock. For high-resolution NMR there are typically three or four different frequency channels (in addition to the lock channel) transmitting RF pulses, whereas for MRI usually only a proton channel is present. In high-resolution NMR a single receive coil is used: in contrast, for MRI there are usually multiple receiver coils and associated receive channels, with 32 being a typical number for commercial systems.

In addition, there are a series of electronic components used to switch the gradients on and off, to produce high power RF pulses, and to amplify and digitize the signal. A simplified block diagram of a generalized MR system is shown in Figure 1.3.

Table 1.1 gives an idea of the characteristics and performance of components in typical commercial NMR and MRI systems. The system performances of each of the components in the table are explained in greater detail in the relevant sections throughout the book.

In the following sections in this chapter the basic phenomena involved in magnetic resonance are explained briefly, with links to the relevant system hardware. There are a large number of MR books dealing with the basic theory of high-resolution liquid-state NMR,[1–5] solid-state NMR[6–9] and MRI,[10–13] and readers are advised to consult these tomes for much more in-depth analyses of different aspects of basic MR theory.

Table 1.1 Characteristics of the magnet, gradients, and RF coils in commercial systems.

	NMR	MRI	MRI
		Human	Animal
Static magnetic field	4.7–23.5 tesla	1.5–9.4 tesla	4.7–21.1 tesla
Proton frequency	200 MHz–1 GHz	63.8–400 MHz	200–900 MHz
Gradient strength	500^a mT m^{-1}	40–80^b mT m^{-1}	1 T m^{-1} (6 cm i.d.) 750 mT m^{-1} (9 cm i.d.) 450 mT m^{-1} (12 cm i.d.)
RF coil diameter	1.3–20 mm	10–60 cm	10–60 mm
RF amplifier power	1 kW	30 kW	4 kW
Shortest pulse	~1 μs	~10–50 μs	5–10 μs

^{a}Single axis diffusion gradients can reach ~25 T m^{-1}.
^{b}300 mT m^{-1} for specialized systems for the human-connectome project.[28,29]

1.2 The Superconducting Magnet and Nuclear Polarization

The role of the magnet is to polarize the nuclei to produce a net magnetization within the sample. For NMR spectroscopy and MRI, almost all magnets are superconducting. The magnetic field should be temporally stable and homogeneous to within parts-per-billion (ppb) throughout the sample. Most magnets are actively shielded, *i.e.* the fringe field does not extend significantly outside the physical dimensions of the magnet itself. Magnet design is considered in detail in Chapter 2 of this book, as well as Appendix 1A at the end of this chapter.

All nuclei with an odd atomic weight and/or an odd atomic number possess a fundamental quantum mechanical property termed "spin" and are termed "spin-active" or "NMR-active". The most important spin-active nuclei include ^{1}H, ^{13}C, ^{15}N, ^{23}Na, ^{17}O, ^{31}P and ^{2}H. Notably *spin-inactive* are nuclei such as ^{16}O and ^{12}C. Considering the proton as the simplest example, the property of spin can be viewed as the proton spinning around an internal axis of rotation giving it a certain value of angular momentum (P). Since the proton is a charged particle, this rotation results in a magnetic moment (μ). This magnetic moment produces an associated magnetic field, which has a configuration similar to that of a bar magnet. The magnitude of P is quantized in terms of the nuclear spin quantum number (I):

$$\left|\vec{P}\right| = \frac{h}{2\pi}\left[I(I+1)\right]^{\frac{1}{2}} \tag{1.1}$$

where h is Planck's constant (6.63 × 10^{-34} Js). In the following analysis a spin 1/2 nucleus (I = 1/2) is considered, corresponding to ^{1}H, ^{13}C, ^{15}N, and ^{31}P in the previous list. In this case:

$$\left|\vec{P}\right| = \frac{h}{2\pi}\frac{\sqrt{3}}{2} \tag{1.2}$$

The magnitudes of the magnetic moment and the angular momentum of the proton are related by:

$$\left|\vec{\mu}\right| = \gamma\left|\vec{P}\right| = \frac{\gamma h}{2\pi}\left[I(I+1)\right]^{\frac{1}{2}} \tag{1.3}$$

where γ is the nuclear gyromagnetic ratio, and has a specific value for different nuclei, with protons having the highest γ (with the exception of tritium). For protons therefore:

$$\left|\vec{\mu}\right| = \frac{\gamma h\sqrt{3}}{4\pi} \tag{1.4}$$

μ contains three components (μ_x, μ_y and μ_z), each of which can have any value within the conditions governed by eqn (1.4): this situation is shown in Figure 1.4(a). However, in the presence of a strong magnetic field, B_0, μ_z is quantized with values given by:

$$\mu_z = \frac{\gamma h}{2\pi} m_{\mathrm{I}} \tag{1.5}$$

where m_{I} is the nuclear magnetic quantum number, and can take values I, $I-1\ldots-I$. In the case of a proton, $m_{\mathrm{I}} = \pm 1/2$ and so:

$$\mu_z = \pm\frac{\gamma h}{4\pi} \tag{1.6}$$

The orientation of μ with respect to B_0 is shown in Figure 1.4(b). The interaction of the static magnetic (B_0) field with μ_z results in Zeeman splitting, producing two energy levels: one in which μ_z aligns parallel to B_0 (the lower energy state) and the other anti-parallel (the higher energy state).

The net magnetization, M_0, of a sample containing N_s protons is proportional to the difference in populations between the two energy levels, which is dictated by Boltzmann's equation:

$$M_0 = \sum_{n=1}^{N_s} \mu_{z,n} = \frac{\gamma h}{4\pi}\left(N_{\text{parallel}} - N_{\text{anti-parallel}}\right) = \frac{\gamma^2 h^2 B_0 N_s}{16\pi^2 kT} \tag{1.7}$$

Eqn (1.7) shows that the net polarization of a sample is proportional to the strength of the main magnetic field. However, since the energy difference between the two levels is very small, so is the population difference. For example, at an operating magnetic field of 11.7 tesla for every one million protons, there is a population difference of only ~40 protons between the parallel and anti-parallel orientations.

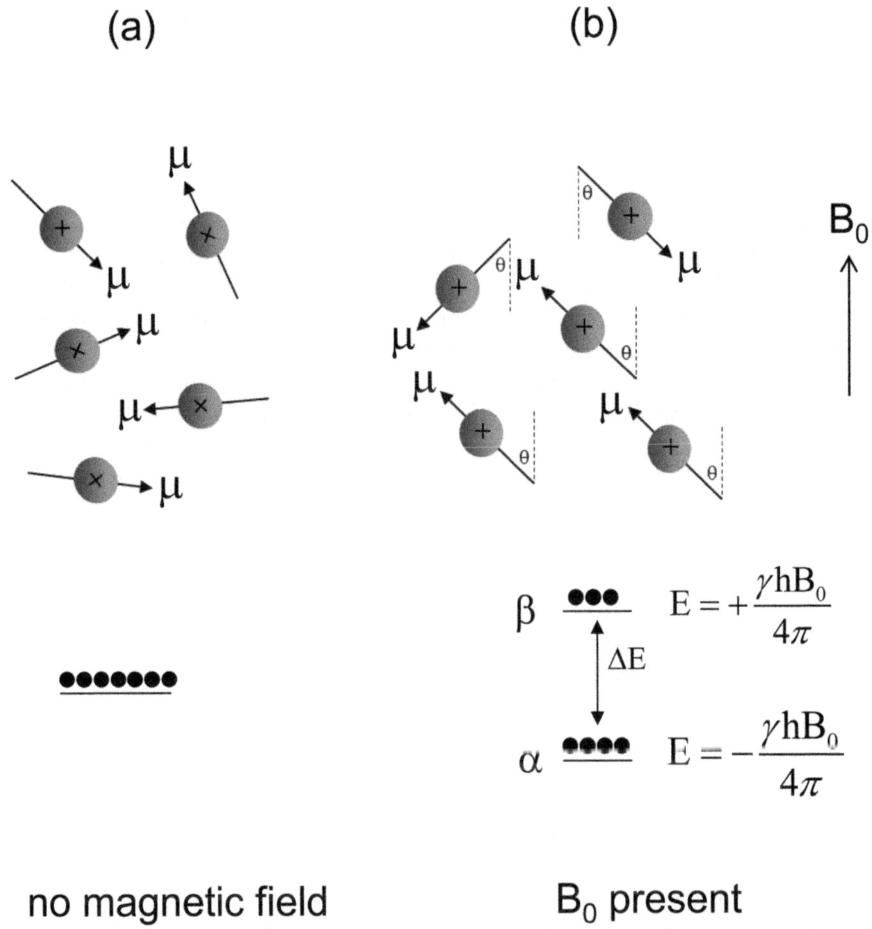

Figure 1.4 (a) In the absence of a magnetic field μ lies in a random direction and the nuclei occupy a single energy level. (b) When a static magnetic field is applied, μ_z becomes quantized at an angle θ of 54.7° with respect to B_0. Parallel (α) and anti-parallel (β) orientations correspond to two energy levels, with a difference $\Delta E = \gamma h B_0/2\pi$. A greater number of nuclei occupy the lower energy, parallel state.

1.3 The Transmitter Coil to Generate Radiofrequency Pulses

In order to detect an MR signal, energy must be applied to the nuclear spin system to stimulate transitions between the two energy levels. A pulse of EM energy is applied at the specific resonance frequency (f_0), which corresponds to the energy difference between the two levels *via*:

$$hf_0 = \Delta E = \frac{\gamma h B_0}{2\pi} \tag{1.8}$$

The resonance frequency in Hz, or resonance angular frequency (ω_0) in radians per second, is therefore given by:

$$f_0 = \frac{\gamma B_0}{2\pi}, \omega_0 = \gamma B_0 \tag{1.9}$$

The pulsed EM wave consists of both magnetic and electric field components, and it is the *magnetic* component that interacts with the nuclear magnetization. The EM energy, termed a radiofrequency (RF) pulse, is transmitted *via* an RF coil. As shown in Figure 1.5, the magnetic field component, B_1^+, of the EM wave from the RF coil must be created in a direction *perpendicular* to B_0 in order to interact with the magnetization.

Using classical mechanics, the action of the RF pulse applied along one axis produces a torque perpendicular to that axis. As shown in Figure 1.5(b), the angle α by which the magnetization is rotated is proportional to the product of the strength of the applied RF field and the time, τ_{B_1}, for which it is applied.

$$\alpha = \gamma B_1^+ \tau_{B_1} \tag{1.10}$$

After application of an RF pulse with tip angle α about the x-axis, the net magnetization can be expressed as three different vectors.

$$M_z = M_0 \cos\alpha, M_x = 0, M_y = M_0 \sin\alpha \tag{1.11}$$

The geometric design of the RF coil is determined by the requirement that the B_1^+ field produced by the coil must be perpendicular to B_0. There are a large number of different geometries, which are discussed in detail in Chapter 3. One example, widely used in high-resolution liquid-state NMR, is the "saddle" coil, shown in Figure 1.6. The long axis of the coil is coincident with B_0, and current flowing through the copper conductor produces a B_1^+ field with the required orientation. The RF coil is tuned and impedance matched to 50 Ω at the Larmor frequency for high efficiency. The RF coil may be tuned to more than one frequency, as also discussed in Chapter 3.

Power to the RF coil is supplied by an RF amplifier, the general specifications of which include:

(i) Amplifiers for liquid-state and solid-state NMR experiments must be able to produce pulses as short as 1 μs, with accurate and reproducible shape, at frequencies up to ~1 GHz. In solid-state NMR applications, very high power pulses with a very high duty cycle may be used for proton decoupling.

(ii) In the case of MRI, up to 30–60 kW of power must be available without the performance varying over time owing to any component heating, as well as the amplifiers being capable of high duty cycles for RF-intensive sequences, such as turbo spin-echoes.

The design of RF amplifiers is described in detail in Chapter 6.

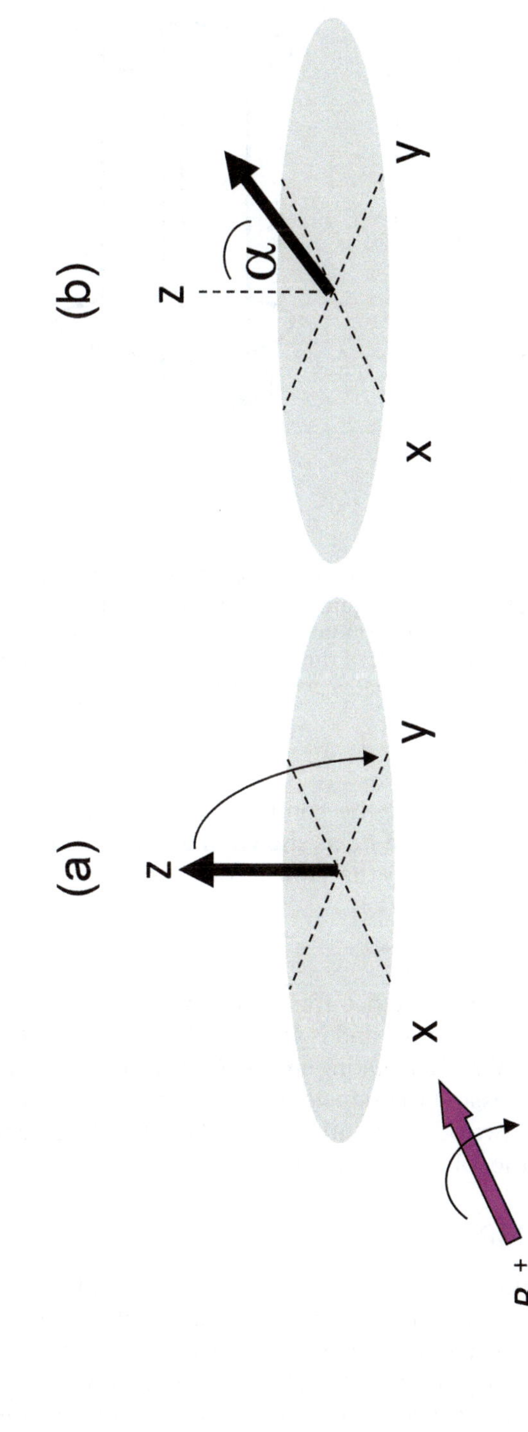

Figure 1.5 (a) The effect of an RF pulse on the net magnetization is to rotate the magnetization about the axis along which B_1^+ is applied. (b) The angle of rotation α is given by eqn (1.10).

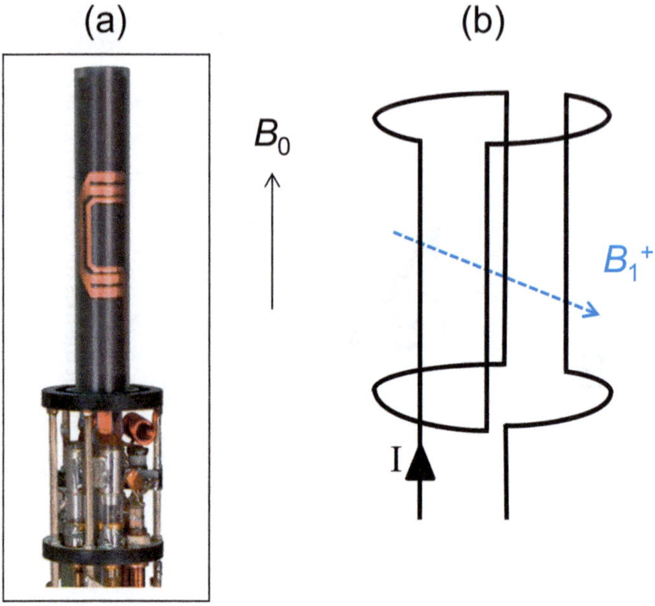

Figure 1.6 (a) Photograph of a saddle coil, formed of an etched copper substrate, used for high-resolution liquid-state NMR. The electrical circuitry used for tuning and impedance matching is situated below the coil. The sample is placed in the centre of the coil. (b) Current I flowing through the coil produces a B_1 field that is perpendicular to B_0.

1.4 Precession

When the RF pulse is turned off, the component of magnetization in the transverse plane precesses around B_0, as shown in Figure 1.7(a). The precession frequency, $\omega_{\text{precession}}$, is exactly the same as the frequency of irradiation:

$$\omega_{\text{precession}} = \omega_0 = \gamma B_0 \tag{1.12}$$

The concept of a rotating reference frame is very useful in analyzing the behaviour of the net magnetization, and is shown in Figure 1.7(b). The rotating reference frame ($x'y'$) is defined as rotating around B_0 at an angular frequency ω_0.

In fact, the exact precession frequencies of different nuclei within a molecule are determined by two other factors, chemical shift and scalar coupling, which are outlined in the following sections.

1.4.1 Chemical Shift

The term "chemical shift" refers to the fact that protons in different chemical environments within a molecule resonate at slightly different frequencies: their precession frequencies are *shifted* (with respect to a reference, discussed below) with the magnitude of the shift depending on their particular *chemical* environment. The cause of this shift is that the exact magnetic field

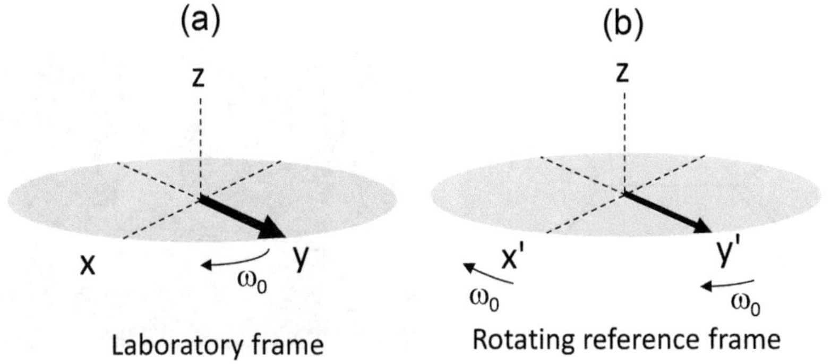

Figure 1.7 (a) Immediately after a 90° pulse about the x-axis, the magnetization lies along the y-axis, and starts to precess at a frequency $\omega_{precession}$, given by eqn (1.12). (b) In the rotating reference frame the magnetization vector appears static.

experienced by each proton in the molecule is slightly lower than B_0 owing to the shielding effects of the electron cloud surrounding each proton. Electrons have a magnetic moment, which is opposite in sign to that of the proton, and so the effective magnetic field experienced by the proton is reduced. As an example, consider the lactic acid molecule shown in Figure 1.8. There are four different proton groups (CH_3, CH, OH, COOH) in this molecule: each of these proton groups has a slightly different precession frequency since they experience a slightly different magnetic field. The effective magnetic field, B_{eff}, experienced by a proton is given by:

$$B_{eff} = B_0(1 - \sigma) \tag{1.13}$$

where σ is called the shielding constant, and is related to the electronic environment surrounding the nucleus. The resonant frequency of the proton is given by:

$$\omega = \gamma B_{eff} = \gamma B_0(1 - \sigma) \tag{1.14}$$

One of the main factors that determines the value of σ is the electronegativity of the atoms connected to the protons: the proton in an –OH group has a lower shielding constant than those in a $-CH_2-$ group since oxygen is more electronegative than carbon and pulls electrons away from the proton, thus reducing the shielding. The resonant frequency of the proton of an –OH group is therefore higher than that of the protons in a $-CH_2-$ group. The chemical shift (δ), in units of parts-per-million (ppm), is defined as:

$$\delta(\text{ppm}) = 10^6 \frac{f - f_{ref}}{f_{ref}} \tag{1.15}$$

where f_{ref} is the resonant frequency of the protons in tetramethylsilane (TMS), which acts as a reference for proton NMR spectra. Figure 1.9 shows approximate chemical shift ranges for protons in different chemical

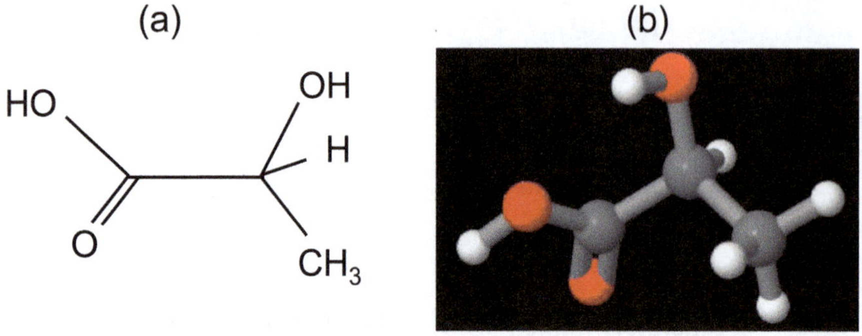

Figure 1.8 (a) Chemical structure of a lactic acid molecule. (b) Ball-and-stick model of the lactic acid molecule with oxygens shown in red, protons in white and carbons in gray.

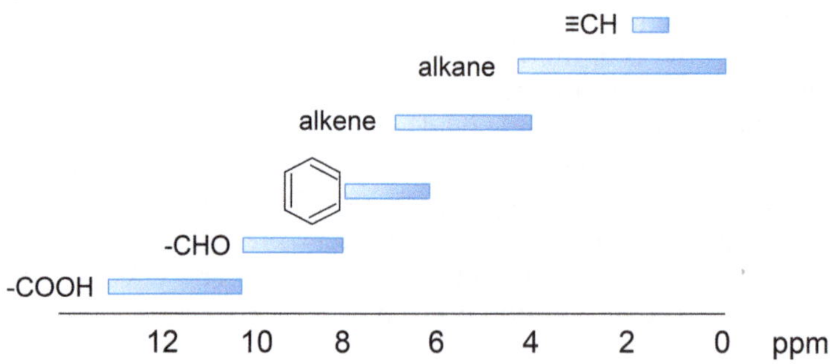

Figure 1.9 The range of proton chemical shifts for different chemical moieties.

environments. For the lactic acid molecule, the order of resonance frequencies is: CH_3 < CH < OH < COOH.

1.4.2 Scalar Coupling

Consider the protons in the molecule shown in Figure 1.10. There are three chemically distinct protons (H_A, H_B, and H_C), which have three different chemical shifts and corresponding resonant frequencies. However, there is also an interaction between the two protons H_A and H_B, which are separated by three chemical bonds (H_A-to-C-to-C-to-H_B). The magnetic field experienced by H_A depends upon whether H_B is in the α (parallel) or β (anti-parallel) state, and *vice versa*: this phenomenon is termed scalar coupling. Therefore, there are four different energy levels for this coupled two-proton system: αα, αβ, βα, and ββ, as shown in Figure 1.10(b). The exact value of the scalar coupling constant J depends upon a number of

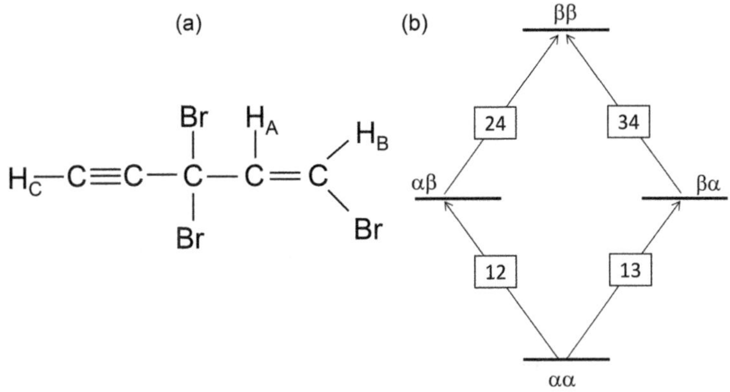

Figure 1.10 (a) A chemical molecule with three different protons. H_A and H_B are scalar coupled, whereas H_C is not coupled. (b) Energy level diagram for the coupled H_A and H_B protons.

different factors, including the particular type of chemical bond (single, double or triple bond), as well as the angle subtended between the C–H bonds (the Karplus angle), but in general the values are between 1 and 7 Hz. The greater the number of bonds between the protons the weaker the coupling, and so for the molecule in Figure 1.10(a) there is effectively no coupling between H_A and H_C since there are five chemical bonds between the protons.

For the coupled H_A–H_B system, the four different transition energies corresponding to the energy level diagram in Figure 1.10(b) are given by:

$$\omega_{12} = \omega_A - \frac{1}{2}J_{AB}, \omega_{34} = \omega_A + \frac{1}{2}J_{AB}$$
$$\omega_{13} = \omega_B - \frac{1}{2}J_{AB}, \omega_{24} = \omega_B + \frac{1}{2}J_{AB} \quad (1.16)$$

Figure 1.11 shows the evolution of the precessing magnetization for the three different protons. From Figure 1.9 the chemical shift of H_C has the lowest value, followed by H_A and the highest value for H_B. H_C precesses at a single frequency, whereas protons H_A and H_B are "split" into two frequencies by the scalar coupling.

1.4.3 Relaxation Processes

The final effect that must be considered in terms of the precession of the net magnetization is relaxation. After the RF pulse has been turned off, each of the magnetization components M_z, M_x and M_y returns to their thermal equilibrium values, with the time-evolution determined by specific time-constants.

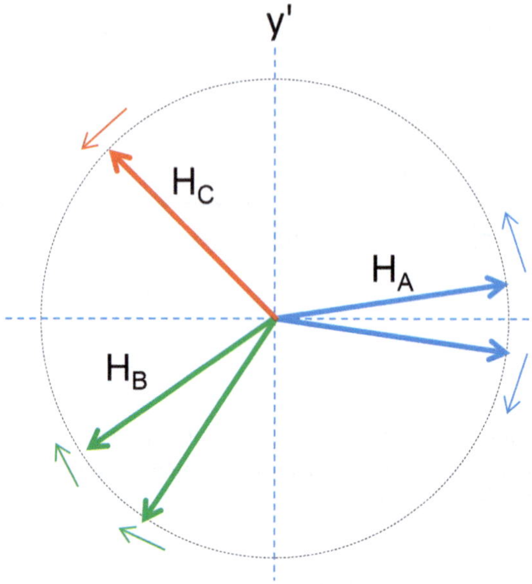

Figure 1.11 Evolution of the magnetization in the molecule shown in Figure 1.10 from the combined effects of chemical shift and scalar coupling. The rotating reference frame ($x'y'$) frequency is set to the chemical shift of H_A.

The time-evolutions of M_z, M_x and M_y are characterized by differential equations known as the Bloch equations:[14]

$$\begin{aligned}\frac{dM_x}{dt} &= \gamma M_y\left(B_0 - \frac{\omega}{\gamma}\right) - \frac{M_x}{T_2}\\ \frac{dM_y}{dt} &= \gamma M_z B_1 - \gamma M_x\left(B_0 - \frac{\omega}{\gamma}\right) - \frac{M_y}{T_2}\\ \frac{dM_z}{dt} &= -\gamma M_y B_1 - \frac{M_z - M_0}{T_1}\end{aligned} \tag{1.17}$$

The return of M_z to its equilibrium value of M_0 is governed by the spin-lattice (T_1) relaxation time, and the return of M_x and M_y to their thermal equilibrium value of zero by the spin–spin (T_2) relaxation time. It should be noted that the relative values of T_1 and T_2 can be very different for different types of sample, but T_1 is always *greater or equal* to T_2. For non-viscous liquids used in high-resolution NMR, the values of T_1 and T_2 are very similar. In contrast, in solid samples T_2 can be up to six orders shorter than T_1, and for human MRI the T_2 of many soft tissues is between one and two orders of magnitude smaller than T_1.

Solving the Bloch equations for the M_x and M_y components of magnetization gives:

$$M_y(t) = M_y(t=0)e^{-t/T_2},\ M_x(t) = M_x(t=0)e^{-t/T_2} \tag{1.18}$$

Therefore, the precessing magnetization vectors shown in Figure 1.11 decay exponentially as a function of time. In practice, there is an addition term that gives rise to the decay of transverse magnetization, which is termed "inhomogeneous line-broadening" and has a time-constant of T_2^+. This term arises from spatial inhomogeneities in the magnetic field within the sample as-a-whole, and can arise from the magnet itself or more commonly with an inhomogeneous sample. The overall relaxation time for transverse magnetization is termed T_2^*, which is given by:

$$T_2^* = \frac{T_2 T_2^+}{T_2 + T_2^+} \tag{1.19}$$

The return of M_z to its thermal equilibrium value of M_0 is governed by the T_1 relaxation time:

$$M_z(t) = M_z(t=0) + \left(M_0 - M_z(t=0)\right)\left(1 - e^{-\frac{t}{T_1}}\right) \tag{1.20}$$

1.5 The Receiver Coil for Detecting the MR Signal

In high-resolution NMR experiments, the same RF coil used to transmit the RF pulses is also used to detect the MR signal. In contrast, in MRI experiments a large number of small coils are typically used to receive the signal, with (usually) a single large coil used to transmit the RF pulses. For simplicity, and without losing generality, Figure 1.12(a) shows the case relevant to

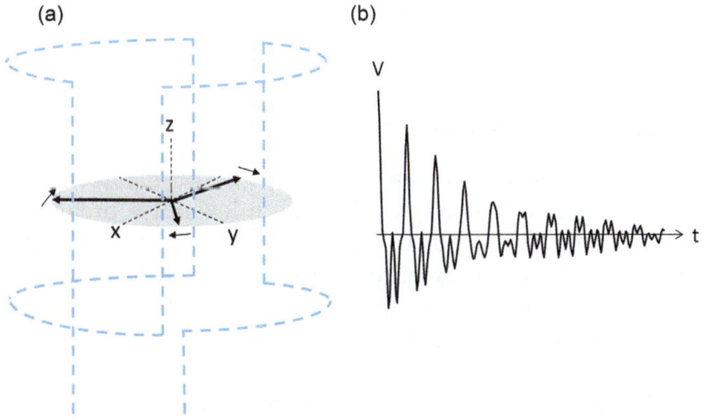

Figure 1.12 (a) Detection of the MR signal occurs *via* the voltage induced across the terminals of an RF coil by the time-varying magnetic flux produced by the precessing magnetization. (b) The induced voltage as a function of time, incorporating chemical shift and scalar coupling, as well as T_2^* relaxation.

high-resolution NMR in which the saddle coil shown in Figure 1.6 is used as the receiver coil. The time-varying magnetic field produced by the precession of the magnetization vectors results in a voltage being induced in this receiver coil. The induced voltage, V, is given by Faraday's law and is proportional to the rate-of-change of magnetic flux:

$$V = -\frac{\partial \phi(t)}{\partial t} = -\frac{\partial}{\partial t}\int_{\text{sample}} \frac{B_1}{I} \cdot M_{xy}\, \mathrm{d}r \tag{1.21}$$

where the coil sensitivity is defined as B_1/I, *i.e.* the B_1 field produced per unit applied current. At higher strengths of the B_0 field, the protons precess at a higher frequency, the value of $\mathrm{d}\phi/\mathrm{d}t$ increases, and so the detected MR signal is higher.

The time-domain signal is shown in Figure 1.12(b). Based on the previous sections the signal contains a number of different frequencies owing to chemical shift and the effects of scalar coupling, and decays exponentially as a function of time. The time-domain signal is usually referred to as the free induction decay (FID).

1.6 The Receiver: Signal Demodulation, Digitization and Fourier Transformation

The analogue time-domain signal shown in Figure 1.12(b) must be converted into a digital signal so that it can be stored and processed. Typically, the magnitude of the signal is on the order of millivolts or even lower, which means that it cannot be digitized directly with a high dynamic range by a commercial analogue-to-digital converter (ADC). It must first pass through the MR receiver, which is covered in detail in Chapter 7 and so only a brief overview is given here.

1.6.1 Receiver Electronics

A block diagram of the electronic circuits in the receiver is shown in Figure 1.13. The signal first passes through a low-noise preamplifier, with a typical gain factor of 20–30 dB and noise figure ~0.5 dB. Commercial ADCs used in MR have a maximum input voltage of 5 volts, with resolution of 16 bits, and so the voltage should be amplified to close to 5 volts before being digitized, otherwise the full dynamic range of the ADC is not utilized.

High-resolution ADCs cannot sample at the very high frequencies associated with MR, and so it is necessary to demodulate the signal to reduce the sampling rate in order to take advantage of the high dynamic range. The first demodulation step uses two phase sensitive detectors (PSDs), with the inputs to the two PSDs phase shifted by 90° to produce real and imaginary

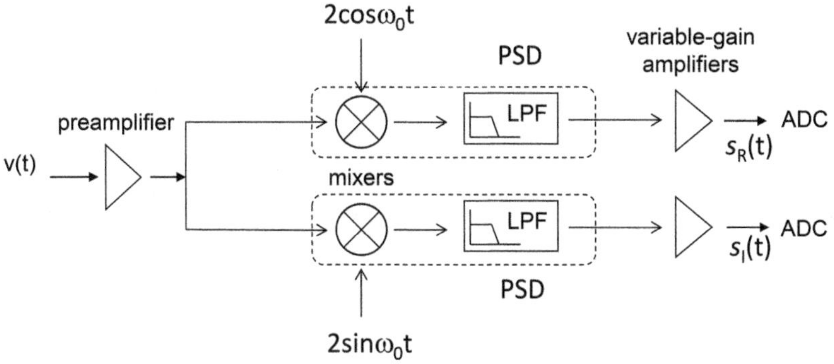

Figure 1.13 Block diagram of the individual components in an MR receiver.

outputs centred at a lower intermediate frequency, ω_{IF}. Each PSD contains a mixer and low pass filter, with the mixer effectively acting as a signal multiplier. The output of the mixer has frequency components at $(\omega_0 + \omega_{IF})$ and $(\omega_0 - \omega_{IF})$: the low pass filter removes the higher frequency components. The real, $s_R(t)$, and imaginary, $s_I(t)$, components of the signal pass through a second amplification stage and are digitized using separate ADCs. In practice, digital sampling is now common in both NMR and MRI,[15] and is discussed further in Chapter 7.

1.6.2 Signal Processing

In NMR experiments, after the signal has been digitized and stored, it is Fourier transformed into the frequency domain (for imaging experiments, covered later in this chapter, the transformation is a multi-dimensional inverse Fourier transform). The frequency-domain signal has both real, $S_R(f)$, and imaginary, $S_I(f)$, components given by:

$$S_R(f) + S_I(f) \propto M_0\left(\frac{T_2^*}{1+(2\pi T_2^* f)^2} - \mathrm{j}\frac{(T_2^*)^2 2\pi f}{1+(2\pi T_2^* f)^2}\right) \tag{1.22}$$

The real part of the frequency-domain signal, $S_R(f)$, is termed the absorption-mode spectrum, as shown in Figure 1.14. Each line in the spectrum is described by a Lorentz function with a full-width-half-maximum (FWHM) given by $(\pi T_2^*)^{-1}$. The imaginary component of the frequency-domain signal is termed the dispersion-mode spectrum, as also shown in Figure 1.14. The real part of the spectrum is displayed after the phase of the spectra is adjusted to account for the finite delay between the RF pulse being switched off and the start of data acquisition. Frequency independent (zero-order) and frequency-dependent (first-order) phasing are applied so that the real part of the spectrum is completely absorptive.

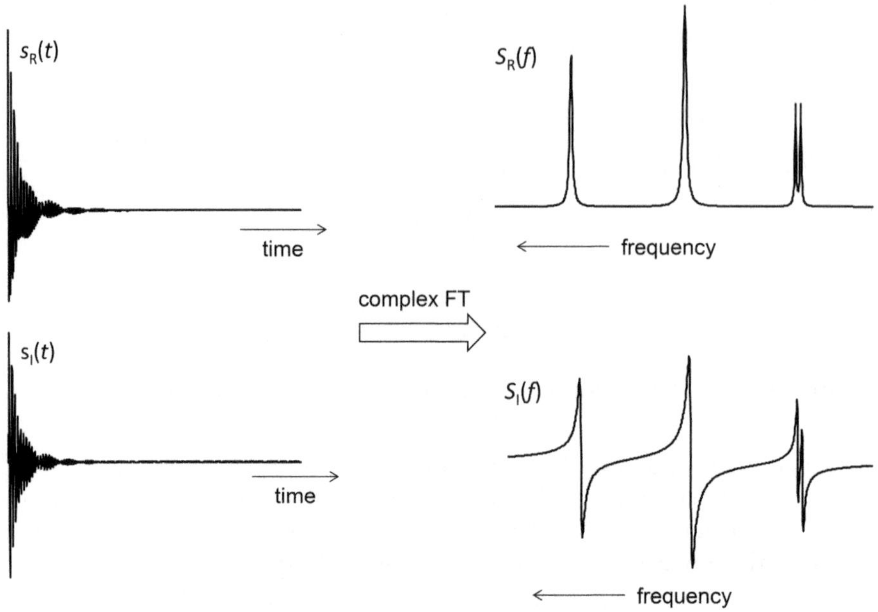

Figure 1.14 Fourier transformation of the complex time-domain digitized signal gives an absorption spectrum (top) and a dispersion spectrum (below).

1.7 Shim Coils

Despite every effort to construct a magnet that produces a completely homogeneous static field throughout the entire sample, this is impossible to achieve in practice. Finite mechanical tolerances in winding and placement of the superconducting wires, and material imperfections within the superconducting wire itself result in a small degree of spatial variation of the B_0 field. Indeed, it is theoretically only possible to produce a completely homogeneous field with an infinitely long magnet, as covered in Chapter 2. Furthermore, unless the sample is a pure liquid, the different magnetic susceptibilities of components within the sample, and in particular the boundaries between these components, produce distortions in the magnetic field: this is the case particularly for MRI experiments in animals or humans, as described in Chapter 4.

In order to counteract this sample-dependent inhomogeneity in the static magnetic field, it is necessary to introduce "shim coils", which are discussed in more detail in Chapter 4. These are "correction" coils, through which current can be passed to produce a spatially-varying compensatory magnetic field for the particular sample being studied. There are two different types of shim coils: (i) superconducting shim coils, which are situated in the cryostat of the magnet, as shown in Figure 1.2, and which are adjusted only during magnet installation, and (ii) room-temperature shims, which are normally placed on the inside of the magnet bore. In high-resolution NMR there may be up to thirty different shim coils through which current can be passed. The reason for such a high number is that it is essential to obtain narrow linewidths in high-resolution NMR

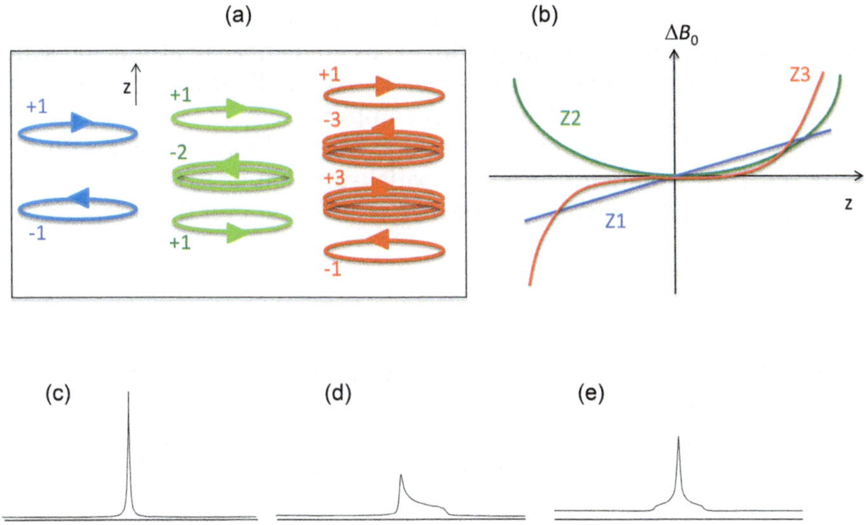

Figure 1.15 (a) Wire geometries for Z1, Z2 and Z3 shim coils with the appropriate relative current values and directions provided by the individual shim DC power supplies. (b) The spatial variation in magnetic field in the *z*-direction for each of the Z1, Z2 and Z3 shim coils. (c) A spectrum showing a well-shimmed line shape. (d) A spectrum showing low-field asymmetry produced by a sub-optimal value of the Z2 shim. (e) A spectrum displaying a symmetric broad baseline produced by a sub-optimal value of the Z3 shim.

spectroscopy in order to separate the resonances from protons in similar chemical environments. In MRI there are far fewer shim coils since: (i) imaging is inherently less sensitive than spectroscopy to B_0 field inhomogeneities, (ii) the effect of the shims is less since they are situated a much greater distance away from the sample than in high-resolution NMR, and (iii) the intrinsic inhomogeneities produced by the sample are much greater.

The geometry of the shim coils is designed to produce different spatial variations in the magnetic field. Figure 1.15(a) shows examples of three sets of shim coils used to produce different spatial compensations to the magnetic field, Figure 1.15(b), as a function of location in the *z*-direction. The effects of mis-setting the shim currents are also shown in Figure 1.15.

Optimization of the currents through the shim coils can be performed in two different ways. In one method, the real component of the deuterium lock signal is monitored, covered later in Section 1.9, in the other method magnetic field gradients are used, as described in Section 1.8.4.

1.8 Gradient Coils

Magnetic field gradient coils, a term usually shortened to simply "gradient coils", are designed to produce a linear dependence of the effective magnetic field on spatial position. They consist of shaped conductors (wires or thin sheets) through which current passes to produce the gradient field. In high-resolution NMR probes, a single *z*-axis gradient coil is usually integrated into the RF probe, although three-axis gradients can also be present. For MRI, three independent orthogonal sets of gradient coils are present, each controlled by independent gradient amplifiers.

Since only the *z*-component of the magnetic field (B_z) interacts with the proton magnetic moments, it is the spatial variation of B_z that is relevant. The magnetic field gradients are designed to be linear over the sample, *i.e.*:

$$\frac{\partial B_z}{\partial z} = G_z \tag{1.23}$$

Figure 1.16 shows a plot of magnetic field *vs.* spatial position for a gradient applied along the *z*-axis.

When the gradient is switched on by passing current through the coil, the magnetic field B_z becomes a function of *z* position and is given by:

$$B_z = B_0 + zG_z \tag{1.24}$$

where G_z has units of tesla (T) per metre. The corresponding precession frequencies (ω_z) of the protons, in the rotating frame, are given by:

$$\omega_z = \gamma z G_z \tag{1.25}$$

Details on gradient coil design, as well as the gradient amplifiers, are covered in Chapter 5. Gradient coils are used for many purposes in MR, the following being the most common:

(i) Dephasing or "crushing" of unwanted transverse magnetization.
(ii) Coherence selection in multiple quantum coherence experiments.[16]

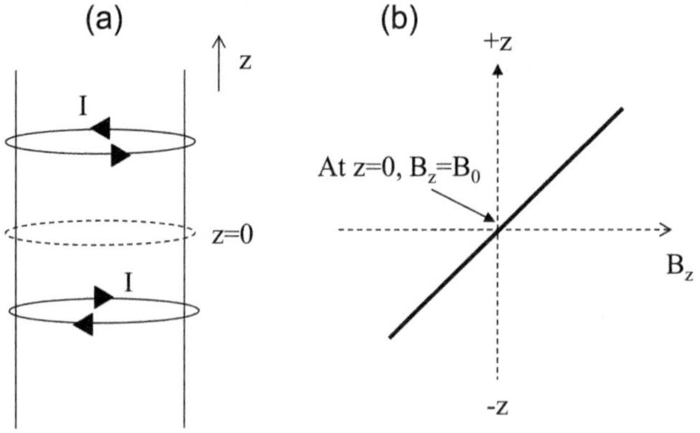

Figure 1.16 A linear gradient in the *z*-direction can be produced by two sets of loops of wire (a), each one carrying current in the opposite direction. (b) Midway between the loops, at $z = 0$, the additional magnetic field from the two sets of loops cancels out, and the magnetic field experienced by the nuclei is B_0.

(iii) Measurements of apparent diffusion coefficients of liquid samples,[17] or tissues in MRI.[18]
(iv) Rapid optimization of shim currents for high-resolution NMR and MRI.
(v) Spatial localization in MRI[19,20] and localized *in vivo* magnetic resonance spectroscopy (MRS).

The following sections give a brief overview of these gradient-based applications.

1.8.1 Crusher Gradients to Dephase Transverse Magnetization

"Crusher" gradients are used to dephase transverse magnetization and are widely used, for example, in localized *in vivo* MR spectroscopy to suppress unwanted signals arising from imperfect RF pulses. In order to calculate the required strength and duration of a gradient pulse for effective dephasing, consider the effect of a gradient, G_z, applied for a time τ as shown in Figure 1.17(a), on transverse magnetization, M_{xy}. The resonance frequency of the protons is given by eqn (1.25). If one assumes a sample with length L in the z-direction, and with uniform proton density, then the average transverse magnetization, as a function of G_z and τ, over the length L is given by:

$$M_{xy}(\tau) = \frac{1}{L}\int_{-\frac{L}{2}}^{\frac{L}{2}} M_{xy}(t=0)\mathrm{e}^{\mathrm{j}\gamma G_z z}\mathrm{d}z = \frac{\sin(\gamma L G_z \tau/2)}{\gamma L G_z \tau/2} \tag{1.26}$$

i.e. a sinc function. Figure 1.17(b) shows a plot of the transverse magnetization as a function of time and gradient strength. If one wishes, as an example,

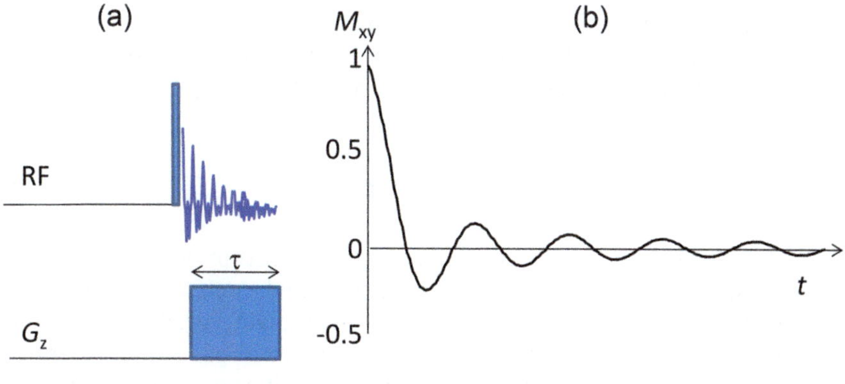

Figure 1.17 (a) Application of a gradient pulse directly after an RF pulse dephases the magnetization. (b) The time-evolution of the transverse magnetization as a function of gradient strength.

to completely dephase the magnetization for a sample of length 1 cm, then a 2 ms long gradient pulse with strength 0.37 T m^{-1} reduces the transverse magnetization by a factor of 1000.

1.8.2 Gradients for Coherence Selection in High-Resolution NMR

Many NMR experiments involve the creation and selection of multiple quantum coherences to simplify the resulting spectra. Examples include homonuclear double quantum and triple quantum coherence selection in correlated spectroscopy (COSY) experiments, as well as heteronuclear multiple quantum coherences created in triple-resonance sequences. The topic of multiple-quantum sequences is covered in many textbooks on high resolution NMR spectroscopy. Gradients can be used to select between different coherence pathways. The coherence order, p, is defined as the difference in the magnetic quantum number, m, of the relevant eigenstates, *i.e.*:

$$p = m_r - m_s \tag{1.27}$$

Longitudinal magnetization corresponds to $p = 0$, single quantum coherence transverse magnetization to $p = \pm 1$, double-quantum coherence to $p = \pm 2$, and so on. In any sequence, the coherence transfer pathway begins with $p = 0$, corresponding to thermal equilibrium magnetization along B_0.

In order to select only the desired coherence pathway, phase cycling can be used, *i.e.* repeating the sequence a number of times with different phases of the RF pulses. This is an efficient way of coherence selection, but limits the minimum total experiment time, especially for sequences requiring extensive phase cycles (16, 32 or 64 steps). An alternative method of selecting specific coherence orders is to use gradients, using the principle that the phase accumulation of the spins during the application of the gradient pulse is proportional to the coherence order. So for a rectangular-shaped gradient pulse with strength G_z, applied for time τ, the phase ϕ is given by:

$$\phi(z,\tau) = p\gamma G_z z\tau \tag{1.28}$$

Eqn (1.28) shows, for example, that double-quantum coherences accumulate phase twice as fast as single-quantum coherence, and triple-quantum coherences three times as fast. Thus, by applying gradient pulses of different strengths or durations it is possible to refocus selective coherence orders, while unwanted coherences are dephased. Simple examples are shown in Figure 1.18 for double-quantum and triple-quantum-filtered COSY sequences.

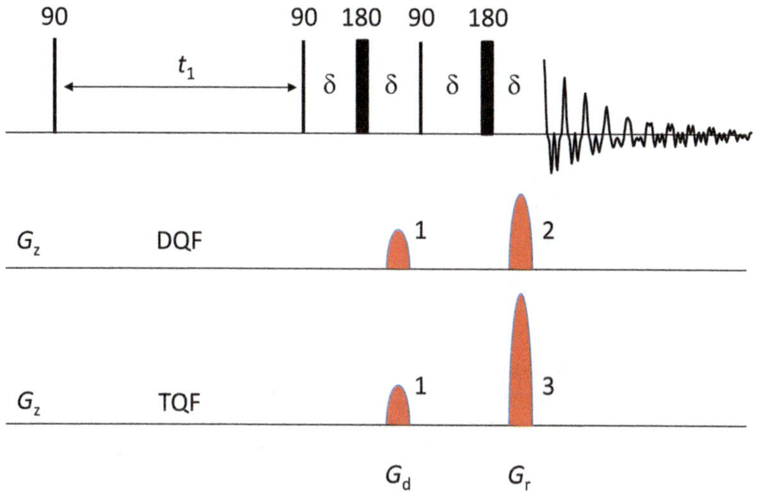

Figure 1.18 Pulse sequences for double quantum filtered (DQF) and triple quantum filtered (TQF) magnitude COSY spectra. The sequence of RF pulses is identical for both experiments. In the DQF experiment, the double quantum coherence created by the second 90° pulse is dephased by gradient G_d and the detected single quantum coherence is rephased by gradient G_r. If $G_r = 2G_d$, then DQ coherence only is refocused, if $G_r = 3G_d$ then TQ coherence only is refocused.

1.8.3 Measurements of Apparent Diffusion Coefficients Using Gradients

The phase evolution of transverse magnetization when a gradient is applied can also be used to measure the molecular diffusion coefficient of a sample. The most common sequence used for this purpose is the Stejskal–Tanner spin-echo, or variations thereof, shown in Figure 1.19.

The phase accumulation during the first gradient is given by:

$$\phi_z = \gamma \int G_z z \mathrm{d}t = \gamma G_z z \delta \tag{1.29}$$

The 180° pulse refocuses B_0 inhomogeneities and chemical shift effects. If the diffusion coefficient of the sample is zero, then there is an equal and opposite phase accumulation produced by the gradient applied after the 180° pulse, and the effects of phase accumulation are completely rephased, the signal simply being attenuated by T_2 relaxation. However, if there is indeed a physical displacement of the molecules, *i.e.* diffusion, during the time interval between the two gradient pulses, then the two phase accumulations are not equal and the signal amplitude is attenuated. The faster the diffusion coefficient (D) the greater the signal attenuation. In the "short-pulse limit", *i.e.* when $\delta \ll \Delta$, the signal intensity is given by:

$$S(G_z) = S(G_z = 0)\mathrm{e}^{-\gamma^2 G^2 \delta^2 (\Delta - \delta/3) D} = S(G_z = 0)\mathrm{e}^{-bD} \tag{1.30}$$

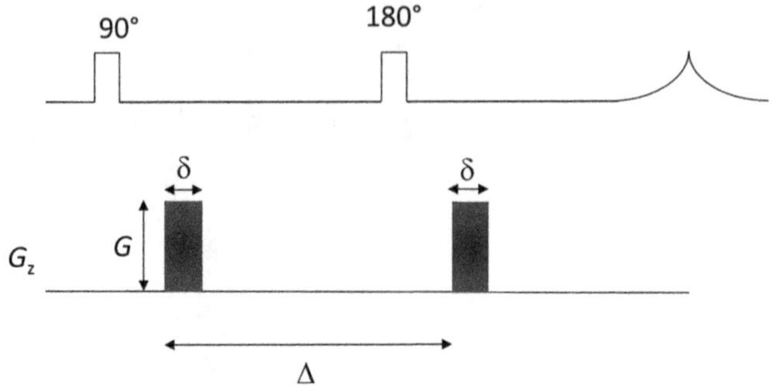

Figure 1.19 The Stejskal–Tanner spin-echo experiment used to measure the apparent diffusion coefficient (D) of a sample. The value of D can be calculated from a plot of the measured signal intensities as a function of the applied gradient strength for a series of different gradient values.

where the b-factor is defined as:

$$b = \gamma^2 G^2 \delta^2 (\Delta - \delta/3) \tag{1.31}$$

By repeating the sequence with a set of different b-values, usually *via* different values of the diffusion-encoding gradient, the resulting signal amplitudes can be fitted to eqn (1.30) to determine the value of D. In addition to measuring diffusion in homogeneous liquid samples, diffusion weighting can be used in high-resolution NMR[21,22] to filter out signals with a very fast diffusion coefficient, for example the water signal in a protein solution. Diffusion encoding sequences are very widely used in MRI in the areas of diffusion weighted imaging (DWI),[18,23] diffusion tensor imaging (DTI),[24] diffusion kurtosis imaging (DKI),[25,26] and fiber tractography.[27]

1.8.4 Gradient-Based Shimming

As outlined in Section 1.7, maximizing the magnetic field homogeneity for each sample involves B_0 shimming, *i.e.* optimizing the currents through each of the individual shim coils. This is a time-consuming process that is not ideal to perform manually, for example, for *in vivo* experiments. Rather than manual optimization, the process can be automated and speeded up by the use of gradients using a pulse sequence, such as the one shown in Figure 1.20(a), which measures the magnetic field inhomogeneity, $\Delta B_0(z)$ as:

$$\Delta B_0(z) = \frac{\Delta\phi(z)}{\gamma(TE_1 - TE_2)} \tag{1.32}$$

where $\Delta\phi$ is the spatially dependent phase difference between the measured projections shown in Figure 1.20(a). In terms of the shim settings, the residual magnetic field $\Delta B_{\text{res}}(z)$ can be defined as:

$$\Delta B_{\text{res}}(z) = \Delta B_0(z) - \sum_j c_j S_j(z) \tag{1.33}$$

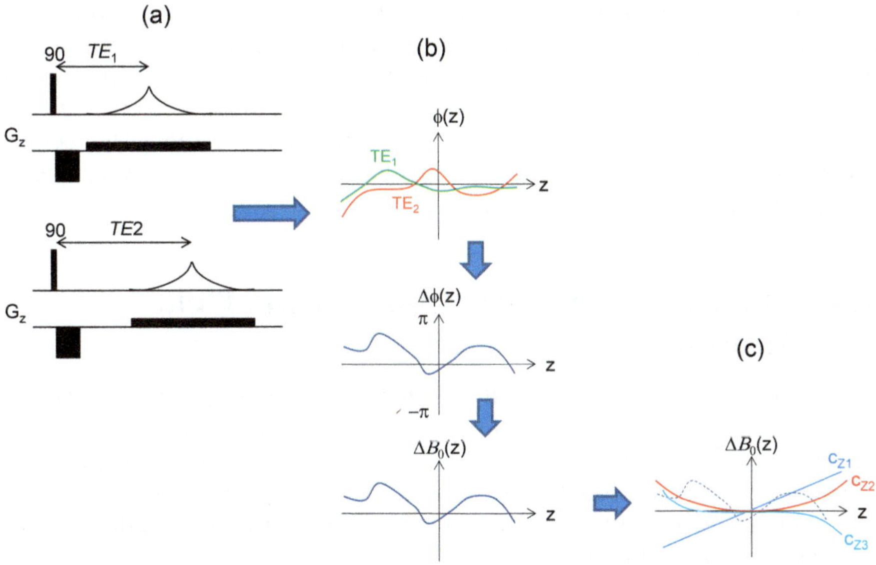

Figure 1.20 Schematic of the processes involved in gradient-based shimming. (a) One dimensional gradient projections of the sample are obtained for two different echo times, TE_1 and TE_2. (b) The difference in the phases ($\Delta\phi$) between the two projections is calculated and scaled to give the B_0 inhomogeneity (ΔB_0) as a function of *z*-position. (c) Knowing the magnetic field profiles of each of the shim coils, a least square fitting routine is implemented to calculate the currents through each of the shim coils which best compensates for the measured B_0 inhomogeneity.

where $S_j(z)$ is the magnetic field distribution of shim j, and c_j represents the corresponding weighting function. This residual is minimized by calculating the values of the coefficients c_j (*i.e.* the currents through each shim coil), which is normally performed in a least-squares sense over the entire one-dimensional *z*-projection:

$$\text{Min}\left[\sum_{k=1}^{N_z}\left|\Delta B_{\text{res}}(z_k)^2\right|\right] \tag{1.34}$$

where N_z is the number of data points acquired in the *z*-projection.

1.8.5 Gradients in MRI

In MRI three orthogonal sets of gradients are used to spatially encode the *x*, *y* and *z* dimensions of the sample. Each of these gradients is designed to provide a linear variation of magnetic field as a function of spatial dimension, *i.e.*:

$$\frac{\partial B_z}{\partial z} = G_z \; \frac{\partial B_z}{\partial x} = G_x \; \frac{\partial B_z}{\partial y} = G_y \tag{1.35}$$

The process of image formation is covered in more detail in Section 1.11 later in this chapter. For human MRI applications, typical gradient strengths of 60–70 mT m^{-1} are used, with some specialized gradients going up to 300 mT m^{-1}.[28,29] High gradient values can also be produced by reduced diameter "head-only insert" gradient coils. MRI is also extensively used in pre-clinical animal imaging systems, and also for microimaging experiments on samples as small as single cells.

1.9 The Deuterium Lock Channel and Field Monitoring

Although the static magnetic field is extremely stable over time, every magnet displays small frequency "drifts" that are typically of the order of 1–10 Hz per hour. Since many NMR experiments require data acquisition times on the order of hours or even days, it is essential to compensate in real-time for this field drift. If not corrected, then spectral linewidths are broadened, and artifacts can appear in the spectrum.

In high-resolution liquid-state NMR, this correction is performed using a deuterium "lock channel". One channel of one of the RF coils in the NMR probe is tuned to the deuterium (^{2}H) resonance frequency. The sample is dissolved in a fully or partially deuterated solvent (*e.g.* D_2O, $CDCl_3$, or C_6D_6). Deuterium is a quadrupolar nucleus with a very short T_1 value, and so the signal can be pulsed and sampled effectively continuously. The imaginary "dispersive" component of the deuterium spectrum is monitored, as shown in Figure 1.21(a). Changes in the static magnetic field produce a shift in the dispersive spectrum, which is analyzed to define an error signal, as shown in Figure 1.21(a). This error signal forms the input to a negative feedback loop, which controls the current applied to the Z_0 shim coil. This coil is a long solenoid, which produces a uniform magnetic field which can add or subtract from the main B_0 field. Since the deuterium signal can be quite noisy (owing to the low power used for pulse transmission, the low gamma of the deuterium nucleus, and the low sensitivity of the deuterium channel of the RF coil), the deuterium signal is averaged over many hundreds or thousands of samples so that the current in the Z_0 shim is only updated slowly. The negative feedback loop is designed such that it integrates the deuterium signals over a long period of time, and thus has a very slow time-constant. The lock channel effectively forms a separate simple "spectrometer", linked by a central clock/master oscillator to the rest of the data acquisition system.

The lock-channel can also be used to monitor the static field homogeneity during shimming. The better the homogeneity the sharper the deuterium line shape and the higher its amplitude. This system is still present on many high-resolution NMR systems but gradient-based shimming, as described in Section 1.8.4, is increasingly being used.

For most *in vivo* MRI and MRS experiments there are not as stringent requirements for the stability of the magnetic field as there are in high-resolution

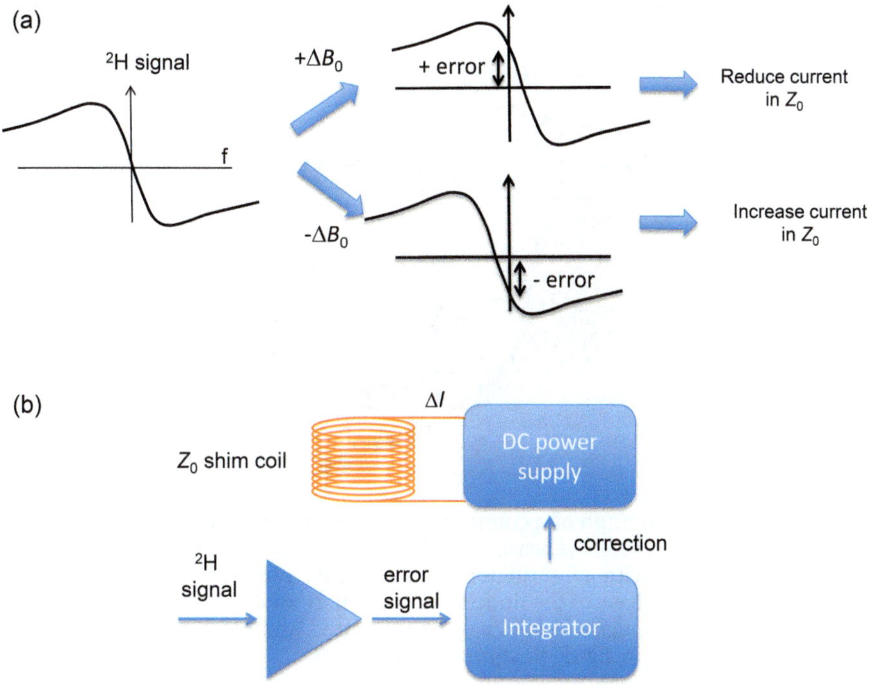

Figure 1.21 Schematic of the operation of the deuterium-based frequency lock on a high-resolution NMR system. (a) A drift in the magnetic field shifts the deuterium resonance to a higher or lower frequency. The dispersive line shape is monitored and a corresponding positive or negative "error" signal recorded as the signal intensity at zero frequency in the rotating reference frame. (b) This error signal is integrated over time and, *via* a negative feedback loop, alters the current (ΔI) fed into the Z_0 shim coil to correct for the magnetic field drift.

NMR. Experiments typically do not take as long, spectral linewidths *in vivo* are much wider owing to the intrinsic inhomogeneity of biological tissue, and imaging *per se* is much more robust with respect to small frequency drifts than spectroscopy. Therefore, in MRI systems there is no need for a separate lock channel. However, in MRI there are other mechanisms that can lead to drifts in the magnetic field. In sequences requiring rapid switching of the gradients, significant temperature changes can occur in the gradient coil, which in turn can heat components such as the passive iron shims and field booster rings. These thermal disturbances change the magnetic susceptibility of the component iron, and hence cause a change in the magnetic field. Changes owing to gradient heating are relatively slow owing to the large heat capacity of the physical former on which the gradient coils are wound, and the large mass of the structure, and so monitoring does not have to be continuous but can be interspersed between scans using very low tip angle spectroscopic measurements of the water resonance frequency, since it is the dominant species for *in vivo* measurements. Based on these measurements,

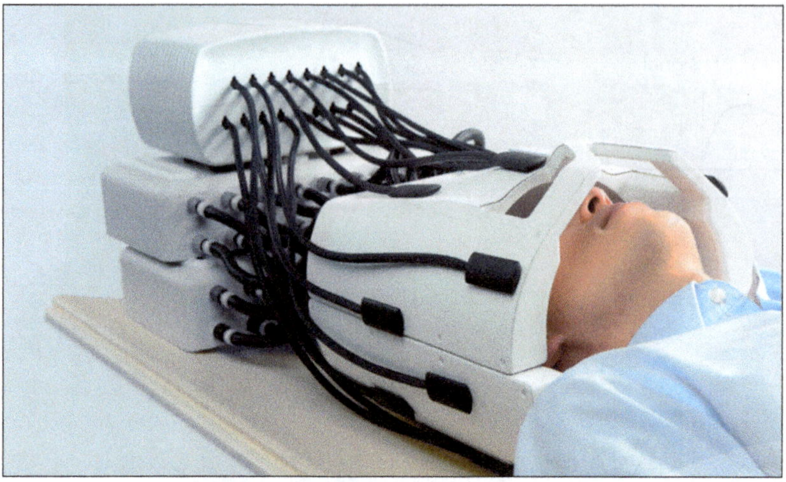

Figure 1.22 A photograph of a commercial field-monitoring system using a series of microcoils (encased in black plastic), tuned to the fluorine resonance, with a fluorinated sample inside. By continuous measurement of the phase of the fluorine MR signal, spatial and temporal variations in the magnetic field can be measured and corrected either in real-time or in signal post-processing.

rather than changing the magnetic field itself as described for the deuterium lock, the system simply alters the frequencies of the transmitter and receiver.

Although this type of monitoring can be used to compensate for global changes in the magnetic field, it cannot correct for spatially dependent changes, which, in addition to gradient heating, can be caused by many other factors, such as patient motion. Correcting for spatially dependent changes in the magnetic field clearly requires spatially dependent measurements. These can be implemented, for example, using multiple small microcoil probes that contain a fluorinated liquid and, similarly to the deuterium lock channel on a high-resolution NMR system, form a separate but linked MR spectrometer.[30] The use of multiple fluorinated probes, shown in Figure 1.22, allows the extraction of spatial variations in the main magnetic field, and the ability to apply real-time corrections to compensate for these fluctuations.

1.10 Magic Angle Spinning Solid-State NMR: Principles and Instrumental Requirements

NMR spectra of liquid samples consist of a series of very sharp resonances, owing to the averaging of anisotropic NMR interactions by rapid random tumbling, resulting in very long T_2 values, typically hundreds of milliseconds. In contrast, NMR spectra of solid or rigid samples are very broad, since anisotropic and orientation-dependent interactions are not averaged: an example is shown in Figure 1.23(a). The two most important interactions

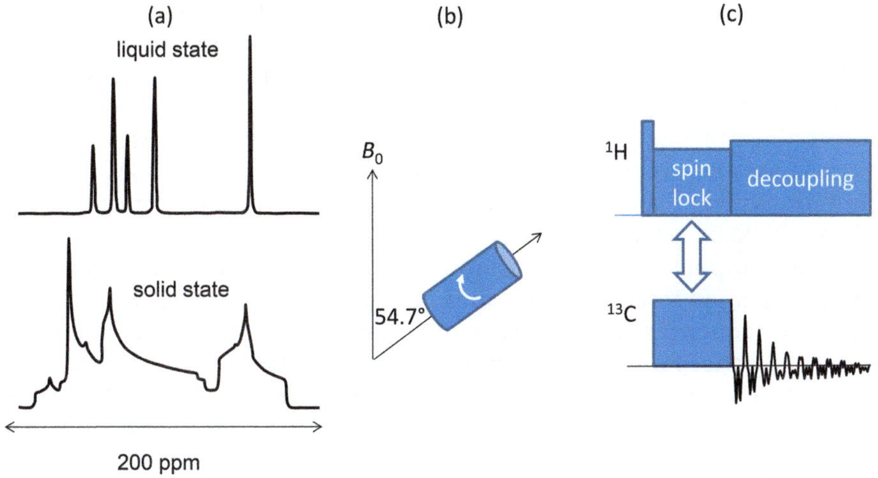

Figure 1.23 (a) Schematics of the respective spectra for ^{13}C in the liquid and solid states. (b) Spinning the sample at the magic angle (54.7°) with respect to B_0 is used to reduce the effect of dipolar coupling and chemical shift anisotropy. (c) A pulse sequence used for heteronuclear solid-state spectroscopy involving cross-polarization and high-power proton decoupling.

that produce broad linewidths in solids are dipolar coupling and chemical shielding, with the dipolar term being dominant especially at low magnetic fields.

Dipolar coupling results from the interaction of one nuclear spin with the magnetic field generated by a second nuclear spin, and *vice versa*. Dipolar coupling represents a direct through-space interaction, the magnitude of which is given by:

$$R_{jk}^{\mathrm{DD}} = \frac{\mu_0 \hbar}{4\pi} \frac{\gamma_j \gamma_k}{\langle r_{jk} \rangle^3} \tag{1.36}$$

where R_{jk}^{DD} is the dipolar coupling between two spins j and k, and r_{jk} is the distance between the spins.

Chemical shielding is related to the chemical shift outlined earlier for liquid samples, and is an anisotropic interaction characterized by a shielding tensor. Anisotropy in the chemical shift leads to broadening of the spectral lines. For molecules which are spatially symmetric in three-dimensions the chemical shift anisotropy (CSA) is very small. For non-symmetric molecules, the CSA is larger, and linewidths correspondingly wider.

Both dipolar coupling and chemical shielding have an angular dependence with respect to B_0, containing multiplicative terms $(3\cos^2\theta - 1)$, where θ is the angle between a line connecting the two nuclei and B_0 for the dipolar coupling, and between the direction with the largest deshielding and B_0 for the chemical shift term. One can therefore effectively average out these components by rotating the sample very rapidly at an angle of 54.74° with respect to B_0 such that $3\cos^2\theta - 1 = 0$, a technique termed magic-angle spinning (MAS),[31] illustrated in Figure 1.23(b). For full averaging the rate of spinning must be greater than the anisotropic interaction. State-of-the-art microprobes are able to spin at a rate of 111 kHz.[32] For regular-sized samples, typical diameter/spinning speeds are 1.3 mm diameter/65 kHz, 2.5 mm/35 kHz, 3.2 mm/23 kHz, 4 mm/8 kHz, and 7 mm/4 kHz. Since proton dipole–dipole coupling constants can be much greater than 100 kHz, full averaging is rarely achieved and MAS is often combined with multiple-pulse sequences, which also reduce the effects of dipolar coupling, based on concepts originally developed as WAHUHA[33] or MREV-8.[34,35] The combination of MAS with multiple-pulse sequences is called CRAMPS (Combined Rotation and Multiple-Pulse Sequence) and requires high-power amplifiers, short RF times, and fast switching between the transmit and receive mode.

For dilute spin-1/2 nuclei, such as ^{13}C, the homonuclear (^{13}C–^{13}C) dipolar coupling is insignificant owing to its low occurrence. The ^{1}H–^{13}C heteronuclear dipolar coupling can be minimized with high-power proton decoupling,[33,36] as shown in Figure 1.23(c). The sensitivity of this type of experiment can be increased using cross-polarization (CP)[37] techniques, which involve polarization transfer from ^{1}H to ^{13}C. Efficient CP requires that the precessional frequencies in the rotating frame are the same for protons and carbon, *i.e.*:

$$\gamma_C B_{1,C}^{+} = \gamma_H B_{1,H}^{+} \tag{1.37}$$

This requirement implies that the B_1^+ fields must be spatially homogeneous at both frequencies, and as discussed in Chapter 3 this requires that the coils be very well balanced, which is challenging, particularly at very high frequencies. Cross-polarization is generally combined with relatively low-speed MAS (5 to 15 kHz) in a combined approach known as CP–MAS.

In a similar fashion to that described previously for liquid-state NMR, gradient coils can be incorporated into the probe in order to perform coherence pathway selection in multiple quantum solid-state experiments. In this case, the gradients must be oriented along the magic angle.[38,39]

The instrumental requirements corresponding to the types of experiments outlined above include:

(i) The ability to spin the sample extremely rapidly with a high degree of mechanical precision.
(ii) Very short (~1 μs), high-power RF pulses must be applied, which place high degrees of demand on the RF amplifiers and components in the RF probe.
(iii) The RF fields for different nuclei must be spatially very well matched and homogeneous for efficient cross-polarization.

(iv) Isolation between the receive chains of the different nuclei, as well as between transmit and receive chains, must be very high since high-power decoupling may be applied during data acquisition.
(v) Magic angle gradients and associated gradient amplifiers must be capable of producing gradient pulses with durations below 100 μs and strengths of 500 mT m^{-1}.

1.11 Magnetic Resonance Imaging: Principles and Instrumental Requirements

In MRI there are a large number of different imaging sequences, which can be used for structural imaging, functional imaging, magnetic resonance angiography, and diffusion-weighted imaging, to name only a few.[12] Two generic sequences, or "building-blocks", are shown in Figure 1.24, namely a gradient-echo and spin-echo sequence.

In the sequences shown in Figure 1.24 there are three "processes" used to produce the two-dimensional image. First, slice-selection uses a frequency-selective RF pulse applied simultaneously with a magnetic field gradient, denoted by G_{slice}. If the selective RF pulse, with an excitation bandwidth of $\pm\Delta\omega_s$, is applied at a frequency ω_s, then protons precessing at frequencies between $\omega_s + \Delta\omega_s$ and $\omega_s - \Delta\omega_s$ are rotated from the z-axis into the transverse plane, whereas protons with precession frequencies outside this range do not experience the RF pulse, as shown in Figure 1.25(a). By changing the centre frequency (ω_s) of the RF pulse, the slice can be centred at different parts of the sample. As indicated in Figure 1.25(b), the thickness of the slice can be decreased by using a larger gradient strength or a longer duration RF pulse that has a smaller frequency bandwidth.

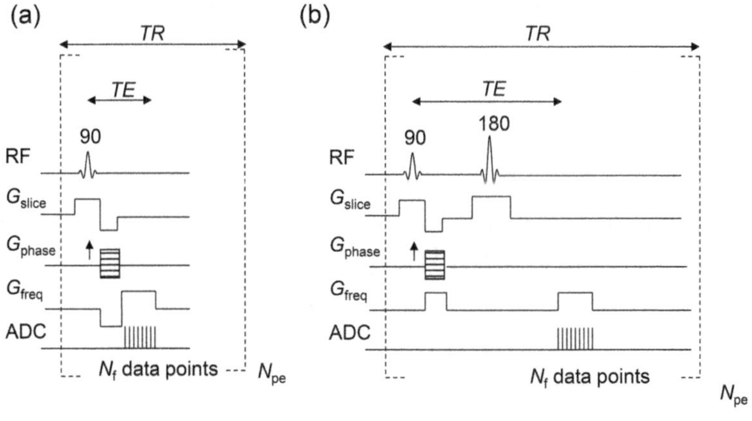

Figure 1.24 (a) Generic two-dimensional gradient-echo imaging sequence. The sequence is repeated N_{pe} times, with the phase encoding gradient being incremented from its maximum negative value to its maximum positive value. (b) Corresponding generic spin-echo imaging sequence. TR represents the repetition time and TE the echo time.

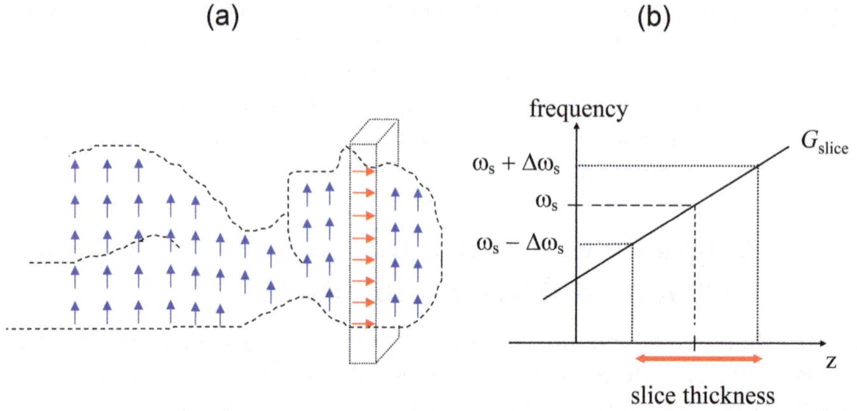

Figure 1.25 The principle of slice selection in MRI. (a) By applying a frequency-selective RF pulse in combination with a gradient, only the nuclei within a thin "slice" experience the RF pulse and are rotated from the longitudinal axis (blue) into the transverse plane (red). (b) The thickness of the slice can be controlled by the strength of the gradient and the bandwidth of the RF pulse.

Having selected a slice, a phase encoding gradient is applied to encode an orthogonal direction *via* the phase of the signal. After the phase encoding gradient has been switched off, the frequency-encoding gradient (G_{freq}) is turned on and data acquisition begins. N_f data points are sampled while the frequency encoding gradient is on, after which a relaxation delay (TR) occurs before the sequence is repeated using the next incremented value of the phase-encoding gradient. A total of N_{pe} phase encoding gradients are used, resulting in a total imaging time of TR × N_{pe}.

Assigning the G_y gradient to phase encoding, and G_x to frequency encoding, the combined effect of the phase-encoding and frequency-encoding gradients gives a signal:

$$s\left(G_y,\tau_{pe},G_x,t\right)\propto\int_{slice}\int_{slice}\rho(x,y)e^{-j\gamma G_x xt}e^{-j\gamma G_y y\tau_{pe}}dx\,dy \tag{1.38}$$

where τ_{pe} is the time for which the phase encoding gradient is applied. Two variables k_x and k_y can be defined as:[40,41]

$$k_x=\frac{\gamma}{2\pi}G_x t\ \ k_y=\frac{\gamma}{2\pi}G_y\tau_{pe} \tag{1.39}$$

and eqn (1.38) can be re-expressed as:

$$S\left(k_x,k_y\right)\propto\int_{slice}\int_{slice}\rho(x,y)e^{-2\pi jk_x x}e^{-2\pi jk_y y}dx\,dy \tag{1.40}$$

The $N_f \times N_p$ data matrix can be visualized as a two-dimensional dataset as a function of k_x and k_y, commonly referred to as the *k*-space domain. Consider

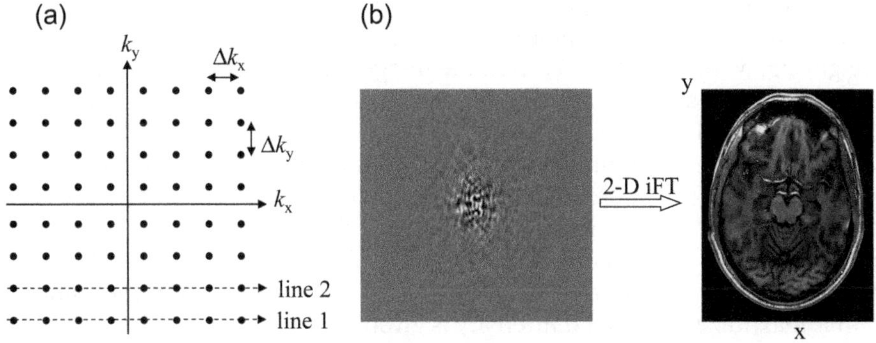

Figure 1.26 (a) Data for MRI are acquired in "k-space", with a Cartesian grid shown in this case. (b) The image is reconstruced by a two-dimensional inverse Fourier transform of the k-space data.

the N_r data points collected when the maximum negative value of the phase-encoding gradient, G_y, is applied. From eqn (1.39) the value of k_y for all N_r data points corresponds to its maximum negative value. When the frequency-encoding gradient is switched on, the first data point collected corresponds to the maximum negative value of k_x due to the negative dephasing lobe shown in Figure 1.24(a), the second data point to a slightly more positive value of k_x and so forth, and so the N_f data points correspond to one "line" in k-space, shown as line 1 in Figure 1.26(a). The second line in k-space corresponds to the next value of the phase-encoding gradient, and so on. The spacing between the k-space points is inversely proportional to the respective spatial dimensions, or fields-of-view (FOV), in the image: $\Delta k_x = 1/\mathrm{FOV}_x$, $\Delta k_y = 1/\mathrm{FOV}_y$.

A two-dimensional inverse Fourier transform of the k-space data $S(k_x, k_y)$ gives $\rho(x,y)$, which is the MR image, as shown in Figure 1.26(b). MR images are conventionally represented as the magnitude of $\rho(x,y)$.

$$\rho(x,y) \propto \int_{-\infty}^{\infty}\int_{-\infty}^{\infty} S(k_x,k_y)\mathrm{e}^{+\mathrm{j}2\pi(k_x x+k_y y)}\mathrm{d}k_x\mathrm{d}k_y \tag{1.41}$$

The signal intensity in each voxel of the image depends on the tissue relaxation times as well as the proton density. In terms of the MR relaxation times, for an axial image acquired using slice-selection in the z-direction, the image intensity of each voxel, $I(x,y)$, is given by:

$$I(x,y) \propto \rho(x,y)\frac{\left(1-\mathrm{e}^{-\frac{\mathrm{TR}}{T_1}}\right)\sin\alpha}{1-\mathrm{e}^{-\frac{\mathrm{TR}}{T_1}}\cos\alpha}\mathrm{e}^{-\frac{\mathrm{TE}}{T_2^*}} \tag{1.42}$$

where $\rho(x,y)$ is the "proton density" and the tip angle of the RF pulse in the gradient echo sequence is $\alpha°$. The degree of T_1-weighting of the image is determined by the respective values of the tissue T_1 and the chosen TR interval between repeated phase-encoding steps: if TR $\ll T_1$ then there is no weighting, whereas if TR $\approx T_1$ then the image intensity is weighted by different tissue T_1 relaxation times. Similarly, the situation of TE $\approx T_2^*$ introduces T_2^*-weighting into the image.

Although eqn (1.42) indicates that the SNR is maximized using a value of $\alpha = 90°$, this case would require a long value of TR to allow full T_1 relaxation to occur, and as a result an impractically long data acquisition time for clinical applications. In order to image more rapidly, the value of α is reduced to a value considerably smaller than 90°. For a given value of TR, the value of α that maximizes the signal intensity is given by:

$$\alpha_{\text{Ernst}} = \cos^{-1} e^{-\frac{\text{TR}}{T_1}} \tag{1.43}$$

For example, if TR is reduced to 0.05 T_1, then the optimum value of α is 8°. Using these parameters, images can be acquired in a few tens of seconds.

Gradient-echo sequences allow very rapid image acquisition. The major disadvantage is that they are weighted by tissue T_2^* values, which are typically much shorter than T_2 *in vivo*. In order to introduce pure T_2-contrast into the image a spin-echo sequence can be used, as shown in Figure 1.24(b). The spatial encoding principles are exactly the same as for the gradient echo image, but now the intensity is weighed by the T_2 rather than T_2^* value:

$$I(x,y) \propto \rho(x,y)\left(1 - e^{-\frac{\text{TR}}{T_1}}\right) e^{-\frac{\text{TE}}{T_2}} \tag{1.44}$$

As mentioned previously there are very many types of MRI sequences. Common instrumental requirements include:

1. RF coils and RF amplifiers must be able to handle/provide up to 30–60 kW of power, with duty cycles up to ~20%. The pulses are both amplitude and phase modulated, with strict requirements for the reproducibility. The amplifiers must be linear over a large range of signal inputs/outputs.
2. For rapid imaging sequences such as echo planar imaging, gradient switching rates in the hundreds of hertz to kilohertz may be necessary. In this case, one also needs to consider peripheral nerve stimulation (PNS), which can occur when the dB/dt is on the order of 50–100 T s^{-1}.
3. Data acquisition involves much higher bandwidths than for spectroscopic experiments: fast imaging sequences may have bandwidths up to 1 MHz for very high gradient strengths.
4. The amount of streaming data is very high: a commercial 32-channel receive array may acquire data essentially continuously over 20 minutes, representing several gigabytes of data that has to be processed in real-time for clinical evaluation.

Appendices

As seen in this first chapter, the hardware components of the MR system share many geometric properties and design goals. A cylindrical geometry is inherent in superconducting magnet design, RF volume coils, shim coils and gradient coils. Maximizing the homogeneity of the magnetic field is critical for superconducting magnet and RF coil design; producing a linear spatial variation in magnetic field is key for gradient coil design; and well-defined higher-order spatial distributions of the magnetic field are required for shim coil design. The basis for each of these component is a set of current-carrying electrical conductors. Despite the fact that the operating frequencies of these different components are quite different (DC for magnets and shim coils, tens to hundreds of kHz for gradients, and tens to hundreds of MHz for RF coils), there are common mathematical tools that can be used to produce the desired magnetic field distributions. Two of the most common formalisms, namely the Biot–Savart law and spherical harmonic decomposition, together with specific examples of their use, are described in these two appendices. These models are further refined, and more sophisticated examples given, in several of the chapters later in the book.

Appendix A. Maxwell's Equations and the Biot–Savart Law

The fundamental bases for calculating both magnetic and electric fields are termed Maxwell's equations, listed in eqn (1.45–1.48).

$$\nabla \bullet E = \frac{\rho_v}{\varepsilon} \text{(Gauss' law)} \tag{1.45}$$

$$\nabla \bullet H = 0 \text{(Gauss' law for magnetism)} \tag{1.46}$$

$$\nabla \times E = -\mu \frac{\partial H}{\partial t} \text{(Faraday's law)} \tag{1.47}$$

$$\nabla \times H = J + \varepsilon \frac{\partial E}{\partial t} \text{(Ampere's law)} \tag{1.48}$$

where ρ_v is the electric charge density, ε the permittivity, μ the permeability, H the magnetic field, E the electric field, and J the electric current density. The operators on the left of eqn (1.45) and (1.46) represent divergence, and those in eqn (1.47) and (1.48) represent curl operators. In physical terms, the divergence of a vector field describes the degree to which it behaves as either a source or a sink at a given point. If the divergence is non-zero at some point then there must be a source or sink at that position: in contrast, if the divergence is zero then it indicates that there is neither. The curl operators on the left of eqn (1.47) and (1.48) represent the spatial-variation of the electric and

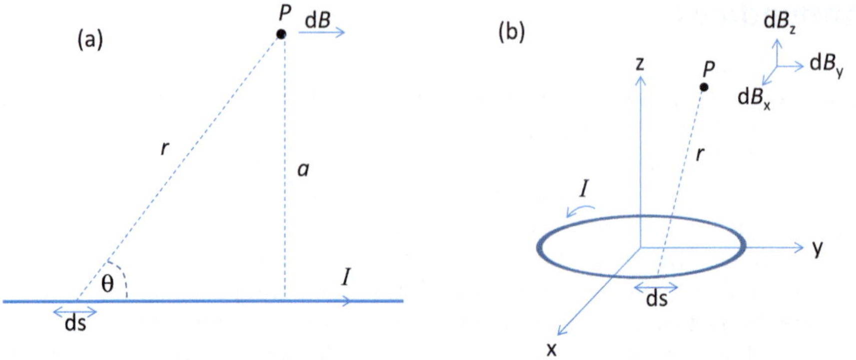

Figure 1.27 Examples of (a) a current-carrying straight wire and (b) a current-carrying circular wire-loop, for calculation of magnetic fields using the Biot–Savart law.

magnetic fields, which are coupled to the time-variation of the magnetic and electric fields on the right hand side. In other words, as an electric field propagates throughout the sample it gives rise to a time-varying magnetic field. This magnetic field, which varies as a function of space, in turn gives rise to a time-varying electric field. Eqn (1.47) and (1.48) show that magnetic and electric fields are intrinsically coupled.

Designs for magnets, gradient coils, shim coils or RF coils can be formulated by simplifying Maxwell's equations under conditions known as "quasi-static", which corresponds to being able to neglect any effects of the finite wavelength of the EM wave within the sample. In this case, the magnetic field can be estimated quite accurately by the Biot–Savart law, which can be derived directly from Maxwell's equations. The Biot–Savart law describes the magnetic field (dB) at a point P distance r from the wire, produced by a current I flowing through a small section (ds) of wire. The value of dB is given by:

$$\mathrm{d}\vec{B} = \frac{\mu_0}{4\pi} I \frac{\mathrm{d}\vec{s} \times \vec{r}}{\left| r^3 \right|} \tag{1.49}$$

where μ_0 is the permeability of free space, equal to 1.257×10^{-6} T m A^{-1}. Two very simple examples are given below, a straight wire and a circular loop, both shown in Figure 1.27.

Example A1. Magnetic Field Produced by a Straight Wire

In Figure 1.27(a), the value of dB for a straight wire can be calculated by simple trigonometry:

$$\mathrm{d}\vec{B} = \frac{\mu_0}{4\pi} I \left(\frac{\sin\theta}{a} \right)^2 \sin\theta \mathrm{d}x \tag{1.50}$$

The total magnetic field produced by the current in the infinitely long wire is then simply given by the integral of this expression:

$$B = \frac{\mu_0 I}{4\pi a}\int_0^{\pi} -\sin\theta \, d\theta = \frac{\mu_0 I}{2\pi a} \tag{1.51}$$

The result shows that the magnetic field decreases as the inverse of distance from the wire.

Example A2. Magnetic Field Produced by a Circular Wire Loop

Consider a loop of wire with radius a, shown in Figure 1.27(b). If one considers the field produced along the central axis of the loop, there is only one non-zero component of the magnetic field, which is oriented along the z-direction and is given by:

$$B_z = \frac{\mu_0 a^2 I}{2\left(z^2 + a^2\right)^{3/2}} \tag{1.52}$$

So at the centre of the loop, the magnetic field is given by $\mu_0 I/2a$, *i.e.* the smaller the wire loop the higher the magnetic field produced per unit current. Along the central axis of the coil, *i.e.* along the z-axis, the magnetic field decreases approximately exponentially with distance.

The off-axis magnetic fields can also be calculated analytically, but are more complicated. They are given (in polar coordinates) by:

$$\begin{aligned} B_r &= \frac{\mu_0 I}{2\pi}\frac{z}{r\left[(a+r)^2 + z^2\right]^{\frac{1}{2}}}\left[\frac{a^2 + r^2 + z^2}{(a-r)^2 + z^2}E(k) - K(k)\right] \\ B_\theta &= 0 \\ B_z &= \frac{\mu_0 I}{2\pi}\frac{z}{r\left[(a+r)^2 + z^2\right]^{\frac{1}{2}}}\left[\frac{a^2 - r^2 - z^2}{(a-r)^2 + z^2}E(k) + K(k)\right] \end{aligned} \tag{1.53}$$

where E and K are complete elliptical integrals of the second and first kinds, respectively, and k is defined as:

$$k = \sqrt{\frac{4ar}{(a+r)^2 + z^2}} \tag{1.54}$$

Using these analytical results, designs using circular loops in particular can be optimized in terms of producing the desired spatial distribution of the magnetic field. The following two examples show the application in designing a linear magnetic field gradient and a homogeneous magnetic field, both with a simple two-loop arrangement.

Example A3. Design of a Two-Loop Coil Geometry to Produce a Linear *z*-Gradient

The simplest configuration for a coil producing a gradient in the *z*-direction is a "Maxwell pair", shown in Figure 1.28(a), which consists of two separate loops of multiple turns of wire, each loop containing equal currents, *I*, flowing in opposite directions. In order to estimate the distance between the two loops that maximizes the linearity of the gradient, the value of B_z is first calculated using eqn (1.52):

$$B_z = \frac{\mu_0 I a^2}{2\left[\left(d/2 - z\right)^2 + a^2\right]^{1.5}} - \frac{\mu_0 I a^2}{2\left[\left(d/2 + z\right)^2 + a^2\right]^{1.5}} \tag{1.55}$$

where μ_0 is the permeability of free space and a is the radius of the gradient set. By applying a Taylor series expansion, and noting that by symmetry the first two differentiable terms are zero, the first term that can give a non-linear contribution is the third derivative, given by:

$$\frac{\mathrm{d}^3 B_z}{\mathrm{d}z^3} = \frac{15\mu_0 I a^2}{2}\left\{\frac{4\left(d/2 - z\right)^3 - 3\left(d/2 - z\right)a^2}{\left[\left(d/2 - z\right)^2 + a^2\right]^{9/2}} + \frac{4\left(d/2 + z\right)^3 - 3\left(d/2 + z\right)a^2}{\left[\left(d/2 + z\right)^2 + a^2\right]^{9/2}}\right\} \tag{1.56}$$

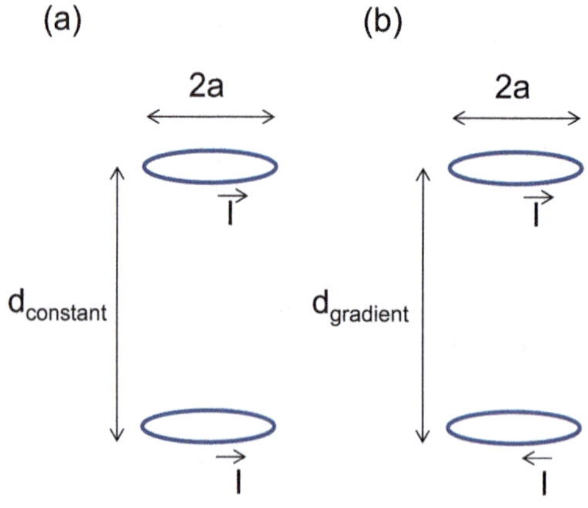

Figure 1.28 Physical arrangements of wire loops for (a) a linear gradient and (b) a constant homogeneous field. The values of d_{gradient} and d_{constant} are derived in terms of the coil radius a in examples A3 and A4.

This term becomes zero at a value of $d = a\sqrt{3}$. The magnetic field produced by this gradient coil is zero at the centre of the coil, and is linearly dependent upon position in the *z*-direction over about one-third of the separation of the two loops. The region over which the gradient is linear can be extended by adding other sets of coils in the axial dimension, as covered later in more detail in Chapters 2 and 4.

Example A4. Design of a Two-Loop Coil Geometry to Produce a Homogeneous Magnetic Field

Using the same arrangement as before, except with the currents now flowing in the same direction shown in Figure 1.28(b), the first-order term in the Taylor expansion is zero, and so the first perturbation arises from the second order term, which should therefore be set to zero, *i.e.*:

$$\frac{\partial^2 B_z}{\partial z^2} = 0 \tag{1.57}$$

Combining eqn (1.55) and (1.57) gives the optimum value of the distance between the coils equal to the radius of each coil. In this case the magnetic field at the centre of the coil is given by:

$$B_{z=0} = \left(\frac{4}{5}\right)^{3/2} \frac{\mu_0 nI}{a} \tag{1.58}$$

Appendix B. Spherical Harmonic Representation of Magnetic Fields

As seen in Appendix A, the Biot–Savart law is an effective method for optimizing the design of simple geometries in terms of their behaviour on-axis. However, it becomes much more cumbersome to use when one needs to consider off-axis terms. In this case, a very widely used formalism is to represent magnetic fields in terms of spherical harmonics, as described in a simplified form below.

Visual Description

The starting point is to consider the magnetic field within a sphere of a certain diameter positioned at the centre of the magnet. If the aim is to produce a homogeneous magnetic field then one can clearly define the desired field by a single vector term B_z with fixed amplitude and direction within the sphere. As will be seen in Chapter 2, this is not possible to achieve in practice (theoretically one can achieve this only by passing current through an infinitely long solenoid) and there will be spatial perturbations in the magnetic field. As in many other areas of engineering one can define these perturbations in terms of increasing orders, *i.e.* first-order, second-order *etc.* In the case of spherical harmonics, there are three first order perturbations, five second order, seven third order, and so on. Within each order, there are a number of degrees, *e.g.* five degrees within the second order, seven degrees within the third order *etc.* The order is given the symbol n, and the degree the symbol m. The zero order term ($m = 0$, $n = 0$) represents a homogeneous magnetic field. The three first order ($n = 1$) perturbations correspond to linear variations in magnetic field, with the three orthogonal directions obviously being x, y and z. The degree (m) of the perturbation corresponds to the particular direction, with $m = +1$ being x, $m = 0$ being z, and $m = -1$ being y. The five second order terms ($n = 2$) correspond to $m = +2, +1, 0, -1$ and -2. Figure 1.29 shows the spatial distributions of the zero, first and second order spherical harmonics.

If one considers only the terms with $m = 0$, one can see that these are rotationally invariant with respect to the z-axis: these are called zonal harmonics. Considering the spheres to have longitudinal and latitudinal axes, analogous to a globe, there are $2|m|$ zeroes in the longitudinal axis and $n - |m|$ zeroes in the latitudinal one.

Mathematical Description of Spherical Harmonics

Starting from Gauss' law of magnetism and Ampere's law, eqn (1.47) and (1.48), one can derive a very useful equation, which is referred to as Laplace's equation.

$$\nabla^2 B_x = \nabla^2 B_y = \nabla^2 B_z = 0 \tag{1.59}$$

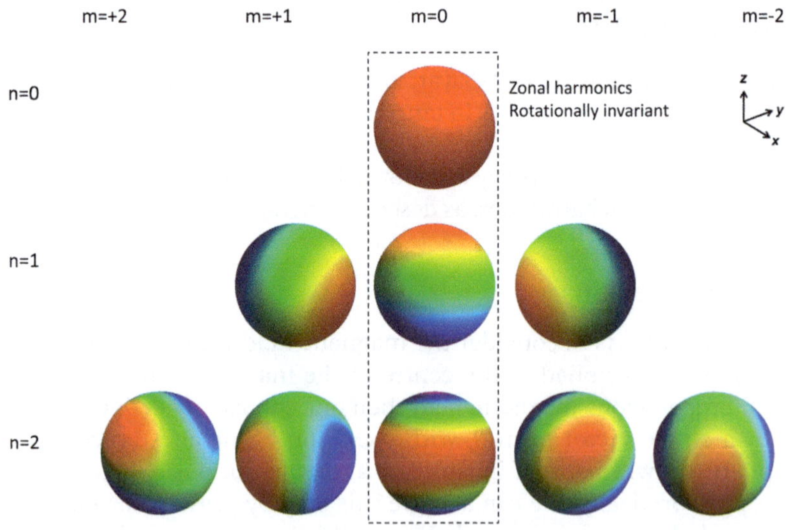

Figure 1.29 Visual representation of lower order spherical harmonics.

where the differential Laplacian operator is defined as:

$$\nabla^2 = \frac{\partial^2}{\partial x^2} + \frac{\partial^2}{\partial y^2}\frac{\partial^2}{\partial z^2} \tag{1.60}$$

In the case of an MR magnet, the magnetic field is aligned along one axis (the *z*-axis), and so one can simplify the equation to give:

$$\nabla^2 B_z = 0 \tag{1.61}$$

This equation is often solved in cylindrical coordinates (the derivation is long and complicated and covered in many textbooks so is not repeated here), shown in Figure 1.30, where θ and ϕ are the polar and azimuthal angles, respectively, and r is the radius.

The solution to Laplace's equation results in a magnetic field $B(r, \theta, \phi)$ which can be described by an expansion of orthogonal spherical harmonic functions Y_{nm} and coefficients $C_{n,m}$.

$$B(r,\theta,\phi) = \sum_{n=0}^{\infty}\sum_{m=-n}^{+n} C_{n,m} r^n Y_{nm}(\theta,\phi) \tag{1.62}$$

where n is the order and m the degree, as outlined in the previous section. Spherical harmonic functions are defined on the surface of a sphere, and are given by:

$$Y_{n,m}(\theta,\phi) = C_{n,m} r^n P_{n,m}(\cos\theta)\cos(m\phi - \phi_{n,m}) \tag{1.63}$$

where $C_{n,m}$ and $\phi_{n,m}$ are constants ($\phi_{n,m} = 0$ for $m \geq 0$ and $\pi/2$ for $m < 0$), and $P_{nm}(\cos\theta)$ are polynomial functions known as Legendre polynomial functions

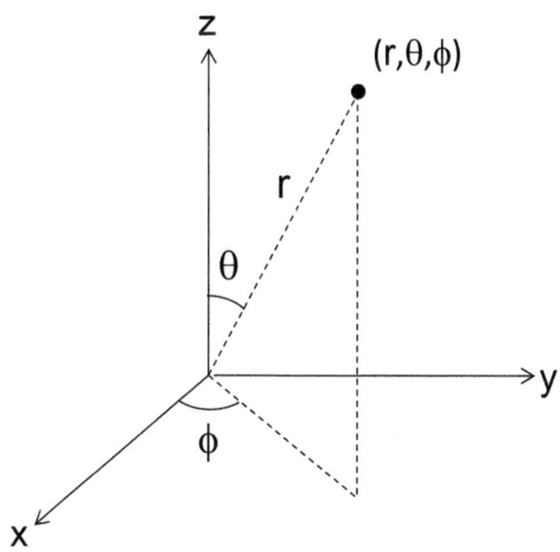

Figure 1.30 Cylindrical axis used for the description of spherical harmonic functions.

for $m = 0$, and associated Legendre polynomial functions for $m \neq 0$. Legendre polynomial functions can be calculated from:

$$P_n(x) = \frac{1}{2^n n!} \frac{\mathrm{d}^n}{\mathrm{d}x^n}\left[\left(x^2 - 1\right)^n\right] \tag{1.64}$$

and associated Legendre polynomials *via*:

$$P_n^m(x) = (-1)^m \left(1 - x^2\right)^{m/2} \frac{\mathrm{d}^m}{\mathrm{d}x^m}\left[P_n(x)\right] \tag{1.65}$$

Combining eqn (1.62) and (1.63), the magnetic field can be expressed as:

$$B(r,\theta,\phi) = \sum_{n=0}^{\infty} \sum_{m=-n}^{+n} C_{n,m} r^n P_{n,m}(\cos\theta)\cos(m\phi - \phi_{m,n}) \tag{1.66}$$

Table 1.2 gives the mathematical values of the lower order spherical harmonics that correspond to those shown in Figure 1.29.

To relate this mathematical description to the visual one presented earlier, one can consider a number of simple cases:

(i) All cases with $m = 0$. From eqn (1.66) the magnetic field is independent of ϕ, *i.e.* it is rotationally invariant around the z-axis. For the particular case that $n = 0$, $m = 0$, the magnetic field is constant. When $m = 0$, the remaining Legendre polynomials are referred to as *zonal* harmonics, and give rise to what is termed a *zonal* magnetic field.

$$B_{z,\text{zonal}} = C_n r^n P_n(\cos\theta) \tag{1.67}$$

If $\theta = 0$, *i.e.* along the z-axis:

$$B_z = C_n z^n P_n(1) = C_n z^n \tag{1.68}$$

Since $P_n(1) = 1$, this can be expanded to give:

$$B_z = C_0 + C_1 z + C_2 z^2 + C_3 z^3 \ldots \tag{1.69}$$

Table 1.2 Legendre and associated Legendre functions $P_{n,m}(\cos\theta)$.

		$P_{0,0}(\cos\theta)$ **1**		
	$P_{1,1}(\cos\theta)$ $\mathbf{\sin\theta}$	$P_{1,0}(\cos\theta)$ $\mathbf{\cos\theta}$	$P_{1,-1}(\cos\theta)$ $\mathbf{\sin\theta}$	
$P_{2,2}(\cos\theta)$ $\mathbf{3\sin^2\theta}$	$P_{2,1}(\cos\theta)$ $\mathbf{3\sin\theta\cos\theta}$	$P_{2,0}(\cos\theta)$ $\mathbf{1/2(3\cos^2\theta - 1)}$	$P_{2,-1}(\cos\theta)$ $\mathbf{3\sin\theta\cos\theta}$	$P_{2,-2}(\cos\theta)$ $\mathbf{3\sin^2\theta}$

This expression is very useful in determining parameters in, for example, gradient coil design as covered in the examples at the end of this Appendix.

(ii) When $m \neq 0$, the magnetic field around the z axis "oscillates" with a frequency $m\phi$ and initial phase $\phi_{n,m}$. For the case that $m = +1$, the field around the circumference should exhibit one cycle, and for $m = -1$ also for one cycle but with a $\pi/2$ phase shift in the origin of the cycle. Similarly for $m = +2$ and -2, there are two cycles circumferentially, as seen in Figure 1.29. Spherical harmonics with $m \neq 0$ are referred to as *tesseral* harmonics.

(iii) For $n = \pm1$ and $m = 0$ the function is rotationally invariant, and there is one cycle as a function of θ.

(iv) For $n = 1$, $m = \pm1$ the number of zeros as a function of θ is zero. These correspond to what are termed as *sectoral* harmonics, corresponding to the case $n = |m|$. Since $|m| = 1$ then there is one cycle around the circumference.

Having given a mathematical description of spherical harmonics, the final step is to show how these can be used in very simple examples of generating specific spatial distributions of the magnetic field. The first two cases are identical to those solved in Appendix A to show the correspondence between the two techniques.

Example B1. Generation of a Homogeneous Magnetic Field

The basic building block to generate zonal harmonics is a current loop with its central axis lying along the z-direction. A declination angle α is defined with respect to a central point $z = 0$ as shown in Figure 1.31.

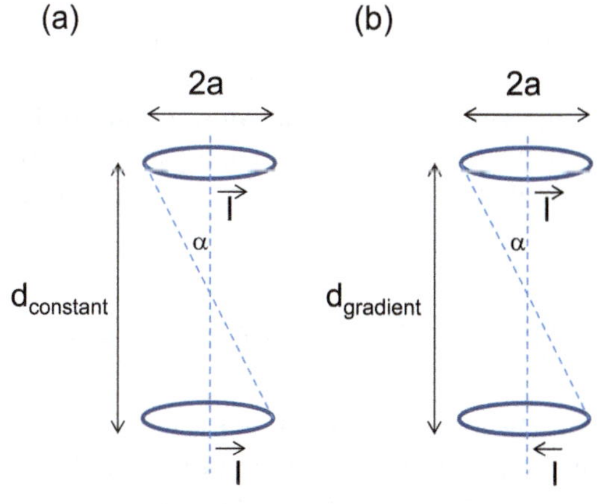

Figure 1.31 Two wire loops producing (a) a constant homogeneous field and (b) a linear gradient. In this case, the declination angle α is solved for using a spherical harmonic approach.

The magnetic field in terms of zonal harmonics is given by:

$$B_z = \frac{\mu_0 I}{2a} \sum_{n=0}^{\infty} \left(\frac{z}{a}\right)^n \sin\alpha P_{n+1,1}(\cos\alpha) \tag{1.70}$$

If a second loop is placed symmetrically about $z = 0$, then one can use the fact that:

$$P_{n+1,1}\left[\cos(\pi - \alpha)\right] = (-1)^n P_{n+1,1}\left[\cos\alpha\right] \tag{1.71}$$

and if equal currents are applied in the same direction in each loop, then eqn (1.71) means that all the odd zonal terms in z, z^3, z^5 *etc.* are cancelled out, leaving the desired term (*i.e.* the homogeneous magnetic field) in z^0 and the first terms that must be cancelled are in z^2. From eqn (1.70) the contribution is determined by the value of $P_{3,1}(\cos\alpha)$. Using Table 1.2, this value is zero when $5\cos^2\alpha - 1 = 0$, from which $\alpha = 63°$ and $d_{\text{constant}} = a$, *i.e.* the separation is equal to the radius, the same value (fortunately!) as derived in example A1.

Example B2. Generating a Linear Magnetic Field Gradient in *z*

Consider the same two loops of wire; if the currents in the two loops are now equal and *opposite*, then even orders cancel, *i.e.* $z^2 = z^4 = z^6 = 0$. Therefore, the first term that needs to be set to zero is the third order one. From Table 1.2, P_{41} contains a term in $(7\cos^3\theta - 3\cos\theta)$, which is equal to zero for $\theta = 49.1°$. This corresponds to a separation given by 1.73 times the radius.

Example B3. Generating a Linear Magnetic Field Gradient in *x* or *y*

As can be appreciated from Figure 1.29, generating a magnetic field gradient in either the *x*- or *y*-direction involves spherical harmonics with $n = 1$, $m = \pm 1$, and since $m \neq 0$ these involve tesseral harmonics rather than zonal harmonics. Unlike zonal harmonics, tesseral harmonics cannot be generated by a current loop, but can be generated instead using arcs of current, as shown in Figure 1.32(a). As shown by Romeo and Hoult,[42] the magnetic field produced by such an arc is given by:

$$B = I\sum_{n=0}^{\infty}\sum_{m=0}^{n} C_{n,m} \frac{r^n}{f^{n+1}} P_{n,m}(\cos\theta)\cos(m\phi - m\psi)\,\mathrm{d}\psi \tag{1.72}$$

where dψ represents the azimuthal width of the arc. In this case the constant terms $C_{n,m}$ contain two associated Legendre polynomials. As was shown previously for rings, if two arcs are placed 180° apart with current flowing in opposite directions, then the even degrees cancel out, as shown in Figure 1.32(a). So, with respect to the degree n there are terms in B_{11}, B_{13}, B_{15} *etc.* and

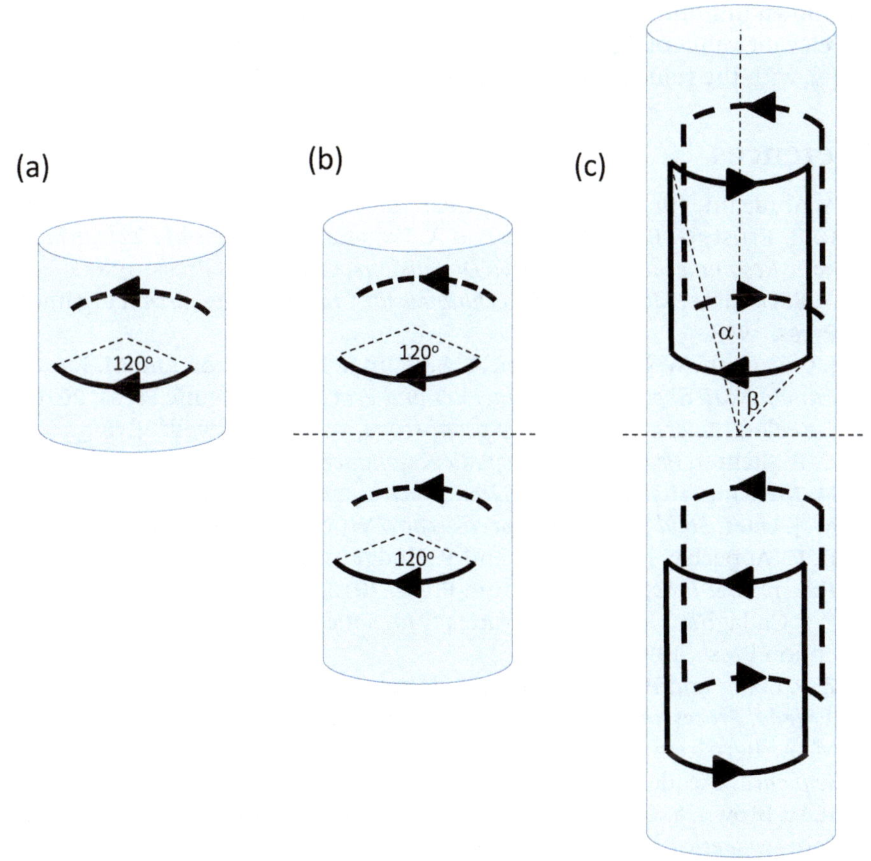

Figure 1.32 (a) An arc of current produces tesseral spherical harmonics. Two arcs placed 180° apart cancel out even degrees of these harmonics (note that the arcs are shown as not being physically connected). The angle 120° is chosen to cancel the B_{13} term. (b) By adding additional arcs symmetrically about the $z = 0$ plane, even orders are cancelled. (c) Physical realization of the gradient coil involves connecting the individual arcs by straight line sections, and optimizing the values of α and β for linearity.

with respect to the order m there are terms in B_{21}, B_{31}, B_{41} *etc.* The term in B_{13} can be cancelled by appropriate choice of the width $\Delta\psi$ of the arc:

$$B \propto \int_{-\Delta\psi/2}^{\Delta\psi/2} \cos(m\phi - m\psi)\mathrm{d}\psi = \frac{2}{m}\cos m\phi \sin\left(\frac{m\Delta\psi}{2}\right) \tag{1.73}$$

This term is zero when the angle subtended by the arc is 120°. The next step is to cancel out the higher order terms. As shown in the previous example, by using a symmetrical arrangement about the $z = 0$ axis, even order terms are cancelled out by extending the number of arcs from two to four, as shown in Figure 1.32(b). The next term is the contribution from B_{31}, which as also shown in Romeo and Hoult,[42] can be made equal to zero by calculating the relevant value of $C_{n,m}$, and results in the arrangement shown in Figure 1.32(c), with the values of α and β given by 21.3° and 68.7°, respectively.

References

1. A. Abragam, *The Principles of Nuclear Magnetism*, Clarendon Press, 1961.
2. R. R. Ernst, G. Bodenhausen and A. Wokaun, *Principles of Nuclear Magnetic Resonance in One and Two Dimensions*, Clarendon Press, 1987.
3. A. E. Derome, *Modern NMR Techniques for Chemistry Research*, Pergamon Press, 1993.
4. J. Cavanagh, W. J. Fairbrother, A. G. Palmer, N. J. Skelton and M. Rance, *Protein NMR Spectroscopy, Principles and Practice*, Academic Press, 2010.
5. J. Keeler, *Understanding NMR Spectroscopy*, Wiley-Blackwell, 2010.
6. C. P. Slichter, *Principles of Magnetic Resonance*, Springer, 2010.
7. M. Mehring, *High Resolution NMR Spectroscopy in Solids*, Springer, 2012.
8. M. J. Duer, *Solid State NMR Spectroscopy*, Wiley-Blackwell, 2001.
9. D. C. Apperley, R. K. Harris and P. Hodgkinson, *Solid State NMR: Basic Principles & Practice*, Momentum Press, 2012.
10. P. T. Callaghan, *Principles of Nuclear Magnetic Resonance Microscopy*, Clarendon Press, 1991.
11. Z. P. Liang and P. C. Lauterbur, *Principles of Magnetic Resonance Imaging: A Signal Processing Perspective*, John Wiley & Sons, 1999.
12. M. A. Bernstein, K. F. Z. King and X. J. Zhou, *Handbook of MRI pulse sequences*, Academic Press, 2004.
13. R. W. Brown, Y. C. Cheng, E. M. Haake, M. R. Thompson and R. Venkatesan, *Magnetic Resonance Imaging: Physical Properties and Sequence Design*, Wiley-Blackwell, 2014.
14. F. Bloch, W. W. Hansen and M. Packard, *Phys. Rev.*, 1946, **69**, 127.
15. D. Moskau, *Concepts Magn. Reson.*, 2002, **15**, 164.
16. R. E. Hurd, *J. Magn. Reson.*, 1990, **87**, 422.
17. E. O. Stejskal and J. E. Tanner, *J. Chem. Phys.*, 1965, **42**, 288.
18. G. E. Wesbey, M. E. Moseley and R. L. Ehman, *Invest. Radiol.*, 1984, **19**, 491.
19. P. C. Lauterbur, *Nature*, 1973, **242**, 190.
20. P. Mansfield and P. K. Grannell, *J. Phys. C: Solid State Phys.*, 1973, **6**, L422.
21. P. J. Hajduk, E. T. Olejniczak and S. W. Fesik, *J. Am. Chem. Soc.*, 1997, **119**, 12257.
22. N. Esturau and J. F. Espinosa, *J. Org. Chem.*, 2006, **71**, 4103.
23. C. Beaulieu, H. D'Arceuil, M. Hedehus, A. de Crespigny, A. Kastrup and M. E. Moseley, *Semin. Pediatr. Neurol.*, 1999, **6**, 87.
24. P. J. Basser, J. Mattiello and D. LeBihan, *J. Magn. Reson., Ser. B*, 1994, **103**, 247.
25. J. H. Jensen and J. A. Helpern, *NMR Biomed.*, 2010, **23**, 698.
26. J. H. Jensen, J. A. Helpern, A. Ramani, H. Z. Lu and K. Kaczynski, *Magn. Reson. Med.*, 2005, **53**, 1432.

27. S. Mori, B. J. Crain, V. P. Chacko and P. C. van Zijl, *Ann. Neurol.*, 1999, **45**, 265.
28. J. A. McNab, B. L. Edlow, T. Witzel, S. Y. Huang, H. Bhat, K. Heberlein, T. Feiweier, K. C. Liu, B. Keil, J. Cohen-Adad, M. D. Tisdall, R. D. Folkerth, H. C. Kinney and L. L. Wald, *Neuroimage*, 2013, **80**, 234.
29. K. Setsompop, R. Kimmlingen, E. Eberlein, T. Witzel, J. Cohen-Adad, J. A. McNab, B. Keil, M. D. Tisdall, P. Hoecht, P. Dietz, S. F. Cauley, V. Tountcheva, V. Matschl, V. H. Lenz, K. Heberlein, A. Potthast, H. Thein, J. Van Horn, A. Toga, F. Schmitt, D. Lehne, B. R. Rosen, V. Wedeen and L. L. Wald, *Neuroimage*, 2013, **80**, 220.
30. C. Barmet, N. De Zanche, B. J. Wilm and K. P. Pruessmann, *Magn. Reson. Med.*, 2009, **62**, 269.
31. E. R. Andrew, A. Bradbury and R. G. Eades, *Nature*, 1958, **182**, 1659.
32. L. B. Andreas, T. Le Marchand, K. Jaudzems and G. Pintacuda, *J. Magn. Reson.*, 2015, **253**, 36.
33. U. Haeberle and J. S. Waugh, *Phys. Rev.*, 1968, **175**, 453.
34. P. Mansfield, M. J. Orchard, D. C. Stalker and K. H. Richards, *Phys. Rev. B*, 1973, **7**, 90.
35. W. K. Rhim, D. D. Elleman and R. W. Vaughan, *J. Chem. Phys.*, 1973, **58**, 1772.
36. J. S. Waugh, L. M. Huber and U. Haeberle, *Phys. Rev. Lett.*, 1968, **20**, 180.
37. A. Pines, M. G. Gibby and J. S. Waugh, *J. Chem. Phys.*, 1973, **59**, 569.
38. W. E. Maas, A. Bielecki, M. Ziliox, F. H. Laukien and D. G. Cory, *J. Magn. Reson.*, 1999, **141**, 29.
39. W. E. Maas, F. H. Laukien and D. G. Cory, *J. Am. Chem. Soc.*, 1996, **118**, 13085.
40. D. B. Twieg, *Med. Phys.*, 1983, **10**, 610.
41. S. Ljunggren, *J. Magn. Reson.*, 1983, **54**, 338.
42. F. Romeo and D. I. Hoult, *Magn. Reson. Med.*, 1984, **1**, 44.

CHAPTER 2

Magnets

RORY WARNER[a] AND SIMON PITTARD*[a]

[a]Tesla Engineering Ltd., Water Lane, Storrington, West Sussex, RH20 3EA, UK
*E-mail: pittard@tesla.co.uk

2.1 Introduction

As covered in Chapter 1, the nuclear polarization and the induced MR voltage both scale linearly with magnetic field. The increase in SNR with magnetic field is one of the driving forces behind the development of ever-higher static magnetic fields in both high resolution NMR spectroscopy and MRI (animal and human). As an example from NMR, Figure 2.1 shows a chronological plot of the SNR of a standard sample (0.1% ethylbenzene in deuterated chloroform) as a function of the highest field strength available. In addition to the increased SNR, the spectral resolution is increased both for high resolution NMR and localized *in vivo* spectroscopy, meaning that spectral overlap becomes much less of a problem, and more resonances can be assigned. In high resolution NMR there are also specific experiments in which a high field has additional advantages: for example transverse relaxation optimized spectroscopy (TROSY)-based experiments show the narrowest linewidths at the "magic fields" of 23.5 tesla for amide protons in polypeptides, and 21.1 tesla for amide nitrogens.[1] In human MRI, several groups have shown that tissue contrast arising from the presence of diffuse iron or changes in myelination is much higher than at lower clinical magnetic fields, providing a major impetus for the use of high field MRI to study neurodegenerative diseases.[2] High field

New Developments in NMR No. 7
Magnetic Resonance Technology: Hardware and System Component Design
Edited by Andrew G Webb

Published by the Royal Society of Chemistry, www.rsc.org

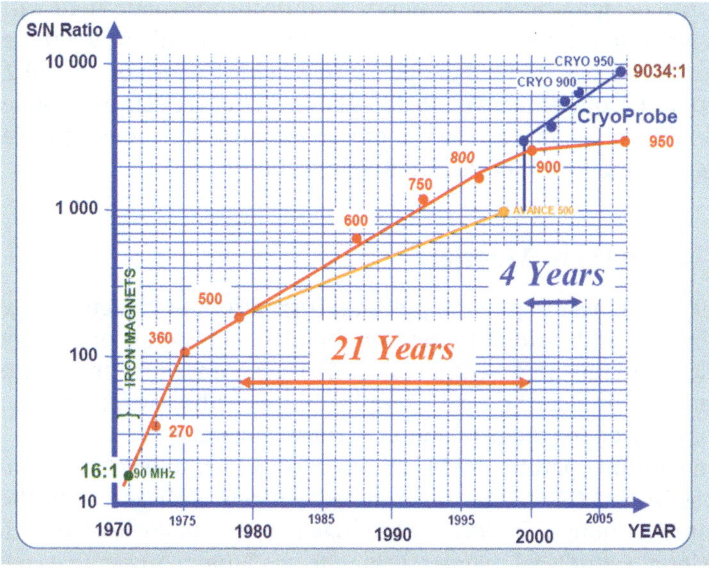

Figure 2.1 Increase in the SNR for the high resolution NMR industry's standard (0.1% ethylbenzene in deuterated chloroform) over the past ~45 years. The major factor is the increase in the static magnetic field. Also shown are the advantages bestowed by using cryoprobe technology, covered in Chapter 3. Figure provided by Bruker Instruments.

systems are also highly promising for *in vivo* MR of low gyromagnetic ratio nuclei, such as ^{31}P and ^{23}Na. Currently, the highest commercially available high resolution NMR systems operate at just over 23.5 T (1 GHz), animal systems at 16.8 tesla (720 MHz), and human MRI systems at 11.7 tesla (500 MHz).

In addition to the absolute value of the magnetic field, other critical design criteria for the magnet include the temporal stability of the field and the useable volume, *i.e.* the volume over which the magnetic field lies within a specified homogeneity (typically a few parts per billion).

2.2 Magnet Types

A number of different technologies can be used to generate the magnetic field for MR. This chapter focuses on the most widespread technology, namely a cylindrical superconducting magnet. However, there are a variety of other types of magnet that offer advantages for particular requirements, such as low operating costs or open access. The various different magnet types are briefly described in this section.

2.2.1 Air-Cored Resistive Magnets

Some of the earliest electromagnets designed for MRI in humans were wound from copper or aluminium and operated at room temperature. A small number of solenoidal coils, usually four, were situated on the surface of a notional sphere: the design aim is to generate a uniform magnetic field within the sphere, while leaving a gap in the mid-plane of the sphere for the subject. Air-cored magnets were able to operate at a field strength of up to 0.2 T with typical power requirements of 40 kW–100 kW. The large amount of heat generated in the coils required the use of water cooling and generally presented challenges to maintain the required field stability.

2.2.2 Permanent Magnets

Permanent magnets require no power, refrigeration, or cryogens to generate the main magnetic field and therefore have the advantage of low operating costs, although the large weight may be a disadvantage in siting. Historically, China has dominated the world supply of rare earths, such as neodymium–iron–boron, used for the production of these magnets.

The maximum operating field is limited to approximately 0.3 T for a whole-body system and the geometry allows a relatively clear access to the imaging volume, although the level of homogeneity is usually less than that of a conventional cylindrical magnet. Owing to the large temperature coefficient of the magnetisation of typical permanent magnet materials, these types of systems are sensitive to changes in temperature and so normally incorporate some form of temperature stabilisation.

2.2.3 Iron-Cored Resistive Magnets

A number of systems incorporate resistive windings into a C-shaped iron pole. The iron core significantly increases the field generated by the resistive coils. The field strength is limited to typically 0.35 T, but the open access geometry has the advantage that it is less claustrophobic and allows access to the patient for treatment during the imaging. Installation costs are low. However, the magnet is driven from a high-power, high-stability power supply and so the operating costs tend to be high. As with resistive and permanent magnets, some sort of temperature stabilisation is required to achieve the field stability required for MRI.

2.2.4 Iron-Cored Superconducting Magnets

Replacing the resistive coils of an iron-cored resistive magnet with superconducting coils allows higher field strengths to be generated and eliminates the difficulties caused by heat generated in the coils. Field strengths of typically 0.5 T to 1.0 T are commercially available using this technology. The main benefit of this type of system is the open access geometry, which reduces claustrophobia and allows easy access to the patient. The higher operating fields result in higher forces between the poles so it becomes difficult to maintain mechanical stability with a simple C-shaped configuration. Therefore, these systems normally have two pillars to support the additional forces.

2.2.5 Superconducting Cylindrical Magnets

The vast majority of the magnets used for both high resolution NMR and MRI are superconducting cylindrical magnets, oriented vertically or horizontally, respectively. These magnets can operate in persistent mode with no need for an external current source, making their field intrinsically stable. A photograph and schematic of a high resolution NMR magnet is shown in Figure 2.2. For human MRI systems, the cylindrical bore allows access for the patient through the circular aperture at the end of the cylinder. This geometry allows the most efficient generation of a high and homogeneous field at clinical field strengths of 1.5 T and 3 T. Whole-body magnets with field strengths up to 10.6 T have been built using this type of magnet. Modern magnets are relatively light compared to iron-cored magnets and are usually built with shielding coils which enable the stray field to be controlled without the requirement of steel shielding. Table 2.1 shows typical specifications for a high resolution NMR magnet and a clinical MRI system.

2.3 Magnetic Field Generation

2.3.1 Basic Physics

As outlined in the previous section, the magnetic field in an MRI system is most commonly provided by an electromagnet, *i.e.* a superconducting magnet. An electromagnet exploits the physical phenomenon whereby a magnetic

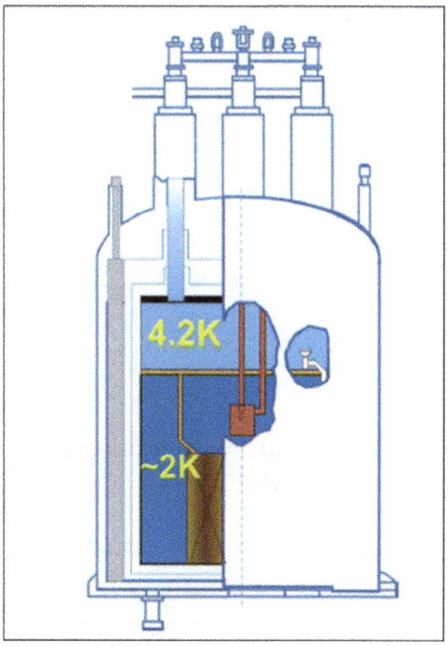

Figure 2.2 (left) Photograph and (right) schematic of a high resolution NMR magnet. The actual superconducting windings of the magnet itself occupy relative little of the total space of the magnet, which is dominated by the cryostat and vacuum vessels necessary to maintain a temperature of 2 K around the superconducting wire. Figure provided by Bruker Instruments.

Table 2.1 Characteristics of a mid-frequency high resolution NMR magnet (left) and three standard clinical MRI systems (right).

	NMR magnet	Clinical MRI magnets		
Field strength	11.7 tesla	3 tesla	1.5 tesla	1.5 tesla
Magnet weight	450 kg	4600 kg	3000 kg	2800 kg
Open bore	54 mm	70 cm	70 cm	60 cm
Homogeneity (55 × 55 × 50 cm)		≤5 ppm	≤5 ppm	≤5 ppm
Homogeneity (50 × 50 × 45 cm)		≤1.8 ppm	≤1.8 ppm	≤1.8 ppm
Homogeneity (25 × 5 × 5 mm)	≤0.1 ppb			
Temporal stability (per hour)	≤10 ppb per hour	<0.1 ppm	<0.1 ppm	<0.1 ppm

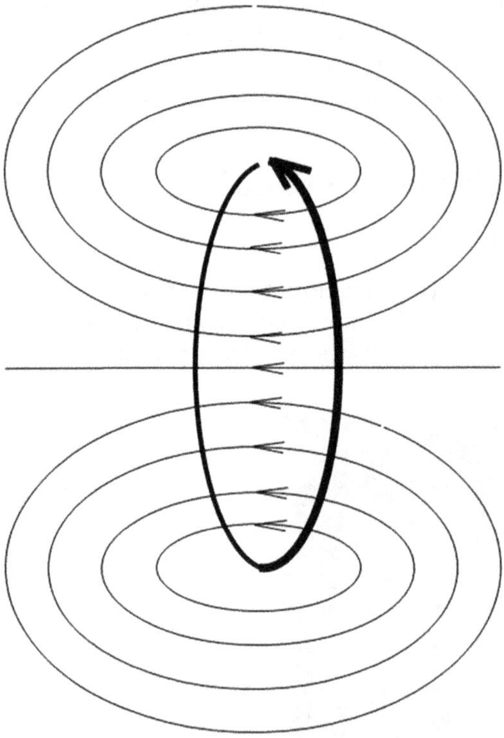

Figure 2.3 Concentration of magnetic field at the centre of a loop of wire. The field drops off inversely with the distance from the wire.

field is generated around a conductor carrying an electric current. The magnetic field is a vector quantity in that it has both magnitude and direction.

The magnetic field can be expressed mathematically by the Biot–Savart law (Chapter 1, Appendix A) which describes the magnetic field at a point in space created by a current carrying element. Compared to a straight-wire, the magnetic field can be concentrated by forming a loop of current, as shown in Figure 2.3.

The strength of the magnetic field can be enhanced by combining multiple loops to form a solenoid, as shown in Figure 2.4. The field at the centre of a

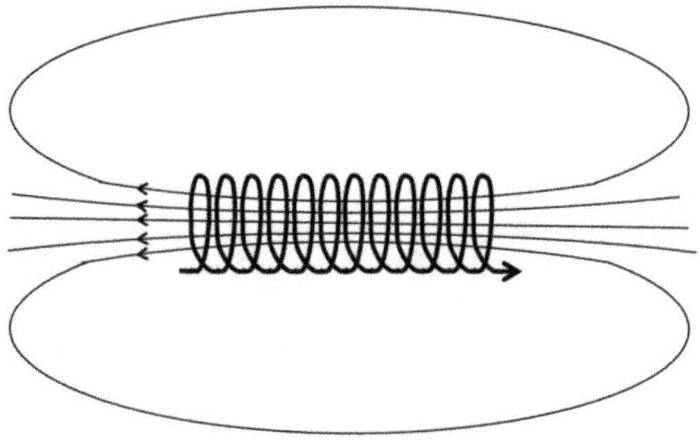

Figure 2.4 Multiple loops of current forming a solenoid produce an enhanced magnetic field strength. The field is very uniform at the centre of the solenoid, becoming weaker towards the ends, and at the very ends is one-half the value in the centre.

very long solenoid is given in eqn (2.1), where N is the total number of turns and L is the length.

$$B = \frac{\mu_0 IN}{2L} \tag{2.1}$$

2.3.2 Field Homogeneity

MRI requires a homogeneous background field, typically 5 parts-per-million (ppm) over the imaging volume. This level of homogeneity could be achieved using a simple long solenoid: however it would result in a magnet length of approximately 18 metres for a human sized imaging system!

As a practical alternative, homogeneous magnetic fields can be produced using discrete sets of coils arranged at optimal positions and with optimal current densities.[3] The simplest example of an arrangement of coils is two identical coils with the axial spacing equal to their radius (Chapter 1, Appendix A, Example A4) as shown in Figure 2.5(a); this arrangement is known as a Helmholtz pair. An arrangement of a single pair of coils is only able to achieve a spherical homogeneity volume of about 60 mm diameter for a human sized magnet. In order to extend the homogeneity volume to a suitable level for whole body imaging (450–500 mm spherical diameter), additional coils need to be added. The positions and number of windings are optimized so as to remove successive orders of harmonics in the spherical harmonic expansion of the magnetic field (Chapter 1, Appendix B). The Helmholz coil corresponds to compensating the second order terms. Successive approximations are shown in Figure 2.5(b) and (c).

Generally, most magnets used for human imaging consist of six or seven discrete coils—the so-called multi-coil approach, compensating up to 10th order terms for the six coil symmetric configuration. The field contours produced by such an arrangement are shown in Figure 2.6. The blue field

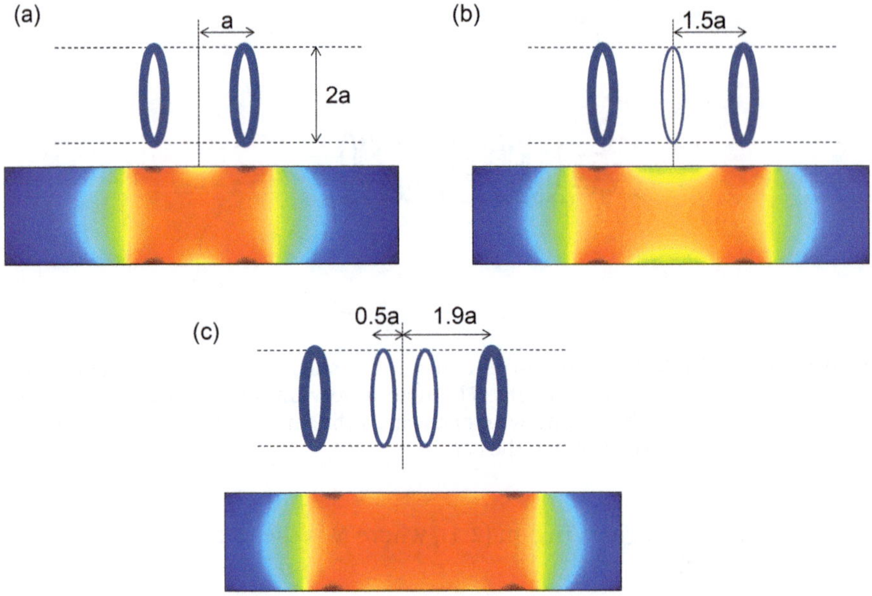

Figure 2.5 Demonstration of the increase in magnetic field homogeneity by using a greater number of coil loops. The magnetic field is calculated using the Biot–Savart law, with red corresponding to a high field and blue to a low one. (a) Two coils with equal current, spaced at a distance equal to the radius of each loop. (b) Three coil arrangement, with four times as much current in the outer loops as the central one. (c) Four coil arrangement with approximately 2.5 times as much current in the outer loops as in the inner loops.

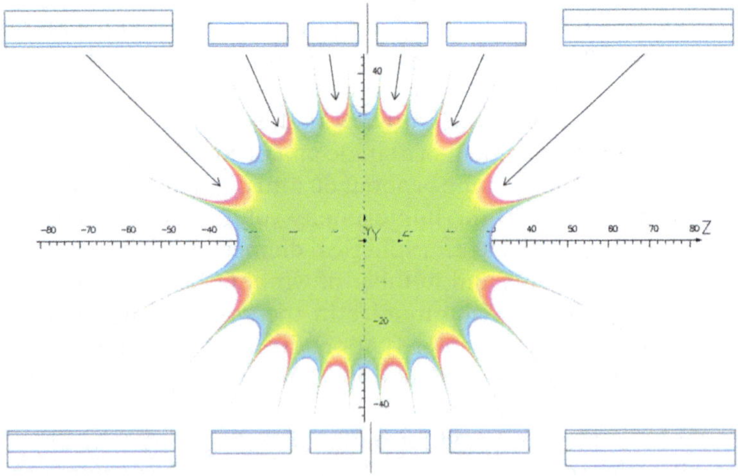

Figure 2.6 Field contours produced by a six coil magnet design.

contour shows regions where the field is lower than the central field and the pink field contour shows regions where the field is higher than the central field.

An alternative method to the multi-coil approach is termed the compensated solenoid, in which a short solenoid forms the inner winding, and compensation coils are added at a slightly larger diameter, as shown in Figure 2.7. The compensated solenoid is typically used for higher magnetic fields, above 3 tesla, whereas the multi-coil approach is used for clinical field strengths.

In terms of industry standards and specifications there are a number of ways of expressing the magnet's homogeneity:

(i) Plotted peak-to-peak: this is the gold standard and states the worst case (the difference between the highest and lowest points) taken on a sampling surface defined by up to 24 planes and 24 angular positions.
(ii) VRMS: this is usually extrapolated from the above analysis and is the root mean square (RMS) of an evenly sampled space within the volume. This measure is typically an order of magnitude less demanding than a peak-to-peak measurement.
(iii) FWHH: the full-width-half-height is an NMR linewidth definition; again it samples the whole volume, often with a flood phantom. The definition makes some assumptions about line shape and is probably the least satisfactory in terms of comparing magnets.

Figure 2.7 Schematic of a compensated solenoid design in which the internal structure is a continuously wound solenoid and the outer four coils provide compensation for field inhomogeneities over the central region.

2.3.3 Magnetic Shielding

Electromagnets generate a stray magnetic field surrounding the device, which can pose a number of problems, particularly in a hospital environment.

- Hazards owing to large forces on nearby ferrous objects, generally for fields exceeding 5 mT (50 gauss).
- Damage to magnetic storage devices for fields exceeding 1 mT (10 gauss).
- Disruption to cardiac pace-makers for fields exceeding 0.5 mT (5 gauss).
- Disruption of sensitive equipment, such as CT scanners, PET scanners, and nuclear medicine instruments, for fields exceeding 0.1 mT (1 gauss).

The stray field of the magnet can be reduced with the use of reverse wound coils, usually two, located outside the main coils of the magnet. Figure 2.8 shows an arrangement of shielding coils.

Figure 2.9 shows how the 0.5 mT (5 gauss) contour is reduced by the shielding coils.

Although all magnets at clinical field strengths are actively shielded, the process for magnets with field strengths exceeding 7 T tends to be expensive and adds to the size and complexity of an already large magnet. Until around 2012, most ultra-high field systems were installed in steel shields. The shield typically consists of a box approximately 5 m wide × 5 m tall × 10 m long with a mass of between 200 000 and 800 000 kg of low-carbon mild steel. The refrigeration devices (cold heads) are situated on tall turrets that penetrate through the roof of the steel shield into a low field region above the shield. A section through the shield of a typical high field site is shown in Figure 2.10. Since ~2012 most 7 tesla human magnets have been actively shielded, mainly for practical purposes, although there remains significant debate as to the relative advantages and disadvantages of active *vs.* passive shielding in terms of patient safety.

2.3.4 System Shielding from External Interference

Ferromagnetism is the process whereby certain special materials, such as iron, nickel and cobalt, become strongly magnetised when an external magnetic field is applied. Ferromagnetic materials have a varying degree of

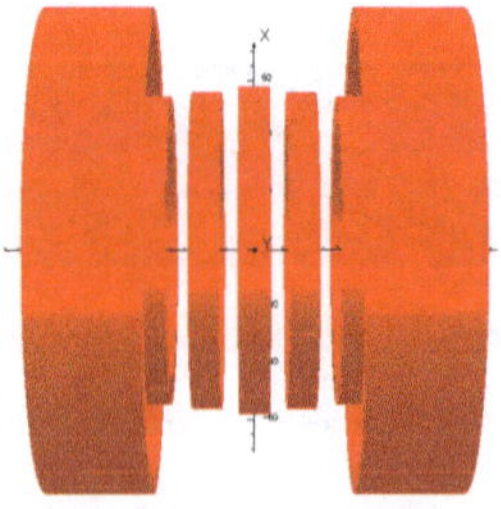

Figure 2.8 Typical arrangement of magnetic field shielding coils, shown in relationship to the six element multi-coil design of Figure 2.6.

residual magnetisation after the external magnetic field is removed. Materials with a large residual field are used as permanent magnets with numerous common applications, such as electric motors.

Ferromagnetism has two main applications in magnet systems for MRI: firstly, small pieces of carbon steel can be located at selected regions within the bore of the magnet to optimise the homogeneity of the magnetic field within the imaging volume; this process is called passive shimming.[4] A second application is to use carbon steel to construct a shield around a sensitive piece of equipment, such as a cryocooler, to reduce the magnetic field inside the shield.

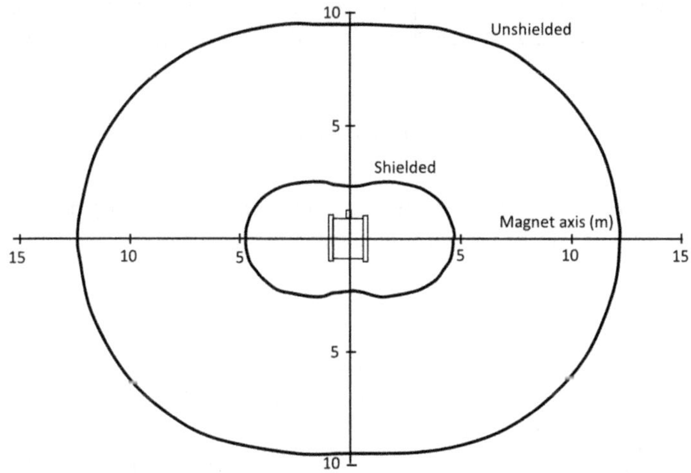

Figure 2.9 Comparison of the stray field (5 gauss line) of a shielded and unshielded magnet. A reduction of an order of magnitude in the space required can easily be achieved.

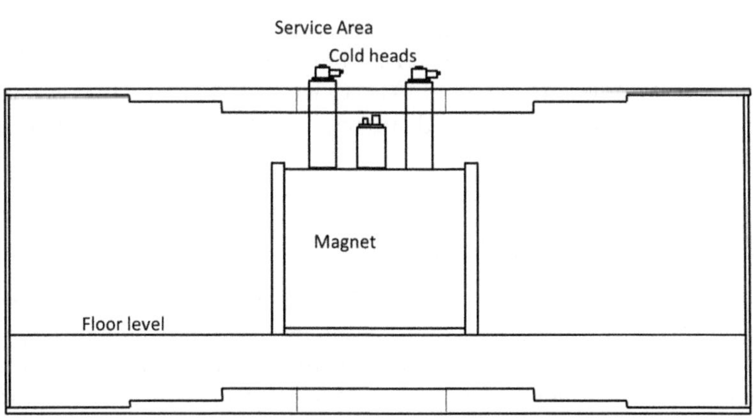

Figure 2.10 Side section though a typical ultra-high magnet shielded room.

One negative impact of ferromagnetism is that objects nearby the magnet become magnetised. If the object is large and moving, such as a car or truck, then it will produce a short-term disturbance to the field stability. Typically, a short-term variation in excess of 0.025 microTesla (0.25 mgauss) is likely to impact on the quality of the image.

The field from a magnetic dipole is given by

$$B=\frac{\mu_0}{4\pi}\left(\frac{3\mathbf{r}(m\cdot\mathbf{r})}{\mathbf{r}^5}-\frac{m}{\mathbf{r}^3}\right) \tag{2.2}$$

where m is referred to as the magnetic moment and $\mathbf{r}$ is the vector from the point of measurement to the dipole. If the object is located on the magnet axis and is magnetised in the direction of the axis, the expression can be simplified to

$$B_z=\frac{2\mu_0 m_z}{4\pi z^3} \tag{2.3}$$

where m_z is the magnetisation and z is the distance along the axis of the magnet. A steel object has a magnet moment given by

$$m = VM \tag{2.4}$$

where M is referred to as the magnetisation and V is the volume of ferrous material. For a carbon steel object in a background field, the value of $\mu_0 M$ can be determined from a B–H curve, where H is the background field ($\mu_0 M = B - \mu_0 H$) and has a saturation field of 2.1 T in a background field in excess of 0.02 T (200 gauss), although the exact value will depend on the specific properties of the steel. A 1 kg carbon steel object will therefore typically have a magnetic moment of ~210 Am^2.

Actively shielded magnets are particularly susceptible to the influence of environmental effects and are usually fitted with a special external interference shield coil. The coil consists of a separate superconducting circuit specifically designed to shield out external field variations. The theory behind the design is described in a paper by Gabrielse and Tan.[5] Practically, a typical factor of 20 reduction to the field disturbance can be obtained with an external interference coil.

2.3.5 Magnetic Field Shimming

The optimum B_0 homogeneity that can be achieved theoretically for a given coil geometry will, in practice, not be achieved owing to small positional errors in the superconducting wire placement caused by manufacturing tolerances, and magnetic components both within the cryostat and the local environment. The sample or patient also generates field inhomogeneities. The disturbance to the homogeneity can be corrected by generating additional magnetic fields designed to cancel out the unwanted field variation; this process is called "field shimming". Field shimming is usually accomplished using either small pieces of iron (passive shims) or coils (shim coils) located in the bore of the magnet, or a combination of both. Large amounts of passive shimming can be problematic as the magnetic properties of the iron change with temperature. This can lead to field drift and homogeneity changes during the course of the scan.

Shim coils operating in the room temperature bore of the magnet are usually contained within the gradient coil assembly and are wound from copper wire or etched (mechanically or chemically) from copper sheet. These are referred to as resistive shims. Up to a maximum of thirty (high resolution NMR) or fifteen (MRI) shim channels might be provided and these will be driven by high-stability power supplies. By careful adjustment of the currents in the coils, field disturbances caused by the patient or sample can be corrected. The topic of shimming is covered in extensive detail in Chapter 4 of this book.

Shim coils are also located inside the helium vessel. These are wound from superconducting wire and then set to a particular current setting during the initial installation of the magnet. The optimum current is determined by mapping the field, usually over the surface of a spherical volume, with a small NMR detector. As with the magnet, the superconducting shims are fitted with a superconducting switch (described later). Once the currents in the superconducting shims are set, the power supply can be removed.

As described in Chapter 1, Appendix B, the variation of the central field of the magnet is usually expressed as a sum of Legendre polynomials and associated Legendre polynomials. These functions are a convenient way of expressing the inhomogeneity of the central field because simple arrangements of circular coils and circular arcs can produce good approximations of each of the most significant harmonics, and this facilitates the shimming process in terms of designing arrangements which can compensate for unwanted higher order terms. Eqn (2.5) shows the expansion of the magnetic field expressed in spherical coordinates.[6]

$$B(r,\theta,\phi)=\sum_{n=0}^{\infty}\sum_{m=0}^{m=n} r^n P_n^m\left(\cos\theta\right)\left[A_m^n \cos(m\phi)+B_m^n \sin(m\phi)\right] \tag{2.5}$$

where $P_n^m(\cos\theta)$ are associated Legendre polynomials; A_n^m and B_n^m are constants that define the field variation; r, θ, and ϕ are the polar coordinates relative to the magnet centre; and m and n are the polynomial order and degree, respectively.

Each shim coil is referenced by the shape of field it produces in terms of Cartesian coordinates, where the magnet axis is defined as the z axis. Table 2.2 lists the Cartesian representation of the most commonly used shim channels. The order (n) and degree (m) given in the table relate to the values of n and m in eqn (2.5).

Table 2.2 Cartesian representation of the lower order harmonics.[a]

Shim	Order, n	Degree, m	Cartesian function
Z1	1	0	z
Z2	2	0	$z^2 - 0.5\rho^2$
Z3	3	0	$z^3 - 1.5\rho^2 z$
Z4	4	0	$z^4 - 3\rho^2 z^2 + 0.375\rho^4$
X	1	1	x
Y	1	1	y
ZX	2	1	$3zx$
ZY	2	1	$3zy$
X2–Y2	2	2	$3(x^2 - y^2)$
XY	2	2	$3(2xy)$
Z2X	3	1	$6(z^2 - 0.25\rho^2)x$
Z2Y	3	1	$6(z^2 - 0.25\rho^2)y$
Z(X2–Y2)	3	2	$15z(x^2 - y^2)$
ZXY	3	2	$15z(2xy)$
X3	3	3	$15(x^3 - 3xy^2)$
Y3	3	3	$15(y^3 - 3yx^2)$

[a]The variable ρ is used to simplify the Cartesian expressions with $\rho^2 = x^2 + y^2$.

Ideally, there would be no spatial variation in the magnetic field within the imaging volume: this baseline is represented by the term corresponding to $n = 0$ and $m = 0$. There are three first order field variations, *i.e.* the field varies linearly with the x, y or z coordinates. In eqn (2.5), the linear variation in the z direction corresponds to $n = 1$ and $m = 0$ with A_1^0 representing the magnitude of this variation. The linear variations in both x and y are obtained when $n = 1$ and $m = 1$ with the magnitude of the variations given by A_1^1 for the x direction and B_1^1 for the y direction. Note that these three linear terms are also produced by the gradient coils that are used to encode spatial information in the MR signal during an imaging sequence.

As the values of n and m increase, the spatial variation of field increases in complexity. However, the magnitude of these more complex variations tends to be smaller and so good levels of field homogeneity can usually be obtained by concentrating shim coil design to remove only the lower order field impurities ($n \leq 4$ and $m \leq 3$).

2.4 Superconductivity

The electrical resistance of most conductors reduces as their temperature decreases. Superconductivity is the physical phenomenon whereby some special conductors' resistance falls to zero at a certain temperature known as the critical temperature.[7] For temperatures less than the critical temperature, the performance of the superconductor is limited by its critical current density and critical field. A graphical representation is shown in Figure 2.11. Superconductivity allows high currents to be carried without heating, enabling the generation of high magnetic fields by relatively compact coils. Superconductivity also allows the construction of a switch that

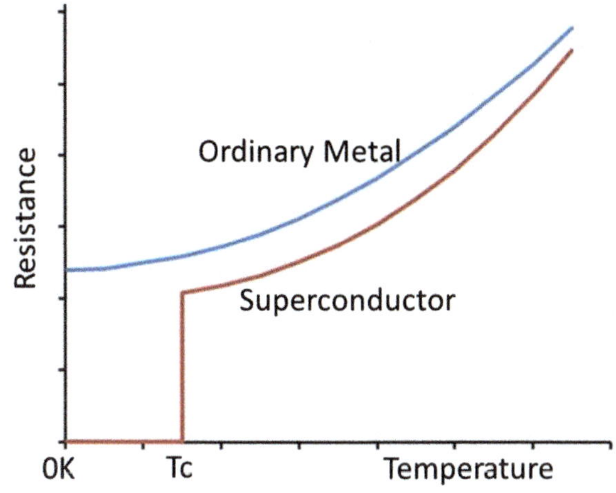

Figure 2.11 Transition to zero resistance in a superconductor at a temperature termed the critical temperature (T_c).

enables the magnetic field to be maintained with high stability without an external power supply.

2.4.1 Superconducting Materials

The superconductor most commonly used in MRI magnets is niobium titanium alloy. This has a critical temperature of 10 K and for practical applications is able to carry significant current with a background field of up to 10 tesla at a temperature of 4.2 K (the boiling point of liquid helium). The maximum practical field can be enhanced to 12 tesla by reducing the temperature to 2.3 K. For higher field or higher temperature applications, a compound of niobium and tin (Nb_3Sn) is most commonly used. Niobium tin conductor has been successfully used in applications in magnets for field strengths just above 20 tesla. The drawback of niobium tin is that it is an order of magnitude more expensive than niobium titanium and is brittle after reaction, resulting in difficulties in manufacturing. Figure 2.12(a) shows a comparison of the current-carrying properties of the two conductors. The superconducting wire is formed into very thin filaments or wires, and these are embedded into a long cylindrical copper structure, which then forms the windings of the magnet, as shown in Figure 2.12(b).

Magnesium di-boride is a relatively new conventional superconductor with a critical temperature of 39 K.[8,9] It has been used in the construction of magnets that have demonstrated the feasibility of this technology for MRI applications.[10–13] At the time of writing, the cost of the conductor and other practical limitations (such as the construction of superconducting joints) means that commercial systems are only just being developed to use this material. However, as cryogen costs increase, it may well become the material of choice for cryo-free systems (magnets not using liquid cryogens to cool the conductors).

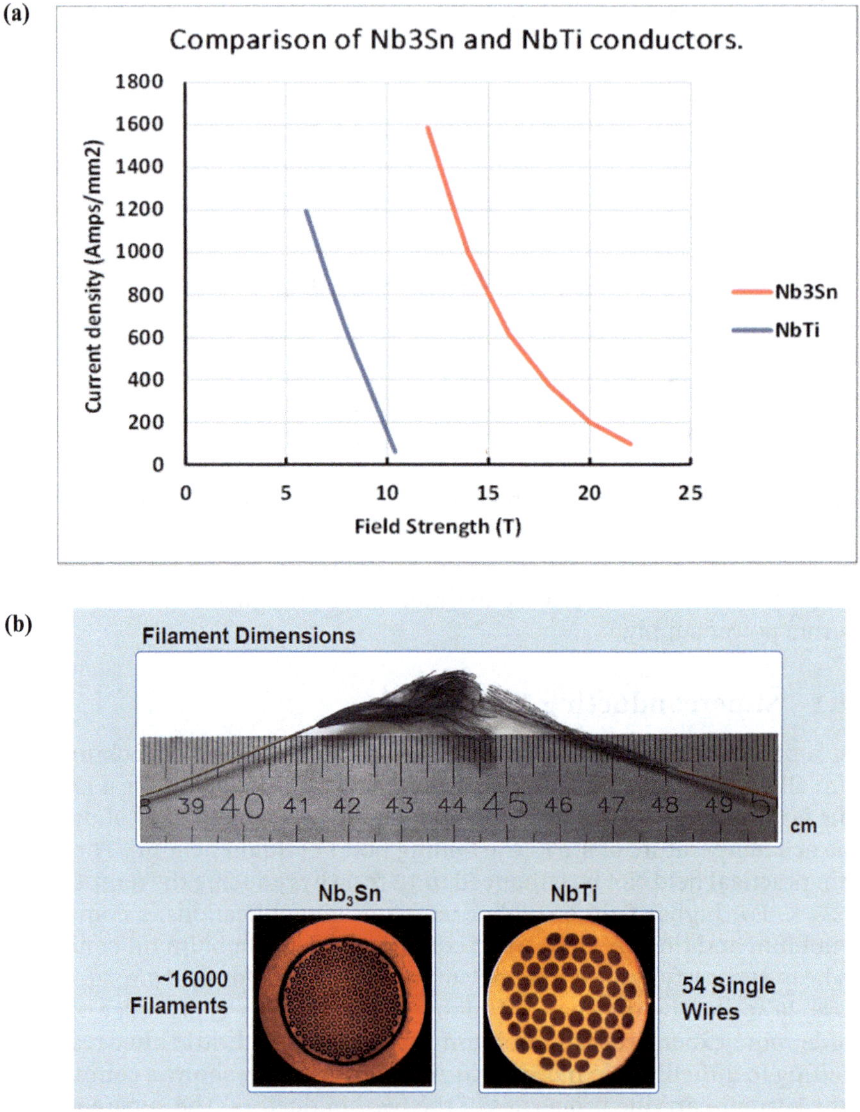

Figure 2.12 (a) Comparison of niobium titanium and niobium tin current carrying properties. (b) Photograph of the multifilamentary nature of superconductors embedded within a copper matrix.

High temperature superconductor (HTS) materials were discovered in 1986[14] in some particular ceramic materials, and critical temperatures in excess of 130 K have been recorded.[15] Although it may eventually be possible to construct high field MR magnets using only HTS materials, this is currently not possible, but HTS inserts can be added inside conventional low temperature superconductor (LTS) magnets. Much of the research is aimed at producing extremely high magnetic fields for high resolution NMR, fields which cannot be reached using LTS technology alone: for example HTS inserts will be series-connected with LTS technology in order to build high resolution NMR magnets up to 35 tesla. However, the technology can also potentially be applied to making much more compact magnets in general, and this aspect is extremely interesting for high field human systems in particular. The most promising materials currently are members of the Bi–Sr–Ca–Cu–O (BSSCO) system. There are three phases of BSSCO, of which Bi-2212 and Bi-2223[16] have been studied the most (the numbers refer to the relative number of Bi, Sr, Ca and Cu atoms). Bi-2201 ($Bi_2Sr_2CaCu_2O_8$) has a transition temperature of 85 K and Bi-2223 110 K. Bi-2212 is particularly favourable since it is the first HTS that can be formed into round wires, and the critical current density (J_c) is not highly dependent on the macroscopic texture (unlike Bi-2223, which requires uniaxial texture formed by deformation and reaction). These properties means that Bi-2212 can potentially be integrated into current LTS production methods. Other possible materials include rare-earth barium copper oxide (REBCO).[17] At the time of writing, these materials have not been used for the commercial construction of coils. However the use of the material in current leads with low thermal heat-load is widespread. Figure 2.13 shows a comparison of the current-carrying capabilities of a variety of different types of superconductor.

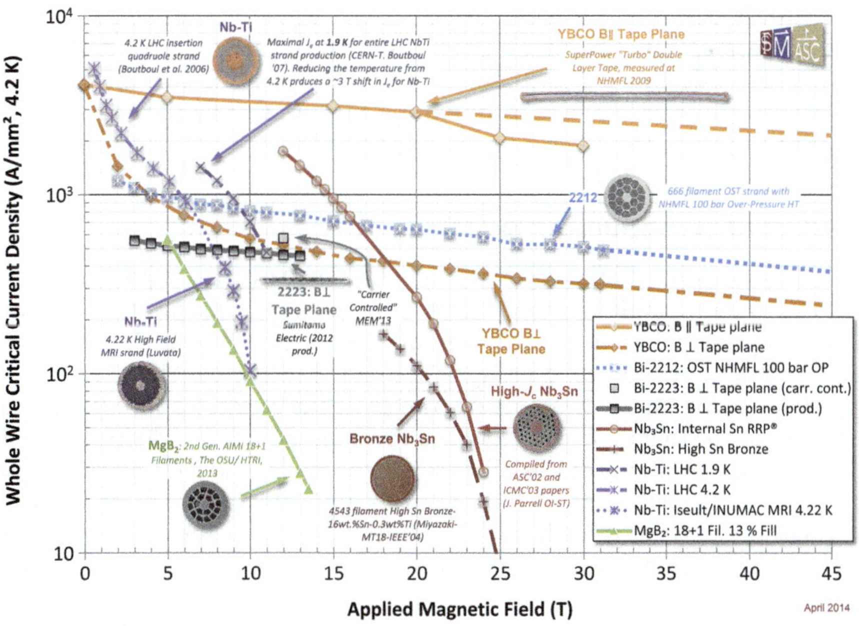

Figure 2.13 Typical performance of various types of superconductor. Courtesy Peter Lee, NHMFL Florida State University.

2.4.2 Energising a Superconducting Magnet

Superconducting magnets for MRI have a large inductance and virtually no resistance. Typical operating currents are between 100 A and 1000 A, but owing to the virtually negligible resistance the magnet power supply only needs to supply a relatively low voltage of typically less than 10 volts: the voltage is only needed to overcome the resistive loss in the charging leads and the inductance. The inductive voltage V during ramping is given by eqn (2.6), where L is the inductance of the magnet and $\mathrm{d}I/\mathrm{d}t$ is the rate of change of current.

$$V = L\frac{\mathrm{d}I}{\mathrm{d}t} \tag{2.6}$$

The time required to ramp the magnet to its operating field is given by eqn (2.7), where I is the operating current.

$$\text{ramptime} = \frac{LI}{V} \tag{2.7}$$

For example, a 3 T magnet has an inductance in the region of 200 H and an operating current of 400 A. Charged at 5 V the magnet could be ramped in approximately 4.5 hours. In practice, the ramp rate is slowed down as the magnet reaches full field. Typically a 1.5 T MRI magnet can be charged in 30 minutes, but an ultra-high field magnet such as a 9.4 T 900 mm magnet may take ~100 hours.

A process of "over-fielding" is typically performed, in which the current is taken slightly higher than the nominal operating current and then slowly brought back down to its operating current, this helps to "settle" the magnet.

2.4.3 Superconducting Switch

In addition to a very homogeneous magnetic field, MRI also requires the field to be extremely stable over the duration of the scanning protocol. This is achieved principally with a superconducting switch. The switch consists of a short length of superconducting wire that is in close contact with a heating element. When the heater is on, the switch is warmed above its critical temperature so that there is no longer a continuous superconducting circuit. Under these conditions the switch is said to be 'open' and current can be run into the magnet windings *via* a DC power supply. When the current flowing through the magnet is such that the central field is at the required value, the switch heater is turned off. The switch quickly cools, becomes superconducting, and so forms a continuous superconducting circuit. The switch is now referred to as being 'closed' and the current can be reduced to zero. The power supply can be removed, leaving the current flowing in the magnet. Figure 2.14 shows the current paths with the switch in the open and closed

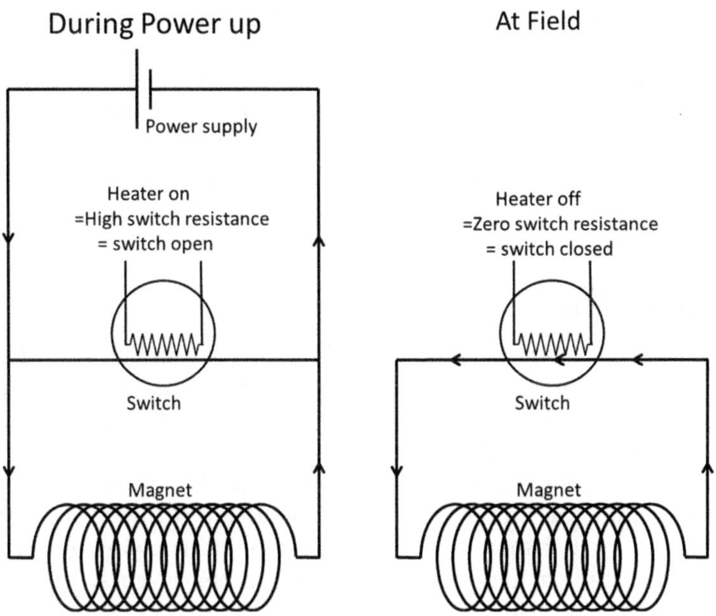

Figure 2.14 Current paths in a magnet with the superconducting switch in the open and closed positions.

configuration. The magnet is now operating in persistent mode and has a field stability of typically better than 0.1 ppm hour^{-1}.

2.4.4 Superconducting Joints

Once the magnet is persistent, the temporal stability is determined by two principal components: the short term drift as current is redistributed within the conductor and the long term drift that is caused by the residual resistance of the superconducting circuit. The latter consists of the very small residual resistance of the wire and the resistance of the joints. A magnet is often constructed from multiple lengths of superconducting wire. These individual lengths of wire need to be joined together and connected to a superconducting switch with a very low resistance joint to achieve the levels of field stability required for MRI, *i.e.* a drift less than 0.1 ppm hour^{-1}.

For a typical 3 T MRI magnet a drift of 0.1 ppm hour^{-1} corresponds to a current decay of 1.1×10^{-8} amps sec^{-1}. The rate of current decay is driven by a resistive voltage as given by eqn (2.6). In this example the resistive voltage will need to be less than 2.2×10^{-6} volts and so the total resistance of the circuit will need to be 6×10^{-9} ohms. Typically, there will be 20 joints in a 3 T magnet, and therefore the resistance of each superconducting joint needs to be less than 3×10^{-10} ohms! A well-soldered room-temperature joint has a resistance of a few parts in 10^{-9} ohms; at least an order magnitude improvement is required for field stabilities suitable for MRI.

There are multiple methods of making superconducting joints. These methods basically involve eliminating non-superconducting materials (such as copper) from the wires and then bringing the superconducting filaments into close proximity, bridging the gap with a superconducting solder such as Woods metal (an alloy of bismuth, lead, tin and cadmium). The detailed jointing process can be critical as oxides layers can significantly degrade the properties of the joints.[18]

2.4.5 Quenching

A magnet will continue in a superconducting state indefinitely. However, local heating (perhaps caused by a small wire movement) can disturb this equilibrium state. The heat capacity of materials used in superconducting wire is very low and so a small amount of energy can result in a temperature rise sufficient to drive the superconductor above its critical temperature.[19] When the conductor moves above its critical temperature its resistance increases, and further increases in temperature occur owing to ohmic (I^2R) heating. The extra heat causes an additional temperature rise which in turn generates higher coil resistance. The locally generated heat then permeates to adjacent conductors and the effect quickly spreads throughout the coil, preferentially propagating along the length of the wire. The effect is shown schematically in Figure 2.15.

2.4.6 Quench Protection

The resistive voltage in the quenching coil drives down the current in the coil. The decrease in current is slowed by the inductance of the coil such that there is a combination of high currents and resistances during the quench. This combination results in voltages that can increase beyond the capabilities of the wire insulation. As the insulation on the coil fails, arcing will occur that causes irreversible damage to the conductor. A second risk in a quenching coil is thermal runaway in a hot spot, where the temperature increases enough to damage the conductor. To prevent damage owing to a quench, a protection circuit is designed to allow some of the magnet current to bypass the section of the coil that is undergoing a quench. By dividing the magnet into a number of protection loops, the entire magnet can be protected from a quench because the subdivided coils have a much smaller inductance than the entire magnet, so current is able to flow more quickly from the coil into the protection circuit. An example of a protection circuit is shown in Figure 2.16. In this example, the protection circuit divides the magnet into four sections; some of the current will flow through the protection resistors as the quench progresses. Note that there is a protection resistor in parallel with the switch; this prevents damage to the switch if it were to quench.

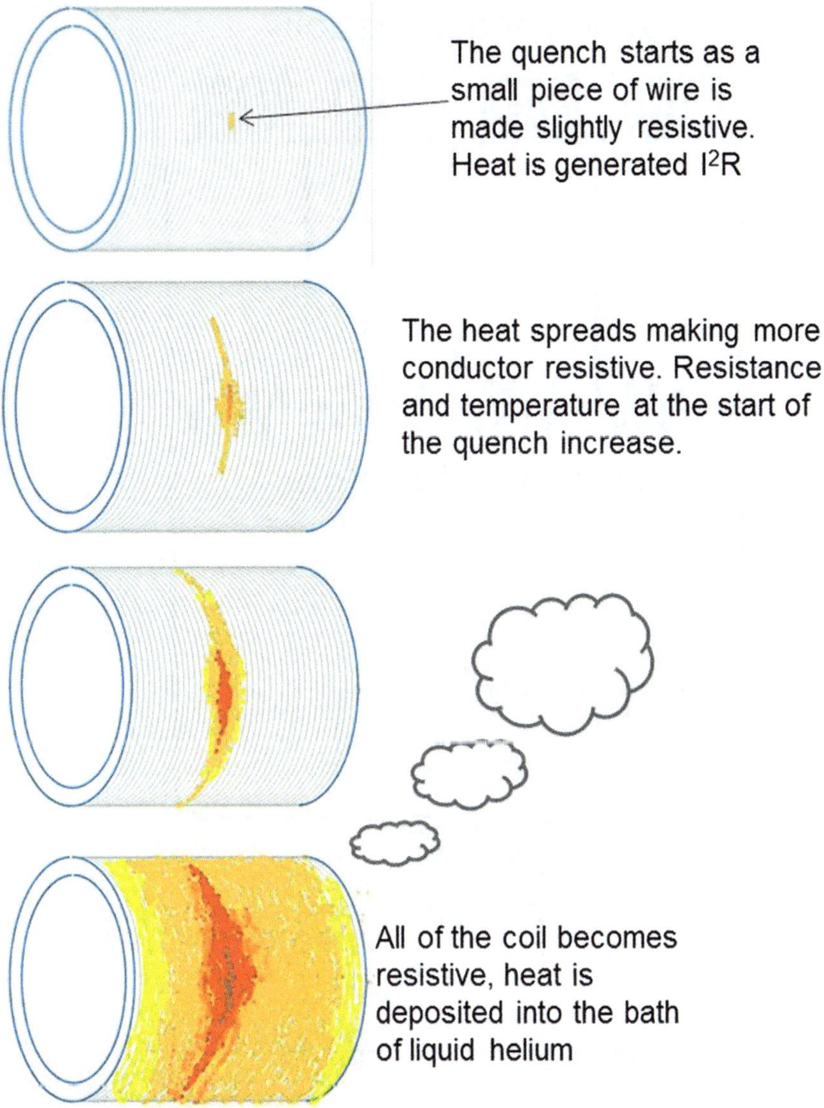

Figure 2.15 Schematic diagram to illustrate how a quench spreads in a coil *via* a positive feedback between the wire resistance and the heating caused by the passage of current through the wire.

2.4.7 Stress Limits

The force on a conductor element **dl**, carrying a current *I* in a magnetic field **B** is given by Lorenz' law in eqn (2.8):

$$\mathbf{F} = \mathbf{B} \times I\mathbf{dl} \tag{2.8}$$

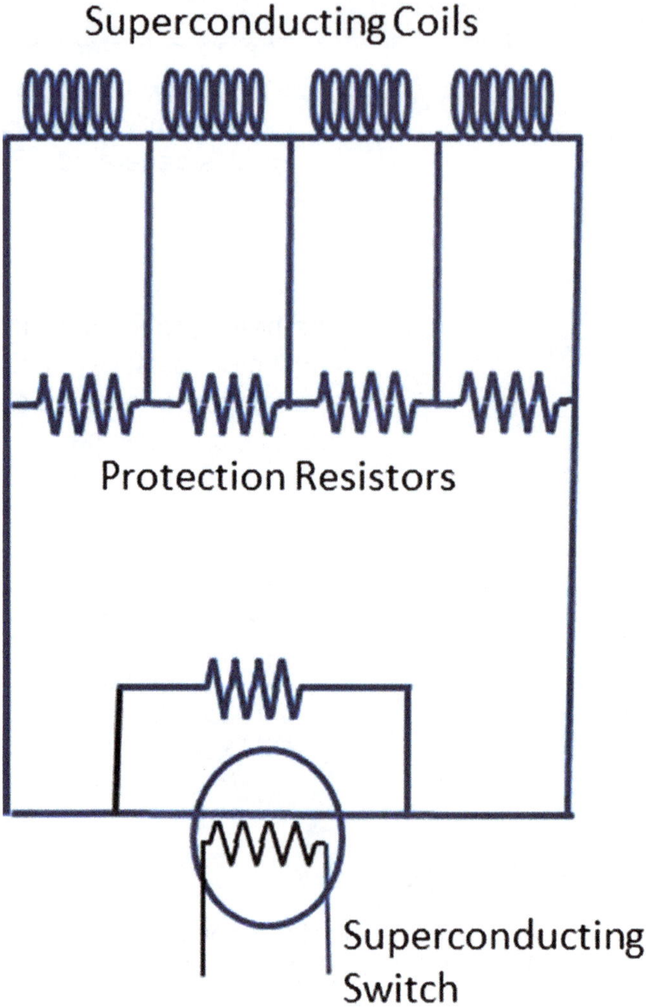

Figure 2.16 Schematic of a quench protection circuit.

where **F**, **B** and **dl** are vector quantities and **B** × **dl** is the cross product of these two vector quantities; **F** is therefore perpendicular to both **B** and **dl**. The direction of force where **B** and **dl** are perpendicular is given by Fleming's left hand rule.[20] For a simple solenoid, the forces on a current element can be resolved into an outward radial force proportional to the *z* component of the magnetic field and an axial force proportional to the radial component of the magnetic field. These resolved forces on a current element are shown in Figure 2.17.

The effect of a radial force integrated around a loop results in hoop (bursting) pressure on the conductor. The geometry of the hoop stress is shown in Figure 2.18.

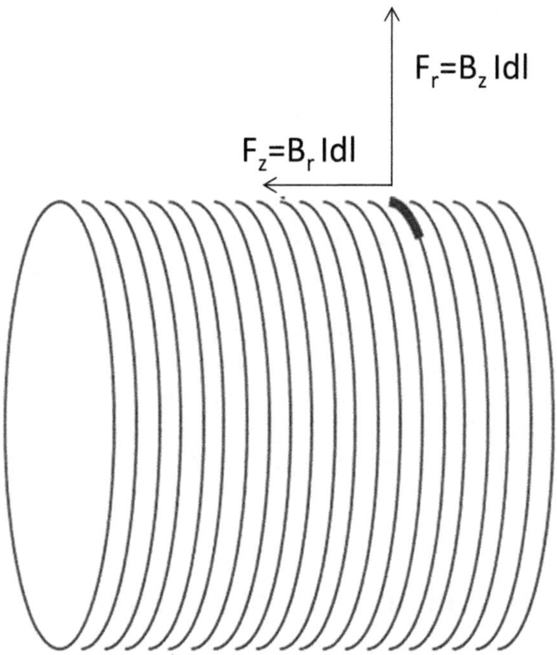

Figure 2.17 The force on a current element in a simple solenoid.

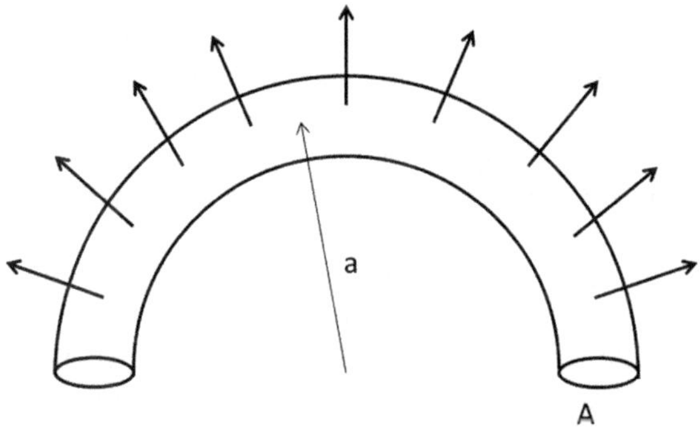

Figure 2.18 Radial forces create a stress within the cross section of the conductor.

The stress in the conductor is given by eqn (2.9):

$$\text{hoopstress} = \frac{B_z I a}{A} \tag{2.9}$$

where a is the radius of the loop and A is the cross section area of the conductor. The stress in a conductor will tend to degrade the superconducting properties of the wire and in the case of Nb_3Sn conductor there is a level of stress that will cause permanent damage. Apart from the stress within a conductor, forces can cause individual wires to move. The heating caused by wire movement can increase the temperature beyond the critical temperature, resulting in a magnet quench. The ability to control the forces on the conductors and prevent such sudden "stick-shift" movement ultimately dictates the maximum field a superconducting magnet can achieve (assuming the critical current of the wire is not exceeded). Filling the small gaps between the coils by impregnating the structure helps to prevent movement. Impregnating the coils with epoxy resin prevents wire movement and allows the forces to be distributed over the whole coil. Additional structures, such as glass fibres or external cylinders, can provide further stiffness, enabling the conductors to operate at higher stresses.

A sufficiently large wire movement will cause the magnet to quench. However, the wire will now be in a new position that may be better supported. When the magnet is brought back to field the wire will be less likely to move and the magnet can be taken to a higher field. The process whereby the operating field is increased by repeated stress-induced quenching is called training.

2.5 Heat Transfer and Cryostat Design

Most MRI magnets are constructed from niobium titanium superconductor. This material needs to be maintained well below its critical temperature of 10 K. The performance of the superconductor increases with decreasing temperature and it is therefore convenient to cool the coils in a bath of liquid helium that has a boiling point of 4.2 K at atmospheric pressure. Liquid helium has a density of 0.125 kg litre^{-1} and a low latent heat of vaporisation: 1 watt of heat will evaporate 1.38 litres hour^{-1} of liquid. A cryostat is required to minimise the heat reaching the helium bath. The cryostat addresses the three principal mechanisms of heat transfer: convection, conduction and radiation.

Convection is the transfer of heat by the circulation of molecules of gas (or liquid). Molecules are excited by a warm surface and then move to the cold surface where they deposit some of their energy. An MRI cryostat eliminates the effect of convection by surrounding the helium vessel with a vacuum. The external structure of the cryostat that contains the vacuum is usually referred to as an outer vacuum case (OVC). The cryostat is evacuated with a high-vacuum pump; the level of vacuum for a typical MRI magnet is less than 10^{-6} mbar. During cooling, any residual molecules of nitrogen and oxygen remaining inside the OVC will freeze on the cold surfaces. An absorber—basically consisting of activated carbon—absorbs any residual atoms of helium. The expected life of a magnet is approximately twenty years. Over this period, small leaks of helium gas will eventually saturate the absorber and this will degrade the cryogenic performance.

Conduction is the process of heat transfer along a structure by the vibration of atoms within the structure. The helium vessel is supported in the OVC with components that have a good ratio between strength and thermal conductivity. Materials used for cryostat supports include carbon fibre, glass fibre, titanium and stainless steel. Power transfer owing to conduction is given by eqn (2.10):

$$Q = \frac{k(\Delta T)A}{L} \quad (2.10)$$

where Q is the heat transferred, k is the thermal conductivity, A is the cross-sectional area of the support, L is the length of the support and ΔT is the difference in temperature at either end of the support. It should be noted that thermal conductivity depends on temperature. The cross-sectional area of the support is determined by the required strength; however, the heat transfer can be minimised by maximising the length of the support. MRI magnet supports typically consist of bands or rods up to 1 m in length. A typical heat load of 0.1 to 0.3 watts will be supplied to the helium vessel owing to conduction down the supports. If the system is cooled with cryogens, then there will be a tube connecting the helium vessel to the outside of the vacuum case. This tube will be sized for pressure relief in a quench and for service activities, and will typically be constructed from thin wall (0.4 mm) stainless steel tube. The tube contributes approximately 0.1 to 0.2 watts of heat to the helium vessel.

Thermal radiation is the process by which electromagnetic radiation (light waves) is emitted from a warm surface and then travels to a cold surface where it is absorbed. Heat transfer by radiation is given by the Stefan–Boltzmann law, as shown in eqn (2.11):

$$Q = \sigma \varepsilon A T^4 \quad (2.11)$$

where σ is the Stefan–Boltzmann constant with a value of 5.67×10^{-8} W m^{-2} K^{-4}, ε is the emissivity, A the area, and T the temperature of the surface. A surface at room temperature (298 K) with a high emissivity ($\varepsilon = 1$) will radiate 447 watts per square metre. Heat transfer owing to radiation is minimised firstly with the choice of reflective surfaces with low values of emissivity, as well as by intercepting radiation with shields at intermediate temperatures, and finally with the use of multi-layers (30–50 layers) of thin aluminised polyester referred to as superinsulation. The radiation shield (or shields) is typically constructed from aluminium alloy and is cooled with a cryo-refrigerator. A schematic of the general construction of a magnet and cryostat is shown in Figure 2.19.

2.5.1 Cryo-Refrigerators

Cryo-refrigerators or cryocoolers are standalone devices that provides cooling at low temperatures. These are used on MRI magnets to cool the thermal radiation shields and recondense gas that has evaporated from the helium bath. There are two types of cryocooler commonly used to cool MRI magnets: one type is based on the Gifford–McMahon method,[21,22] the other is based

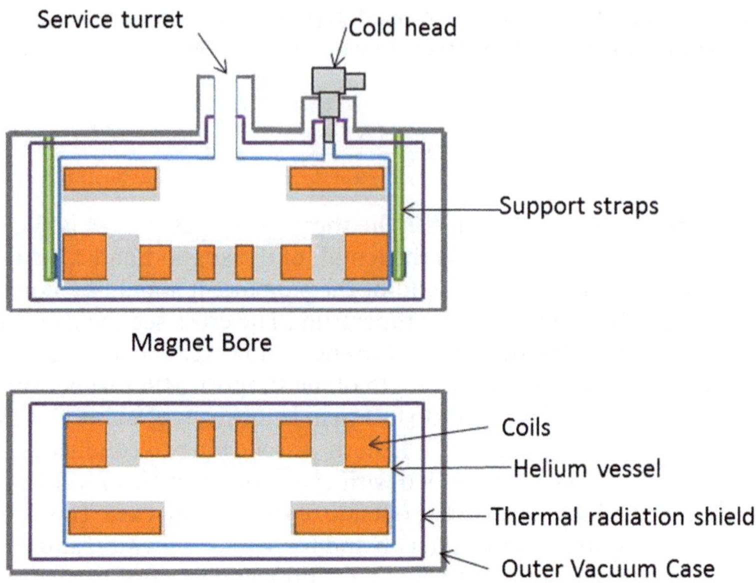

Figure 2.19 Schematic of the general assembly of a magnet and cryostat.

on pulsed tube cooling. Both systems use a closed high-pressure helium gas cycle and comprise of an expansion chamber often referred to as a "cold-head" connected *via* high pressure gas lines to a compressor.

A Gifford–McMahon system works by pushing gas backwards and forwards with a displacer within an expansion chamber. As the displacer moves to increase the volume of the chamber, the helium gas expands and cools. The cold gas cools a heat exchanger consisting of a porous material with a high heat capacity known as a regenerator. The helium gas is then circulated to the compressor and heat is removed from the system as the gas is recompressed. Figure 2.20 shows an example of a cold-head using the Gifford–McMahon cycle.

A pulse tube cooler works in a similar way; in this case, expansion is achieved by pulsing the gas rather than using a displacer. As there are no moving parts inside this type of cold-head the service life tends to be longer. Figure 2.21 shows an example of a pulse tube cold-head.

The cryo-refrigerator usually has two stages, with greater cooling power available at the warmer (40 K) first stage. The second stage will often have approximately one watt of cooling power at a second stage temperature of 4.2 K and in this case it is used to re-liquefy evaporated gas from the helium vessel and effectively achieve a "zero boil-off" cryogenic system.

2.5.2 Sub-Atmospheric Operation

The performance of superconducting wire depends on the operating temperature. Generally it is convenient to operate at the boiling point of liquid helium (4.2 K), however an increase of approximately 20% in the operating

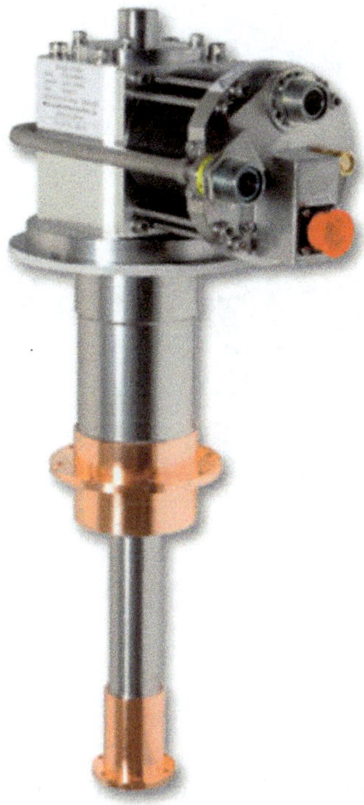

Figure 2.20 Photograph of a cold head using a Gifford–McMahon cycle. Photo courtesy of Sumitomo Heavy Industries (SHI).

field can be obtained with NbTi conductor if the temperature is reduced to 2.3 K. This enables magnets constructed from NbTi conductor to operate at fields up to 11.75 T. The temperature reduction is achieved by reducing the pressure in the helium vessel with the use of a vacuum pump. A temperature of 2.3 K is achieved by reducing the pressure in the vessel to 0.067 atm (6700 Pa). Approximately one-third of the helium volume is lost during pumping as the liquid cools from 4.2 K to 2.3 K and this extra consumption is one of the drawbacks of operating at lower temperatures. The other drawback of operating at reduced pressure is the increased risk of ice accumulating in the neck of the helium vessel.

2.5.3 Gradient-Induced Heating

During an imaging sequence, a time-varying magnetic field is generated by the gradient coils. The magnitude of the field which leaks into the magnet space depends on the effectiveness of the shielding of the gradient coil (see Chapter 5 for details on gradient coil shielding). Any leaked field generates eddy currents in the thermal radiation shield of the cryostat and the radiation shield will therefore effectively shield the helium vessel. However, at particular frequencies the thermal radiation shield may resonate, and its

Figure 2.21 Photograph of a cold head of a pulse tube cryocooler. Photo courtesy of Cryomech Inc.

ability to shield stray magnetic field from the gradient will be reduced. The magnetic field now penetrates the helium vessel and the resultant eddy currents will generate heat in the helium vessel bore as they decay. This heat can result in a significantly increased helium consumption (~20 litres hour^{-1}) and in extreme cases can quench the magnet.[23]

2.6 Practical Considerations

2.6.1 Safety

There are a number of practical considerations that need to be taken into account when planning to site a large MRI magnet. The first, of course, is the stray magnetic field. Public areas need to have a field less than 5 gauss; controlled areas can be higher with most modern electronics able to operate within a field of 50 gauss. It is important to also check the vertical direction as the stray magnet field usually extends into the interstitial space between floors and sometimes into the room above.

Active shield magnets can also suffer from "bloom field". During a quench, the relative distribution of current between the shield and main coils goes out of balance, resulting in either excess current running in the main coils or reduced current running in the shield coils. In these circumstances the shielding is compromised and it is possible that the stray magnetic field will momentarily bloom. Most modern magnet quench circuits take this into account and minimise the amount of bloom field.

The use of liquid helium to cool the magnet results in a number of safety issues. Firstly, the very low temperature can result in severe burns during service processes, particularly if cold gas is exhausted at high pressure onto unprotected skin. During a service procedure involving cryogens, protective clothing including face marks are recommended. The second risk associated with the use of helium is that of asphyxiation; helium gas is a colourless, odourless gas that will cause asphyxiation without any noticeable physical warnings. It is recommended that a superconducting magnet is installed in a well-ventilated room with an oxygen monitor. During a quench or emergency run down, up to 50 litres s^{-1} of liquid helium will evaporate and be expelled from the cryostat. This quantity of gas exhausting into the scan room poses a significant risk of asphyxiation: therefore it is necessary to provide ducting (the quench pipe) to route the gas directly from the magnet to the outside of the building. The back pressure from the duct will depend on the length, diameter and number of bends in the duct. In general, the number of bends should be minimised. The duct should be designed so the back pressure does not exceed the pressure rating of the helium vessel. In normal operation, the quench pipe is sealed from the helium vessel by a burst disk. The burst disk is usually made from graphite and needs to be checked annually for cracks. In the event of a quench, the burst disk will rupture and allow the helium vapour to escape. Once a quench has finished, the magnet coils are still very cold and it is important to reseal the helium can as soon as possible to avoid cryopumping.

The other major risk associated with cryogens relates to the large expansion as the cold liquid evaporates and warms up to room temperature; the fast expansion can result in the risk of explosions. Liquid helium expands by a factor of 748 as it changes from liquid to gas at room temperature. Air leaks into the helium vessel or into a storage vessel can be extremely dangerous as air will tend to freeze in the neck of the vessel causing an ice block. Pressure will then build-up inside the vessel with the inevitable consequence that it will explode with potentially catastrophic consequences. Good cryogenic housekeeping and regular checks for leaks and the presence of ice are essential for safe operation of cryogenic systems. In the event of an ice blockage, expert advice should be sought immediately to address this potentially dangerous situation.

2.6.2 Installation Issues

The level of mechanical vibration of the proposed site needs to be checked. The magnet itself is very robust against vibration, being designed to withstand freighting on trucks to site. However, when the magnet is on-field very low levels of mechanical vibration will cause the radiation shields inside the

Figure 2.22 A 7 T magnet entering the site through a knock-out panel. Photo courtesy of CMRR, University of Minnesota.

cryostat to move and as these conducting surfaces move within the magnetic field they can generate perturbations in the magnetic field resulting in ghosting in the MR image. The site survey should assess the magnitude and frequency of any vibration and have the system vendor confirm that the vibrations are within the operating parameters of the MRI system. In extreme circumstances it is possible to incorporate an isolated slab with anti-vibration pads into the floor of the building. High resolution NMR systems are routinely supplied with pneumatic anti-vibration stands, but these are not normally used for MRI systems.

Building engineers will usually need to check on the suitability of the site to take delivery and commission the magnets. This will need to include the route into the building for the magnet which may be along building corridors, or possibly craning the magnet in through a large window or other opening of the building. It is also worth considering a knockout panel in a wall to allow the magnet to be easily removed at some future date. An example of a magnet delivery is shown in Figure 2.22.

2.7 Future Developments

2.7.1 High and Ultra-High Field Magnets

In the past decade, the number of whole body 7 tesla systems has increased to over 50 worldwide. Until ~2012 these were entirely unshielded systems, with the transition to actively shielded systems occurring soon thereafter. For an unshielded magnet, a compensated solenoid is used, with a magnet length of 3.37 m, diameter 2.38 m, weight 32 tonnes, bore size of 90 cm, field drift 0.05 ppm per hour, and field uniformity of 5 ppm over a 45 cm DSV. Improved gradient technology (see Chapter 5) enabled the bore of the magnet to be reduced from 90 cm to 82 cm, and this was one of the key developments in the design of an actively shielded 7 T magnet.

The highest field human MRI magnet (head-only) being manufactured at present operates at a field strength of 11.7 T. This field is as high as is practical using niobium titanium conductor. To reach these fields, the conductor performance is enhanced by operating at 2.3 K using super-cooled liquid helium. The most ambitious 11.7 T magnet is the 900 mm bore system being built by the CEA in France: a schematic of the magnet is shown in Figure 2.23.[24–26] The magnet is constructed from niobium-titanium conductor and operates at a temperature of 1.8 K in a helium bath that is permanently connected to a helium reliquifier. The main coils are constructed from a series of double pancakes positioned to optimise the homogeneity. The magnet incorporates standard winding techniques (wires wound onto a former) for the shielding coils used to control the stray field. Owing to the high level of stored energy, about 340 MJ, and a relatively high nominal current, about 1500 A, the magnet will be operated in a non-persistent mode with a stabilized power supply.

For pre-clinical applications, very high field magnets are being manufactured using niobium-tin conductor, with the highest field currently being 17.2 tesla. This conductor is significantly more expensive than niobium-titanium

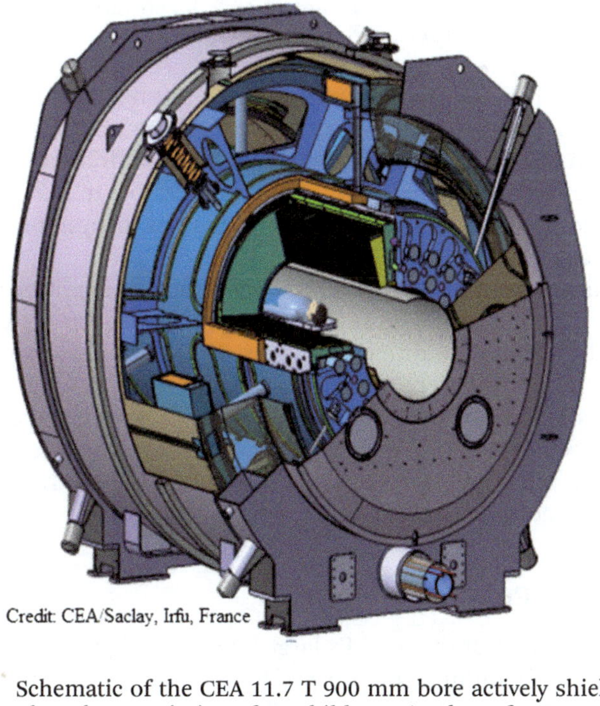

Figure 2.23 Schematic of the CEA 11.7 T 900 mm bore actively shielded magnet. Photo by permission of T. Schild, CEA/Saclay, Irfu, France.

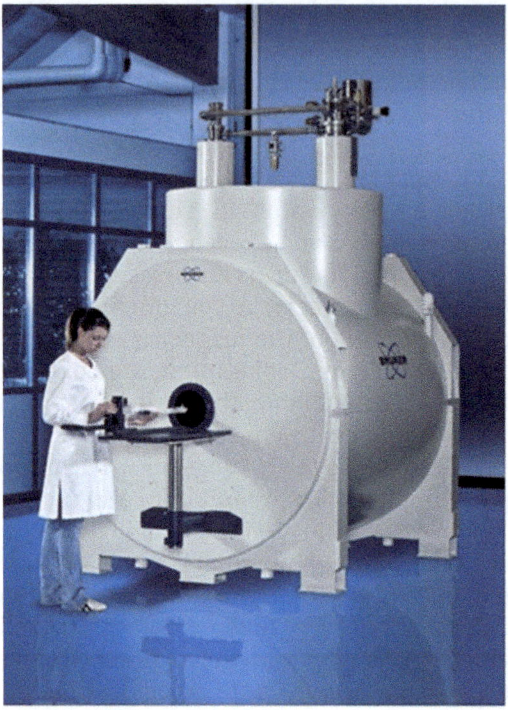

Figure 2.24 An ultra-high field pre-clinical 17.2 T 260 mm bore magnet. Photo courtesy of Bruker Corporation.

and so far has only been used in narrow bore NMR magnets and small bore animal magnets. Figure 2.24 shows an example of an ultra-high field pre-clinical magnet. In the future, using a combination of a niobium-tin conductor and reduced temperature operation, it may be possible to build a magnet of 14 T suitable for imaging humans; however, this type of system is likely to be prohibitively expensive.

A high performance tin conductor has been used on large bore magnets for particle beam accelerators. The authors of this chapter are currently exploring the use of this accelerator technology in ultra-high field magnet designs that are very much more compact than current systems. The niobium tin conductor has an important advantage in that it remains superconducting at higher temperatures. In principle, this would result in a more thermally robust magnet and could offer benefits in minimising the risk of a quench owing to high gradient usage.

2.7.2 Helium-Free Technology

MRI magnets are one of the world's largest users of liquid helium. Up until the turn of the millennium, the price of helium had been depressed by the release of the US strategic reserve in Texas; previous to this, alternative

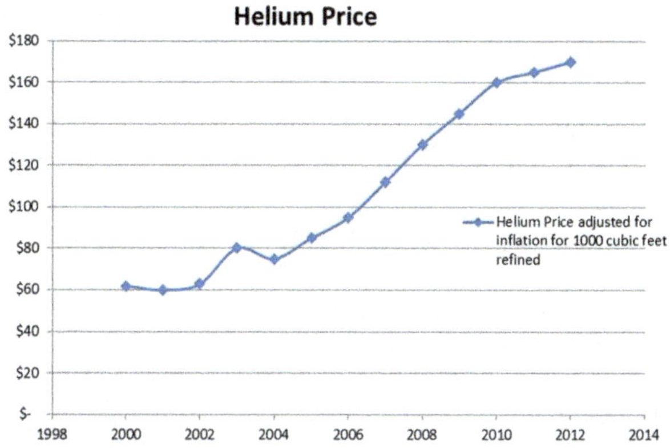

Figure 2.25 Helium price since 2000.

sources had not been developed as it was uneconomic to do so. However, over the last decade, the price of helium has been allowed to increase and this has allowed other alternative sources to be developed in Qatar, Poland and Russia. The increasing helium price is shown in Figure 2.25.

The resulting increase in helium price has driven a number of developments, including higher capacity and more efficient cryocoolers. The majority of magnets today use cryocoolers that condense the helium vapour resulting in virtually no helium consumption during the magnet's operating life. By thermally attaching the cryocoolers directly to the coils, it is possible to operate magnets "dry" (*i.e.* entirely free from helium). As an example, ASG Superconductors built a 0.5 T magnet for Paramed Medical Systems that is entirely cryo-free by utilising a magnesium diboride superconductor operating at a temperature of 20 K. The system demonstrates the feasibility of creating certain types of future MRI systems that will be completely independent of liquid cryogens,[13,27] allowing the modality to become widespread in parts of the world where the cost of cryogens is prohibitively expensive and the availability and transport is severely limited.

References

1. K. Pervushin, R. Riek, G. Wider and K. Wuthrich, *Proc. Natl. Acad. Sci. U. S. A.*, 1997, **94**, 12366.
2. J. Duyn, *J. Magn. Reson.*, 2013, **229**, 198.
3. M. W. Garrett, *J. Appl. Phys.*, 1951, **22**, 1091.
4. D. I. Hoult and D. Lee, *Rev. Sci. Instrum.*, 1985, **56**, 131.
5. G. Gabrielse and J. Tan, *J. Appl. Phys.*, 1988, **63**, 5143.
6. D. Rayner, P. J. Feenen and R. J. Warner, in *Methods in Biomedical Magnetic Resonance and Spectroscopy*, ed. I. R. Young, Wiley, Chichester, 2000.
7. A. Mann, *Nature*, 2011, **475**, 280.

8. J. Nagamatsu, N. Nakagawa, T. Muranaka, Y. Zenitani and J. Akimitsu, *Nature*, 2001, **410**, 63.
9. P. C. Canfield, D. K. Finnemore, S. L. Bud'ko, J. E. Ostenson, G. Lapertot, C. E. Cunningham and C. Petrovic, *Phys. Rev. Lett.*, 2001, **86**, 2423.
10. A. Yamamoto, A. Ishihara, M. Tomita and K. Kishio, *Appl. Phys. Lett.*, 2014, **105**, 032601.
11. Y. Iwasa, J. Bascunan, S. Hahn, M. Tomita and W. J. Yao, *IEEE Trans. Appl. Supercond.*, 2010, **20**, 718.
12. J. Y. Ling, J. Voccio, Y. Kim, S. Hahn, J. Bascunan, D. K. Park and Y. Iwasa, *IEEE Trans. Appl. Supercond.*, 2013, **23**, 6200304.
13. M. Razeti, S. Angius, L. Bertora, D. Damiani, R. Marabotto, M. Modica, D. Nardelli, M. Perrella and M. Tassisto, *IEEE Trans. Appl. Supercond.*, 2008, **18**, 882.
14. J. G. Bednorz and K. A. Muller, *Z. Phys. B: Condens. Matter*, 1986, **64**, 189.
15. A. Schilling, M. Cantoni, J. D. Guo and H. R. Ott, *Nature*, 1993, **363**, 56.
16. Y. Kim, J. Bascunan, T. Lecrevisse, S. Hahn, J. Voccio, D. K. Park and Y. Iwasa, *IEEE Trans. Appl. Supercond.*, 2013, **23**, 6800704.
17. H. W. Weijers, W. D. Markiewicz, A. J. Voran, S. R. Gundlach, W. R. Sheppard, B. Jarvis, Z. L. Johnson, P. D. Noyes, J. Lu, H. Kandel, H. Bai, A. V. Gavrilin, Y. L. Viouchkov, D. C. Larbalestier and D. V. Abraimov, *IEEE Trans. Appl. Supercond.*, 2014, **24**, 4301805.
18. V. C. Strivastava, Method and apparatus for making a superconducting joint, *US Pat.* 4901429A, 1990.
19. M. N. Wilson, *Superconducting magnets*, Clarendon Press, Oxford, 1998.
20. J. A. Fleming, *Magnets and Electric Currents*, E.&F.N.Spon, London, 1902, p. 173.
21. W. E. Gifford and R. C. Longsworth, *Adv. Cryog. Eng.*, 1966, **11**, 171.
22. W. E. Gifford and R. C. Longsworth, *Adv. Cryog. Eng.*, 1966, **10B**, 69.
23. G. Ries, Magnetic resonance apparatus having a basic field magnet with damping of mechanical oscillations, Patent US6707302 B2, 2004.
24. L. Quettier, C. Berriaud, A. Bourquard, G. Gilgrass, R. Leboeuf, M. Nusbaum, J. L. Oudot, T. Schild, M. Schweitzer, V. Stepanov and P. Vedrine, *IEEE Trans. Appl. Supercond.*, 2014, **24**, 4401304.
25. T. Schild, S. Bermond, P. Bredy, A. Donati, O. Dubois, J. M. Gheller, J. J. Goc, J. C. Guillard, H. Lannou, R. Leboeuf, D. Medioni, F. Nunio, F. Molinie, L. Scola, A. Sinanna, V. Stepanov and P. Vedrine, *IEEE Trans. Appl. Supercond.*, 2014, **24**, 4402205.
26. P. Vedrine, G. Aubert, J. Belorgey, C. Berriaud, A. Bourquard, P. Bredy, A. Donati, O. Dubois, F. Elefant, G. Gilgrass, F. P. Juster, H. Lannou, F. Molinie, M. Nusbaum, F. Nunio, A. Payn, L. Quettier, T. Schild, L. Scola and A. Sinanna, *IEEE Trans. Appl. Supercond.*, 2014, **24**, 4401206.
27. B. J. Parkinson, R. Slade, M. J. D. Mallett and V. Chamritski, *IEEE Trans. Appl. Supercond.*, 2013, **23**, 4400405.

CHAPTER 3

Radiofrequency Coils

ANDREW WEBB*[a]

[a]C.J.Gorter Center for High Field MRI, Department of Radiology, Leiden University Medical Center, Leiden, The Netherlands
*E-mail: a.webb@lumc.nl

3.1 Introduction

As elegantly put by Vaughan,[1] "Radiofrequency coils are the lenses of the MR system", designed for efficient transmission of electromagnetic energy into the sample and for high sensitivity signal detection. The same coil may be used both to transmit RF pulses and receive the MR signal, or the transmit and receive coil(s) may be separate physical entities. RF coils range in geometric complexity from a single-frequency surface coil to multiple-frequency nested volume resonators, receive arrays containing more than 100 elements[2] and transmit arrays of up to 32 elements.[3] RF coils have been constructed for experiments in the Earth's magnetic field[4] (Larmor frequency ~2 kHz) all the way up to pulsed field NMR above 2 GHz.[5] RF coils for whole body MRI are on the metre size scale, whereas "microcoils" with dimensions below 100 μm[6] are used for detection of trace elements as well as imaging and spectroscopy of single neurons.[7,8] In high-resolution NMR, superconducting coils[9] have provided tremendous increases in sensitivity over the past decade, and magic angle probes for solid-state NMR spin the sample at rates above 100 kHz[10] with manufacturing tolerances in the micrometer

New Developments in NMR No. 7
Magnetic Resonance Technology: Hardware and System Component Design
Edited by Andrew G Webb

Published by the Royal Society of Chemistry, www.rsc.org

range.[11] Figure 3.1 illustrates the wide variety of RF coils currently used in NMR and MRI experiments.

Despite the enormous ranges in operating frequency, physical size, and specific application, certain basic principles underlie all RF coil design, and Sections 3.2 and 3.3 cover these principles. Sections 3.4–3.8 describe the different types of RF coil(s) that are commonly used in MR experiments. The remainder of the sections in this chapter describe specific coils used for high resolution liquid- and solid-state NMR spectroscopy, MR microimaging/microscopy and human MRI.

In addition to the references in this chapter, recent specialized publications contain extended descriptions of the process of RF coil design[12] and different types of RF coil.[13] Review articles on probes for high resolution NMR of proteins,[14] probes for small animal imaging,[15] solid-state NMR probes[11] and microcoils[16] have also been published.

3.2 General Electromagnetic Principles for RF Coil Design

At the risk of oversimplification, one can summarize the general goals of RF coil design as:

(i) Obtaining the highest transmit magnetic (B_1^+) field within the sample per unit input current to the coil;
(ii) Producing the minimum electric (E) field within the sample per unit input current to the coil; and
(iii) For volume coils, giving as homogeneous a B_1^+ field throughout as large a volume of the sample as possible.

Treating each of these criteria in turn, a high magnetic field per unit current means that short RF pulses can be produced according to (Chapter 1):

$$\alpha = \gamma B_1^+ \tau \tag{3.1}$$

where α is the tip angle, γ is the gyromagnetic ratio, and τ the duration of the pulse. Shorter RF pulses are desirable in all types of MR experiments since they result in reduced artifacts arising from off-resonance effects, and also minimize transverse relaxation during the RF pulse. In localized *in vivo* MR spectroscopy, short pulses minimize the chemical shift displacement artifact and in MRI they allow shorter echo times. In addition, as covered later in this chapter when the principle of reciprocity is discussed, the higher the transmit efficiency of an RF coil, the higher the SNR produced by the same coil when it is used in receive mode.

The electric field component is important since it produces heating in conductive samples, whether these be salt-containing biological samples for

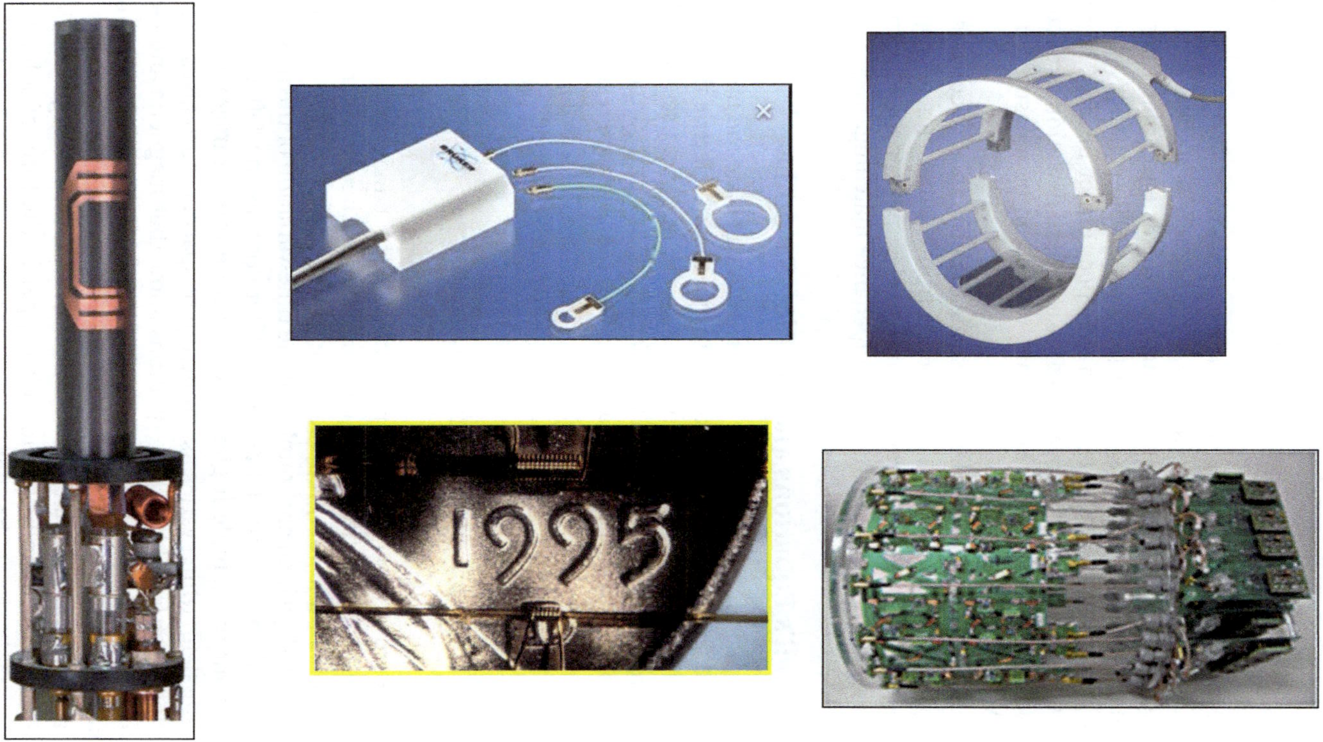

Figure 3.1 An illustration of the many different types of RF coil used in high resolution NMR, animal and human MRI, and MR microscopy.

liquid- or solid-state NMR, or an animal or human in MRI studies. The power absorbed, P_{abs}, by the sample is given by:

$$P_{abs} = \frac{1}{2}\int \sigma |E|^2 \, dv \tag{3.2}$$

where σ is the conductivity, and v is the sample volume.

As covered in Section 3.4, many RF coils are designed to produce a homogeneous magnetic field throughout the sample. This is particularly important when many RF pulses are used in a particular MR sequence or when quantitative information is required from the spectrum or image. The uniformity can be expressed as the covariance (% standard deviation divided by the mean) of the B_1^+ field within the sample, which can be measured using specialized MRI sequences.

In the following sections, in order to illustrate the principles of coil design, a simple 10 cm diameter loop coil constructed from 1 mm diameter copper wire is analyzed in terms of the transmit and receive magnetic field components, the electric field components, and impedance matching.

3.2.1 Maxwell's Equations and the Biot–Savart law

As covered in Appendix A in Chapter 1, in order to calculate the magnitude, orientation, and spatial distribution of the magnetic and electric fields produced within the sample by the particular RF coil, one can use Maxwell's equations. Here only the two equations, Faraday's law and Ampere's law, which are directly related to electromagnetic (EM) radiation are reshown:

$$\nabla \times E = -\mu \frac{\partial H}{\partial t} \text{ (Faraday's law)} \tag{3.3}$$

$$\nabla \times H = J + \varepsilon \frac{\partial E}{\partial t} \text{ (Ampere's law)} \tag{3.4}$$

where ε is the permittivity, μ is the permeability, H is the magnetic field, E is the electric field, and J is the electric current density. The curl operators on the left of eqn (3.3) and (3.4) represent the spatial-variation of the electric and magnetic fields, which are coupled to the time-variation of the magnetic and electric fields on the right hand side. Therefore, as the E-field propagates throughout the sample it gives rise to a time-varying magnetic field. This magnetic field, which varies as a function of space, in turn gives rise to a time-varying electric field. These equations show that magnetic and electric fields are intrinsically coupled. The propagation of these fields through space is illustrated in Figure 3.2.

The initial design stage for an RF coil is to determine into what geometrical structure the electrical conductor (usually copper in the form of solid wire, hollow tube, or foil) should be shaped to produce the desired magnetic field distribution within the sample. EM simulations based on Maxwell's equations can be performed, but these usually require sophisticated commercial

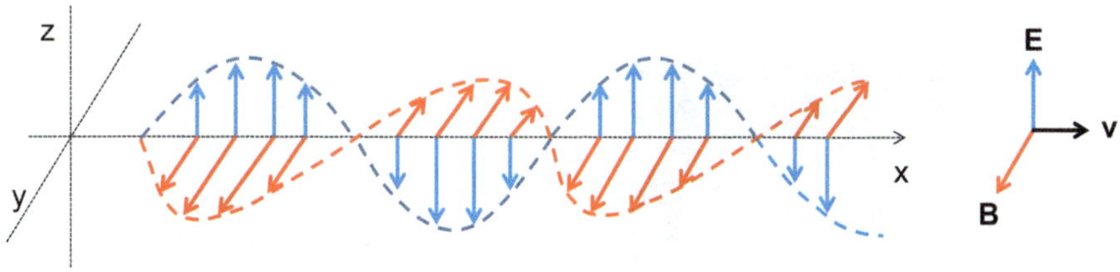

Figure 3.2 Propagation of electric (blue) and magnetic (red) fields through space at velocity v in the x-direction. The fields propagate at an angle of 90° with respect to one another.

software packages, covered briefly later in this chapter and in much more depth in Chapter 8. An initial design can be formulated by simplifying Maxwell's equations under conditions known as "quasi-static", which corresponds to being able to neglect any effects of the finite wavelength of the EM wave within the sample. In this case, the magnetic field can be estimated quite accurately by the Biot–Savart law, as shown in Appendix A in Chapter 1, which can be derived directly from Maxwell's equations.

For a loop of wire radius a the magnetic field, B_y, produced along the central axis of the loop, was derived in Chapter 1, Appendix A as:

$$B_y = \frac{\mu_0 a^2 I}{2\left(y^2 + a^2\right)^{3/2}} \tag{3.5}$$

Applying eqn (3.5) to the example of the 10 cm diameter surface coil the magnetic field produced by unit current in the centre of the coil ($y = 0$) is approximately 12.5 μT. As the distance away from the coil increases to 1 cm, 2 cm, 5 cm (one radius) and 10 cm (one diameter), the magnetic field decreases to 12 μT, 10 μT, 4.4 μT and 1 μT, respectively.

As also covered in Appendix A, the off-axis magnetic fields contain terms in complete elliptical integrals. Figure 3.3 shows plots of the transmit magnetic field produced by a loop of wire oriented perpendicular to the direction of the main magnetic field. The field is strongest close to the wires, but also extends significantly outside the loop. Beyond one coil diameter the magnetic field is reduced by over 90% of its maximum value.

3.2.2 Transmit (B_1^+) and Receive (B_1^-) Magnetic Fields

As mentioned earlier, RF coils can be used either in transceive mode (*i.e.* the same coil is used both as a transmitter and as a receiver) or one coil can be used to transmit the RF pulses and another (or multiple others) used to receive the signal. As outlined in Chapter 1, the transmit B_1 field producing the RF pulses is an oscillating magnetic field, which can be represented mathematically as:

$$\overrightarrow{B_1} = 2B_1 \cos(\omega t + \phi) \tag{3.6}$$

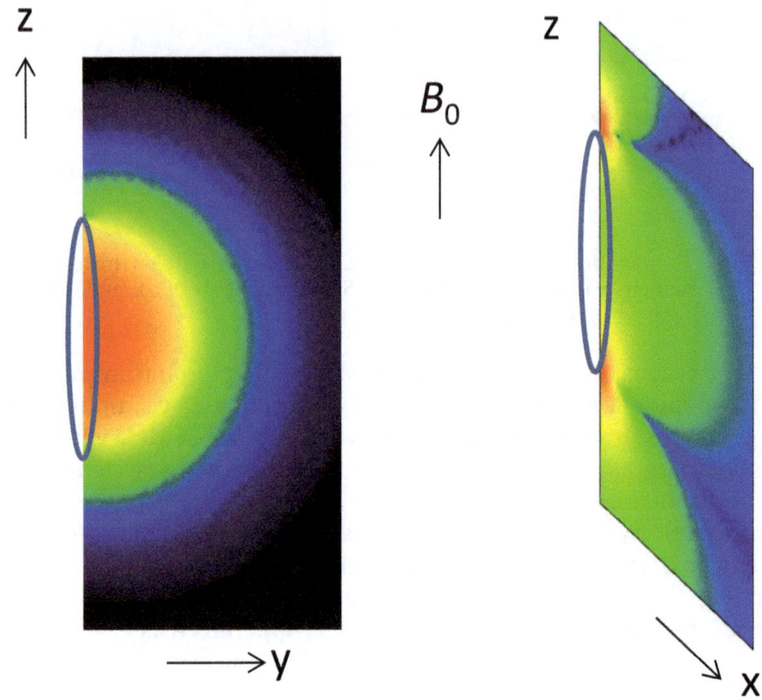

Figure 3.3 Images of the transmit and magnetic field through the centre of two orthogonal planes produced by a loop of wire.

where ω is the Larmor frequency, and ϕ represents the fact that there is an arbitrary phase of the transmit pulse that is dependent upon many instrumental factors. This field can be decomposed into two circularly polarized fields, which rotate in opposite directions:

$$\vec{B_1} = B_1\left[\cos\left(\omega t + \phi\right) - \sin\left(\omega t + \phi\right)\right] + B_1\left[\cos\left(\omega t + \phi\right) + \sin\left(\omega t + \phi\right)\right] \quad (3.7)$$

This decomposition is shown schematically in Figure 3.4.

The first term in eqn (3.7) rotates in the same direction as the precessing spins and therefore interacts with the net magnetization. The second term rotates in the opposite sense and has no interaction. The symbol B_1^+ is assigned to the component that rotates in the same direction as the nuclear precession and the symbol B_1^- to the counter-rotating component. So the relevant B_1 field for transmitting RF pulses is given by:

$$\vec{B_1} = B_1\left[\cos\left(\omega t + \phi\right) - \sin\left(\omega t + \phi\right)\right] \quad (3.8)$$

which has x- and y-components given by:

$$B_{1,x} = B_1 \cos(\omega t + \phi),\ B_{1,y} = -B_1 \sin(\omega t + \phi) \quad (3.9)$$

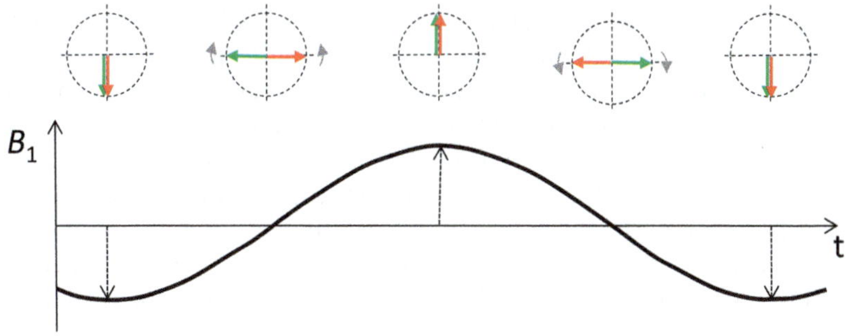

Figure 3.4 Illustration of the decomposition of a linear wave (black line) into two counter-rotating components, shown at the top in red and green. One component (red) rotates in the same sense as the nuclear magnetization and therefore interacts with it, whereas the other component (green) rotates in the opposite direction and has no interaction.

The transmit (B_1^+) and receive B_1^- fields can be expressed as:

$$B_1^+ = \frac{B_{1x} + jB_{1y}}{2} \tag{3.10}$$

$$B_1^- = \frac{\left(B_{1x} - jB_{1y}\right)^*}{2} \tag{3.11}$$

In terms of the relationship between the B_1^+ and B_1^- fields the principle of reciprocity, as originally applied to magnetic resonance by Hoult in 1976,[17] is a very useful way to link the coil transmit efficiency to the coil receive sensitivity. The initial formulation dealt with direct current, but is equally valid at MRI frequencies and indeed, as pointed out by Hoult in 2000,[18] is an inviolable physical law. Applying the principle of reciprocity the received NMR signal, S, is proportional to $(B_1^-)^*$, *i.e.* the complex conjugate of the counter-rotating field induced in the coil per unit current. The signal intensity from a simple one-pulse experiment, with a tip angle of α degrees, is given by:

$$S \propto (\sin\alpha)[\alpha] \propto \left(\sin\gamma B_1^+\tau\right)\left[\gamma\left(B_1^-\right)^*\tau\right] \tag{3.12}$$

For non-conducting samples, the B_1^+ and B_1^- fields are spatially identical, but as the sample conductivity increases the fields become asymmetric and mirror images of one another, as covered later in this chapter.

3.2.3 Linear and Circular Polarization

So far the EM field from a single loop has been considered. This EM field is linearly polarized, *i.e.* it only has one component that interacts with the net magnetization, as seen in the previous section. The concept of circularly polarized EM fields is familiar from many fields such as optics and radar. Linear and circular polarization in MRI were first described in detail in the work of Glover *et al.*[19] As can be appreciated from Figure 3.4, half the energy of the transmitted RF pulse is effectively wasted, since it produces a component that is rotating in the "wrong" sense and so does not interact with the net magnetization. Instead, suppose that the magnetic field is split into two components, with a 90° phase shift introduced into one component, as shown in Figure 3.5. In this case the rotating vector in the "correct" sense (shown in red) in the two channels adds up constructively, whereas the component rotating in the wrong sense (shown in green) in the two channels cancels out. Therefore, all of the energy is used to produce a rotating vector which interacts with the nuclear magnetization. This means that transmission is twice as efficient, *i.e.* only one-half the power is needed to attain a given pulse tip angle, or alternatively the pulse duration can be reduced by a factor-of-two for the same input power.[20]

How might one implement this principle in practice using the loop coil design as a simple example? Two loops must be used, and the magnetic fields they produce must be orthogonal to one another. One way to achieve this is to place the two coils at 90° to one another, as shown in Figure 3.6.

As shown in Figure 3.6, in order for a quadrature coil to function properly in both transmission and reception, the signal from the coil that leads by 90°

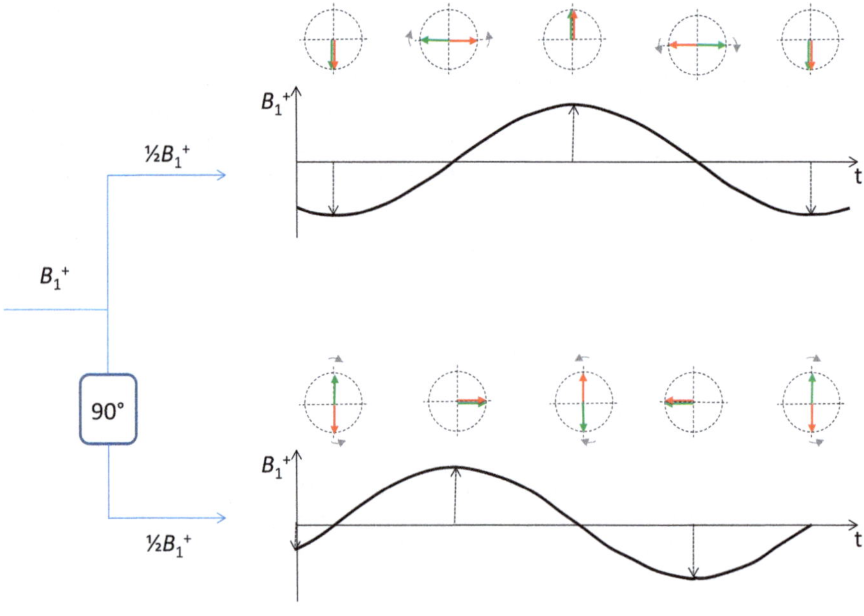

Figure 3.5 Illustration of a circularly polarized electromagnetic field produced by splitting the transmission field into two separate halves, with a phase difference of 90°. The component rotating in the same sense as the nuclear magnetization, shown in red, adds constructively, whereas that rotating in the opposite sense, in green, cancels.

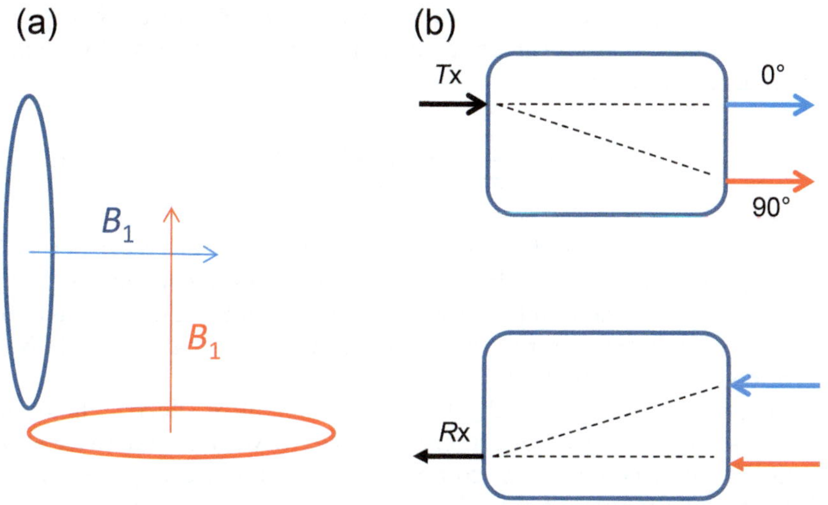

Figure 3.6 (a) Two loops placed orthogonally to one another produce orthogonal B_1 fields, which can be used for quadrature transmission and reception of a circularly polarized field. (b) Schematic of a quadrature hybrid coupler used to introduce a +90° phase difference between the two channels in transmission, and a −90° phase difference in reception. During signal transmission, the extra path-length in the signal transmitted to the coil shown in red introduces a phase lag. During reception, the extra path length is now in the channel connected to the blue coil.

during transmission lags by 90° during reception. Eqn (3.14) shows that the received signal is proportional to the complex conjugate of the B_1^- field. From Figure 3.5, one can see that the signal cancels out if the same 90° phase shift is applied during signal reception as is applied during pulse transmission. However, by applying a 90° phase shift in the opposite sense, the B_1^- components add constructively.

3.2.4 Conservative and Non-Conservative Electric Fields

So far only the *magnetic* field component produced by an RF coil has been considered. However, it is also important to consider the *electric* fields produced by the coil, particularly those present within a conducting sample, such as the human body, animal or a protein sample with significant salt content, since the electric field induces Joule heating *via* the power deposited, as shown in eqn (3.2). The electric field produced by the RF coil has two components. The first is the magnetically induced component of the *E*-field (E_i) produced by a changing magnetic flux as dictated by Maxwell's equations. This is a *non-conservative* electric field, where a field is defined as conservative if its line integral around any closed path is zero. The second component is a *conservative* electric field (E_c), which results from the voltage across the RF coil required to drive current through the coil. In exact analogy to the Biot–Savart law for calculating the magnetic field, the electric field generated by stationary charges can be calculated from Coulomb's law for a continuous charge distribution:

$$E(r) = \frac{1}{4\pi\varepsilon_0}\int \rho(r')\frac{r-r'}{|r-r'|}\mathrm{d}^3 r' \tag{3.13}$$

Figure 3.7 shows a plot of the electric field produced in the central axes of a wire loop. The electric field is lowest at the centre of the coil, and highest right next to the wire itself. In practice, this means that it is not desirable to place the sample directly next to the wires, but to put a small spacer in-between the coil and the sample.[21]

Joule heating associated with the non-conservative electric field is referred to as *inductive loss*, and increases linearly with conductivity and quadratically with frequency. The component E_i cannot be changed without changing the RF magnetic field (B_1^+), and so essentially this loss is inherent to the MR experiment. In terms of power deposition arising from the conservative electric field, a high voltage across the terminals of the coil results in a high E_c: in practice, this is the case if the inductance of the coil is high. Strong conservative electric fields are also typically found close to capacitors in the coils, and any areas where there is a large voltage drop.

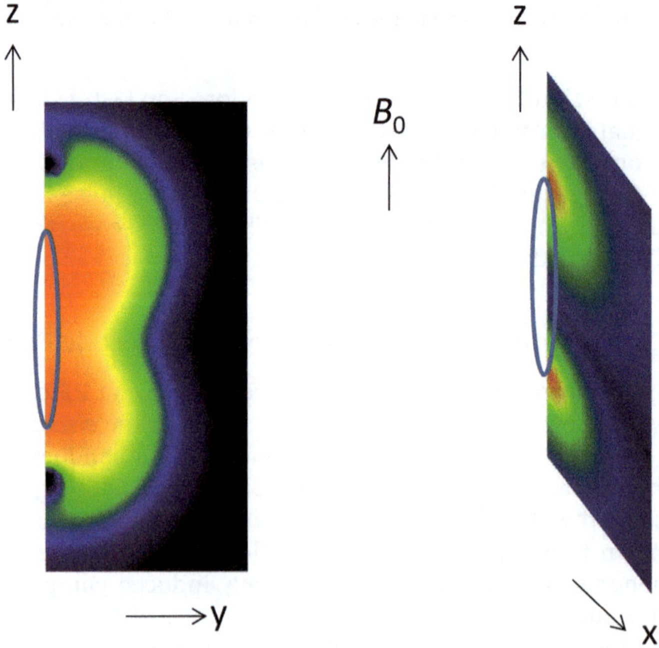

Figure 3.7 Plots of the electric field through the centre of two orthogonal planes produced by a loop of wire (see Figure 3.3 for the corresponding magnetic field plots).

The loss associated with the conservative electric field is termed the *dielectric loss*. In some cases, E_c can be a significant component of the total electric field and can be responsible for the majority of heating in the sample.[22] Since the dielectric loss *does* depend upon the coil geometry, one of the aims of coil design is to minimize the conservative electric field within the sample.[23–26]

3.2.5 Electromagnetic Simulations

Calculation of magnetic and electric field distributions can be performed analytically only for very simple geometries such as a single loop or straight wire segments under quasi-static conditions. For more complicated coil geometries and higher operating frequencies (where "high" refers to the fact that wavelength effects cannot be ignored and so quasi-static assumptions are no longer valid), Maxwell's equations must be solved numerically. This is most commonly performed using commercial electromagnetic software packages. Examples of such packages include X finite difference time domain (XFdtd), Computer Simulation Technology (CST) Microwave Studio, Sim4Life, high frequency structural simulator (HFSS) and FEldberechnung für Körper mit beliebiger Oberfläche (FEKO). These packages use different techniques such as finite difference time domain, finite element or finite integration. Although many of these software packages were originally developed for radar applications or the cellphone industry, MRI plays an increasingly important role in their applications, and more MRI-specific parameters (such as explicit calculation of B_1^+ and B_1^- as well as inter-coil interactions) are being incorporated. Chapter 8 in this book covers the topic of electromagnetic simulations in extensive detail.

3.3 Electrical Circuit Analysis

As shown in Figure 3.8, the RF coil is interfaced to the MR system *via* a transmit/receive switch. Ideally, all of the power from the RF amplifier should be delivered to the coil during pulse transmission and all of the MR signal should be delivered to the preamplifier during reception. In order to ensure these conditions, a number of electrical circuits have to be integrated into or close to the RF coil. The two major circuits are:

(i) An impedance matching circuit, which consists of elements(s) to resonate the RF coil at the Larmor frequency with an input impedance of 50 Ω. The concepts of impedance, resonance and impedance matching are covered in Sections 3.3.1–3.3.5.
(ii) Incorporation of balanced-to-unbalanced (balun) circuits and cable traps to ensure that common mode currents on the outside of the coaxial cables connected to the RF coils are minimized. This is covered in Section 3.3.6.

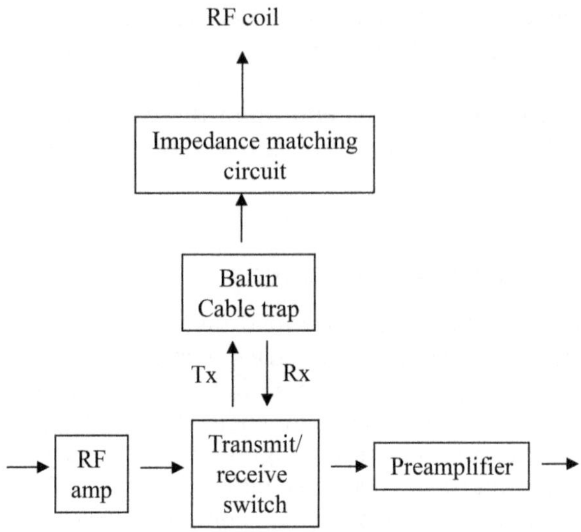

Figure 3.8 Schematic of the transmit and receive RF chain in an MR system.

3.3.1 RF Coil Impedance

Impedance (Z) can be considered as a frequency-dependent resistance. The impedances of an inductor, L, and a capacitor, C, are given by:

$$Z_L = j\omega L$$
$$Z_C = -j\frac{1}{\omega C} = \frac{1}{j\omega C} \tag{3.14}$$

where j represents $\sqrt{-1}$. These equations show that the impedance of an inductor increases with frequency, whereas that of a capacitor decreases. The combination of impedances in series or in parallel are exactly analogous to the expressions for resistors in the DC case, *i.e.* for two impedances Z_1 and Z_2:

$$Z_{\text{series}} = Z_1 + Z_2$$
$$Z_{\text{parallel}} = \frac{Z_1 Z_2}{Z_1 + Z_2} \tag{3.15}$$

Consider the example of the 10 cm diameter circular loop of wire connected *via* a coaxial cable (which has a 50 Ω characteristic impedance) to the transmit/receive switch, as shown in Figure 3.9(a).

The inductance (in nanohenries) of a circular loop of wire is given by:[27]

$$L(nH) = \frac{\pi}{5} d_{\text{coil}} \left(\ln\left(\frac{8 d_{\text{coil}}}{d_{\text{wire}}} \right) - 2 \right) \tag{3.16}$$

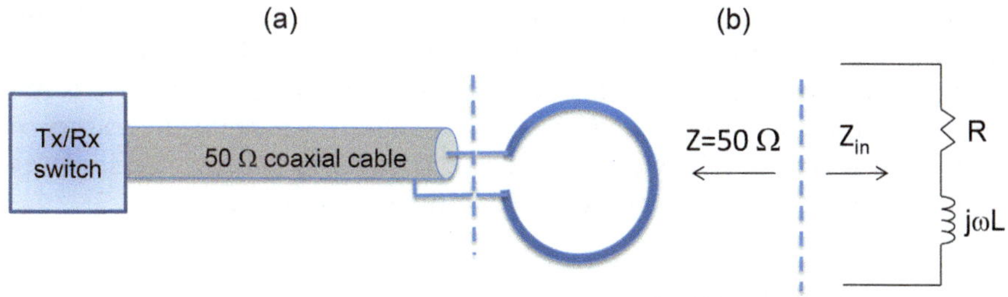

Figure 3.9 (a) Schematic of a wire loop connected to a coaxial cable. (b) Corresponding electrical circuit in which the RF coil is represented by an inductor in series with a resistor, and has an impedance given by Z_{in}.

where the coil and wire diameters, d_{coil} and d_{wire}, respectively, are measured in mm. A 10 cm diameter loop with a wire diameter of 1 mm has an inductance of 294 nH. Since the copper has finite conductivity (~6×10^7 S m^{-1}), there is also a resistance, R_{wire}, associated with the coil. This resistance is given by:

$$R_{wire} = \rho_{wire} \frac{d_{coil}}{\delta d_{wire}} \tag{3.17}$$

where ρ is the resistivity and δ is the skin depth, in turn given by:

$$\delta = \sqrt{\frac{2\rho_{wire}}{\omega\mu_0}} \tag{3.18}$$

For copper wire the resistivity is approximately 2.2×10^{-8} Ωm.[28] If the loop is designed to operate on a 3 tesla magnet at a proton Larmor frequency of 127.8 MHz the wire resistance is approximately 0.03 Ω, and so the impedance of the coil at that frequency is given by:

$$Z = R + j\omega L = 0.03 + j236 \tag{3.19}$$

At the interface/connection between the cable and the coil the amount of power that is reflected from the connection depends upon the "impedance match" between the cable and coil. This is defined in terms of a reflection coefficient, Γ, given by:

$$\Gamma = \frac{Z_{cable} - Z_{coil}}{Z_{cable} + Z_{coil}} \tag{3.20}$$

Ideally, $\Gamma = 0$, in which case no power is reflected from the interface and all the power is transmitted to the RF coil. However, for the loop coil $\Gamma = -0.9 - j0.4$, corresponding to $|\Gamma| > 0.999$. The fraction of the incident power from the amplifier which is delivered to the coil is therefore much less than 0.1%!!

In order to minimize the reflected power, the impedance of the RF coil has to be matched to that of the coaxial cable. It is well-known that maximum power transfer in a circuit is achieved when the load impedance is the complex conjugate of the source impedance. So in this case it is necessary to make the impedance of the coil in eqn (3.21) equal to 50 Ω, which involves the design of a resonant circuit, covered in the next section.

3.3.2 Resonant Circuits

The concept of resonance occurs in many different aspects of everyday life. It refers to the phenomenon in which the output of a particular device or system is a maximum at a certain input frequency, with this frequency being referred to as the resonance frequency. For example, acoustic and mechanical resonance occurs in the human ear, where the pinna and ear canal have dimensions that amplify frequencies associated with the human voice more than other frequencies. Similar concepts are present in wind and string musical instruments in which the air-filled structures produce resonances. An example of optical resonance is using specifically-sized cavities in the production of lasers.

For MR, the relevant parameter is the output voltage induced in the RF coil by the precessing magnetization. In order to use the loop as an RF coil it should be designed such that the resonance frequency is the Larmor frequency. An equivalent definition of resonance in electrical circuits is that the imaginary part of the input impedance is zero, *i.e.* there is only a real part. From eqn (3.21) the loop has a positive imaginary impedance owing to its inductance and therefore a capacitor with a compensating negative imaginary impedance must be placed in series with the loop. The impedance of the coil plus capacitor, shown in Figure 3.10(a), is given by:

$$Z_{\text{in}} = \frac{R}{(1-\omega^2 LC)^2 + \omega^2 R^2 C^2} + \mathrm{j}\,\frac{\omega L\left[1 - \dfrac{R^2 C}{L} - \omega^2 LC\right]}{(1-\omega^2 LC)^2 + \omega^2 R^2 C^2} \tag{3.21}$$

The imaginary component is zero at a resonant frequency, ω_0, given by:

$$\omega_0 = \sqrt{\frac{1}{LC} - \frac{R^2}{L}} \tag{3.22}$$

The first term in the square root is much larger than the second, and therefore the following approximation is commonly made:

$$\omega_0 = \frac{1}{\sqrt{LC}} \tag{3.23}$$

For the 10 cm diameter loop with inductance 294 nH, a capacitance of 5 pF is needed to achieve a resonance frequency of 127.8 MHz. Figure 3.10(b)

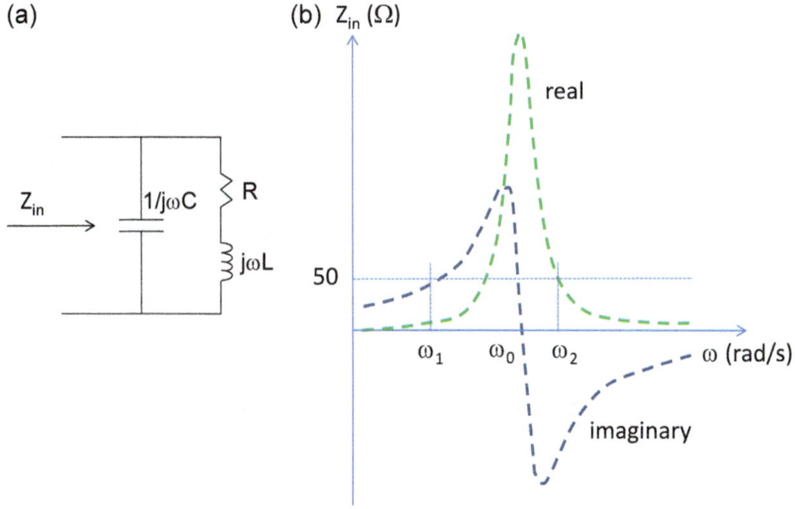

Figure 3.10 (a) A capacitor is placed across the terminals of the RF coil in order to resonate it. (b) Plots of the real and imaginary components of the input impedance as a function of frequency. The resonant frequency is indicated by ω_0.

plots the real and imaginary components of the input impedance as a function of frequency. At ω_0 the real component is maximum and the imaginary component zero.

Although the coil is now resonant, the impedance is not equal to 50 Ω at ω_0 which is required for maximum power transfer. From Figure 3.10(b), it is clear that if the coil is tuned to a frequency ω_1, which is slightly lower than ω_0, then the real part of the impedance is 50 Ω, and there is an inductive (positive) imaginary component. Similarly, if the coil is tuned to a frequency ω_2, the real part of the impedance is also 50 Ω, and there is a capacitive (negative) imaginary component. Cancelling out the imaginary component in either of these cases is the second stage of impedance matching. There are three basic methods of performing this task, capacitive impedance matching, inductive impedance matching, and using transmission lines, each of which is discussed in turn in the following sections.

3.3.3 Capacitive Impedance Matching

As outlined above, if the coil is tuned to a frequency ω_1 a capacitor can be added in series such that its negative impedance is equal and opposite to the positive impedance of the coil, with the result that the input impedance is 50 Ω, *i.e.* it will be "impedance matched". The most simple impedance matching circuit is shown in Figure 3.11(a).

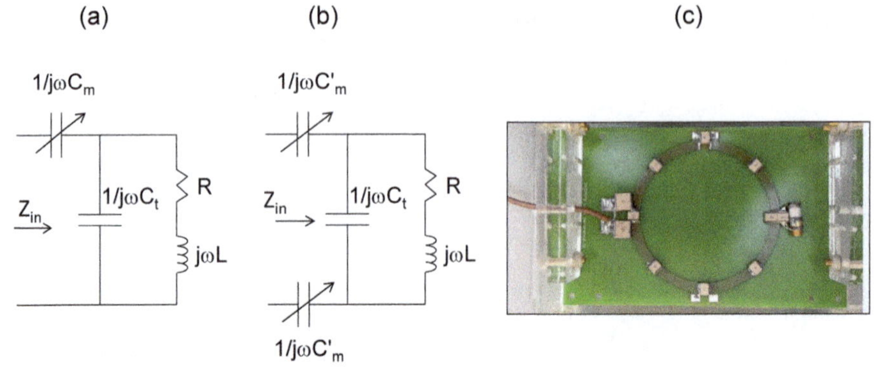

Figure 3.11 (a) A simple capacitive "L-network" to impedance match the RF coil to 50 Ω. A capacitor is connected to the central conductor of the coaxial cable: the other side of the coil is connected to the outer braid (ground) of the cable. (b) An improved balanced "pi-network" in which the value of *C'*m is twice that of C_m. (c) Photograph of a capacitively matched loop coil: the loop coil has many segmenting capacitors (as discussed in Section 3.3.7) and a variable capacitor for fine tuning.

The impedance of the circuit in Figure 3.11(a) is given by:

$$Z_{in} = \frac{R}{\left(1 - \omega^2 LC_t\right)^2 - \omega^2 R^2 C_t^2} + j\left[\frac{\omega L\left(1 - \frac{R^2 C_t}{L} - \omega^2 LC_t\right)}{\left(1 - \omega^2 LC_t\right)^2 + \omega^2 R^2 C_t^2} - \frac{1}{\omega\, C_m}\right] \tag{3.24}$$

For a given value of ω, there is a unique solution for C_t and C_m such that $Z_{in} = 50 + j0$. Often a balanced matching design is used with two capacitors in series with the coil, a configuration that has been shown to minimize the dielectric losses for a conducting sample.[29] There are a large number of alternative and more advanced networks that can be used for impedance matching, with improved balancing properties,[12] but the pi-network shown in Figure 3.11 (b) represents a simple and practical solution. A photograph of a loop with a pi-network, and one variable capacitor for fine-tuning the resonance frequency, is shown in Figure 3.11(c).

3.3.4 Inductive Impedance Matching

A second method of impedance matching the RF coil is to use inductive matching *via* an inductively coupled, physically separated electrical circuit. A circuit diagram is shown in Figure 3.12(a), an equivalent circuit in Figure 3.12(b), and a photograph of the arrangement in Figure 3.12(c).

The input impedance in Figure 3.12(a) and (b) is given by:

$$Z_{in} = R_1 + j\left(\omega L_1 - \frac{1}{\omega C_1}\right) + \frac{\omega^2 M^2}{R_2 + j\left(\omega L_2 - \frac{1}{\omega C_2}\right)} \tag{3.25}$$

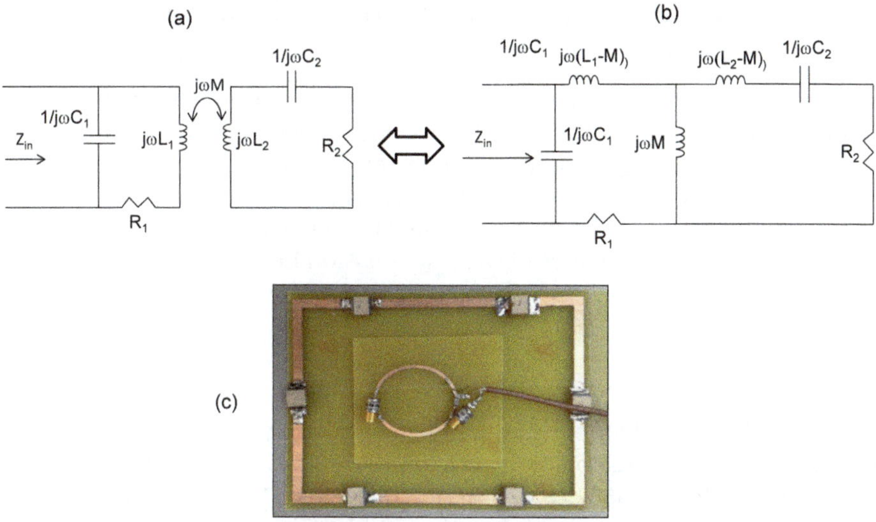

Figure 3.12 (a) Circuit representation of inductive impedance matching using the mutual inductance (M) between the RF coil ($R_2C_2L_2$) and a smaller loop (L_1, R_1). (b) The equivalent electrical circuit used to derive the value of Z_{in}. (c) Photograph of an inductively matched loop coil.

The mutual inductance, M, measured in henries, between the two circuits can be expressed in terms of a coupling constant, k:

$$M = k\sqrt{L_1L_2} \tag{3.26}$$

k has values between 0 and 1. If the two circuits are tuned to similar frequencies and are physically very close, then the coupling constant is high. If the circuits are physically separated by a large distance then the coupling constant is close to zero. This impedance matching circuit can be implemented in different ways. If one can mechanically vary the distance between the primary and secondary circuits, thus changing the value of M, then the loop can be fixed tuned at the Larmor frequency with the capacitor C_1. If, on the other hand, the positioning of the two loops is fixed, then the capacitor C_1 should be made variable.

It should be noted that, of course, the current in the small loop creates a magnetic field, which interacts with that produced by the RF coil. Ideally this additional magnetic field should be as small as possible. The ratio of the currents in the two circuits is given by:

$$\frac{I_1}{I_2} = \sqrt{\frac{R_2}{50}} \tag{3.27}$$

Therefore, as long as the resistance of the coil (plus sample) is much lower than 50 Ω, the additional magnetic field from the coupling loop is very small. Extensive analysis of inductive coupling can be found in the publications of Bilgen[30] and Mispelter.[12]

3.3.5 Impedance Matching Using Transmission Line Elements

A third method of impedance matching is to use transmission lines. The main advantages of transmission lines over lumped elements occurs when very high power is required, most commonly in solid-state NMR probes. In its simplest form, a transmission line consists of an inner conductor and an outer conductor separated by a material with relative dielectric constant ε_r. The characteristic impedance of the transmission line, Z_0, is given by:

$$Z_0 \approx 60 \ln\left(\frac{b}{a}\right)\sqrt{\frac{1}{\varepsilon_r}} \tag{3.28}$$

where a is the outer diameter of the inner conductor and b is the inner diameter of the outer conductor. Many transmission lines are designed to have a Z_0 of 50 Ω (*e.g.* RG8, RG58, RG 174, RG178, RG223) or 75 Ω (RG6, RG11, RG59, RG179), although there are many others with higher values (RG63–125 Ω, RG180–95 Ω).

The input impedance, Z_{in}, of a transmission line of length L and characteristic impedance Z_0 that is terminated in an impedance Z_{load}, as shown in Figure 3.13(a), is given by:

$$Z_{in} = Z_0 \frac{jZ_0 \tan(\beta l) + Z_{load}}{jZ_{load} \tan(\beta l) + Z_0} \tag{3.29}$$

where $\beta = 2\pi/\lambda$. Figure 3.13(b) show an example of how a transmission line can be used to impedance match an RF coil. The two parameters $D1$ and L determine the transmission line impedances, which correspond to their lumped element equivalents for tuning and matching as covered earlier. Typically the "stub" with length L is short-circuited as shown, although an open circuit can also be used, but this latter arrangement has the disadvantage that it may radiate energy away from the circuit, particularly if the length of the transmission line is a significant fraction of a wavelength.

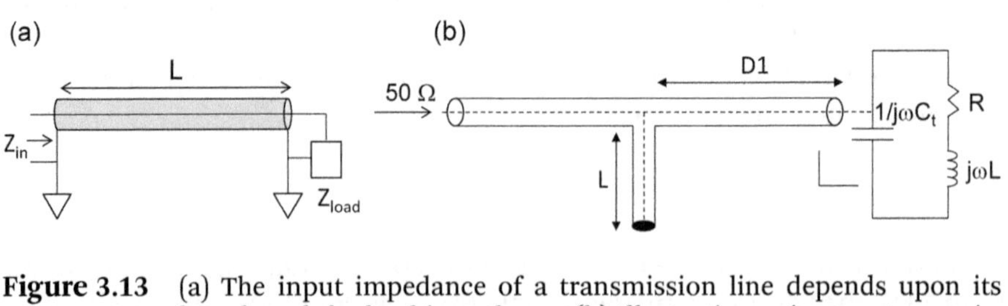

Figure 3.13 (a) The input impedance of a transmission line depends upon its length and the load impedance. (b) Illustration using two transmission line elements to impedance match a coil.

3.3.6 Baluns and Cable Traps

As outlined earlier, a coaxial cable is used to connect the RF coil to the transmit/receive switch. This coaxial cable has an inner conductor and outer shield, as shown in Figure 3.14(a). However, in practice the cable actually has three conductors: the inner conductor, the inside of the shield and the outside of the shield, as shown in Figure 3.14(b). Equal and opposite currents run in the central conductor and the inside of the shield, forming the desired differential mode signal. However, there can also be common-mode currents, which represent "unwanted" currents that flow along the outer shield of the cable, as shown in Figure 3.14(c). These currents are unwanted since they can produce an MR signal from objects far away from the RF coil, they can interact with the patient to detune the coil, and can also potentially be dangerous if stray electric fields interact with the patient. One way to recognize these currents is to touch the coaxial cable when the coil is connected to a measuring device. Any shifts in the coil impedance indicate that there are significant cable currents present. The fundamental cause of these currents is that the shield of the cable is not at true electrical "ground". This can be caused by an imbalance when the cable is connected to an RF coil, which results in one-half of the coil being electrically asymmetric with respect to the other half.

There are two classes of methods that can be used to reduce the magnitude of the common-mode current: the first is *balanced* impedance matching networks and the second is *cable traps*. It should be noted that sometimes these two terms are used interchangeably since the net aim of the two approaches is essentially the same.

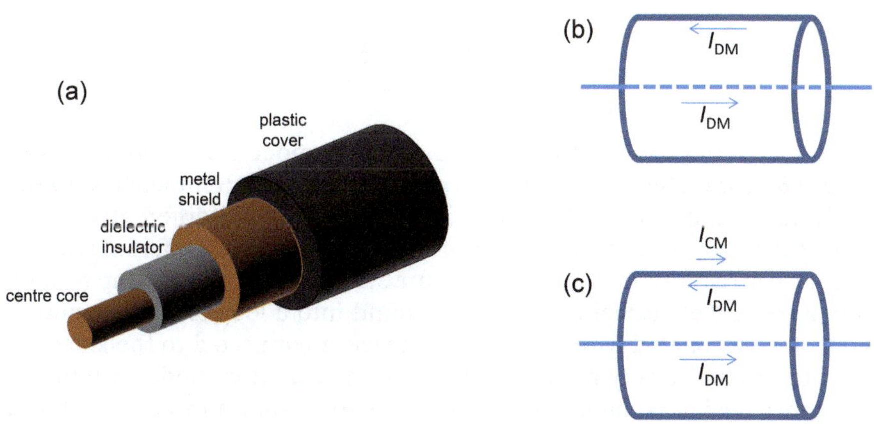

Figure 3.14 (a) Schematic of a standard coaxial cable used to connect the RF coil with the transmit/receive switch. (b) In the ideal case there are only differential mode currents, meaning that no RF leaks outside the cable and there is no influence on the currents from outside phenomena. (c) In practice, there is an additional common-mode current present on the shield of the cable.

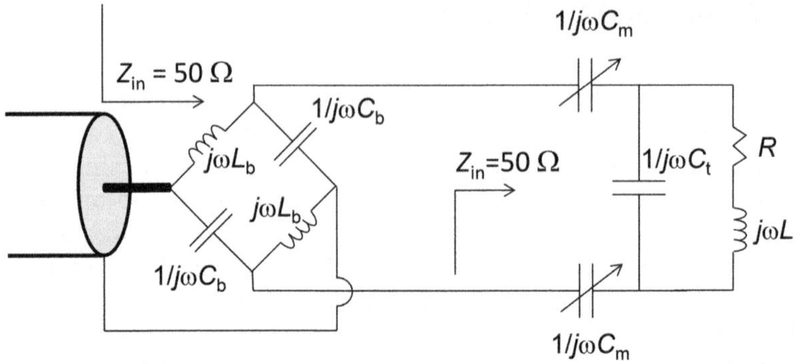

Figure 3.15 Circuit schematic of an RF coil with matching circuit connected to the coaxial cable by an LC-balun circuit.

Balanced matching networks, such as a pi-network or inductive coupling, have already been discussed. However, even if these schemes are used, asymmetric loading of the coil can lead to an unbalanced network. There are a large number of methods for transforming an unbalanced circuit into a balanced circuit, using what are termed in the antenna literature as *baluns* (balanced-to-unbalanced). Many of these use transmission line elements, which can be quite large at the relatively low (with respect to communications devices) MR frequencies. Therefore, balun circuits using lumped elements are often preferred. One example is shown in Figure 3.15, an LC-balun, which is added in front of the matching circuit. This balun comprises a four-element network that isolates the coil circuitry from the ground of the cable. The values of the balun inductor, L_b, and balun capacitor, C_b, are given by:

$$L_b = \frac{50}{\omega_0}, C_b = \frac{1}{50\omega_0} \tag{3.30}$$

The second method of reducing the common mode shield currents is to use a *cable trap*. This is a device that presents a very high impedance to the shield current, but a very low impedance to the desired differential mode signal carried by the central conductor of the coaxial cable. Different types of cable trap are shown in Figure 3.16. Figure 3.16(a) shows a cable trap formed from a section of coaxial cable that is wound into a loop and the resulting inductance resonated with an external capacitor connected to the shield in order to produce a very high impedance to the common-mode current. An alternative design is the floating sleeve cable trap shown in Figure 3.16(b), which has the advantage that it can easily be moved along the length of the coaxial cable to find the most effective position for reducing cable currents.[31] This design consists of a splittable plastic cylinder, covered inside and outside with a metal conductor. Capacitors are soldered between the inside and outside conductors on each half of the cylinder, and the structure resonated at the Larmor frequency. There are two modes of this cable trap owing to the coupling between the two identical halves of the cylinder. The low frequency

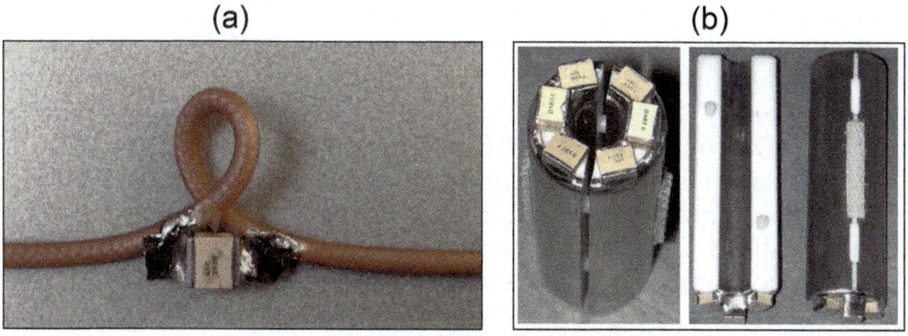

Figure 3.16 (a) A solenoidal cable trap in which the capacitance connected across the shield resonates with the inductance of the loop to produce a very high impedance to shield currents. (b) A floating cable trap comprising two splittable halves. A number of cables can be fed through the central hole in this type of cable trap.

mode is the desired one and this can be fine-tuned by changing the separation between the two halves of the cylinder.

In general, it has been found that for coils that are used in transmit and receive mode, a balancing network such as the LC-balun alone is sufficient to reduce cable currents, but for coils that are used as receive-only the use of cable traps is required.[32] Ideally one cable trap should be placed as close to the coil as possible, with additional cable traps placed at approximately one-quarter wavelength distances along the cable.

3.3.7 RF Coil Loading—The Effect of the Sample

The circuit model for the RF coil considered so far includes a simple resistor, which represents the equivalent series resistance (ESR) of the copper wire, solder joints and capacitor. The quality factor (Q) of the coil is defined as the ratio of the total magnetic field energy stored in the RF coil to the energy loss, integrated over one cycle of the transmitted RF, and is given by:

$$Q = \omega L / R_{\text{eff}} \tag{3.31}$$

where R_{eff} is the total resistance of all the elements. The Q factor of the coil can be measured very easily,[18] allowing a good estimation of the coil resistance. However, the Q value decreases significantly when the sample is introduced owing to the losses introduced by the sample. The degree to which the value changes is a good indication as to the efficiency of the RF coil. Overall, there are four different important loss mechanisms: coil losses outlined above, inductive and dielectric losses, which are both associated with the sample, and finally radiation losses. The latter three are described below.

In general, the interactions between the coil and the sample are determined by the relative permittivity and conductivity of the sample. There are two sources of noise from the sample, *inductive* and *dielectric*. From Maxwell's equations, the alternating B_1^+ field from the RF coil induces magnetically induced (inductive) losses, also known as eddy current losses, in a conducting sample. These losses can be represented as an effective resistance R_m in series with the receiving coil, as shown in Figure 3.17 using a model originally suggested by Gadian and Robinson.[33] The value of R_m is proportional to the square of the operating frequency and the conductivity.

Dielectric losses result from electrical lines of force, associated with the distributed capacitance of the RF coil, passing through the sample. Gadian and Robinson[33] showed that these dielectric losses could be modelled as a parallel stray capacitance (C_1) between the coil and the sample, with the conducting sample represented as a parallel resistor (R_e) and capacitor (C_2), as shown in Figure 3.17. At low frequencies, R_e is proportional to the fourth power of frequency, in the high frequency limit to the square of the frequency. The higher the sample conductivity the higher the value of R_e and the higher the losses. Since R_e is proportional to the square of the inductance of the coil, dielectric losses can be minimized by designing a coil to have low inductance. A number of alternative, though related, models for the sample/coil interactions have been published,[34–39] but all can essentially be analyzed in terms of inductive and dielectric losses.

One other potential source of loss is radiation loss, which represents energy radiated away from the coil rather than into the sample. Radiation loss can be correlated to an equivalent resistance, $R_{radiation}$. In general, depending upon the particular geometry of the coil:

$$R_{radiation} \propto \omega^4 a^4 \tag{3.32}$$

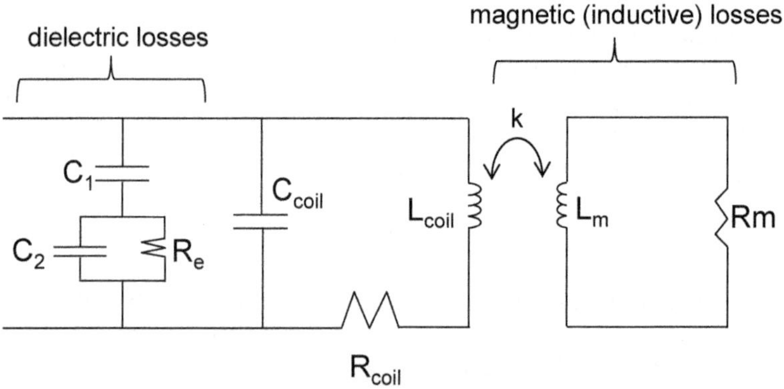

Figure 3.17 A simple circuit that models the interactions between the coil (left) and the sample (right). The capacitive coupling of the electric fields in the air gap between the coil and the sample is modelled by C_1. The reactive and dissipative effects of the electric field within the sample are represented by C_2 and R_e, respectively. Magnetic field coupling between the coil and sample is represented by the coupling constant k, with values between 0 and 1. R_m corresponds to the magnetic losses in the sample.

where a is the radius of the coil. Except for very large coils at very high frequencies, $R_{coil} \gg R_{radiation}$, and so this latter term is generally neglected.

One can define an overall effective "system" resistance, R_{eff}, given by:

$$R_{eff} = R_{coil} + R_{radiation} + R_{inductive} + R_{dielectric} \tag{3.33}$$

With a well-designed coil the radiation and dielectric losses are much lower than the other two components, and the simplification can be made that the overall noise voltage is given by:

$$V_{noise} = \sqrt{4kT\left(R_{coil} + R_{sample}\right)BW} \tag{3.34}$$

where BW is the measurement bandwidth, and R_{sample} is predominantly $R_{inductive}$. It is clear that the ideal coil design results in the sample noise being dominant. One common method of ensuring this is to use extensive capacitive segmentation within the coil. The principle is shown in Figure 3.18. Rather than using a single capacitor, C_t, to resonate the loop, multiple capacitors are used. In the example shown in Figure 3.18 eight capacitors, each with value $8C_t$, are used. Capacitive segmentation reduces the voltage across each capacitor, and therefore also reduces dielectric losses. It also reduces the radiation resistance by effectively making the coil electrically smaller. From a design point-of-view the number of capacitors should be increased up to the point at which the Q-value of the coil loaded with the sample reaches a maximum.

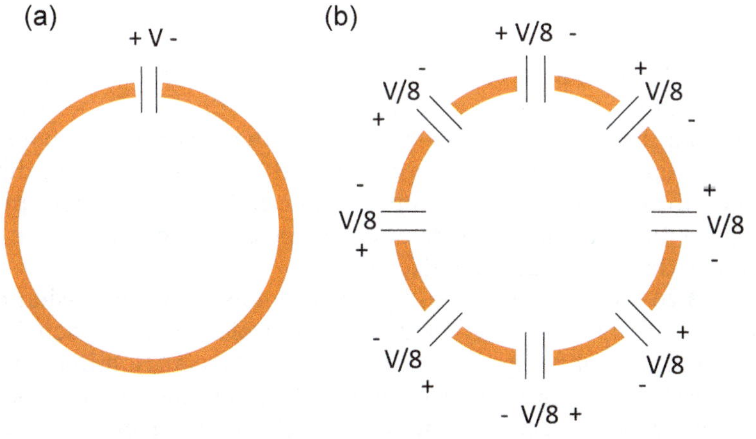

Figure 3.18 Comparison of a coil with a single capacitor (a) and one with eight capacitors in series (b). If a single capacitor only is used, then the voltage across the capacitor is eight times higher than the case of capacitive segmentation.

It should also be noted that the relative noise contributions in eqn (3.35) are frequency dependent. Therefore, for a particular frequency there is a loop size below which the noise from the coil, rather than from the sample, becomes dominant: this topic is covered in Section 3.5.

This completes the sections outlining the basic principles behind RF coil design. The following sections consider specific types of coil and their properties.

3.4 RF Coils Producing a Homogeneous Magnetic Field (Volume Coils)

Most RF coils for NMR and MRI studies are constructed on a cylindrical former, with the long axis of the coil coincident with that of the magnet, *i.e.* vertical for NMR or horizontal for MRI. The B_1^+ field must be produced in a direction perpendicular to the long axis of the coil in order to excite the nuclei as covered in Chapter 1. For many applications the coil geometry is designed to produce a spatially uniform B_1^+ field within the sample. In high resolution NMR a uniform transmit field is particularly important in multi-pulse sequences; in solid-state NMR the efficiency of cross-polarization depends upon having spatially uniform fields at more than one frequency; and in MRI spatial uniformity is critical since image contrast depends upon the tip angle and many clinical diagnoses are performed based upon subtle changes in image contrast.

It can be shown by elementary electromagnetic analysis that a perfectly uniform transverse B_1 field can be produced by an infinitely long cylinder carrying a surface current, J_s, in the z-direction given by:

$$J_s = J_0 \sin\phi, \text{ or } J_s = J_0 \cos\phi \tag{3.35}$$

where ϕ is the azimuthal angle subtended at the centre of the cylinder, as shown in Figure 3.19. Of course, this configuration cannot be realized in practice, but the basic tenet of designing coils that produce a homogeneous transverse magnetic field is to approximate these surface current distributions over the finite length of the coil. The sine and cosine functions produce orthogonal magnetic fields, which offers the possibility of quadrature operation by creating a circularly polarized field.

Figure 3.20 shows a simplified (quasi-chronological) progression of coil designs based upon approximating the sinusoidal current density. The Alderman–Grant coil design[40] is shown in Figure 3.20(a). Nominally, this is a very crude approximation to the desired current distribution, with only two conductor elements, each subtending an angle of ~80°. However, it was the first coil to use segmented capacitors to reduce the electric field entering the sample (thus reducing sample heating), and to minimize radiation loss. The Alderman–Grant design has a very low inductance and is widely used in high-resolution solution-state NMR and solid-state NMR, owing in particular

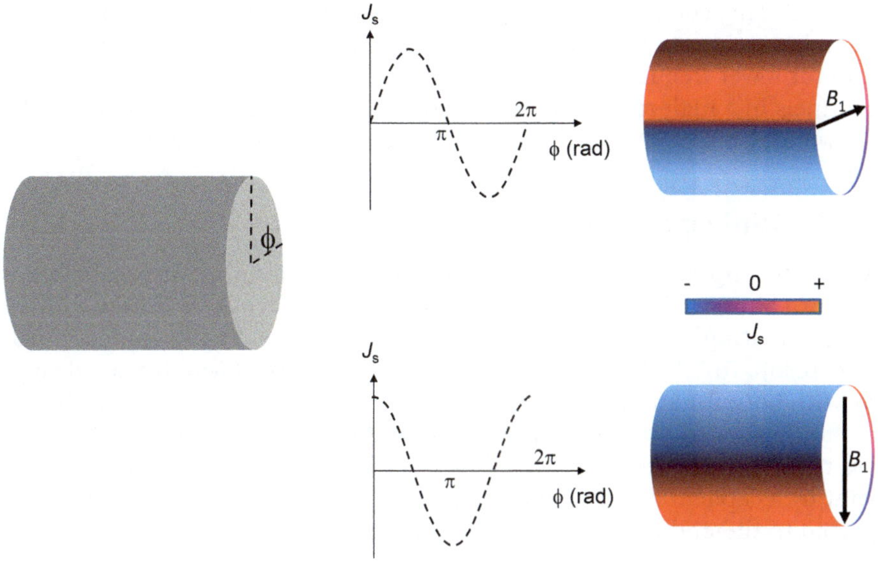

Figure 3.19 Schematic of the current density (either sine or cosine) on an infinitely long cylinder that produces a perfectly homogeneous RF field orthogonal to the long axis of the cylinder.

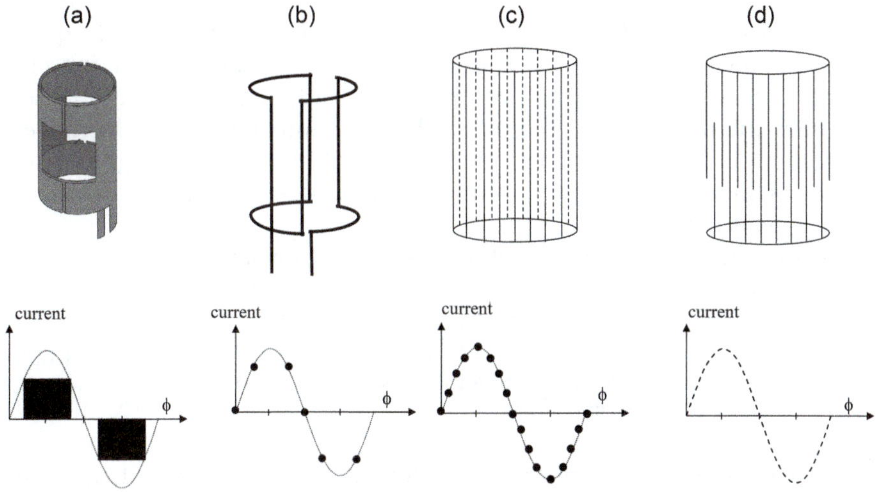

Figure 3.20 Various geometries of cylindrical coils producing a homogeneous RF transmit field, and the corresponding current densities as a function of azimuthal angle. (a) Alderman–Grant, (b) saddle, (c) birdcage, and (d) millipede.

to the reduced sample heating. The saddle coil, shown in Figure 3.20(b), can be considered as a "six-point" approximation to the sinusoidal current distribution.[41] This coil also has a very low inductance and can be constructed as a single continuous loop of wire for high frequencies or as a multi-turn structure for lower frequencies. It is typically used for the proton and X-nucleus channels in a high-resolution NMR probe. The birdcage and millipede coils shown in Figure 3.20(c) and (d) are considered in the next section.

3.4.1 Birdcage Coils

As can be appreciated intuitively, increasing the number of parallel conductor elements in the coil increases the B_1 uniformity by more closely approximating the ideal sinusoidal current density. The so-called "birdcage resonator",[42] shown in Figure 3.20(c), is the most widely used design in MRI, both for animal and human studies. It consists of a series of equidistantly spaced "rungs", typically 16 or 32, and two "end-rings". There are two types of birdcage coil: a low-pass geometry in which each rung is split by a capacitor and a high-pass geometry in which the capacitors are placed in the end-rings between each strut: both structures are shown in Figure 3.21.

The birdcage coil has multiple resonant modes, with one set of frequency-degenerate modes (mode 1) producing two orthogonal homogeneous transverse B_1 fields, corresponding to the sine and cosine current densities described previously: the coil, therefore, can produce a circularly polarized B_1 field. For a low-pass birdcage, mode 1 occurs at the *lowest* frequency with the higher order modes at successively higher frequencies, as shown in Figure 3.21. These higher order modes have an increasingly larger "null point" at the centre of the field, as also shown in Figure 3.21. For the high-pass birdcage, the homogenous mode 1 is the *highest* frequency peak, with the magnetic field distributions of the higher order modes, which occur at lower frequencies, being very similar to those of the low-pass birdcage.

The frequencies of the different modes can be calculated as:[43]

$$\omega_k = \sqrt{\frac{2}{\sum_{n=0}^{N-1} M_n e^{-j2\pi kn/N}}\left[\frac{1}{C_1} + \frac{1}{C_2}\left(1 - \cos\frac{2\pi k}{N}\right)\right]} \tag{3.36}$$

where k is the mode number, $1/C_2 = 0$ for a high-pass birdcage, $1/C_1 = 0$ for a low-pass birdcage, and M_n is the mutual inductance between loops that are n loops apart. The birdcage coil is typically surrounded by an RF shield constructed from thin copper, segmented copper elements, or a copper sheet with capacitive segmentation to prevent eddy currents.

There are several variations on the birdcage geometry. One such is the millipede coil,[44] which consists of hundreds of legs, intercalated as shown in

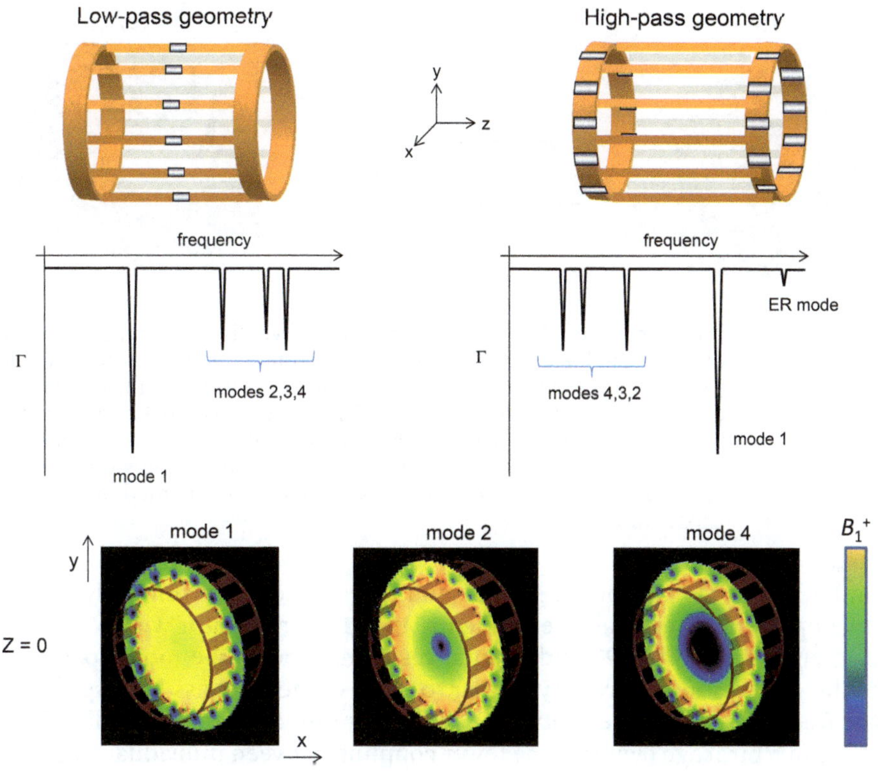

Figure 3.21 Characteristics of birdcage coils. (upper) Two different configurations, low-pass and high-pass, showing conductor elements and segmenting capacitors (white blocks). (centre) Frequency spectrum of the different modes. Mode 1 corresponds to the homogeneous transmit field in both cases. (bottom) Plots of the transmit magnetic field produced at the centre of the coil ($z = 0$). In mode 1 the field is homogeneous, whereas for higher modes there is zero field at the centre, with the field becoming more inhomogeneous the higher the mode number.

Figure 3.20(d). This type of coil is used for some high field microimaging and solid-state-applications. The major alternative to the birdcage for MRI studies is the transverse electromagnetic mode (TEM) resonator, described in the following section.

3.4.2 Transverse Electromagnetic Mode (TEM) Resonators

Although appearing somewhat similar in geometry, the transverse electromagnetic mode (TEM) resonator represents a fundamentally different design from the birdcage, and is derived from a re-entrant cavity.[45–56] It has been used for body coils for MRI at 3 and 4 tesla, as well as for head and knee coils at many different field strengths. The basic form of the re-entrant cavity is shown in Figure 3.22: the largely inductive coaxial line (cavity) is shorted

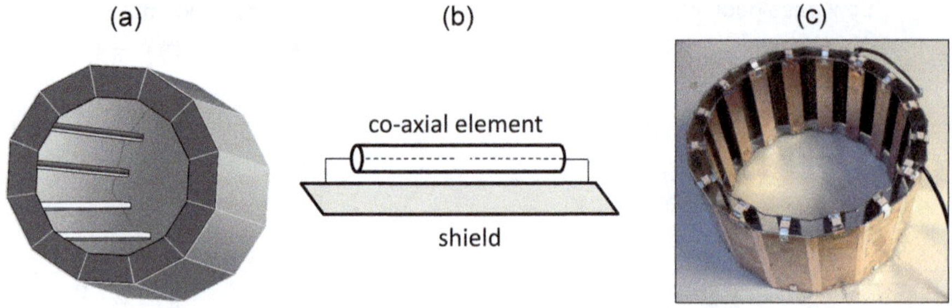

Figure 3.22 (a) Schematic of a TEM resonator, in which each rung consists of a co-axial element connected at both ends to the RF shield. (b) Capacitance is formed by the overlap of the inner conductor with a dielectric material surrounding it. The capacitance can be changed by varying the distance between the two parts of the split inner conductor. (c) Photograph of a TEM resonator in which the individual rungs are realized using conductor strips, which are segmented by discrete capacitors.

on both ends (as part of the shield), and is resonated with primarily-capacitive open-ended coaxial line elements. In a TEM resonator with N coaxial elements, there are $N/2$ fundamental resonant modes: the modal patterns of the TEM resonator are very similar to those of the low-pass birdcage, with the frequency differences between the different modes usually smaller than those of a birdcage owing to the lower coupling between individual loop elements, which are not physically connected. The major advantage of the TEM resonator over the birdcage is that there are no end ring currents in the TEM resonator, which therefore produces a field that extends axially some way outside the coil. The main disadvantage is that the smaller mode splitting in a TEM resonator means that for certain geometries and numbers of elements, different modes may overlap.

3.4.3 Partial-Volume Coils

Often in MRI scans only part of the subject needs to be scanned, for example the calf muscle in the leg, or the back of the brain for functional imaging of the visual cortex. In this case it makes sense to use a half-coil or partial-volume coil, as shown in Figure 3.23, to transmit energy only into the relevant part of the body. Since the homogeneous volume coils considered in the sections above are circularly symmetric, a partial-volume coil can be designed as one-half of the full coil. The advantages of a partial-volume coil over a volume coil are improved patient comfort (since the coil does not surround the entire head/leg), increased power efficiency, and increased SNR close to the coil. There are many different versions of partial-volume coils, including a linear half-birdcage[57,58] and a quadrature half-TEM resonator:[59] an alternative design, covered in Section 3.5.2, is to use a quadrature surface coil wrapped around a half-cylinder.

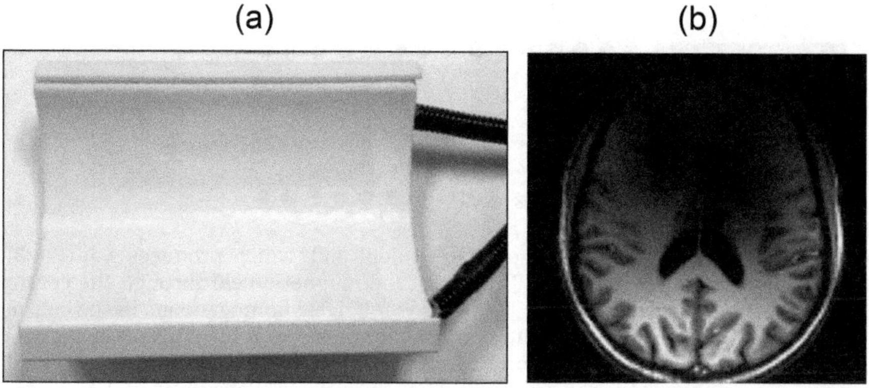

Figure 3.23 (a) Photograph of a partial volume resonator formed on a half-cylinder. (b) Image acquired with a partial volume coil positioned at the back of the head of the subject.

3.4.4 Solenoids and Loop Gap Resonators

In certain MR experiments, the sample can be oriented with its long axis *perpendicular* to B_0, and a different RF coil geometry must be used to produce a homogenous B_1 field, now in the axial direction. The most simple and commonly-used geometry is termed a solenoidal coil, shown in Figure 3.24(a). Solenoidal coils are used in many different applications, including solid-state NMR, microimaging,[7] high resolution NMR of mass-limited samples[126–133] and to couple NMR detection with hyphenated chemical microseparations.[134–141]

The on-axis magnetic field created by a solenoidal coil can be calculated analytically by integrating the Biot–Savart law. For a coil of radius a and length h with n turns of wire:

$$B^{+}_{1,\text{on-axis}} = \frac{\mu_0 nI}{2}\left[\frac{x + h/2}{\sqrt{a^2 + (x + h/2)^2}} - \frac{x - h/2}{\sqrt{a^2 + (x - h/2)^2}}\right] \tag{3.37}$$

where x is the distance along the axial dimension of the solenoid, centred at $x = 0$, and I is the current through the coil. The magnetic field at the centre of the coil, corresponding to $x = 0$, is given by:

$$B^{+}_{1,x=0} = \frac{\mu_0 \mathrm{n} I}{\sqrt{4a^2 + h^2}} \tag{3.38}$$

Solenoids are extremely easy to design, and have a relatively homogeneous B_1^+ field, as shown in Figure 3.24(b), particularly if the length-to-diameter ratio is high. The optimum spacing between the wires is one-half the wire diameter,[17] and for the optimum SNR the length and diameter should be

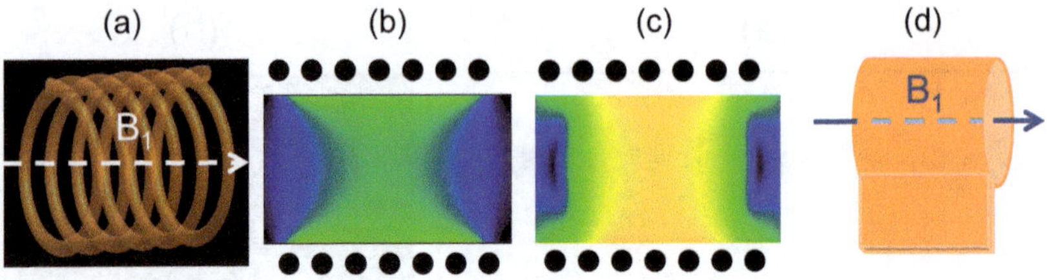

Figure 3.24 (a) Basic structure of a solenoidal coil, which produces a linear B_1^+ field along its axis. (b) Plot of the magnetic field through the centre of a solenoidal coil. The field is relatively homogeneous in the centre and drops to one-half of its maximum value at the ends of the coil. (c) Plot of the corresponding electric field, showing a high electric field throughout most of interior of the coil. (d) Schematic of a loop gap resonator, which essentially forms a single-turn solenoid. Reproduced from ref. 23., B. Park, T. Neuberger, A. G. Webb, D. C. Bigler and C. M. Collins, Faraday shields within a solenoidal coil to reproduce sample heating: numerical comparison of designs and experimental verification, *J. Magn. Reson*, **202**, 72–77. Copyright (2010) with permission from Elsevier.[23]

equal, although there is an obvious trade-off with respect to B_1 homogeneity. One of the main advantages of the solenoid over other designs is that it has an intrinsic higher sensitivity, between two- and three-times that of the saddle/Alderman–Grant/birdcage coil design.[115]

The major disadvantages of a solenoid relate to its very high inductance, given by:

$$L = \frac{4n^2a^2}{92a + 100h} \tag{3.39}$$

where L has units of microhenries, and a and h units of centimetres. The interturn capacitance is given by[60]:

$$C = 2a\left(0.1126\frac{h}{2a} + 0.08 + \frac{0.27}{\sqrt{h/2a}}\right) \tag{3.40}$$

where C is measured in picofarads. The high inductance means that the coil becomes self-resonant at a relatively low frequency, *i.e.* the intrinsic interturn capacitance resonates with the inductance of the coil, so there is an upper limit to the frequency at which a solenoid can be operated. There are ways to overcome this limitation, including series capacitive segmentation, as described previously. The second problem is that the high inductance produces a high conservative electric field within the sample, shown in Figure 3.24(c), which can lead to heating problems with highly conductive samples.[22] Again, there are potential solutions including the use of Faraday shields[23] between the RF coil and the sample, but this reduces the filling factor of the coil and increases the design complexity.

A variation on the solenoid is termed a loop gap resonator (LGR),[61–63] which is essentially a one-turn solenoid with a very wide conductor strip used to form the resonator, as shown in Figure 3.24(d). As such it has a very low inductance and so can be used at high frequencies and does not produce as high an electric field as the solenoid. Originally developed for electron paramagnetic resonance (EPR) applications, the LGR can be used in a variety of different shapes and geometries for MRI[64,65] and NMR,[26,66] with different degrees of capacitive segmentation, or using separated loops to concentrate the electric field away from the sample.

3.5 Surface Coils

Surface coils are used extensively in MRI, both for animal and human applications. As the name suggests, the coil is used to image organs or features very close to the skin surface. Surface coils can be almost any shape, including circular, elliptical, square, rectangular, hexagonal or pentagonal, and can lay flat or be bent to mold around the sample. Surface coils can be used in transmit/receive mode, or as a single-element or part of a multi-element receiver. A general rule-of-thumb is that the surface coil gives an enhanced sensitivity, compared to a volume coil, up to a distance equal to the radius (circular coil) or half-the-width (square coil). The B_1 field patterns from two common forms of surface coil, a loop and a butterfly geometry, are shown in Figure 3.25(a) and (b). Special geometries of surface coils can also be designed to restrict the imaging field-of-view very close to the surface. One example is a meanderline coil,[67] shown in Figure 3.25(c), in which the B_1^+ field drops off very rapidly with distance owing to cancellation of the fields from the currents running in opposite directions in the coil.

3.5.1 Transmit/Receive Surface Coils

Despite the intrinsically inhomogeneous transmit and receive fields associated with a surface coil, there are situations in which using it both for transmitting and receiving is desirable or necessary. For example, when using small-diameter high-power gradients for small animal imaging there is often too little space

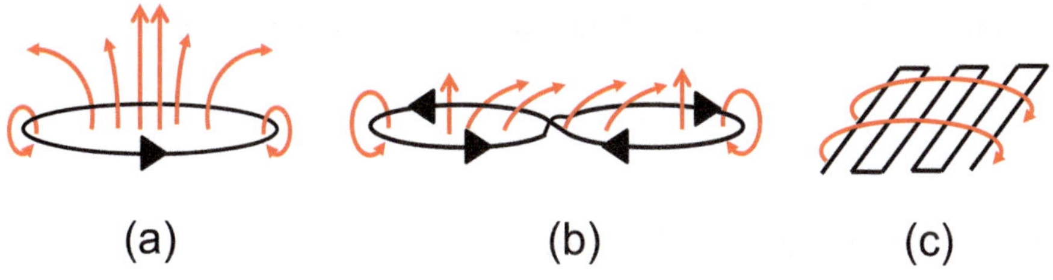

Figure 3.25 B_1^+ fields from three different types of surface coil: (a) a simple circular loop coil, (b) a "butterfly" coil, and (c) a meanderline coil.

between the animal and the inside bore of the gradients to fit a volume coil, and instead a thin surface coil can be used, particularly if only one part of the animal is to be imaged. In human studies, the use of a surface coil transmit means that power is deposited very locally. In very high field human MRI a transmit body coil is not available, and transmit arrays of surface coils can be used instead. If the surface coil is used both to transmit the RF pulse and also to receive the signal, then the acquired signal has a steep drop-off with distance. If desired, this can be mitigated by the use of an adiabatic excitation pulse, which produces a B_1-independent tip angle at the cost of a longer RF pulse.[68]

In terms of designing simple surface coil geometries, there are analytical expressions for the coil inductance, which allows calculation of the appropriate capacitance to resonate the structure. The equation for the inductance of a circular loop was given in eqn (3.16): the equivalent expression for a rectangular loop is given by:

$$L(nH) = 0.4\left[(w+l)\ln\left(\frac{4lw}{d_{\text{wire}}}\right) - l\ln(a+t) - w\ln(l+t) + 2\left(t+\frac{d_{\text{wire}}}{2}\right) - 2(w+l)\right] \tag{3.41}$$

where w and l are the width and length of the coil, respectively, and $t = \sqrt{(w^2 + l^2)}$. As with all coil design, capacitive segmentation should be used to ensure that the length of each conductor element is less than ~$\lambda/10$ to ensure that phase shifts along the length of the coil are small, to minimize the conservative electric field entering the patient, and to reduce frequency shifts between the unloaded and loaded coils. If L_{loop} is the inductance of the unsegmented loop, and n_{seg} the number of segmentation capacitors, then the capacitance for each segment is given by:

$$C_{\text{seg}} = \frac{n_{\text{seg}}}{L_{\text{loop}}\omega_0^2} \tag{3.42}$$

The dimensions of the coil are related to the depth of the imaging region-of-interest (ROI). The coil must be large enough to have adequate B_1 field strength throughout the ROI, but if the coil is too large then it picks up thermal noise from a larger section of the sample than necessary, and so the SNR is degraded. As analyzed by Kumar and Bottomley,[28] for a desired depth d within the tissue, the optimum radii of a loop coil and a butterfly coil are given by:

$$r_{\text{loop}} = \frac{d}{\sqrt{5}},\ r_{\text{butterfly}} = 0.6d \tag{3.43}$$

3.5.2 Quadrature Surface Coils

Surface coils can also easily be designed to operate in quadrature mode. The two coils usually lie in the same plane, although they can also be placed on a curved surface for imaging the back of the head or the calf muscle.

(a) (b)

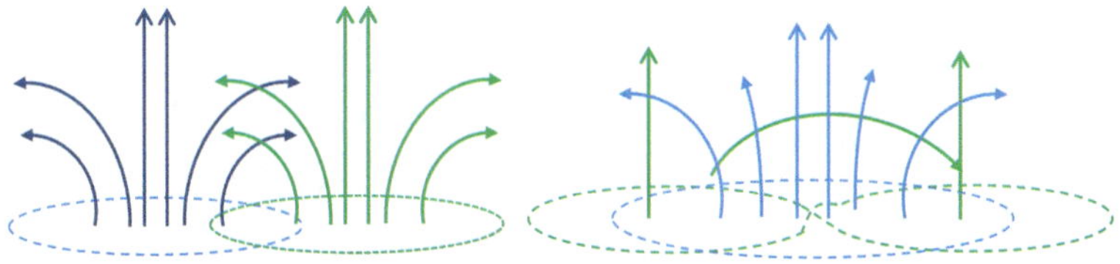

Figure 3.26 Two different types of quadrature surface coils: (a) two overlapping single loops, (b) one butterfly and one single loop. Fields from the two coils are shown in green and blue. Regions in which the field lines cross at an angle of 90° produce a circularly polarized field. Regions in which the angle is between 0 and 90° produce an elliptically polarized field, and where there is essentially no overlap the field is linear.

Quadrature surface coils can be formed from circular/elliptical or square/rectangular loops. Two of the most common geometries are shown in Figure 3.26. In each case, the two different coils (represented by blue and green shapes) are driven with a 90° phase difference. Optimal dimensions of different quadrature coil combinations have also been presented by Kumar and Bottomley.[69]

The major requirement for effective quadrature operation is to electrically isolate one coil from the other. The identical resonant frequencies of the two coils are shifted owing to the mutual inductance between the coils. One shifted frequency is higher, and the other lower, than the Larmor frequency. In order to electrically isolate the two coils, and therefore ensure that they both operate at the Larmor frequency, the mutual inductance must be eliminated. The most common method of ensuring zero mutual inductance is to overlap the two coils, as shown in Figure 3.27.[70] At the point of "critical overlap" the field induced by coil 1 in coil 2 is exactly opposite to that produced by coil 2 itself. For two circular loops or two rectangular coils, there is a slightly different critical overlap criterion, which results in zero mutual inductance, as shown in Figure 3.27(a) and (b), respectively. In the case in Figure 3.26(b), the mutual inductance between the single loop and the "butterfly" coil is zero owing to the intrinsic anti-symmetric nature of the overlap, provided that the geometric centres of the butterfly and loop/square coils coincide.

The coupling between two coils can be quantified in terms of a mutual impedance, Z_{12}, shown in Figure 3.28(a):

$$Z_{12} = \frac{V_1}{I_2} = \mathrm{j}\omega M + R_{12} \tag{3.44}$$

This impedance has a real part (mutual resistance) and imaginary component (mutual inductance). In addition to the overlap method, Figure 3.28(b) and (c) show two alternative methods of minimizing the mutual inductance using a capacitor to "resonate out" the mutual inductance. None

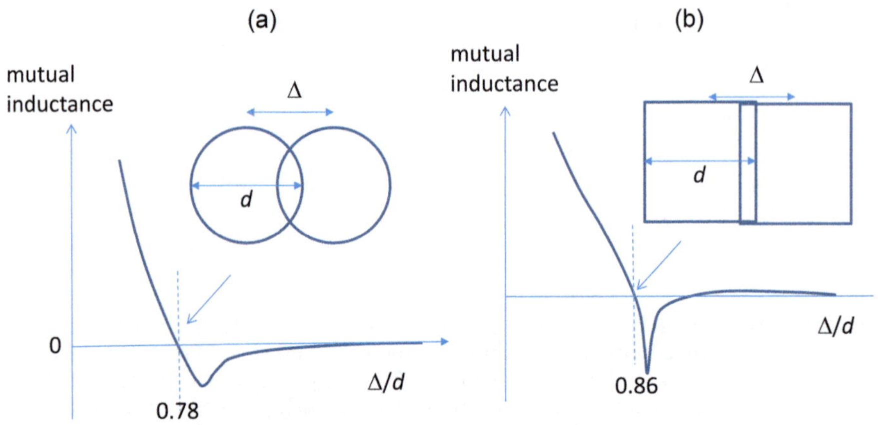

Figure 3.27 Plots of mutual inductance between two coils of the same geometry as a function of the distance between the two coils. (a) Circular surface coils and (b) rectangular surface coils. In each case, there is a critical overlap that results in zero mutual inductance and the two coils being effectively "isolated" from one another.

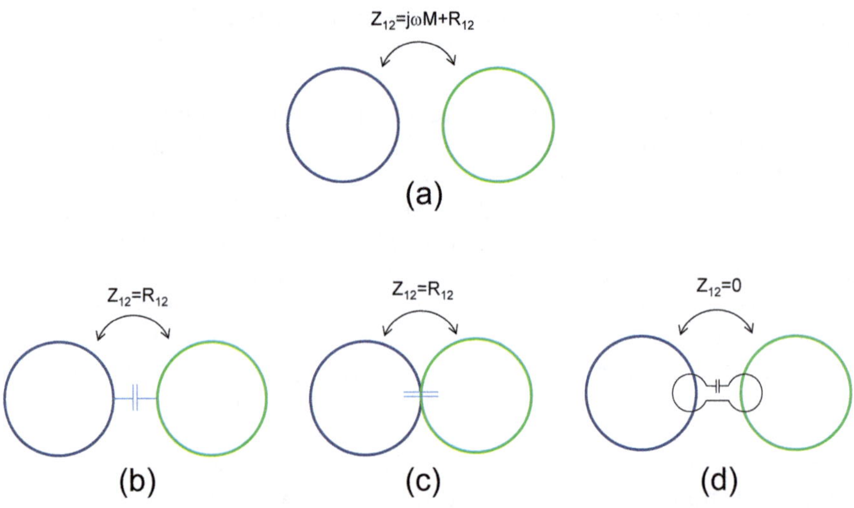

Figure 3.28 (a) Two surface coils placed close to one another have real and imaginary components of mutual impedance. (b and c) Two methods to minimize the mutual inductance by resonating it with a capacitor, which can either be placed (b) between the separated coils, or (c) in a common leg if the two coils are connected. (d) By using an overlapping circuit (thin black lines), both the mutual impedance and mutual resistance can be minimized.[71]

of these three approaches eliminates the mutual resistance term in eqn (3.46), but in practice the mutual inductance term is usually much larger than the mutual resistance. However, there are also circuits that can be used to minimize both the mutual inductance and mutual resistance terms,[71] as shown in Figure 3.28(d).

3.6 Detuning Circuits for Transmit-Only Volume Coils and Receive-Only Surface Coils

The most common setup for MRI is that a homogeneous volume coil is used to transmit the RF pulses with a surface coil (or array of many surface coils) to receive the signal. The transmit and receive coils are both tuned to the same Larmor frequency, and in order to avoid any coupling between them it is necessary to detune the receive coil during pulse transmission and to detune the transmit coil during signal reception. Both of these functions are performed using diodes, with a combination of both active and passive diodes normally being incorporated into the RF coils.

The electrical characteristics and properties of an active PIN diode are summarized in Figure 3.29. A DC voltage is applied to the diode, as shown in Figure 3.29(a). If the voltage is above +0.7 volts, the diode is "turned on" and exhibits a very small impedance, which can be approximated as a short circuit, Figure 3.29(b). Below this voltage, the diode is in the off-state and appears as a very high impedance. In practice, a negative voltage is applied to the diode to switch it into the off state. The positive or negative DC voltages are provided by the MR scanner, with typical values in the order of ±24 volts, although some manufacturers use values as small as ±5 volts. The DC voltage is fed into the coil through large inductors which act as RF "chokes", *i.e.* they present a high impedance at the RF frequency and very low impedance at DC, so that no RF signal leaks into the PIN-diode driving circuit. Passive diodes can also be used in RF coils. These are placed back-to-back in a crossed-diode configuration, shown in Figure 3.29(c), and are included mainly as a safety mechanism in case any damage occurs to the actively driven PIN diodes. Passive diodes are used mainly in the receive coil and are switched to the conducting state by the high intensity of the RF pulse. After the pulse is switched off they return to the

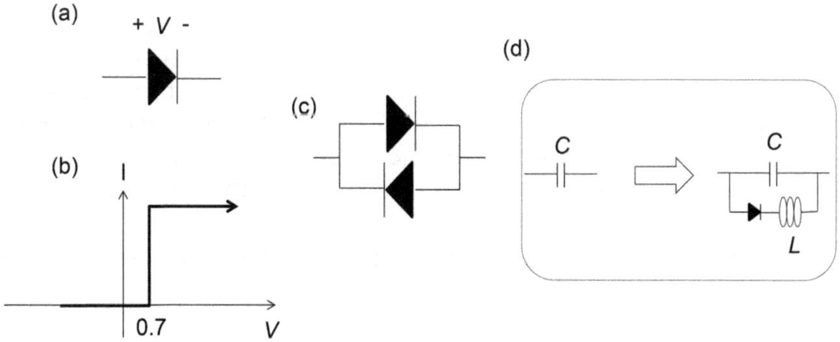

Figure 3.29 (a) Circuit symbol for a PIN-diode with a DC voltage applied across its terminals. (b) An idealized plot of the current passing through a diode as a function of driving voltage for a silicon-based diode. (c) Passive crossed-diodes. (d) Using a PIN-diode to detune a coil consists of replacing one or more of the tuning capacitors with a parallel LC-diode combination.

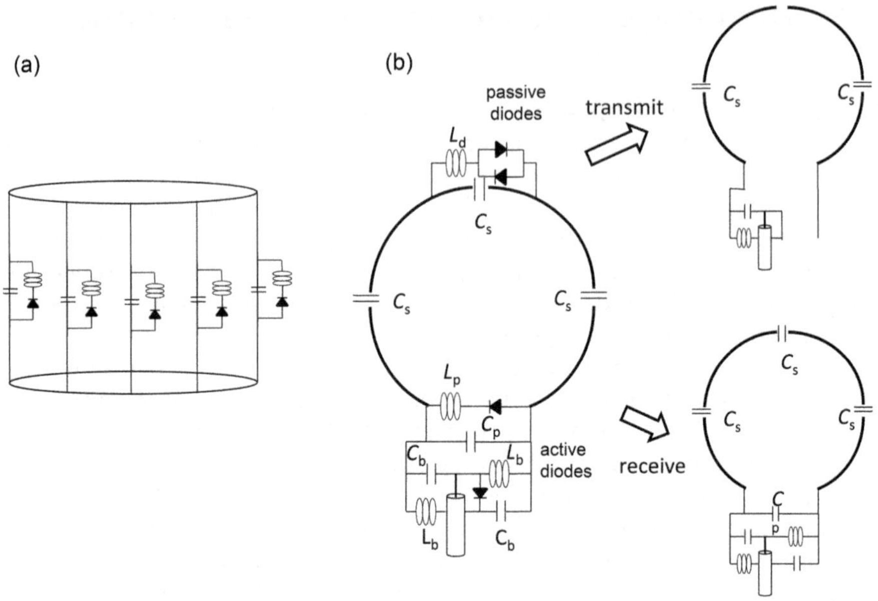

Figure 3.30 (a) Incorporation of active PIN diodes into a low-pass birdcage coil. (b) A receive surface coil decoupling scheme. The active decoupling comprises of a two-stage PIN diode switching design. A parallel inductor (L_p) is switched across C_p and a lattice balun (L_b, C_b) is effectively open circuited across the coil input at C_p. Passive decoupling with high-speed switching crossed-diodes and a detuning inductor (L_d), switched in parallel with C_s, is used as a precautionary decoupling mechanism in the event of a PIN diode failure.[72]

non-conducting state. A back-to-back configuration is necessary since the RF pulse consists of an alternating voltage with positive and negative values.

There are a number of different circuit configurations incorporating PIN diodes that can be used for switching between the tuned and detuned coil modes. One of the basic "modules" is to replace the tuning capacitor (or multiple tuning capacitors in the case of structures such as the birdcage or TEM resonator) with a capacitor in parallel with an actively switched diode and an inductor, as shown in Figure 3.29(d). The inductor value is chosen such that when it is switched into the circuit, it resonates with the capacitor to form a very high impedance, *i.e.* $L = (\omega^2 C)^{-1}$.

Examples of incorporating active and passive diodes into transmit and receive coils are shown in Figure 3.30.[72]

3.7 Receive Arrays

All modern clinical scanners use receive arrays, *i.e.* a collection of (typically) between eight and sixty-four surface coils that must be electrically isolated from each other, and also decoupled from the body transmit coil (see section above). There are two basic advantages of using an array of small coils, rather than one single large coil with an equivalent volumetric coverage. The first is that the sensitivity of each small coil is greater than that of the large coil, and each small coil picks up noise only from a local region of the body, as opposed to the large coil, which picks up noise from the entire imaging field-of-view. Providing that the overall noise is dominated by the sample (Section 3.3.7) this means that, after the combination of the signals from each element of the array, the image SNR is increased for a coil array compared to using the single larger coil. The increase in SNR is greatest for organs close to the surface of the patient. The second advantage of coil arrays is that they enable the use of parallel imaging techniques to increase the imaging speed by reducing the amount of *k*-space data required for image reconstruction. The basic principle of parallel imaging[73] is that signals detected by different receive coils already have spatial information intrinsically encoded by the spatial sensitivity profiles of the coils, *i.e.* the coil closest to a particular organ receives a higher signal from that organ than does a coil positioned further away. This additional spatial information means that a partial undersampling of *k*-space can be performed, reducing the total acquisition time.[74–76] A number of different methods have been implemented, with different acronyms given to each, the most common being based on generalized autocalibrating partially parallel acquisitions (GRAPPA)[77] and sensitivity encoding (SENSE).[78]

3.7.1 Array Optimization

For a fixed overall size of the surface coil array, one key element in the design process is to determine how many individual coils should be used, *i.e.* what should be the size of the individual coils.

The relative noise contributions from the coil and the sample have been studied by Kumar *et al.*[28] as a function of frequency. Figure 3.31 shows a plot

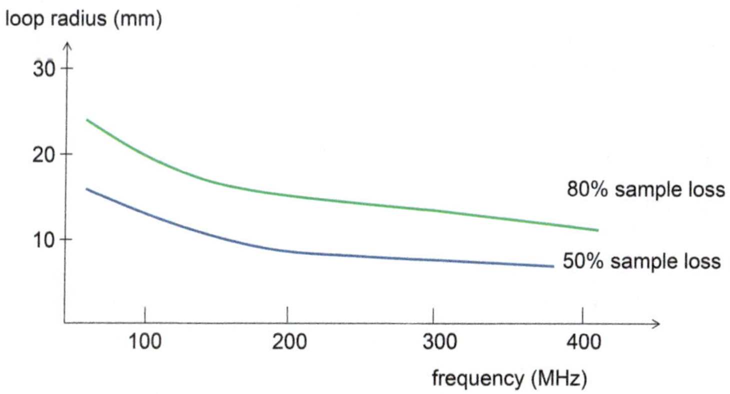

Figure 3.31 Graph of the contribution of sample loss to the overall system loss, as a function of loop size and operating frequency.

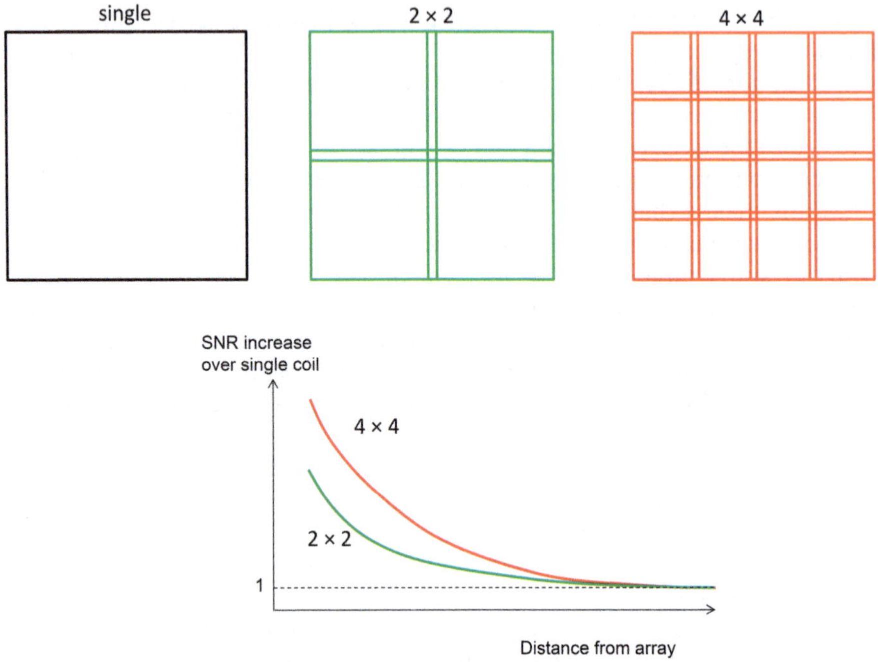

Figure 3.32 The SNR performance of different numbers of individual elements in an array of the same overall size as a single loop. At a distance greater than one-half the dimension of the single loop, the SNR performance is independent of the number of elements.

of loop radius *vs.* frequency for wire coils constructed with a wire diameter of 3.2 mm for loops with radius greater than 15 mm, and 2 mm for loops smaller than this. The graph shows that at higher frequencies coils can be smaller before coil noise becomes an important factor.

If the sample noise is dominant, then the SNR gain is maximum very close to the array, and at a distance equal to the radius (for a circular) or half-the-width (for a square) of the array, the SNR is equivalent to that obtained using a single large surface coil, as shown in Figure 3.32.

3.7.2 Preamplifier Decoupling in Receive Arrays

In an array comprising a large number of elements it is extremely challenging to be able to achieve a high degree of decoupling between all of the individual elements using either the coil overlap or capacitive-decoupling methods outlined previously. Fortunately, there is a relative simple, independent method for increasing the decoupling, which is termed *preamplifier decoupling*. This method works by reducing the current flowing in each coil in the array, thus also reducing the effective inter-coil interactions.[70,79] As discussed earlier in the chapter, preamplifiers typically have an input impedance of 50 Ω, so that they form a conjugate match with the impedance matched coil

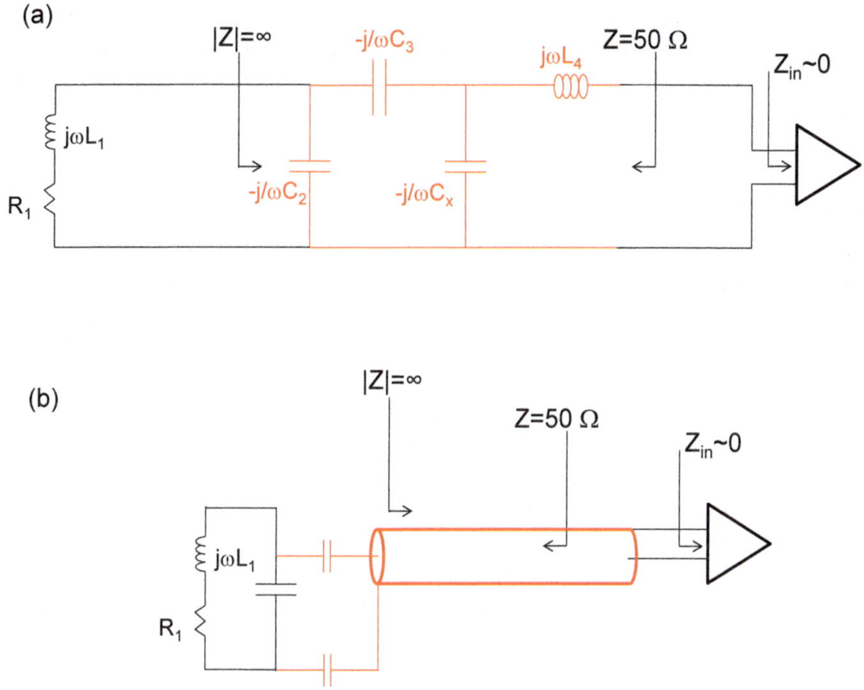

Figure 3.33 Two circuits used for preamplifier decoupling of individual elements in a coil array: the networks are shown in red. (a) A discrete element matching network, which allows flexibility in choosing realistic values over a wide range of coil input impedances. (b) A half-wavelength transmission line is used to convert the very low input impedance of the preamplifier to a very high impedance at the coil.

and coaxial cables for maximum power transfer. For a single coil this is an efficient approach, but for coil arrays it has the disadvantage that, if there is significant mutual impedance between coils, a voltage in one coil produces a significant current in an adjacent receive coil. In contrast, if there is effectively an infinite impedance at the input of the coil, then zero current will flow in the coil and the mutual inductance will be zero. One method of achieving this, while maintaining a 50 Ω match (which produces the lowest noise figure) at the input to the preamplifier is to use a very low input impedance preamplifier, typically with a real part of 1 Ω or less. Two circuits commonly used for preamplifier decoupling are shown in Figure 3.33. These types of circuits minimize the effective mutual inductance between different coil elements in the array, although mutual resistance effects are still manifested as correlated noise between different channels.

Having designed the RF coil array with the minimum mutual impedance between the individual elements of the array, it is important to combine the signals from the individual receive channels in an optimal way.[70,80] Essentially each voxel in an image can be reconstructed optimally by knowing the coil sensitivity at that particular voxel and the coupling between the different coils of the array. Different complex weighting functions are then applied to the signal from each coil for each voxel, and these weighted signals are finally summed.

3.8 Multiple-Frequency Circuits

So far only RF coils tuned and impedance matched to a single frequency have been described. However, many experiments require either a single coil to be tuned to more than one frequency, or several coils tuned to different frequencies to operate together. For example, in high resolution NMR, almost all experiments require the use of multiple frequencies. Multi-dimensional correlation experiments between ^1H, ^{13}C, ^{31}P, ^{15}N are universal in biomolecular research, and a deuterium lock circuit must also be incorporated, as covered in Chapter 1. Similar requirements pertain to solid-state NMR. In MRI, there are also many research protocols that, in addition to proton scans, acquire data from nuclei such as sodium, phosphorus or carbon. There are many designs for RF coils that can produce B_1^+ fields at more than one frequency. The principles behind some of these designs are described in this section.

3.8.1 Multiple-Pole Circuits

One of the most simple methods of double tuning a single RF coil such as a loop or solenoid is to insert a secondary L_1C_1 circuit into the primary *RLC* circuit, as shown in Figure 3.34.[81,82] The impedances of the individual circuits, as well as the combined circuit, are also plotted in Figure 3.34. If the secondary L_1C_1 circuit is considered by itself, at frequencies below its resonant frequency, $\omega_1 = \sqrt{(L_1C_1)}$, the circuit can be represented by an equivalent inductance, L_{eq}. The entire circuit therefore resonates at a frequency given by:

$$\omega_{\text{low}} = \frac{1}{\sqrt{\left(L_{\text{eq}} + L\right)C}} \tag{3.45}$$

where ω_{low} is lower than ω_0, the resonance frequency of the primary circuit alone. At frequencies above ω_1, the parallel L_1C_1 circuit looks like an equivalent capacitor, C_{eq}, and the corresponding circuit resonant frequency is:

$$\omega_{\text{high}} = \frac{1}{\sqrt{L\left(\dfrac{C_{\text{eq}}C}{C_{\text{eq}} + C}\right)}} \tag{3.46}$$

which occurs at a frequency higher than ω_0, since the series combination of capacitors results in a reduced equivalent capacitance.

In many cases, including the design of high resolution NMR probes, it is not possible to insert a secondary circuit into the main RF coil itself, and

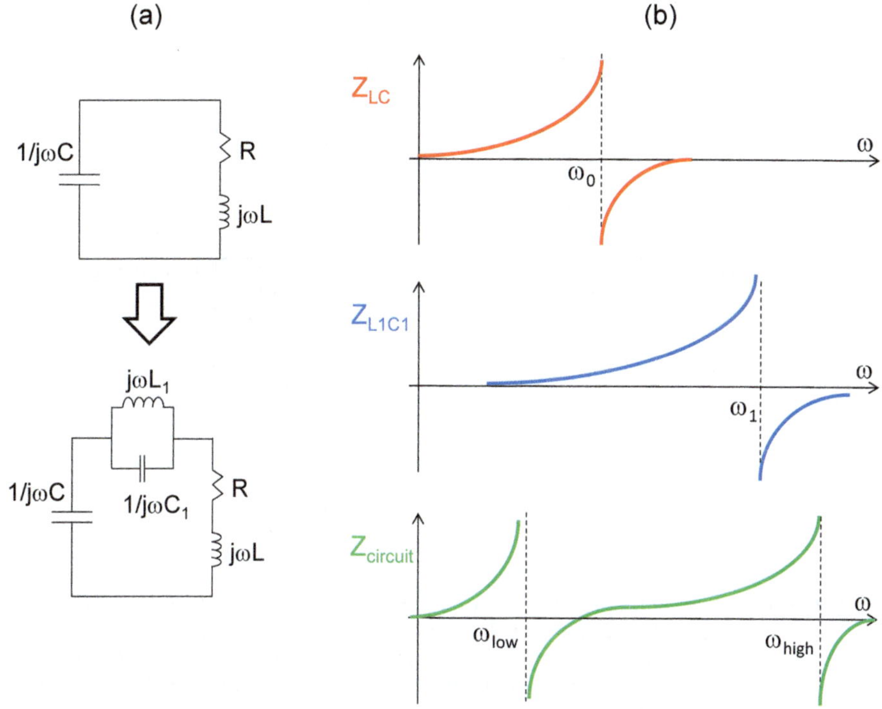

Figure 3.34 (a) Insertion of a secondary L_1C_1 circuit into the primary *RLC* circuit means that the overall circuit resonates at two different frequencies, ω_{low} and ω_{high}, which can be tuned to the Larmor frequencies of two different nuclei. (b) Plots of the impedances of the primary circuit alone (red), the secondary circuit alone (blue), and the combined circuit (green).

so the double-tuning must be performed externally *via* additional electrical circuits. There are many realizations of such circuits, one of which is shown in Figure 3.35. The X-nucleus can be ^{13}C, ^{31}P, ^{15}N or ^{2}H, and has a Larmor frequency significantly lower than that of protons. The simplest way to analyze this type of circuit is to consider the equivalent circuit at high frequency (the proton frequency) and low frequency (the X-nucleus frequency), as also shown in Figure 3.35.

3.8.2 Transformer Coupled Circuits

An alternative approach to double-tune a single coil is to use two RF coils, placed very close to one another, in order to produce RF fields with essentially the same physical coverage at two different frequencies. As seen previously, if two RLC circuits are brought into close proximity to one another, there is a coupling between the circuits that depends upon the coupling constant k, as shown in Figure 3.27. The effect of the mutual coupling is to reduce the total inductance of the higher frequency circuit, and to increase the inductance of

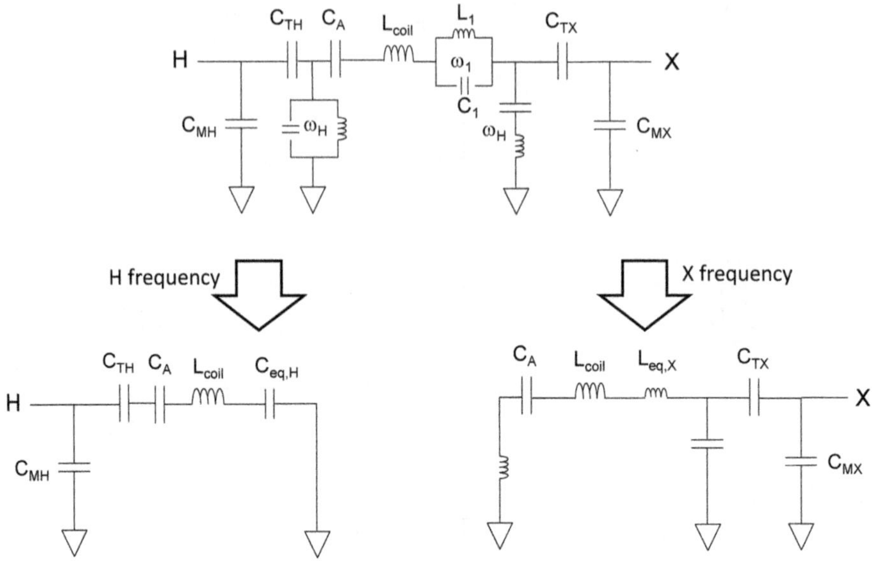

Figure 3.35 Schematic of a double-tuned circuit for a single RF coil, showing the equivalent circuits at the proton (high) frequency and X-nucleus (low) frequency. The parallel L_1C_1 circuit resonates as a frequency between that of ^{1}H and the X-nucleus, and therefore appears as a capacitor ($C_{eq,H}$) at the proton frequency and an inductor ($L_{eq,X}$) at the X-nucleus frequency.

the lower frequency circuit. The relationship between the frequencies of the isolated high frequency (f_h) and low frequency (f_l) circuits, and the combined high frequency primary (f_p) and secondary low frequency (f_s) resonances is given by:

$$f_p^4\left(\frac{1}{f_h^2}-\frac{1}{f_l^2}\right)+f_p^2\left[\frac{f_h^2}{f_l^2}-\frac{f_l^2}{f_h^2}\right]\left(1-k^2\right)+\left(f_l^2-f_h^2\right)\left(1-k^2\right)=0 \tag{3.47}$$

This principle can be used to design a set-up of two coils resonating at the Larmor frequencies of two different nuclei by choosing suitable values of f_h, f_l and k,[83,84] as shown in Figure 3.36.

3.8.3 Multiple-Tuned Volume Coils

There are a number of different approaches to double- or multiple-tuning multi-mode volume coils, *i.e.* birdcage or TEM resonators.[46,85–88] Figure 3.37 shows three different approaches that can be used to double-tune a single birdcage coil.

The mutual inductance between two tightly coupled birdcage coils can also be used to produce a "double-tuned" coil. This approach was shown

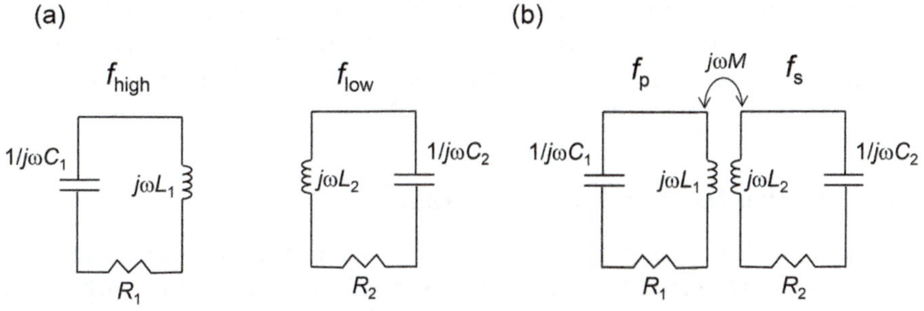

Figure 3.36 (a) Circuit diagram of two uncoupled circuits resonating at different frequencies f_{high} and f_{low}. (b) If the two circuits are brought close together then the mutual coupling M shifts the resonant frequencies to f_p and f_s, respectively, where $f_p > f_{high}$ and $f_s < f_{low}$. The frequency shift in the high frequency channel is greater than that in the low frequency one.

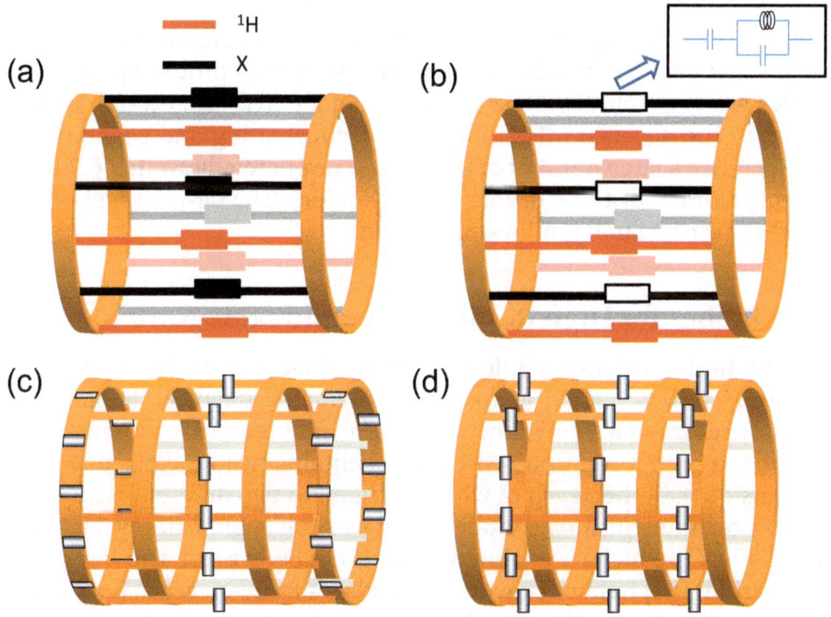

Figure 3.37 (a) A simple variation of a double-tuned birdcage in which alternate rungs are tuned to the ^{1}H and X nuclei using different capacitors. The B_1^+ distribution at the ^{1}H frequency is less homogeneous than that for a single-tuned resonator, but at the X nucleus frequency it is homogeneous. (b) An improved design in which ^{1}H trap circuits are included in the legs which are tuned to the X-nucleus frequency. (c and d) A four-ring birdcage resonator introduced by Murphy-Boesch *et al.*,[187] which consists of a low-pass central birdcage, and two either high-pass (c) or low-pass (d) outer birdcages.

initially by Fitzsimmons *et al.*[83] at 1.5 tesla. The transformer-coupled design[83] involves placing one birdcage coil inside the second, with the smaller diameter coil tuned to the lower Larmor frequency. Figure 3.38 shows schematics, a photograph and *in vivo* results from a recent version of this coil, designed for ^{1}H MRI and ^{31}P MRS of the leg at 7 tesla.[89]

3.8.4 Multiple-Tuned Surface Coils

A number of different methods can also be used to double-tune a surface coil. Poles, also referred to as trap circuits, can be inserted into a single coil, as shown in Figure 3.39(a), or two tightly coupled coils can be used, as shown in Figure 3.39(b), with a proton trap circuit inserted into the X-nucleus loop. Circuits derived from high resolution NMR, such as shown in Figure 3.35, can also be used.

3.9 RF coils for NMR Spectroscopy

Almost every chemistry and physics department has at least one NMR spectrometer that can run liquid or solid samples. Typical proton operating frequencies are between 200 MHz (4.7 tesla) and 1 GHz (23.5 tesla), with the majority in the 400–600 MHz range for practical siting and economic reasons. There are a large number of RF probes commercially available for these systems, provided by three or four major vendors. The basic design of these probes is described in the following sections.

3.9.1 Probes for High Resolution Liquid-State NMR

Probes for high resolution liquid-state NMR must give sub-hertz linewidths and short RF pulses, operate at up to four different frequencies, be stable over periods of several days and function over a wide temperature range. Most commonly, a high resolution NMR probe contains two RF coils, one nested closely inside the other, each of which is tuned to two different frequencies. Each coil operates in linear mode. Owing to the higher filling factor, the inner coil has the higher sensitivity. One of the coils must also be tuned to the deuterium (^{2}H) frequency, which is used as a frequency lock (and also for decoupling in experiments involving partially deuterated proteins). The two most common geometries are a saddle coil and Alderman–Grant coil, one placed within the other, as shown in Figure 3.40. The proton coil is usually the Alderman–Grant design owing to the low associated electric field, and therefore low sample heating during proton decoupling. For the second coil, the resonant frequencies of more than one X nuclei may be covered by having rods with different values of capacitors, which can be swapped in and out.

Commercial NMR probes have nomenclature that describes whether the coil is designed to have maximum sensitivity on the proton or on the X-nucleus channel. For example, in the TXI configuration the T refers to triple resonance, the X to X-nucleus, and the I to an inverse design, which means

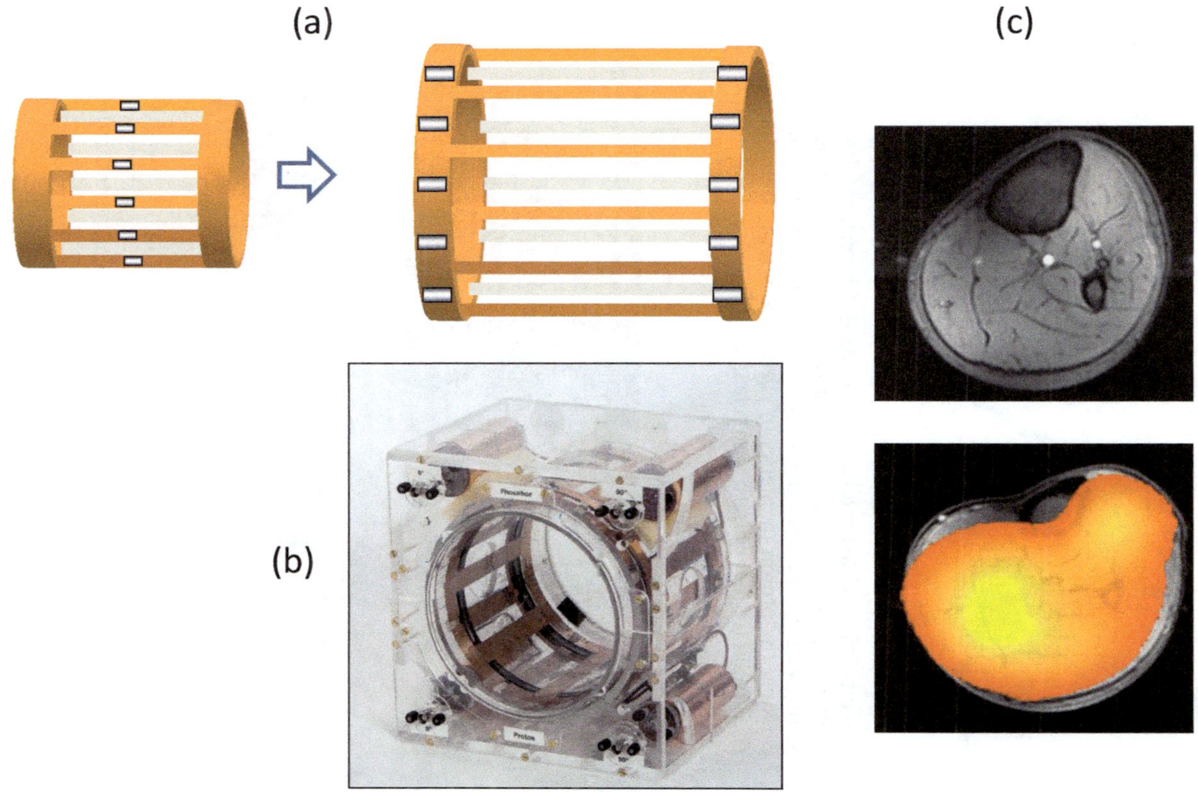

Figure 3.38 (a) Schematic of an imbricated birdcage with the low-pass X-nucleus structure placed inside the high-pass proton resonator. (b) Photograph of an imbricated birdcage constructed for proton and phosphorus MRI and MRS of the human calf muscle at 7 tesla. (c) A proton image and an overlaid map of the phosphocreatine distribution obtained using a ^{31}P chemical shift imaging sequence.[89]

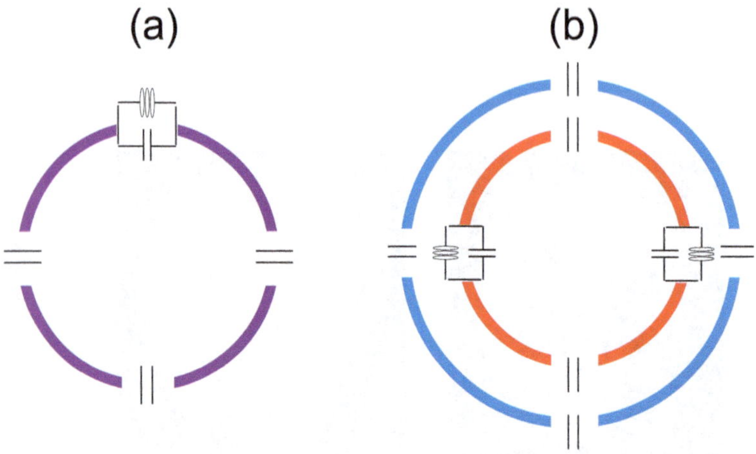

Figure 3.39 Circuits for double-tuned surface coils. (a) Insertion of a pole/trap circuit. (b) Two nested coils, blue-proton, red-X nucleus, coupled strongly through the mutual inductance. The parallel LC circuits resonate at the proton frequency to prevent counter-currents flowing at the proton frequency in the X-nucleus coil.[84]

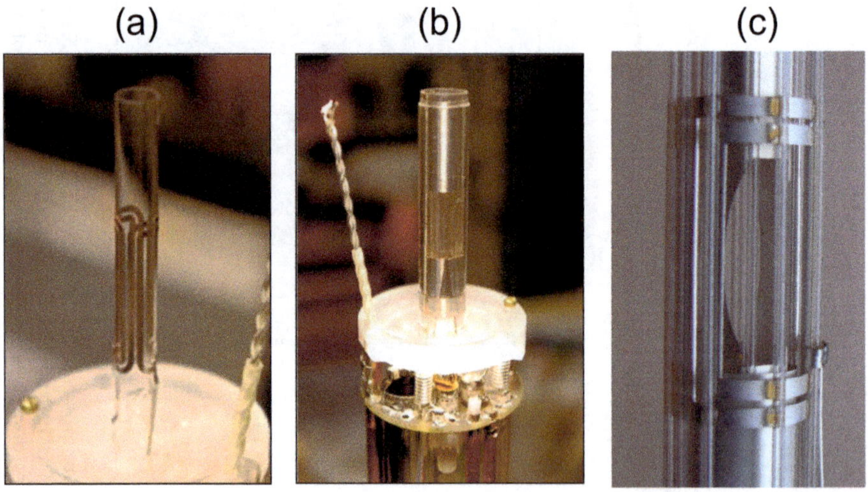

Figure 3.40 Photographs of coils and coil setups used in high resolution NMR spectroscopy. (a) A multi-turn saddle coil, (b) an Alderman–Grant design, and (c) an Alderman–Grant and saddle coil nested within each other.

that the proton channel corresponds to the inner coil. In this configuration, the inner coil is tuned to ^{1}H and ^{2}H with the circuit optimized for ^{1}H sensitivity. The outer coil operates, for example, at ^{13}C and ^{15}N frequencies, with the circuit designed to give higher efficiency on the ^{13}C channel. This means that the ^{13}C pulses are shorter than those for ^{15}N, which is necessary since the chemical shift range of ^{13}C resonances is much larger than that of ^{15}N resonances. Direct detection probes swap the configuration, with the inner coil tuned to the lower gamma nuclei, and the outer one used for proton decoupling. Depending upon the manufacturer, slightly different circuits are used to double tune each coil. High resolution probes are characterized by the maximum tube diameter that they can accommodate: 10 mm, 5 mm, 3 mm and 1.7 mm are standard sizes.

The spectral linewidths required for solution-state NMR of small organic molecules are sub-hertz, which corresponds to less than a 1 part-per-billion (ppb) variation in the magnetic field within the sample. Linewidths and line shapes are quoted at the 50%, 0.55% and 0.11% level of the full height of the spectral peak: typical values for non-spinning liquid probes are ~0.5/5/15 Hz using an industry standard sample of chloroform in acetone-d6. As covered in Chapter 4, there are a large number of room temperature shim coils that can be used to optimize the magnetic field homogeneity within the sample. Nevertheless, it is critical that the materials from which the RF coil is constructed are chosen so as to have as small a perturbing effect on the magnetic field as possible. The former on which the RF coil is mounted is usually a quartz cylinder. The most common metals used for RF coils are copper and silver with magnetic susceptibilities of −9.6, and −24 cgs, respectively.[11,90–92] If the magnetic susceptibility of the metal can be made close to zero, this minimizes the perturbing influence of the coil. The most commonly used method to achieve this is to coat the diamagnetic copper or silver wire with a thin layer of paramagnetic metal, either palladium and rhodium, which have magnetic susceptibilities of +840 and $+168 \times 10^{-6}$ cgs, respectively.[92,93]

In addition to the deuterium lock used to ensure a stable magnetic field *via* a negative feedback loop as described in Chapter 1, it is also important that the RF coil itself is stable over time, *i.e.* that over a period of days the coil remain impedance matched. One relatively recent innovation in high resolution NMR probes is "automatic matching and tuning" (ATM), which means that adjustments in the tuning and matching capacitors can be performed at set time intervals during a long experiment. This is achieve by using an internal reflection bridge and capacitors which have rods controlled by piezoelectric pistons to make the physical adjustments based on the values from the reflection bridge.[94]

High resolution NMR probes also incorporate magnetic field gradients, either in a single-axis (directed along the *z*-axis) or full triple-axis configuration. As described in Chapter 1, these can be used for coherence selection in multiple-quantum experiments, as diffusion-sensitizing gradients in experiments such as diffusion ordered spectroscopy (DOSY),[95] or as diffusion filters in protein NMR dynamic studies.[96]

3.9.2 Microprobes for High Resolution NMR and Hyphenated Microseparations

Although NMR spectroscopy is a powerful analytical method for structural elucidation it does have a very low inherent sensitivity compared to many other analytical techniques such as mass spectroscopy and fluorescence.

Sensitivities for analytical techniques are typically quoted in terms of limits-of-detection (LOD), corresponding to an SNR of 3 : 1 for a single measurement. Mass spectroscopy, for example, has an LOD in the 10^{-19} mole range, Fourier-transform infrared spectroscopy and Raman spectroscopy 10^{-12}–10^{-15} moles, laser induced fluorescence 10^{-13} moles, and Fourier-transform ion cyclotron resonance ~10^{-20} moles. In stark contrast, NMR spectroscopy using conventional 5 mm coils has a value in the 10^{-9} moles range. This low sensitivity means that mass-limited samples either require extremely long data acquisition times or simply cannot be studied. There are two, ultimately complementary, methods to design RF coils capable of studying very small sample masses: one is the design of "microcoils", RF coils with diameters much less than 1 mm,[16,97,98] and the other is to construct cryogenically cooled coils, covered in Section 3.9.4.

Given that that the sensitivity of an NMR probe is inversely proportional to its diameter, the optimal (theoretical) approach to acquiring spectra from a mass-limited sample is to dissolve the sample to the maximum possible concentration in the minimum amount of solvent, and to use a very small RF coil that just accommodates this volume. For truly mass-limited samples this might mean an RF coil with a volume in the nanoliter or hundreds of picoliter volume range, many orders of magnitude less than used in conventional NMR.[99] Although simple in concept, there are many challenges in this approach, both in terms of sample handling of such small volumes, and designing an RF coil that can produce high resolution NMR spectra despite the coil conductors being so close to the sample.

The most efficient RF coil, which is also relatively easy to produce at dimensions less than 1 mm, is a solenoid. Assuming that the length-to-diameter ratio is kept constant, the SNR per unit volume of sample increases as the inverse of the coil diameter down to a diameter ~100 μm, below which it increases as the inverse square root of coil diameter,[100] as shown in Figure 3.41(a). Details on constructing solenoids with dimensions in the hundreds of micrometres have been described by many authors.[97,100] One of the key developments was to be able to obtain high resolution spectra by using a susceptibility matching fluid, a perfluorocarbon, which is placed around the coil. The magnetic susceptibility of the perfluorocarbon is very similar to that of copper, which effectively makes the coil appear (magnetically) as an infinite cylinder and so minimizes the spatial perturbation of the local magnetic field.[98] The perfluorocarbon has no background signal and has a low loss tangent, so does not reduce the coil sensitivity. Another approach to magnetic susceptibility matching is to use a solid matrix, constructed from epoxy glue doped with appropriate concentrations of lanthanides, in which the coil is embedded.[101,102]

A second form of microcoil that can be produced in most laboratories without the need for sophisticated fabrication equipment is a planar surface coil, shown in Figure 3.41(c). Simple etching techniques can also be used for planar coils based upon striplines,[103] as shown in Figure 3.41(d). There are several other types of microcoils that require more sophisticated manufacturing

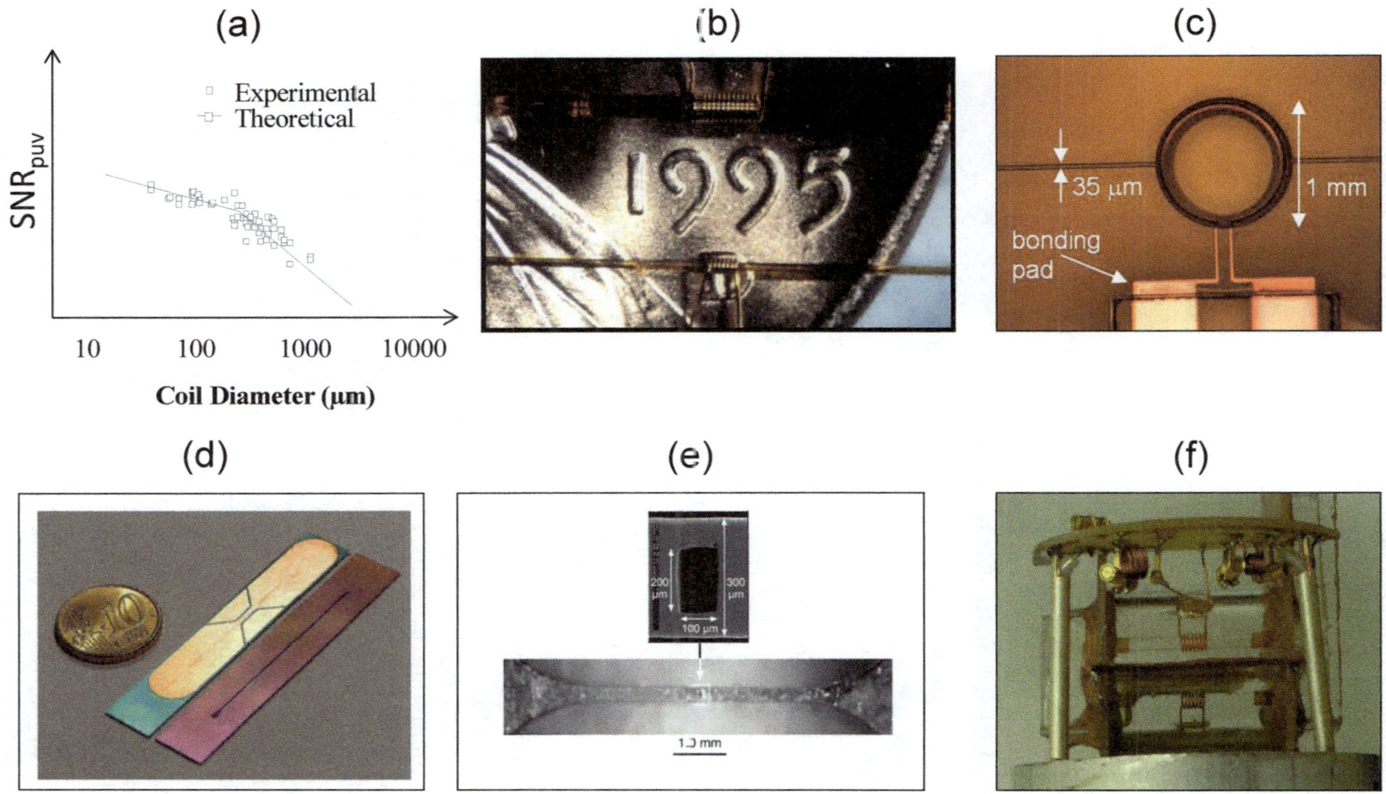

Figure 3.41 (a) SNR per unit volume as a function of coil diameter for a solenoidal microcoil.[100] (b) Photograph of two solenoidal microcoils: the upper with a diameter of 350 μm, the lower 150 μm. (c) Photograph of a lithographically etched surface coil. (d) A stripline resonator that can easily be integrated with microfluidic devices.[103] (e) A microslot coil lithographically etched in copper.[104] (f) A two-coil probe used for $^{1}H/^{15}N$ protein experiments.[97]

techniques. Three-dimensional solenoids can be produced, for example, by laser-etching or microcontact printing. Microslot probes,[104] in which a small slot is introduced into a conductor and a strong B_1 field is produced in the slot by "current-crowding", can be produced using planar ion-beam etching, as shown in Figure 3.41(e).

In addition to being able to study mass-limited samples, microcoils have many advantages, including the very wide excitation bandwidths associated with the very short RF pulses that can be applied, the ability to incorporate more than one coil into the RF probe for multi-sample investigations, as shown in Figure 3.41(f), and the direct coupling to chemical microseparations. There are a number of extensive reviews[6,16,97,105–108] of the applications of MR microcoils in both spectroscopy and imaging.

3.9.3 Probes for Solid-State NMR

Most solid-state NMR experiments are performed on large biomolecules, although there are a number of applications in areas such as solid-state catalysis and material characterization. The linewidths of solid samples are extremely wide (kHz to tens of kHz) compared to liquid samples, since line-broadening mechanisms are not averaged out by rapid molecular tumbling as they are in a liquid. The large linewidths mean that the excitation bandwidths of the RF pulses produced by solid-state probes much be much higher than for liquids, *i.e.* the RF pulses must be much shorter (~1 μs), and the coils must be capable of handling high power (several kW). As discussed briefly in Chapter 1, magic angle spinning (MAS) probes spin the sample at 54.7° with respect to the main magnetic field,[109,110] which averages out susceptibility broadening, chemical shift anisotropy and (to some degree) proton dipolar broadening. Residual dipolar coupling can be reduced by high powered proton decoupling during signal acquisition, and the combined effects of spinning and decoupling can produce X-nucleus (usually ^{13}C or ^{15}N) spectra with relatively narrow linewidths. In samples that are not spun, such as ordered membrane proteins, high powered proton decoupling is applied to reduce the effects of dipolar coupling. The major challenges in solid-state probe design are the very high powers that the capacitors must be able to handle, the problems associated with heating of conductive samples by the high power proton decoupling, and the very accurate machining tolerances necessary to spin at very high rates over a long period of time.

Figure 3.42(a)–(c) show a MAS probe for solid-state NMR. The rotors and stators in an MAS probe are typically made from partially stabilized zirconia, a high strength ceramic. The smaller the rotor diameter, the higher the possible rotation speed: typical numbers are 1.3 mm diameter/65 kHz, 2.5 mm/35 kHz, 3.2 mm/23 kHz, 4 mm/8 kHz, and 7 mm/4 kHz. Extensive details on various aspects of probe and stator design and construction can be found in work by Doty *et al.*[111] Pressurized air or nitrogen is supplied to the magic angle rotors, and data acquisition is synchronized to the spinning speed *via* optical tracking. Variable temperature experiments are often performed, with temperature

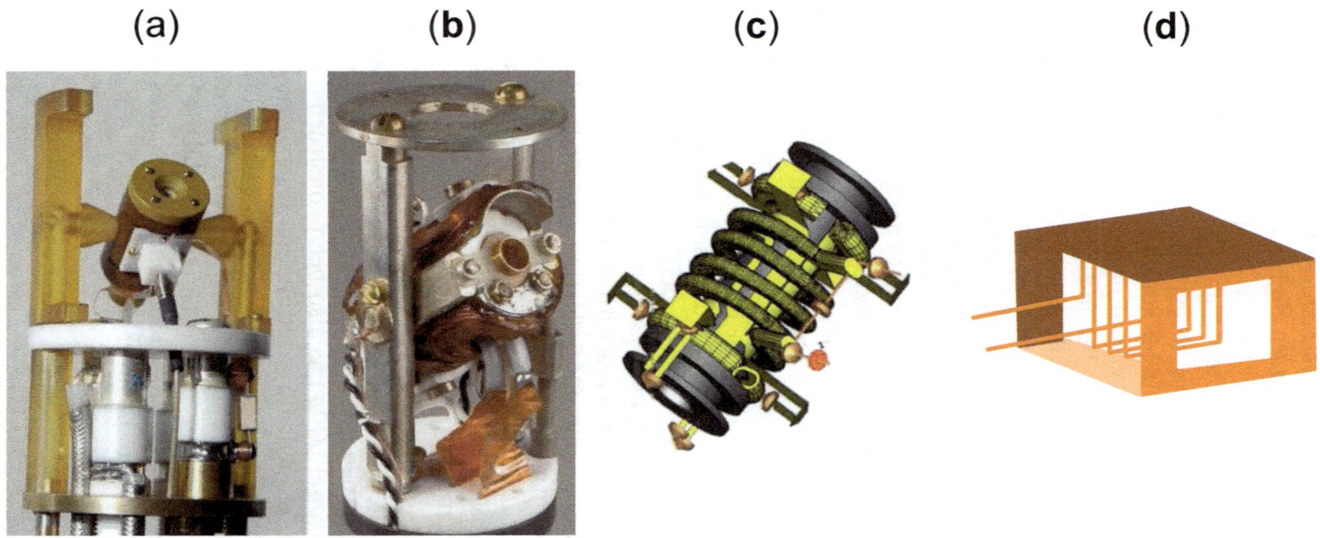

Figure 3.42 (a) A magic angle spinning solid-state NMR probe with the sample inserted into the hole indicated by the arrow. (b) Magic angle gradients (arrow) are often placed around the RF coil[113] in an MAS probe. (c) A schematic of a cross-coil solid-state probe with inner segmented cross-coil and outer solenoid.[25] (d) Schematic of a low *E*-field probe used for oriented membranes. The outer segmented Alderman–Grant coil is used for high power decoupling and the inner solenoid for X-nucleus detection.[26]

ranges for typical commercial probes between −120 °C and +160 °C (although much lower and higher temperatures can be used with specialized probes), controlled by the temperature of a gas stream. As mentioned earlier, the power ratings for variable capacitors are much higher than for liquid-state probes and, since these must also operate over a large temperature range, their design is more critical than in liquid-state probes, and this topic has also been extensively covered by Doty.[111] Gas-variable capacitors tend to be used much more than the air–dielectric devices used in MRI or liquid-state NMR.

In terms of the RF coils, the most common design is a triple-resonance configuration, which covers the ^{1}H, ^{13}C and ^{15}N frequencies. For non-conducting samples, a triple-tuned solenoid coil can be used. However, for conducting samples, the heating induced by high power proton decoupling can be problematic, since as noted previously (Section 3.4.4) the conservative electric field produced by a solenoid is very high. Instead, a two-coil arrangement is used in which the outer coil is designed as a solenoid, and operates at the $^{13}C/^{15}N$ frequencies, and the inner coil is designed to have a very low inductance and therefore very low conservative electric field. Most designs are based on a saddle or Alderman–Grant coil, with the so-called segmented cross-coil[25,112] being used for MAS probes. In order to achieve high spectral resolution, copper wire can be substituted by a zero-susceptibility copper/aluminium hybrid in which the copper forms the outer layer surrounding an inner core of aluminium.[11,91] So-called "magic angle gradients" (MAGs) are incorporated into many solid-state probes.[113,114] These comprise a set of three-axis gradients (*e.g.* one Maxwell pair and two orthogonal Golay coils) that are aligned at the magic angle. As with liquid-state probes, these gradients can be used to perform coherence selection, introduce diffusion weighting/filtering, and reduce artifacts in multi-dimensional experiments.

Non-spinning NMR studies of mechanically-oriented samples typically use rectangular solenoids, as shown in Figure 3.42(d), in order to accommodate slides on which membrane proteins have been deposited. Typically an inner solenoid is used for the X-nucleus, and a very low inductance loop-gap resonator used as the outer proton coil.[26] The two B-fields are orthogonal and so excellent decoupling between the two coils can be achieved. A related probe design with two orthogonal LGRs has also been constructed for operation at proton and fluorine frequencies.[115] A different design that also produces a low conservative electric field is termed a scroll coil, and can be considered as a multi-layer LGR.[116] In this geometry, the conservative electric field is concentrated in the thin dielectric between the conductive turns. A review of different geometries of RF coils used for solid-state studies of membrane proteins has been published by Grant *et al.*[66]

3.9.4 Cryoprobes

The main sources of noise in an NMR experiment are the coil, the preamplifier and the sample itself. Cooling the coil and the preamplifier reduces the noise from these components, thus increasing the SNR, until a limit is reached in which the noise from the sample is dominant. The use of cooled RF coils, usually termed "cryoprobes", has become widespread in high resolution NMR spectroscopy. The coils are constructed either from high temperature superconductors (HTS) or conventional conductors, such as copper or aluminium, and cooled to temperatures between 15 K and 30 K.[117–119]

The first high resolution NMR probe incorporating coils patterned from thin-film yttrium-barium-copper-oxide ($Yba_2Cu_3O_{7-\delta}$ or YBCO) was produced by Anderson *et al.* in 1995.[120]

In terms of the three noise sources, the SNR can be expressed as:[121]

$$\mathrm{SNR} \propto \frac{1}{\sqrt{T_{\mathrm{coil}} R_{\mathrm{coil}} + T_{N,\mathrm{preamp}} \left[R_{\mathrm{coil}} + R_{\mathrm{sample}} \right] + T_{\mathrm{sample}} R_{\mathrm{sample}}}} \tag{3.48}$$

where T_{coil} and T_{sample} are the respective actual physical temperatures, and $T_{\mathrm{N,preamp}}$ is the noise temperature of the preamplifier (which is not the same as the actual physical temperature since it is an active device). In cryogenic probes the temperature of the coil is in the range of 15–30 K (making the resistance of the coil much smaller than that of conventional room-temperature probes), and the preamplifier noise temperature is in the range of 10–15 K, corresponding to it being cooled to liquid nitrogen temperatures. Substantial increases in SNR, compared to equivalently sized room temperature probes, of up to a factor-of-four for molecules dissolved in non-conducting organic solvents have been achieved.[9,122–124] Currently, all of the major NMR vendors of high resolution probes have cryoprobe models at almost all magnetic field strengths: Figure 3.43 shows a typical setup including cryocooler and compressor.

The major challenge in cryoprobe design is clearly to establish an extremely high thermal gradient between the coil and the sample, the latter being at room temperature. The probe body is evacuated with high vacuum to provide the primary method of thermal isolation. A thin single-walled dielectric tube separates the sample space from the coil, which allows the coil to be placed in close proximity to the sample, ensuring a high filling factor. The RF coils are cooled *via* thermal conduction from a coldhead, as shown in Figure 3.44(a), and are not in direct physical contact with the cooling gas. The coldhead is usually made from copper and the coils are mounted onto a sapphire sheet, which is used since sapphire has a high thermal conductivity at cryogenic temperatures. The superconducting coils are soldered to small metal mounts, which are attached to the coldhead with machine screws to reduce the chance of the sapphire substrates cracking. The sample is placed inside the vacuum tube at the centre of the coil, and heated nitrogen gas is blown over the sample tube to counter radiant heat loss.

The superconducting coil is a self-resonant circuit consisting of an inductive loop and interdigitated capacitance, as shown in Figure 3.44(a) and (b), and is etched from an epitaxial film of yttrium-barium-copper compounds, between 300 and 650 nm in thickness, deposited on a planar substrate of lanthanum

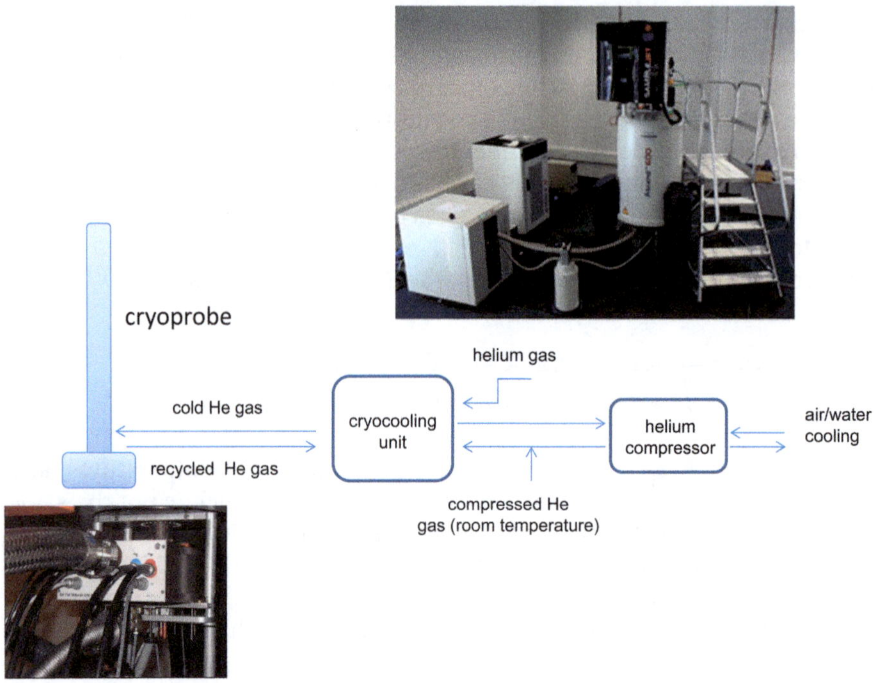

Figure 3.43 Schematic of the subsystems required for cryoprobe operation. Air or water cooling is used for operation of the helium compressors which pressurize room temperature helium gas. A Gifford–McMahon (GM) coldhead is used with two-stage cooling. A secondary flow loop is cooled *via* thermal contact to the interface surface of the GM coldhead. This secondary loop circulates a small fraction of the gas from the compressor within a small capillary tubing and this gas is used to cool a coldhead on which the RF coils are mounted.

aluminate or sapphire. The properties of this very thin superconducting layer are effectively those of a single crystal, which avoids many of the loss mechanisms that are present in the bulk material. Adjustment of the resonant frequency is achieved by laser trimming of the capacitive fingers. Unloaded *Q* values of between 10 000 and 20 000 can be achieved. Fine-tuning of the probe for different samples uses a conductive "paddle" and impedance matching is implemented *via* a moveable coupling loop, as shown in Figure 3.44(a).

In addition to standard-sized (3/5 mm diameter) cryoprobes, cryogenic cooling can also be applied to microcoils to obtain the highest possible sensitivity and lowest LODs. The first example of this concept was a 1 mm HTS probe with ^{1}H, ^{13}C, and ^{15}N channels, in addition to the deuterium lock,[125] shown in Figure 3.44(c). The proton coil was placed closest to the sample for highest sensitivity, and consisted of a "racetrack" design with the electric field constrained between the fingers of the interdigitated capacitors to reduce the power deposited in the sample. The ^{15}N, ^{13}C and ^{2}H coils were all designed as spiral inductors.[125,126] The mass sensitivity of this probe was

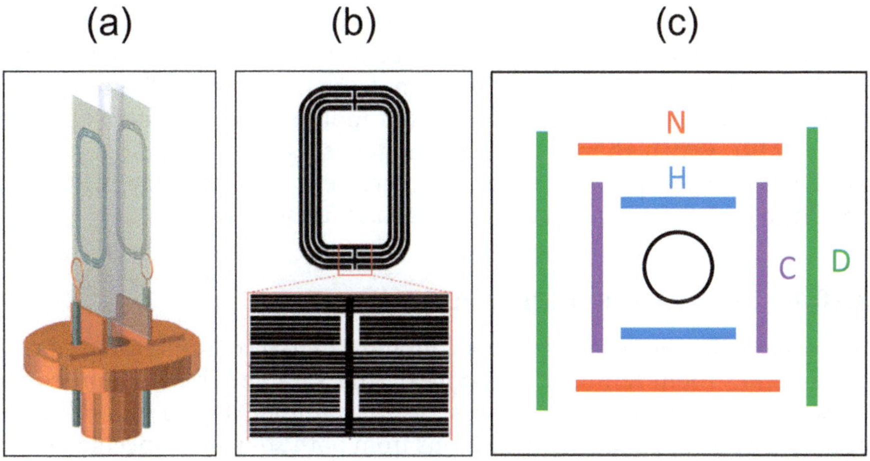

Figure 3.44 (a) The arrangement of a single pair of HTS coils supported by a cold-head and placed around an inner dielectric cylinder, into which the sample tube is loaded. The loops on the outside are for inductive impedance matching. (b) A racetrack design with interdigitated capacitance, typically used for the proton channel. (c) Four nested coils used in the first cryo-microprobe, with sensitivity optimized for proton detection.[125] Reproduced from ref. 125, W. W. Brey, A. S. Edison, R. E. Nast, J. R. Rocca, S. Saha and R. S. Withers, Design, construction, and validation of a 1 mm triple-resonance high-temperature-superconducting probe for NMR, *J. Magn. Reson.*, 2006, **179**, 290–293. Copyright (2006) with permission from Elsevier.[125]

measured to be more than four times greater than a 5 mm cryoprobe. A probe design optimized for ^{13}C detection, in which the positions of the ^{13}C and ^{1}H coils were interchanged, has also been described by the same group.[127]

The principle of cryogenic cooling can also be applied to solid-state MAS probes, although the challenges are even greater than for high resolution liquids probes. For most solid samples the coil noise is dominant, and so even cooling only the RF circuitry used for impedance matching and multiple-frequency operation can reduce the overall noise level.[128] In the CryoMAS probe from Doty Scientific the vacuum insulated RF coils and specially-designed capacitors are cooled to 23 K. The probe includes a triple resonance (^{1}H/^{13}C/^{15}N) RF circuit and a 3 mm insulated ceramic spinner, and is operated with a commercial closed loop GM cryocooler. SNR gains up to a factor-of-four over room-temperature probes have been reported with this setup.[128]

3.10 RF Coils for Small Animal Imaging and MR Microscopy

Most small animal MRI is performed on mice and rats on systems operating between 4.7 and 17.2 tesla for horizontal magnets, and 9.4 and 21.1 tesla for vertical magnets. Coil design concepts for small animal MRI represent an overlap between the design criteria for high resolution NMR and those for clinical MRI. For example, the use of cryogenic resonators developed for high resolution NMR is now being actively implemented into small animal MRI coils, and it is a similar story for arrays of multiple receivers developed for human MRI. Very high resolution imaging on small samples, often referred to as MR microscopy, uses coils which are very similar to the microcoils described earlier for high resolution NMR spectroscopy.

3.10.1 Small Animal Imaging Coils

Whereas coil losses usually dominate for most high resolution NMR samples (considering room temperature probes), and sample losses dominate for most clinical MRI studies, the losses for small animal MRI lie somewhere in between, with significant contributions from both sources.[15] Specific design constraints for RF coils for small animal MRI include:[15]

(i) There is usually a very limited amount of space within small, high powered gradients, meaning that RF shields often have to be placed very close to the transmit coil.
(ii) Capacitor values are much smaller than for clinical coils, meaning that stray capacitance becomes much more of an issue.
(iii) There are a wide variety of loading conditions: even when imaging mice, differences in the size and age of the animals impose very different sample loads, and the relatively small capacitors used in the coils mean that the frequency range over which the coil can be impedance matched is limited if the coil is to be kept electrically balanced.
(iv) It is often necessary to implement a remote tune-and-match scheme in which the impedance matching circuit is physically remote from the RF coil, and if this is not implemented correctly it can lead to high losses.

The majority of volume resonators for animal MRI are based on the birdcage design. As outlined above, the challenges lie in the much smaller capacitor values necessary to resonate the coils, particularly for very high magnetic fields, and the proximity of the RF shield to the coil itself, which reduces the strength of the magnetic field owing to mirror currents being induced in the shield. Variations on the basic birdcage design include: (i) using two struts in parallel rather than a single strut,[129] which has been shown to increase the usable [frequency × diameter] product and also to increase the coil quality factor, (ii) the millipede coil,[44] which consists of a very large number of intercalated struts, which produces a coil with high B_1 homogeneity, but a limited range of loads over which the coil can be matched, and (iii) the Litzcage,[130] shown in Figure 3.45(a) and (b), which has been demonstrated to have certain advantages over the birdcage design in terms of improved B_1 homogeneity and the range of loads over which the coil can be matched.

(a) (b)

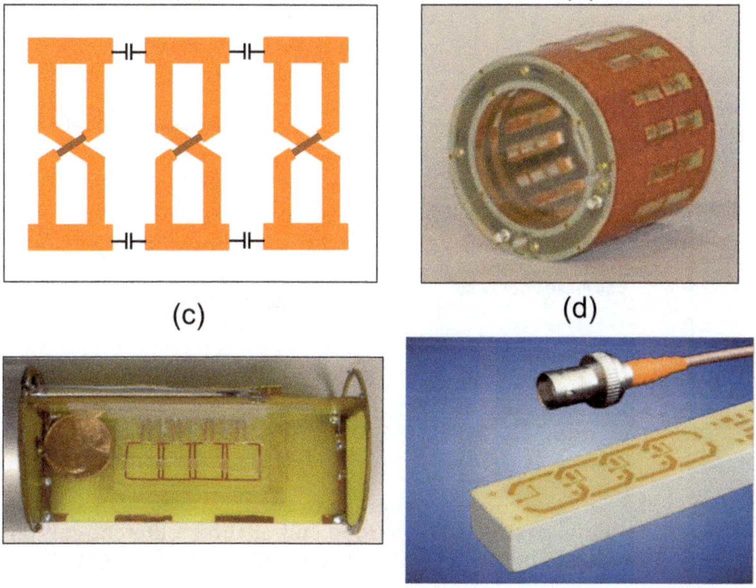

(c) (d)

Figure 3.45 (a) Schematic of the conductor layout for a high-pass Litzcage volume resonator for small animal imaging.[130] A single strut of an equivalent birdcage is split into two conductors, which have a crossover in the centre of the struts. (b) An assembled Litzcage resonator. (c) A four element phased array used in a vertical bore magnet for imaging mice.[164] (d) A commercial four element receive array for imaging mice in a horizontal bore magnet.

As mentioned above, space requirements and the need for tuning the RF coil to each and every load mean that the variable impedance matching capacitors may have to be placed remotely from the coil, connected by a transmission line.[131,132] This case has been analyzed by a number of authors, who have shown that it is critical to have the coil tuned within 1–2% of the operating frequency, in which case there is negligible loss in connecting variable capacitors *via* a half-wavelength cable. However, if the resonant frequency of the loaded coil is significantly shifted from the operating frequency, then a large amount of energy can be lost in standing waves in the transmission line, and the transmit and receive sensitivities will be greatly reduced. Kodibakgar and Conradi have shown that overcoupling of the line to the tuned circuit is key to obtaining a large tuning bandwidth.[132]

Over the past few years, multiple receive channels have become standard on animal imaging systems, and a number of phased array receive coils have been designed: examples are shown in Figure 3.45(c) and (d). Electrical isolation of individual components of the array follows standard techniques described earlier in this chapter, but isolation is typically more challenging than for human applications owing to the much smaller physical sizes of the coils.

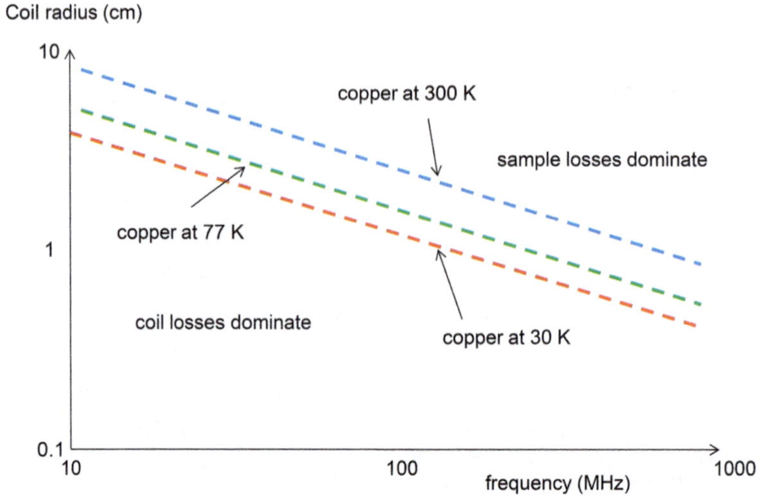

Figure 3.46 Plots of the lines corresponding to equal contributions from coil and sample losses for a single loop coil at three different coil temperatures, as a function of coil radius and operating frequency.[139]

Since small animal imaging lies in the domain in which both sample and coil losses are significant it makes sense to incorporate cryogenic technology where possible.[133–139] Figure 3.46 shows the crossover between coil-dominated and sample-dominated noise at room temperature for a single loop coil as a function of coil radius, temperature and operating frequency. This "crossover point" corresponds to a loop with a 2.5 cm radius at 100 MHz and to a loop with a radius of 0.9 cm at 800 MHz. For a copper coil cooled to 30 K or an HTS coil the respective radii are 1.2 and 0.45 cm. The design and fabrication of HTS coils for small animal MRI is very similar to that described earlier for high resolution NMR probes, with self-resonant structures being formed and then fine-tuned using mutual coupling to external coils. The preamplifier is cooled to liquid nitrogen temperature, and is placed directly outside the magnet. Increases in SNR over conventional room temperature probes of close to a factor-of-three have been reported for *in vivo* studies,[140] and cryogenic phased-array coils are increasingly becoming available from commercial vendors.

3.10.2 RF coils for MR Microscopy and Combined MR/Optical Histology

MR microscopy has traditionally been somewhat loosely defined as corresponding to an imaging regime in which the spatial resolution in one or more dimensions is less than 100 μm, although in practice the resolution is usually far superior to this. Early work by Aguayo[141] and Jacobs[142] using solenoidal coils established the utility of MR microscopy in biological systems. In addition to a small and efficient RF coil, very strong gradients are used in MR microscopy since the fundamental limit to spatial resolution in MR microscopy is the molecular diffusion coefficient of water in the particular sample being imaged.[143–145] Although strong gradients can be used to minimize the full-width-half-maximum (FWHM) of the blurring function caused by diffusion, there are practical limits to the maximum gradient strength that can be achieved. Using microcoils, a number of researchers have reported very high spatial resolutions in the single digit micrometers.[145–147]

Many MR microscopy experiments have been performed using solenoids with sub-millimetre diameters, very similar to those described in the section on microcoils for high resolution NMR. These types of coil are very suitable for imaging single cells or neurons:[7,8] however, sample positioning is extremely challenging for many other types of biological samples. Therefore, most current MR microscopy is performed using planar microfabricated surface microcoils,[148] examples of which are shown in Figure 3.47(a) and (b). Using these surface microcoils cellular-level morphological and diffusion-weighted images have been obtained from a number of different samples, including mammalian nervous tissue and drosophila brain,[149–151] as shown in Figure 3.47(c). Arrays of overlapping planar microcoils have also been produced using microfabrication techniques: an example of a seven-coil array is shown in Figure 3.47(d) with images acquired using this array shown in Figure 3.47(e).

In many MR microscopy experiments, fluorescence and histological images are also obtained from the same sample in order to obtain information complementary to the MRI data. Post-processing of such types of multi-modal data is very challenging: in particular, the task of precise co-registration of the data from histological sections and MRI. This is primarily owing to the deformations and shrinkage that occur when the sample is processed for histology. It is also very difficult to cut histological sections in exactly the same orientation as the acquired MRI data. Finally, the very different spatial resolution of MRI and histological sections makes direct registration a cumbersome task. In order to alleviate these difficulties and possible registration errors, a number of groups have developed RF coils that can obtain MR images from the same histological slice that is used for subsequent staining. The first such setups used a small self-resonant microcoil placed around the sample, and inductively coupled this coil to the much larger volume resonator that is standardly provided with microimaging systems. The approach of using inductively coupled self-resonant coils for MR microscopy was first shown by Banson,[152] with the approach extended by Glover *et al.*[153] to show co-registered optical and MR images of epidermal cells of *Allium cepa*. In a study of *ex vivo* tissue from Alzheimer's patients performed at 9.4 tesla, Nabuurs *et al.* placed the self-resonant coil shown in Figure 3.48(a) and (b) on the back of a standard microscope slide so it completely covered a 60 μm thick tissue sample.[154,155] Using these easily-produced and replaceable coils, high resolution images of histological samples, as shown in Figure 3.48(c), were acquired with a ~15 fold reduction in data acquisition time compared to using a commercial resonator: a histological image from the same sample, stained for iron, is shown in Figure 3.48(d).

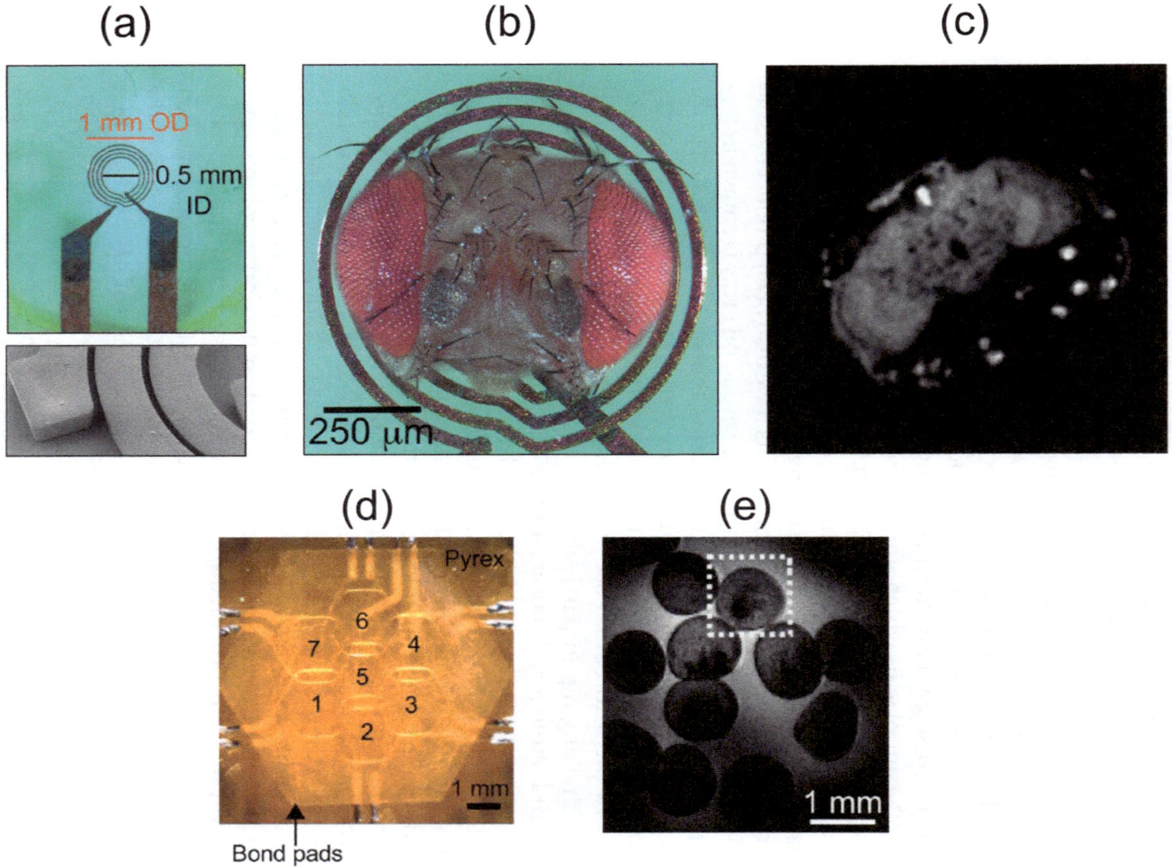

Figure 3.47 (a) Photograph of a spiral inductor planar microcoil and zoomed-in view of the copper conductor paths (Bruker Instruments). (b) One slice from a drosophila brain placed on top of the planar microcoil.[151] (c) One slice from a diffusion-weighted image of a drosophila brain acquired at 10 × 10 × 10 μm^3 spatial resolution.[151] (d) Photograph of a seven-element planar microcoil array.[188] (e) Images of oocytes acquired using the seven-element phased array.[188] Reproduced from ref. 188 with permission from the Royal Society of Chemistry.

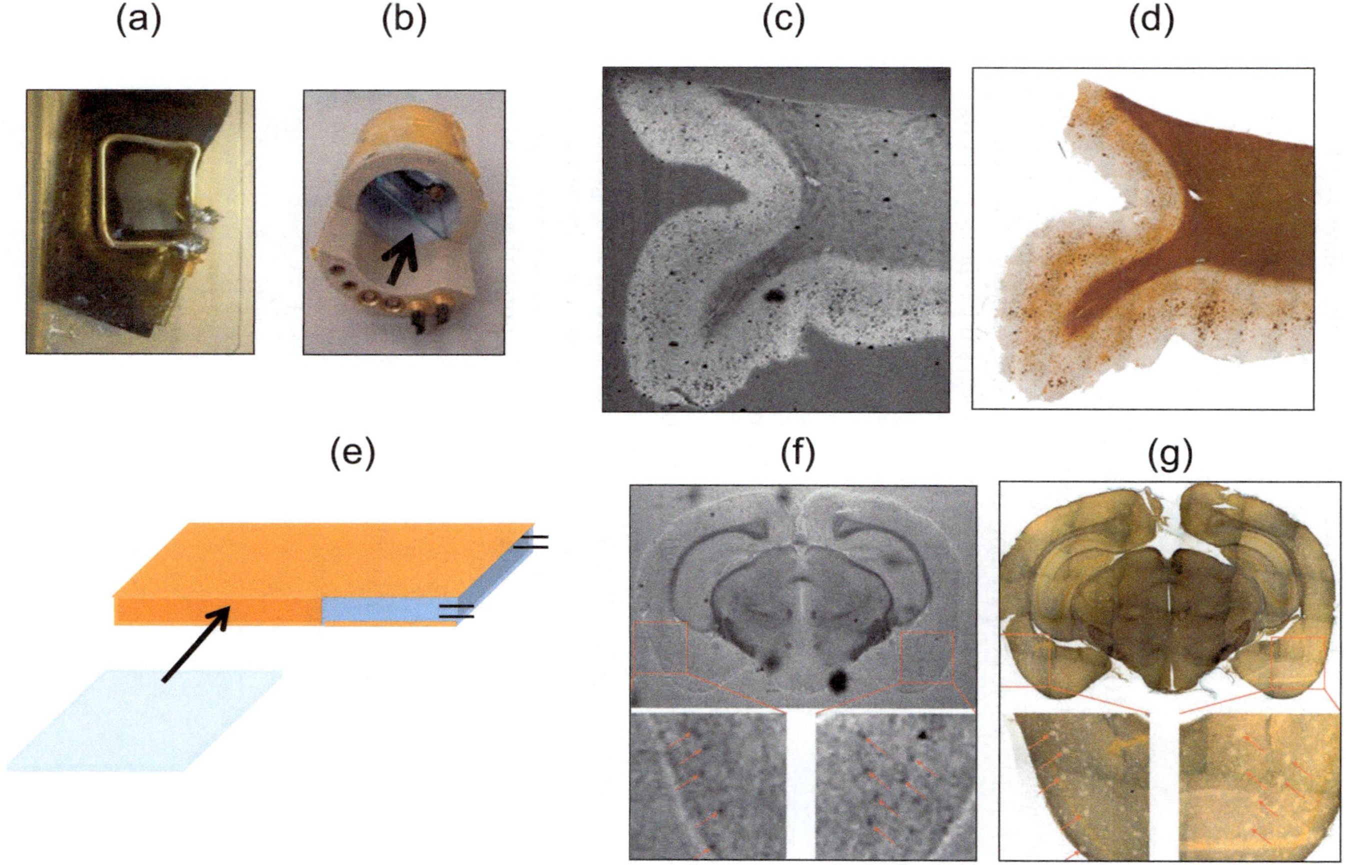

Figure 3.48 (a) A self-resonant microcoil is placed on the back of a standard microscope slide, and (b) the assembly is placed inside a birdcage volume resonator.[155] (c) A T_2^*-weighted image and (d) corresponding iron-stained histological slide of a slice from a human brain with Alzheimer pathology. (e) A specialized RF coil constructed as a loop-gap resonator from copper foil and a Teflon spacer (blue).[157] (f and g) MR images acquired with the loop-gap resonator and correlative histology acquired from a single 60 μm thick sample.[157]

Specialized resonators have also been constructed to image histological slides. Meadowcroft *et al.*[156,157] have designed and built such a setup for horizontal-bore 3 tesla and 7 tesla scanners. The basic design, shown in Figure 3.49(e), is based upon a loop-gap resonator, covered earlier in this chapter. A Teflon spacer is placed close to the feed-point of the coil, and the volume into which the microscope slide is placed has a very homogeneous B_1^+ field. Images obtained from rat brain samples are shown in Figure 3.48(f) and (g).

3.11 RF Coils for Clinical Imaging Systems

The static magnetic field strengths most commonly used for clinical whole body MRI systems are 1.5 tesla and 3 tesla. In the past decade so-called "ultra-high field" whole-body human scanners have become commercially available at 7 tesla and 9.4 tesla, with several clinical studies being performed at 7 tesla. In general, the design of RF coils for human studies becomes more complicated the higher the field strength, and in this section the transition between single-channel (1.5 tesla), dual-channel (3 tesla) and multi-channel (>3 tesla) transmit coils is outlined.

3.11.1 Single-Channel and Dual-Channel Transmit Coils for Clinical Systems

On 1.5 tesla and 3 tesla systems a birdcage "body coil" is used to transmit a homogeneous B_1^+ field. However, the presence of the patient inside the RF coil disturbs the uniformity of the B_1^+ field, with the perturbation being greater on a 3 tesla system. The major factor causing this B_1^+ inhomogeneity within the patient is the high relative permittivity (ε_r) of the body. The RF wavelength in tissue is inversely proportional to the square root of ε_r and if the dimensions of the body are a substantial fraction of the RF wavelength in tissue, the RF energy propagating through the body from the birdcage coil undergoes constructive and destructive interferences. The second factor that can cause B_1^+ inhomogeneity is the spatially dependent conductivity of different tissues,

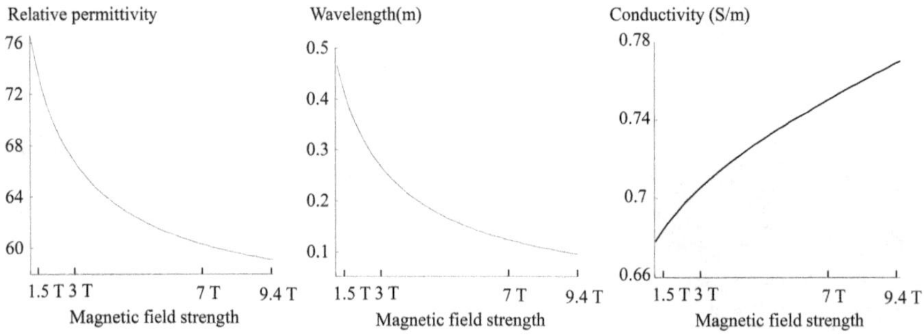

Figure 3.49 Plots of relative permittivity, RF wavelength, and conductivity for muscle tissue at different magnetic field strengths.

with a higher conductivity reducing RF penetration through tissue. Figure 3.49 shows the relative permittivity, RF wavelength, and conductivity of muscle as a function of field strength.

As shown in Figure 3.49, at 1.5 tesla the RF wavelength in muscle is ~40 cm, which means that wavelength effects are relatively minor. A standard birdcage coil can be used as the transmit coil, driven in quadrature mode, as shown in Figure 3.50(a), to produce a circularly polarized EM wave. At 3 tesla the wavelength is ~25 cm, and as a result abdominal images in particular show areas of lower and higher transmit field, and therefore lower and higher SNR. The presence of such "dielectric artifacts" at 3 tesla has resulted in the recent introduction of *dual-channel transmit* systems, in which the birdcage body coil is effectively split into two independent channels, each of which can be driven with an independently-controlled magnitude and phase, as shown in Figure 3.50(b). The additional degrees-of-freedom of the dual-channel system can produce significant increases in RF transmit homogeneity. The

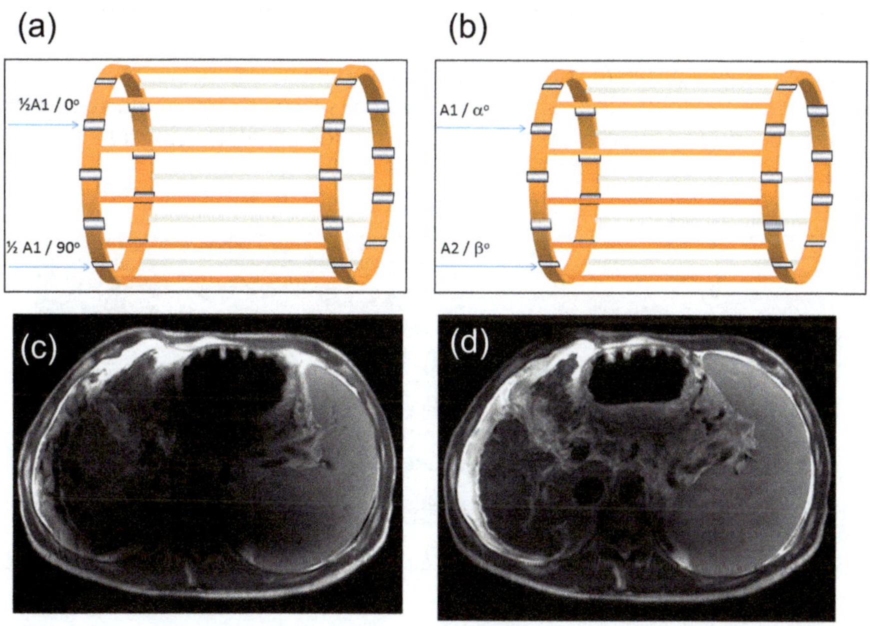

Figure 3.50 (a) Schematic of a single-channel transmit birdcage body coil operating in quadrature mode. There is a fixed phase angle of 90° between the two ports, and the amplitudes of the signal fed into each of the two ports are the same. (b) Corresponding schematic of a dual-channel transmit body coil. Using two separate transmitters, the amplitudes and phases of the two channels can be set independently, giving additional degrees of freedom. (c and d) Example of image improvement from a patient with ascites using a dual-channel transmit system.[189] A T_2-weighted turbo spin echo sequence was run with spatial resolution of 1.2 × 1.4 × 7 mm within a single breath-hold of 14 seconds. (c) Single-channel transmit and (d) dual-channel transmit.

process of optimizing image quality by adjusting the amplitudes and phases of the two channels is referred to as "B_1-shimming". A short additional calibration scan (typically lasting less than a minute) is added to the start of the clinical protocol to map the B_1^+ field within the patient, and based upon this scan the optimum relative phases and magnitudes feeding the two ports of the dual-channel transmit coil are calculated. Figure 3.50(c) shows one clinical example in which areas of low image intensity occur within the liver owing to the low transmit efficiency from a single-channel system, whereas the image in Figure 3.50(d) show significantly higher signal intensity when the amplitudes and phases of the two channels are optimized on a dual-channel system.

3.11.2 Receive Arrays for Clinical Systems

Almost all clinical systems use receive arrays consisting of multiple loops, as covered previously in this chapter. Each loop contains circuitry for detuning during the transmission of RF pulses by the body coil, and is also decoupled from each other element in the array. These arrays can form an overall cylindrical shape, for example for brain imaging as shown in Figure 3.51(a), or can be shaped into a semi-flexible "shell" which can be placed on top of the area of the body to be imaged, for example for cardiac or abdominal imaging, as shown in

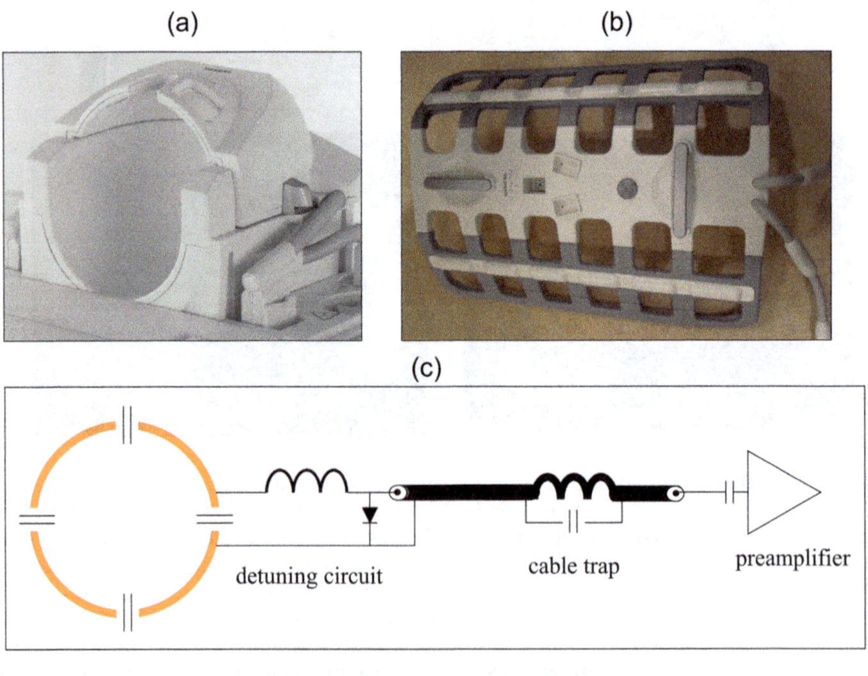

Figure 3.51 (a) A 32 element receive loop array for imaging the brain. (b) A twelve element receive array for imaging the body. (c) A schematic of each element of a receive array.

Figure 3.51(b). Many commercial systems also have rigid arrays that are built into the patient table. Individual elements of the array can be activated or deactivated *via* the scanner software depending upon the particular organ being scanned. Figure 3.51(c) shows a schematic of each element of the array, with capacitive segmentation, a PIN-diode based detuning circuit, a cable trap to reduce the common mode current, and preamplifier decoupling.

3.12 RF Coils for Very High Field Human Imaging

Over the past decade there has been a rapid increase in the number of very high field (7 tesla and above) human scanners installed worldwide. The major challenges associated with these systems are the intrinsically lower B_1^+ homogeneity compared to 3 tesla or 1.5 tesla systems owing to wavelength effects, as described previously, and the higher power deposited in the patient. In addition, if one attempts to make a body coil for such a high field system using a conventional quadrature birdcage or TEM design, the EM energy couples very efficiently to waveguide modes in the magnet bore and a significant amount of energy is transported away from the body. These effects are shown in Figure 3.52, which represents the first *in vivo* body imaging study performed on a 7 tesla system.[158]

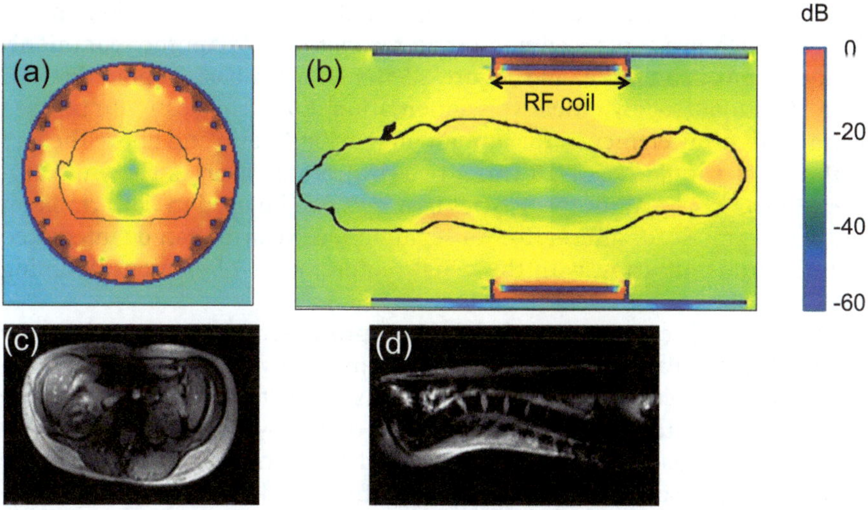

Figure 3.52 (a and b) Finite difference time domain simulations of the B_1^+ field generated by a TEM body coil *in vivo* at 7 tesla in (a) central transverse view and (b) central sagittal view. A significant amount of the magnetic field couples to waveguide modes within the magnet bore and "leaks" into the head and lower abdomen. (c and d) Abdominal images acquired at 7 tesla using a body coil in transmit and receive mode: data were acquired using a low-tip angle gradient echo sequence. Reprinted from ref. 158 with permission of John Wiley & Sons, Inc. © 2008 Wiley-Liss, Inc.[158]

3.12.1 Multi-Channel Transmit Arrays for High Field Imaging

Section 3.11.1 showed that one way of dealing with the intrinsic B_1^+ inhomogeneities in a human subject at 3 tesla is to introduce two RF channels that can drive the transmit coil with independent phases and amplitudes. From Figure 3.52 it is clear that at 7 tesla or 9.4 tesla these inhomogeneities are much increased, and so a transmit array with additional (>2) elements is required for body imaging, and may also be useful for neuroimaging. State-of-the-art systems at 7 tesla and above currently have the capability of driving 8, 16 or 32 element transmit arrays.[159] The basic architecture of such a system is shown in Figure 3.53. Based either on measurements or simulations, the signal input (phase and magnitude) to each element of the RF coil can be optimized to maximize the uniformity of the transmit field within the patient, minimize local and global tissue heating, or a combination of the two.[160]

A number of different types of designs have been used to form a transmit array: a few of these are illustrated in Figure 3.54. Historically, the first widely-adopted design was a microstrip resonator, shown in Figure 3.54(a), which comprises a narrow conductor strip placed on top of a dielectric material, such as Teflon, with a wide ground plane on the other side of the dielectric.[161] The multiple stripline elements in a transmit array can be decoupled either by using specific dimensions[162,163] or *via* capacitive decoupling.[164,165] Microstrip-based arrays with up to 32 elements have been designed for imaging the body at 7 tesla and 9.4 tesla. Figure 3.54(b) shows a variation on this design, which is a centre-fed microstrip resonator.[158,166] Another commonly used element is a dipole, and Figure 3.54(c) shows a basic dipole that has been "fractionated" by inductive segmentation. Loop coils can also be used as transmit array elements: for human imaging at 7 tesla and above the size of the loops should be ~8–10 cm. A two-row rotated array is shown in Figure 3.54(d).[80,167–169] All of these types of arrays can be used in both transmit and receive mode, or alternatively a separate receive array can be integrated with the transmit array, with the usual requirements of element detuning and decoupling.

Figure 3.55 gives one example of the improvement in image uniformity and quality that can be achieved using a transmit array.[170] In this case a sixteen element microstrip transmit array was used in transmit and receive mode in a study to assess the feasibility of renal angiography at 7 tesla. In this particular case the phase of the individual channels was optimized by using a fast pre-scan to map the B_1^+ fields from each of the individual elements of the array: subsequently equal magnitudes but optimized phases were applied to each channel.

3.13 Dielectric Resonators

All of the coil designs considered so far have been constructed from metallic conductors, with lumped elements or distributed capacitance being used to resonate the structure at the desired frequency. An alternative approach to

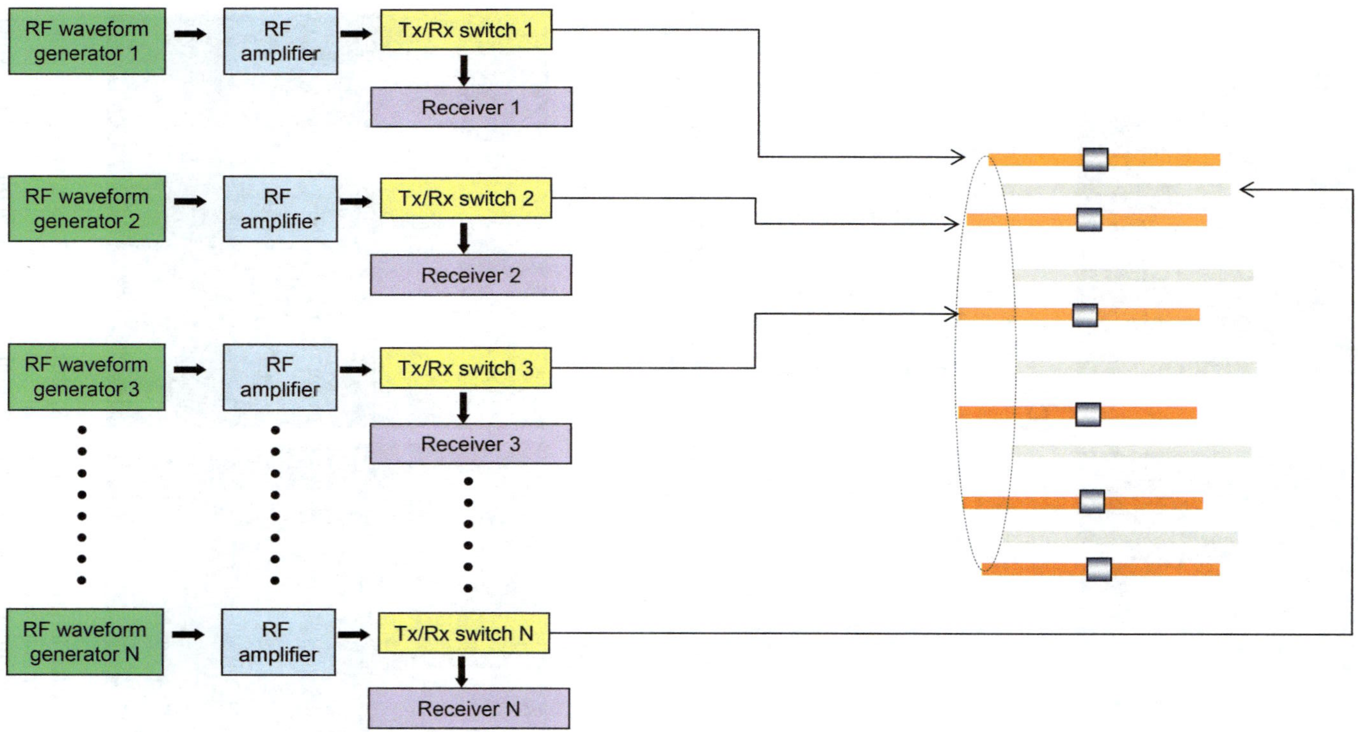

Figure 3.53 Schematic of the hardware platform required to implement an *N*-element transmit/receive array with variable magnitude and phase of the input signal to each element of the array. Each array element may be a microstrip, dipole or loop element, or a combination of different types of element.

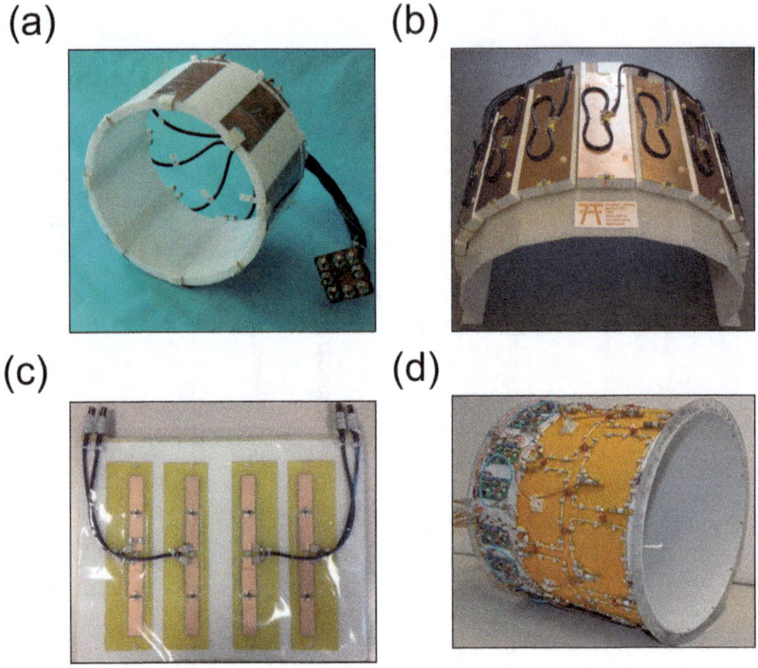

Figure 3.54 Four different coil designs that have been used in transmit arrays: (a) microstrip, (b) centre-fed microstrip, (c) fractionated dipole, and (d) multi-row loop.

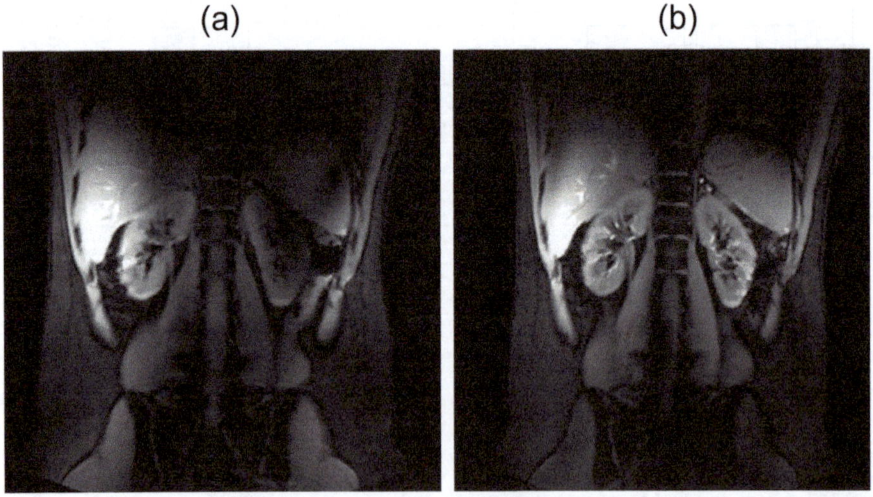

Figure 3.55 Illustration of the benefits of B_1-shimming at 7 tesla using a 16-channel transceiver array. Coronal anatomic images from a volunteer using (a) default transmit phase settings and (b) a local B_1 phase shim, which optimized the B_1^+ over both kidneys.[170] Reprinted from ref. 170 by permission of John Wiley & Sons, Inc. © 2012 Wiley Periodicals, Inc.[170]

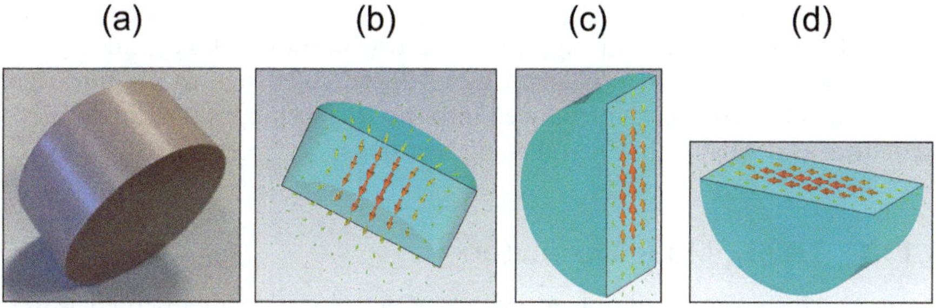

Figure 3.56 (a) Photograph of a cylinder formed of high permittivity ceramic material, barium strontium titanate. (b) Magnetic field distribution corresponding to the TE_{01} mode. (c and d) The two orthogonal frequency-degenerate HEM_{11} modes of a cylindrical resonator.

coil design, which becomes practical at high frequencies, is to use a so-called dielectric resonator (DR). This is a solid structure, typically cylindrical or rectangular in shape, which is formed from a very high permittivity material: an example is shown in Figure 3.56(a). There are several different resonant modes of these structures: the particular mode structure and associated resonant frequency depends on the relative permittivity of the material as well as the size and shape of the resonator. These resonant modes of a DR refer to time-invariant electric- and magnetic-field patterns that are formed within the resonator. For example, cylindrical DRs can have transverse electric (TE), transverse magnetic (TM), and hybrid electromagnetic (HEM) modes. Different modes are denoted by numerical subscripts, *e.g.*, TE_{01} and HEM_{11}: the two subscripts denote the number of half-wavelength field variations in the azimuthal and radial directions, respectively.

Figure 3.56(b)–(d) shows the distribution of magnetic fields for the most useful modes, TE_{01} and HEM_{11}, of a cylindrical DR for MR experiments. The magnetic field in the TE mode is linearly polarized with the main axis coincident with that of the DR. In contrast, the HEM mode supports a circularly polarized magnetic field, with the main axis perpendicular to that of the cylinder. Although the exact frequency for a given resonant mode can only be calculated by rather complicated numerical procedures, empirical expressions have been derived for the TE_{01} and HEM_{11} modes[171] for a simple cylinder:

$$f_{\mathrm{TE}_{01\delta}} = 2.921\frac{c\varepsilon_r^{-0.465}}{2\pi a}\left[0.691 + 0.319\frac{a}{2h} - 0.035\left(\frac{a}{2h}\right)^2\right] \tag{3.49}$$

$$f_{\mathrm{HEM}_{11\delta}} = 2.735\frac{c\varepsilon_r^{-0.436}}{2\pi a}\left[0.543 + 0.589\frac{a}{2h} - 0.05\left(\frac{a}{2h}\right)^2\right] \tag{3.50}$$

where c is the speed of light, a is the radius and h the height of the cylinder in centimetres, and f_r has units of MHz.

As can be appreciated from Figure 3.56(b)–(d), the magnetic field for the TE_{01} and HEM_{11} modes is strongest in the centre of these solid cylinders, and so in order to use these in place of metal-based coils such as the birdcage resonator, a hole must be introduced into the DR, turning it into an annular structure, as described in the following sections.

3.13.1 HEM_{11} Mode Resonators

Figure 3.57 shows details of an annular DR consisting of distilled water ($\varepsilon_r = 78$), which was designed to operate in degenerate quadrature HEM_{11} modes for human imaging at 298.1 MHz (7 tesla). Since the hole is filled with tissue, in this case the knee, which has a lower permittivity ($\varepsilon_r \sim 60$) than water, the height and diameter of the DR are slightly larger than those values calculated from eqn (3.50). EM simulations are very useful for designing the exact dimensions of such a resonator. In this case, the outer diameter was 255 mm, the inner diameter 154 mm, and height 89 mm. *In vivo* images acquired at 7 tesla using a gradient echo sequence are shown in Figure 3.57(c).

Smaller water-based DRs have also been designed for the wrist.[172] HEM resonators have also been designed using much higher permittivity materials which reduces the dimensions of the resonator.[173] Methods for PIN-diode based detuning of DRs have also been introduced in the literature.[172]

3.13.2 TE_{01} Mode Resonators

The TE_{01} mode of a DR can also be used for MRI: in this case the resonator must be aligned with its principal axis perpendicular to B_0. Barium strontium titanate ($\varepsilon_r = 323$) has been used to produce DRs for microimaging on a vertical bore system at 14.1 T (600 MHz)[174,175] and calcium titanate ($\varepsilon_r = 156$) has also been used to construct resonators for microimaging at magnetic fields up to 21.1 tesla (900 MHz). In each of these cases a small hole is bored in the centre of the high permittivity material, into which the specimen can be placed. Since the electric field is zero at the centre of the resonator, introducing a small hole has very little effect on the current distributions in the DR and therefore the B_1^+ distribution is virtually identical to that of a solid resonator.

Transmit arrays for high field human imaging can also be produced using ceramic resonators working in the $TE_{01\delta}$ mode.[176] Figure 3.58 shows an eight element array consisting of ceramic elements ($\varepsilon_r = 170$) designed to operate in transmit/receive mode for cardiac imaging at 7 tesla.

3.14 Antennae for Travelling Wave MRI

As shown in Figure 3.52, the RF energy can travel down the bore of the magnet in a very high field whole body MRI system by coupling to waveguide modes, meaning that energy can be distributed far away from

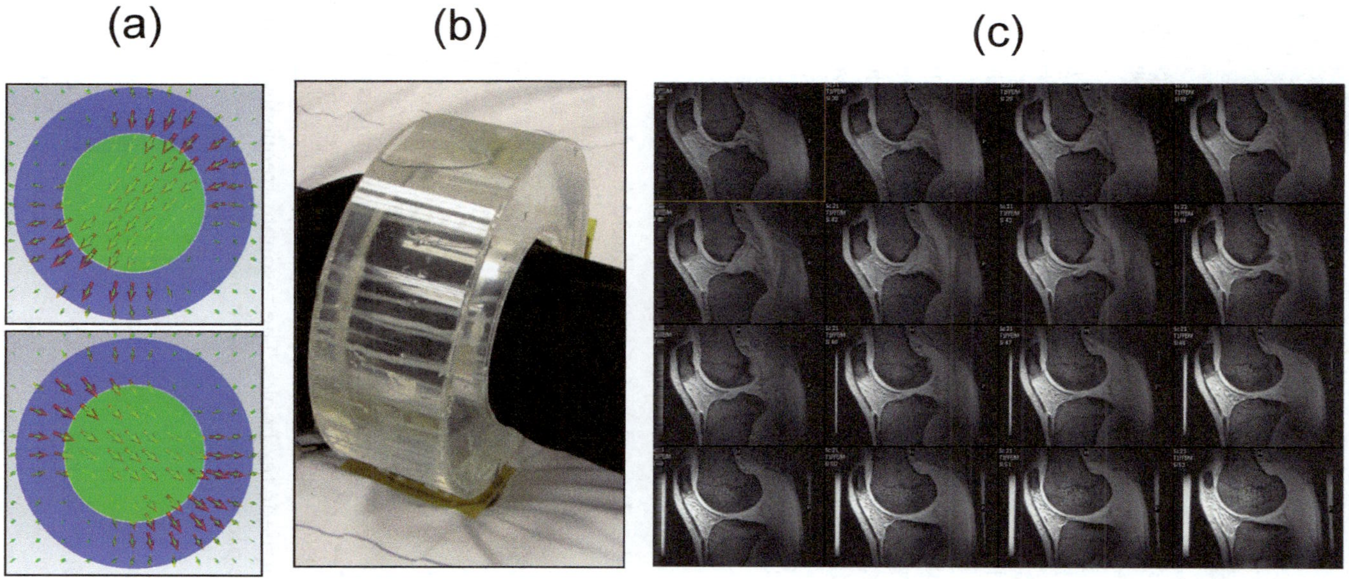

Figure 3.57 (a) Electromagnetic simulations of the B_1^+ fields corresponding to the two frequency-degenerate HEM_{11} modes of a water annulus (blue) filled with tissue (green). (b) Photograph of the annular DR with two critically coupled tuned loops for a 50 Ω impedance match. (c) *In vivo* images of a volunteer's knee acquired with the DR at 7 tesla.

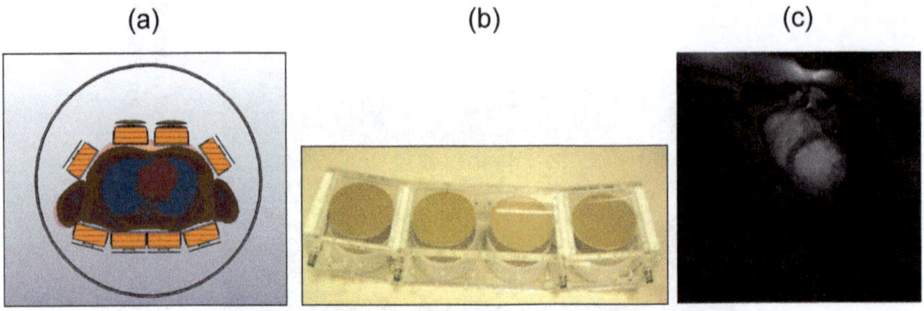

Figure 3.58 (a) Positioning of eight cylindrical DRs around the body for cardiac imaging at 7 tesla. (b) Photograph of an eight-element transmit/receive array constructed from high permittivity discs. (c) *In vivo* cardiac images acquired from a healthy volunteer with the DR array.

the physical location of the coil. This phenomenon forms the basis of a method for large field-of-view imaging[177] using a remote RF antenna specifically designed to produce such a "travelling" wave that propagates through the magnet bore. The cut-off frequencies for the two lowest waveguide modes, the TE_{11} and TM_{01} modes, of the empty cylindrical magnet bore are given by:

$$f_{\text{cutoff},TE_{11}} = \frac{1.8412}{2\pi r}c, f_{\text{cutoff},TM_{01}} = \frac{2.408}{2\pi r}c \quad (3.51)$$

where r refers to the diameter of the RF shield inside the magnet bore. Although different manufacturers have slightly different configurations of the RF shield and gradients bore, a typical diameter of the RF shield in a whole-body magnet is ~58 cm, which corresponds to a cutoff frequency of the TE_{11} mode of 301 MHz, and that of the TM_{01} mode of 396 MHz. When a human subject is placed in the magnet, the cut-off frequencies decrease owing to the increase in the effective permittivity within the bore.

In order to excite the TE_{11} mode, a quadrature patch antenna can be used, with the design shown in Figure 3.59(a).[177–179] The antenna is typically placed at the entrance of the magnet bore, as shown in Figure 3.59(b). The size of the patch antenna can be determined according to design equations found in many electromagnetic textbooks. The resonant frequency of the relevant TM_{110} mode of the patch antenna is given by:

$$f_{TM_{110}^{z}} = \frac{1.8412c}{2\pi a_{\text{eff}}\sqrt{\varepsilon_r}} \quad (3.52)$$

where ε_r is the relative permittivity of the dielectric (usually plastic) between the patch and the ground plane, and a_{eff} is the effective radius of the circular patch, given by:

$$a_{\text{eff}} = a\left\{1 + \frac{2h}{\pi\varepsilon_r a}\left[\ln\frac{\pi a}{2h} + 1.7726\right]\right\}^{\frac{1}{2}} \quad (3.53)$$

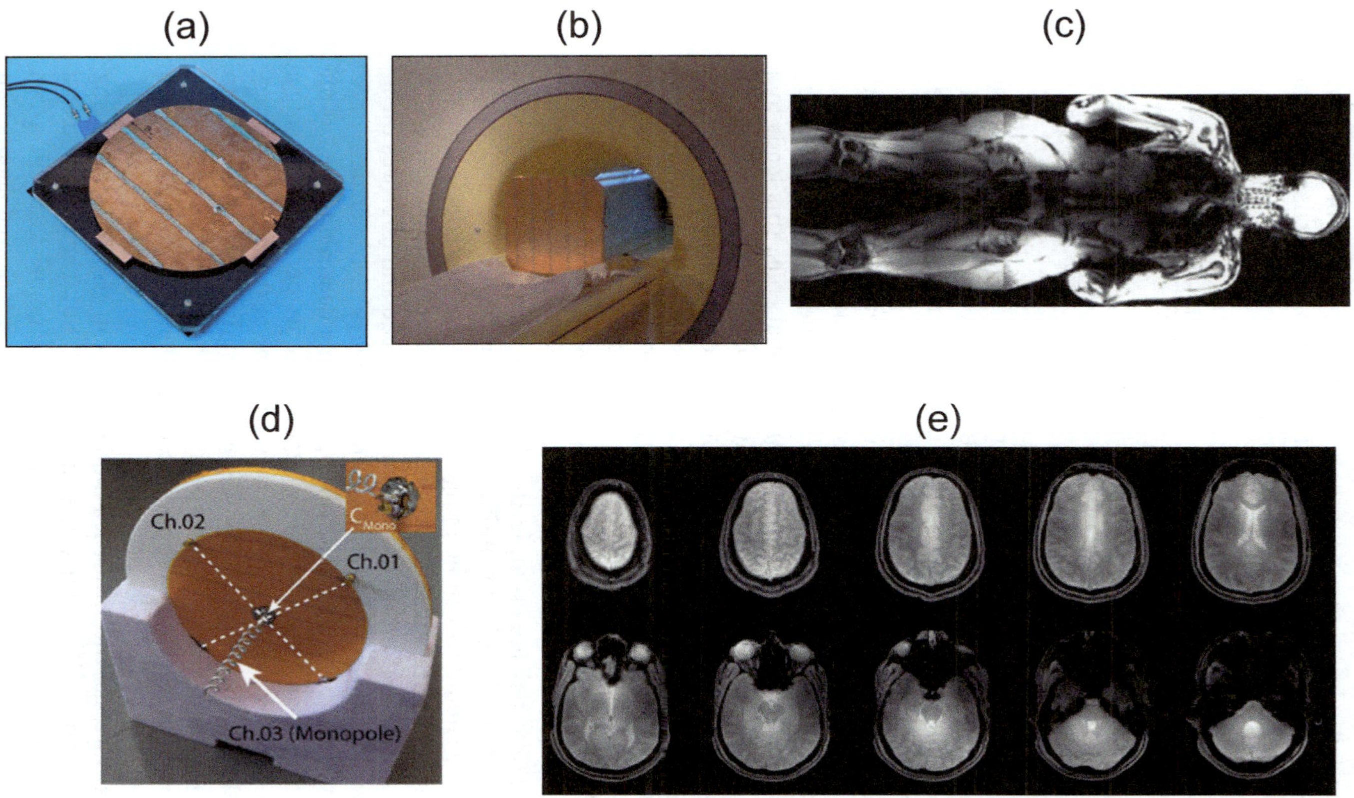

Figure 3.59 (a) Photograph of a quadrature patch antenna used for traveling wave excitation at 7 tesla. (b) Positioning of the patch antenna at the entrance of the magnet bore. (c) Multi-station low tip angle gradient echo images acquired at 7 tesla with the patch antenna in transmit/receive mode.[179] (d) A three-channel assembly consisting of quadrature patch antenna plus monopole used at 9.4 tesla. (e) Low tip angle gradient echo images acquired using the three-channel antenna and the TIAMO technique. Reprinted from ref. 185 by permission of John Wiley & Sons, Inc. © 2015 Wiley Periodicals, Inc.[185]

in which h is the thickness of the dielectric between the patch and the ground plane. For a given geometry there is a specific location at which the coaxial cable is connected, which corresponds to an input impedance of 50 Ω. This point can be derived from equations relating the input impedance to the radial distance ($\rho' = \rho_0$) from the centre of the patch:

$$R_{\text{in}}(\rho' = \rho_0) = R_{\text{in}}(\rho' = a_e)\frac{J_m^2(k\rho_0)}{J_m^2(ka_e)}$$
$$R_{\text{in}}(\rho' = a_e) = \frac{1}{G_t} \tag{3.54}$$

where G is the total conductance owing to radiation, conduction and dielectric losses, and $J_{\text{m}}(x)$ represents a Bessel function of order m. By attaching two cables at 90° to one another, the patch antenna can be driven in quadrature.

As the travelling wave generated by the patch antenna interacts with the different tissue permittivities and conductivities in the body, the pure TE mode structure of the traveling wave is altered and the energy is attenuated by the conducting tissue. As with conventional coils, this leads to a low image intensity in the centre of the body. A whole body image, shown in Figure 3.59(c), acquired using a patch antenna as both transmitter and receiver, illustrates this effect.[179]

There are many challenges with the basic travelling wave approach, which typically has a poor sensitivity compared to local coil excitation,[179] as one might expect. One problem is that abrupt changes in wave impedance between empty regions of the magnet bore and the body cause large amounts of energy to be reflected. For example, the shoulders cause strong reflections, which result in a longitudinal standing wave in the head. The impedance mismatch at the shoulders can be mitigated by putting material with a relative permittivity of ~40, similar to that of tissue, on top of the shoulders. This moves the region of mismatch away from the head and improves the B_1 homogeneity within the brain. Several other approaches have been developed to increase the sensitivity of the technique, including decreasing the cut-off frequency of the magnet bore by increasing the effective permittivity using a dielectric liner,[180] the addition of an extra waveguide at the end of the magnet,[181] and using multiple patch antennas as a transmit array.[182] A method to limit the absorption of energy in parts of the body that are not being imaged is to use a coaxial waveguide section placed between the antenna and the region of interest.[183]

At 400 MHz (9.4 tesla) the lowest TM mode, TM_{01}, can also be excited using a monopole antenna,[184,185] as illustrated in Figure 3.59(d). The patch antenna and monopole can effectively be used as a three channel transmit array. Using an imaging sequence known as TIAMO (time-interleaved acquisition of modes)[186] Hoffmann *et al.* were able to acquire relatively homogeneous images of the human brain using a combination of the three modes, as shown in Figure 3.59(e).

Appendix A

Coil Workbench Measurements Using a Network Analyzer

The network analyzer is the most common piece of equipment used for characterizing RF coils: simple examples are for frequency tuning and impedance matching, and for decoupling adjacent coils in an array. There are a number of ports (2–8 typically) that can be used for measurements, where one coil can be connected to each port. In this section, a simple two-port network analyzer, shown in Figure 3.60, is considered since it is the most commonly used device owing mainly to economic reasons. The network analyzer essentially measures transmitted and reflected voltages and powers as a function of frequency for each of the two ports. The coil is connected to the network analyzer with a coaxial cable with characteristic impedance Z_0 (almost always 50 Ω) and the input impedance of the measurement port of the network analyzer is also Z_0. Common measurements include:

Reflection coefficient (Γ):

$$\Gamma = \frac{Z_{\text{coil}} - Z_0}{Z_{\text{coil}} + Z_0} \tag{3.55}$$

Return loss (RL):

$$\text{RL} = 10 \log \frac{P_{\text{refl}}}{P_{\text{in}}} = -20 \log \Gamma \tag{3.56}$$

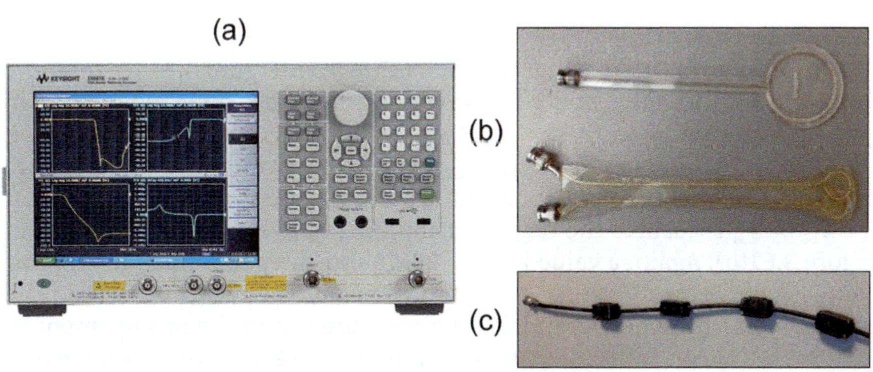

Figure 3.60 (a) Photograph of a two port network analyzer. (b) Pick-up loops used to measure the characteristics of different coils. (c) Coaxial cable to connect the network analyzer and RF coil: the use of external ferrites eliminates common mode currents on the outside of the cable shield.

Voltage standing wave ratio (VSWR):

$$\mathrm{VSWR} = \frac{1 + 10^{\frac{-\mathrm{RL}}{20}}}{1 - 10^{\frac{-\mathrm{RL}}{20}}} = \frac{1 + \Gamma}{1 - \Gamma} \tag{3.57}$$

If a single coil is connected to port 1 then a one-port measurement can be made to see how closely the coil is tuned and impedance matched to 50 Ω at the required frequency. If the impedance of the coil is very close to 50 Ω, for example 51 Ω, then the respective values are: $\Gamma = 0.0099$, RL = −40 dB, and VSWR = 1.02 : 1. In contrast, if the impedance matching is poor, for example the input impedance of the coil is 3 Ω, then the numbers become: $\Gamma = -0.89$, RL = −1 dB, and VSWR = 16.7 : 1.

The most commonly reported measurements made with a network analyzer are called *S*-parameters. These are covered in great detail in many electrical engineering textbooks, and so only a very brief description is given here. There are four basic measurements that can be made on a 2-port network analyzer, namely S_{11}, S_{22}, S_{12} and S_{21}, defined respectively on a log scale as:

$$\begin{aligned}
S_{11} &= 20 \log \frac{V_{1,\text{reflected}}}{V_{1,\text{incident}}} \\
S_{22} &= 20 \log \frac{V_{2,\text{reflected}}}{V_{2,\text{incident}}} \\
S_{12} &= 20 \log \frac{V_{2,\text{transmitted}}}{V_{1,\text{incident}}} \\
S_{21} &= 20 \log \frac{V_{1,\text{transmitted}}}{V_{2,\text{incident}}}
\end{aligned} \tag{3.58}$$

If only a single coil is being tested, as shown in Figure 3.61(a), then the S_{11} effectively measures how well the coil is impedance matched to 50 Ω: in general a value of lower than −20 dB is considered acceptable. A second coil can be attached to port 2 and the S_{22} measures how well this is impedance matched in exactly the same way. If the two coils are to form an array, then the S_{12} or S_{21} measures the degree of coupling between them, as shown in Figure 3.61(b). Again, a value below −20 dB indicates that there is very low coupling between the coils. If the properties of the coil, rather than the coil plus matching network, are to be measured then the arrangement in Figure 3.61(c) can be used, using two shielded pick-up loops, as shown in Figure 3.60.

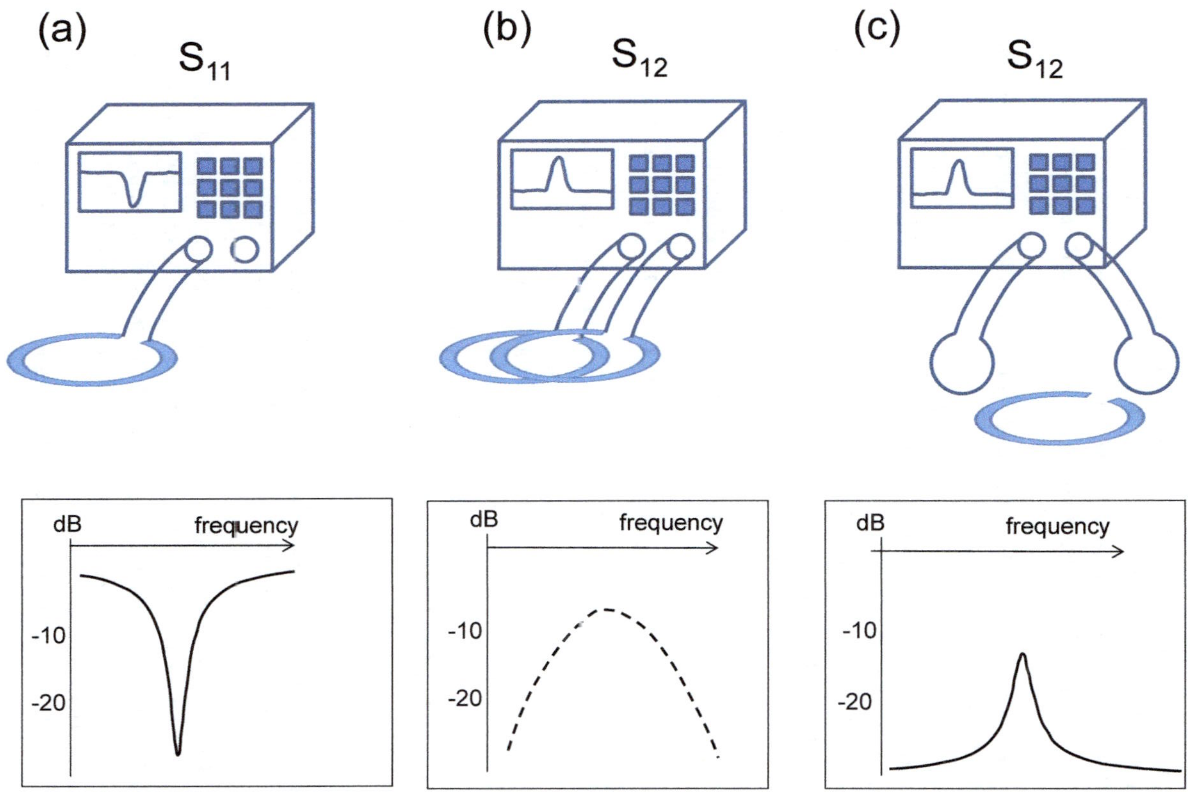

Figure 3.61 Schematics of simple measurements using a 2-port network analyzer. (a) S_{11} measurement on a single coil, (b) S_{12} or S_{21} measurement to assess the degree of coupling between two coils. (c) S_{12} measurement using two untuned pick-up loops to characterize the coil without the matching network.

References

1. J. T. Vaughan, *NMR Biomed.*, 2009, **22**, 907.
2. M. Schmitt, A. Potthast, D. E. Sosnovik, J. R. Polimeni, G. C. Wiggins, C. Triantafyllou and L. L. Wald, *Magn. Reson. Med.*, 2008, **59**, 1431.
3. A. Graessl, W. Renz, F. Hezel, M. A. Dieringer, L. Winter, C. Oezerdem, J. Rieger, P. Kellman, D. Santoro, T. D. Lindel, T. Frauenrath, H. Pfeiffer and T. Niendorf, *Magn. Reson. Med.*, 2014, **72**, 276.
4. P. T. Callaghan, C. D. Eccles and J. D. Seymour, *Rev. Sci. Instrum.*, 1997, **68**, 4263.
5. M. B. Kozlov, J. Haase, C. Baumann and A. G. Webb, *Solid State Nucl. Magn. Reson.*, 2005, **28**, 64.
6. A. G. Webb, *J. Magn. Reson.*, 2013, **229**, 55.
7. S. C. Grant, N. R. Aiken, H. D. Plant, S. Gibbs, T. H. Mareci, A. G. Webb and S. J. Blackband, *Magn. Reson. Med.*, 2000, **44**, 19.
8. S. C. Grant, D. L. Buckley, S. Gibb, A. G. Webb and S. J. Blackband, *Magn. Reson. Med.*, 2001, **46**, 1107.
9. H. Kovacs, D. Moskau and M. Spraul, *Prog. Nucl. Magn. Reson. Spectrosc.*, 2005, **46**, 131.
10. L. B. Andreas, T. Le Marchand, K. Jaudzems and G. Pintacuda, *J. Magn. Reson.*, 2015, **253**, 36.
11. F. D. Doty, G. Entzminger and Y. A. Yang, *Concepts Magn. Reson.*, 1998, **10**, 239.
12. J. L. Mispelter, M. Lupu and A. Briguet, *NMR Probeheads for biophysical and biomedical experiments*, Imperial College Press, 2nd edn, 2015.
13. J. T. Vaugan and J. R. Griffiths, *RF coils for MRI*, Wiley, 2012.
14. A. G. Webb, *Annu. Rep. NMR Spectrosc.*, 2006, **58**, 1.
15. F. D. Doty, G. Entzminger, J. Kulkarni, K. Pamarthy and J. P. Staab, *NMR Biomed.*, 2007, **20**, 304.
16. A. G. Webb, *Prog. Nucl. Magn. Reson. Spectrosc.*, 1997, **31**, 1.
17. D. I. Hoult and R. E. Richards, *J. Magn. Reson.*, 1976, **24**, 71.
18. D. I. Hoult, *Concepts Magn. Reson.*, 2000, **12**, 173.
19. G. H. Glover, C. E. Hayes, N. J. Pelc, W. A. Edelstein, O. M. Mueller, H. R. Hart, C. J. Hardy, M. Odonnell and W. D. Barber, *J. Magn. Reson.*, 1985, **64**, 255.
20. D. I. Hoult, C. N. Chen and V. J. Sank, *Magn. Reson. Med.*, 1984, **1**, 171.
21. B. H. Suits, A. N. Garroway and J. B. Miller, *J. Magn. Reson.*, 1998, **135**, 373.
22. B. Park, A. G. Webb and C. M. Collins, *J. Magn. Reson.*, 2009, **199**, 233.
23. B. Park, T. Neuberger, A. G. Webb, D. C. Bigler and C. M. Collins, *J. Magn. Reson.*, 2010, **202**, 72.
24. A. Krahn, U. Priller, L. Emsley and F. Engelke, *J. Magn. Reson.*, 2008, **191**, 78.
25. F. D. Doty, J. Kulkarni, C. Turner, G. Entzminger and A. Bielecki, *J. Magn. Reson.*, 2006, **182**, 239.
26. P. L. Gor'kov, E. Y. Chekmenev, C. G. Li, M. Cotten, J. J. Buffy, N. J. Traaseth, G. Veglia and W. W. Brey, *J. Magn. Reson.*, 2007, **185**, 77.

27. F. W. Grover, *Inductance Calculations*, Dover, Mineola, New York, USA, 2009.
28. A. Kumar, W. A. Edelstein and P. A. Bottomley, *Magn. Reson. Med.*, 2009, **61**, 1201.
29. J. Murphy-Boesch and A. P. Koretsky, *J. Magn. Reson.*, 1983, **54**, 526.
30. M. Bilgen, *Biomed. Eng. Online*, 2006, **5**, 3.
31. D. A. Seeber, I. Jevtic and A. Menon, *Concepts Magn. Reson., Part B*, 2004, **21B**, 26.
32. D. M. Peterson, B. L. Beck, C. R. Duensing and J. R. Fitzsimmons, *Concepts Magn. Reson., Part B*, 2003, **19B**, 1.
33. D. G. Gadian and F. N. H. Robinson, *J. Magn. Reson.*, 1979, **34**, 449.
34. M. D. Harpen, *Phys. Med. Biol.*, 1989, **34**, 1229.
35. M. D. Harpen, *Med. Phys.*, 1989, **16**, 781.
36. M. D. Harpen, *Phys. Med. Biol.*, 1989, **34**, 843.
37. M. D. Harpen, *Med. Phys.*, 1989, **16**, 234.
38. J. G. van Heteren, R. M. Henkelman and M. J. Bronskill, *Magn. Reson. Imaging*, 1987, **5**, 93.
39. T. W. Redpath and J. M. Hutchison, *Magn. Reson. Imaging*, 1984, **2**, 295.
40. D. W. Alderman and D. M. Grant, *J. Magn. Reson.*, 1979, **36**, 447.
41. D. M. Ginsberg and M. J. Melchner, *Rev. Sci. Instrum.*, 1970, **41**, 122.
42. C. E. Hayes, W. A. Edelstein, J. F. Schenck, O. M. Mueller and M. Eash, *J. Magn. Reson.*, 1985, **63**, 622.
43. M. C. Leifer, *J. Magn. Reson.*, 1997, **124**, 51.
44. W. H. Wong, S. Unno and W. Anderson, Millipede imaging coil design, *US Pat.* 6420871, 2002.
45. J. T. Vaughan, G. Adriany, C. J. Snyder, J. Tian, T. Thiel, L. Bolinger, H. Liu, L. DelaBarre and K. Ugurbil, *Magn. Reson. Med.*, 2004, **52**, 851.
46. J. T. Vaughan, H. P. Hetherington, J. O. Otu, J. W. Pan and G. M. Pohost, *Magn. Reson. Med.*, 1994, **32**, 206.
47. N. I. Avdievich, K. Bradshaw, A. M. Kuznetsov and H. P. Hetherington, *Magn. Reson. Med.*, 2007, **57**, 1190.
48. N. I. Avdievich, K. Bradshaw, J. H. Lee, A. M. Kuznetsov and H. P. Hetherington, *J. Magn. Reson.*, 2007, **187**, 234.
49. N. I. Avdievich and H. P. Hetherington, *NMR Biomed.*, 2009, **22**, 960.
50. N. I. Avdievich, H. P. Hetherington, A. M. Kuznetsov and J. W. Pan, *J. Magn. Reson. Imaging*, 2009, **29**, 461.
51. N. I. Avdievich, V. N. Krymov and H. P. Hetherington, *Magn. Reson. Med.*, 2003, **50**, 13.
52. L. L. Wald, G. C. Wiggins, A. Potthast, C. J. Wiggins and C. Triantafyllou, *Appl. Magn. Reson.*, 2005, **29**, 19.
53. G. C. Wiggins, A. Potthast, C. Triantafyllou, C. J. Wiggins and L. L. Wald, *Magn. Reson. Med.*, 2005, **54**, 235.
54. T. S. Ibrahim, C. Mitchell, R. Abraham and P. Schmalbrock, *NMR Biomed.*, 2007, **20**, 58.
55. M. G. Erickson, K. N. Kurpad, J. H. Holmes and S. B. Fain, *Magn. Reson. Med.*, 2007, **58**, 800.

56. C. S. Wang and G. X. Shen, *J. Magn. Reson. Imaging*, 2006, **24**, 439.
57. D. Ballon, M. C. Graham, S. Miodownik and J. A. Koutcher, *J. Magn. Reson.*, 1990, **90**, 131.
58. A. M. Hudson, W. Kockenberger and R. W. Bowtell, *MAGMA*, 2000, **10**, 69.
59. A. S. Peshkovsky, R. P. Kennan, M. E. Fabry and N. I. Avdievich, *Magn. Reson. Med.*, 2005, **53**, 937.
60. R. Medhurst, *Wireless Eng.*, 1947, **35**, 80.
61. G. A. Rinard and G. R. Eaton, *Biol. Magn. Reson.*, 2005, **24**, 19.
62. J. S. Hyde, W. Froncisz and T. Oles, *J. Magn. Reson.*, 1989, **82**, 223.
63. W. Froncisz, T. Oles and J. S. Hyde, *J. Magn. Reson.*, 1989, **82**, 109.
64. M. F. Koskinen and K. R. Metz, *J. Magn. Reson.*, 1992, **98**, 576.
65. J. P. Hornak, J. Szumowski, D. Rubens, J. Janus and R. G. Bryant, *Radiology*, 1986, **161**, 832.
66. C. V. Grant, C. H. Wu and S. J. Opella, *J. Magn. Reson.*, 2010, **204**, 180.
67. T. Nakada, I. L. Kwee, T. Miyazaki, N. Iriguchi and T. Maki, *Magn. Reson. Med.*, 1987, **5**, 449.
68. A. Tannus and M. Garwood, *NMR Biomed.*, 1997, **10**, 423.
69. A. Kumar and P. A. Bottomley, *Magn. Reson. Mater. Phys.*, 2008, **21**, 41.
70. P. B. Roemer, W. A. Edelstein, C. E. Hayes, S. P. Souza and O. M. Mueller, *Magn. Reson. Med.*, 1990, **16**, 192.
71. N. I. Avdievich, J. W. Pan and H. P. Hetherington, *NMR Biomed.*, 2013, **26**, 1547.
72. E. A. Barberi, J. S. Gati, B. K. Rutt and R. S. Menon, *Magn. Reson. Med.*, 2000, **43**, 284.
73. D. K. Sodickson and W. J. Manning, *Magn. Reson. Med.*, 1997, **38**, 591.
74. D. K. Sodickson and C. A. McKenzie, *Med. Phys.*, 2001, **28**, 1629.
75. M. Blaimer, F. Breuer, M. Mueller, R. M. Heidemann, M. A. Griswold and P. M. Jakob, *Top. Magn. Reson. Imaging*, 2004, **15**, 223.
76. R. M. Heidemann, O. Ozsarlak, P. M. Parizel, J. Michiels, B. Kiefer, V. Jellus, M. Muller, F. Breuer, M. Blaimer, M. A. Griswold and P. M. Jakob, *Eur. Radiol.*, 2003, **13**, 2323.
77. M. A. Griswold, P. M. Jakob, R. M. Heidemann, M. Nittka, V. Jellus, J. Wang, B. Kiefer and A. Haase, *Magn. Reson. Med.*, 2002, **47**, 1202.
78. K. P. Pruessmann, M. Weiger, M. B. Scheidegger and P. Boesiger, *Magn. Reson. Med.*, 1999, **42**, 952.
79. S. M. Wright, R. L. Magin and J. R. Kelton, *Magn. Reson. Med.*, 1991, **17**, 252.
80. S. M. Wright and L. L. Wald, *NMR Biomed.*, 1997, **10**, 394.
81. M. D. Schnall, V. H. Subramanian, J. S. Leigh and B. Chance, *J. Magn. Reson.*, 1985, **65**, 122.
82. M. D. Schnall, V. H. Subramanian, J. S. Leigh, L. Gyulai, A. Mclaughlin and B. Chance, *J. Magn. Reson.*, 1985, **63**, 401.
83. J. R. Fitzsimmons, H. R. Brooker and B. Beck, *Magn. Reson. Med.*, 1987, **5**, 471.
84. M. Alecci, S. Romanzetti, J. Kaffanke, A. Celik, H. P. Wegener and N. J. Shah, *J. Magn. Reson.*, 2006, **181**, 203.

85. A. R. Rath, *J. Magn. Reson.*, 1990, **86**, 488.
86. G. X. Shen, F. E. Boada and K. R. Thulborn, *Magn. Reson. Med.*, 1997, **38**, 717.
87. G. B. Matson, P. Vermathen and T. C. Hill, *Magn. Reson. Med.*, 1999, **42**, 173.
88. S. Amari, A. M. Ulug, J. Bornemann, P. C. van Zijl and P. B. Barker, *Magn. Reson. Med.*, 1997, **37**, 243.
89. R. C. Brand, A. G. Webb and J. W. M. Beenakker, *Concepts Magn. Reson., Part A*, 2013, **42**, 155.
90. L. F. Fuks, F. S. C. Huang, C. M. Carter, W. A. Edelstein and P. B. Roemer, *J. Magn. Reson.*, 1992, **100**, 229.
91. F. D. Doty, G. Entzminger and Y. A. Yang, *Concepts Magn. Reson.*, 1998, **10**, 133.
92. N. Soffe, J. Boyd and M. Leonard, *J. Magn. Reson., Ser. A*, 1995, **116**, 117.
93. F. O. Zelaya, S. Crozier, S. Dodd, R. Mckenna and D. M. Doddrell, *J. Magn. Reson., Ser. A*, 1995, **115**, 131.
94. F. Hwang and D. I. Hoult, *Magn. Reson. Med.*, 1998, **39**, 214.
95. K. F. Morris and C. S. Johnson, *J. Am. Chem. Soc.*, 1993, **115**, 4291.
96. G. Wider, R. Riek and K. Wuthrich, *J. Am. Chem. Soc.*, 1996, **118**, 11629.
97. A. G. Webb, Microcoils, in *RF Coils for MRI*, ed. J. T. Vaughan and J. R. Griffiths, Wiley, Chichester, UK, 2012, p. 225.
98. D. L. Olson, T. L. Peck, A. G. Webb, R. L. Magin and J. V. Sweedler, *Science*, 1995, **270**, 1967.
99. D. L. Olson, J. A. Norcross, M. O'Neil-Johnson, P. F. Molitor, D. J. Detlefsen, A. G. Wilson and T. L. Peck, *Anal. Chem.*, 2004, **76**, 2966.
100. T. L. Peck, R. L. Magin and P. C. Lauterbur, *J. Magn. Reson., Ser. B*, 1995, **108**, 114.
101. C. Barmet, N. De Zanche and K. P. Pruessmann, *Magn. Reson. Med.*, 2008, **60**, 187.
102. N. De Zanche, C. Barmet, J. A. Nordmeyer-Massner and K. P. Pruessmann, *Magn. Reson. Med.*, 2008, **60**, 176.
103. A. P. M. Kentgens, J. Bart, P. J. M. van Bentum, A. Brinkmann, E. R. H. Van Eck, J. G. E. Gardeniers, J. W. G. Janssen, P. Knijn, S. Vasa and M. H. W. Verkuijlen, *J. Chem. Phys.*, 2008, **128**, 052202.
104. Y. Maguire, I. L. Chuang, S. G. Zhang and N. Gershenfeld, *Proc. Natl. Acad. Sci. U. S. A.*, 2007, **104**, 9198.
105. K. Takeda, *Solid State Nucl. Magn. Reson.*, 2012, **47–48**, 1.
106. O. Gokay and K. Albert, *Anal. Bioanal. Chem.*, 2012, **402**, 647.
107. R. M. Fratila and A. H. Velders, *Annu. Rev. Anal. Chem.*, 2011, **4**, 227.
108. F. C. Schroeder and M. Gronquist, *Angew. Chem., Int. Ed.*, 2006, **45**, 7122.
109. E. R. Andrew, A. Bradbury and R. G. Eades, *Nature*, 1958, **182**, 1659.
110. E. R. Andrew and R. A. Newing, *Proc. Phys. Soc., London*, 1958, **72**, 959.
111. F. D. Doty, *eMagRes*, 2007, DOI: 10.1002/9780470034590.emrstm0515.pub2.
112. S. A. McNeill, P. L. Gor'kov, K. Shetty, W. W. Brey and J. R. Long, *J. Magn. Reson.*, 2009, **197**, 135.

113. W. E. Maas, A. Bielecki, M. Ziliox, F. H. Laukien and D. G. Cory, *J. Magn. Reson.*, 1999, **141**, 29.
114. W. E. Maas, F. H. Laukien and D. G. Cory, *J. Am. Chem. Soc.*, 1996, **118**, 13085.
115. P. L. Gor'kov, R. Witter, E. Y. Chekmenev, F. Nozirov, R. Fu and W. W. Brey, *J. Magn. Reson.*, 2007, **189**, 182.
116. J. A. Stringer, C. E. Bronnimann, C. G. Mullen, D. H. Zhou, S. A. Stellfox, Y. Li, E. H. Williams and C. M. Rienstra, *J. Magn. Reson.*, 2005, **173**, 40.
117. P. Styles, N. F. Soffe, C. A. Scott, D. A. Cragg, F. Row, D. J. White and P. C. J. White, *J. Magn. Reson.*, 1984, **60**, 397.
118. M. Jeroschherold and R. K. Kirschman, *J. Magn. Reson.*, 1989, **85**, 141.
119. T. M. Logan, N. Murali, G. S. Wang and C. Jolivet, *Magn. Reson. Chem.*, 1999, **37**, 762.
120. W. A. Anderson, W. W. Brey, A. L. Brooke, B. Cole, K. A. Delin, L. F. Fuks, H. D. W. Hill, M. E. Johanson, V. K. Kotsubo, R. E. Nast, R. S. Withers and W. H. Wong, *Bull. Magn. Reson.*, 1995, **17**, 98.
121. A. E. Kelly, H. D. Ou, R. Withers and V. Dotsch, *J. Am. Chem. Soc.*, 2002, **124**, 12013.
122. A. T. Dossey, M. Gottardo, J. M. Whitaker, W. R. Roush and A. S. Edison, *J. Chem. Ecol.*, 2009, **35**, 861.
123. A. T. Dossey, S. S. Walse, J. R. Rocca and A. S. Edison, *ACS Chem. Biol.*, 2006, **1**, 511.
124. T. F. Molinski, *Curr. Opin. Biotechnol.*, 2010, **21**, 819.
125. W. W. Brey, A. S. Edison, R. E. Nast, J. R. Rocca, S. Saha and R. S. Withers, *J. Magn. Reson.*, 2006, **179**, 290.
126. V. Ramaswamy, J. W. Hooker, R. S. Withers, R. E. Nast, A. S. Edison and W. W. Brey, *eMagRes*, 2013, **2**, 215.
127. V. Ramaswamy, J. W. Hooker, R. S. Withers, R. E. Nast, W. W. Brey and A. S. Edison, *J. Magn. Reson.*, 2013, **235**, 58.
128. F. D. Doty, US 7498812 B2, 2009.
129. S. Crozier, K. Luescher, L. K. Forbes and D. M. Doddrell, *J. Magn. Reson., Ser. B*, 1995, **109**, 1.
130. F. D. Doty, G. Entzminger and C. D. Hauck, *J. Magn. Reson.*, 1999, **140**, 17.
131. A. R. Rath, *Magn. Reson. Med.*, 1990, **13**, 370.
132. V. D. Kodibagkar and M. S. Conradi, *J. Magn. Reson.*, 2000, **144**, 53.
133. R. D. Black, T. A. Early and G. A. Johnson, *J. Magn. Reson., Ser. A*, 1995, **113**, 74.
134. R. D. Black, T. A. Early, P. B. Roemer, O. M. Mueller, A. Mogrocampero, L. G. Turner and G. A. Johnson, *Science*, 1993, **259**, 793.
135. R. D. Black, P. B. Roemer, A. Mogrocampero, L. G. Turner and K. W. Rohling, *Appl. Phys. Lett.*, 1993, **62**, 771.
136. S. E. Hurlston, W. W. Brey, S. A. Suddarth and G. A. Johnson, *Magn. Reson. Med.*, 1999, **41**, 1032.
137. L. Darrasse and J. C. Ginefri, *Biochimie*, 2003, **85**, 915.
138. A. C. Wright, H. K. Song and F. W. Wehrli, *Magn. Reson. Med.*, 2000, **43**, 163.

139. S. Junge, Cryogenic and superconducting coils for MRI, in *RF coils for MRI*, ed. J. T. Vaughan and J. R. Griffiths, Wiley, Chichester, UK, 2012, p. 233.
140. C. Baltes, N. Radzwill, S. Bosshard, D. Marek and M. Rudin, *NMR Biomed.*, 2009, **22**, 834.
141. J. B. Aguayo, S. J. Blackband, J. Schoeniger, M. A. Mattingly and M. Hinterman, *Nature.*, 1986, **322**, 190.
142. R. E. Jacobs and S. E. Fraser, *Science.*, 1994, **263**, 681.
143. Z. H. Cho, C. B. Ahn, S. C. Juh, H. K. Lee, R. E. Jacobs, S. Lee, J. H. Yi and J. M. Jo, *Med. Phys.*, 1988, **15**, 815.
144. P. T. Callaghan and C. D. Eccles, *J. Magn. Reson.*, 1988, **78**, 1.
145. L. Ciobanu, A. G. Webb and C. H. Pennington, *Prog. Nucl. Magn. Reson. Spectrosc.*, 2003, **42**, 69.
146. L. Ciobanu, D. A. Seeber and C. H. Pennington, *J. Magn. Reson.*, 2002, **158**, 178.
147. S. C. Lee, K. Kim, J. Kim, S. Lee, J. H. Yi, S. W. Kim, K. S. Ha and C. Cheong, *J. Magn. Reson.*, 2001, **150**, 207.
148. T. L. Peck, R. L. Magin, J. Kruse and M. Feng, *IEEE Trans. Biomed. Eng.*, 1994, **41**, 706.
149. J. J. Flint, B. Hansen, M. Fey, D. Schmidig, M. A. King, P. Vestergaard-Poulsen and S. J. Blackband, *NeuroImage*, 2010, **52**, 556.
150. J. J. Flint, B. Hansen, S. Portnoy, C. H. Lee, M. A. King, M. Fey, F. Vincent, G. J. Stanisz, P. Vestergaard-Poulsen and S. J. Blackband, *NeuroImage*, 2012, **60**, 1404.
151. C. H. Lee, S. J. Blackband and P. Fernandez-Funez, *Sci. Rep.*, 2015, **5**, 8920.
152. M. L. Banson, G. P. Cofer, R. Black and G. A. Johnson, *Invest. Radiol.*, 1992, **27**, 157.
153. P. M. Glover, R. W. Bowtell, G. D. Brown and P. Mansfield, *Magn. Reson. Med.*, 1994, **31**, 423.
154. R. J. A. Nabuurs, I. Hegeman, R. Natte, S. G. van Duinen, M. A. van Buchem, L. van der Weerd and A. G. Webb, *NMR Biomed.*, 2011, **24**, 351.
155. R. J. A. Nabuurs, R. Natte, F. M. de Ronde, I. Hegeman-Kleinn, J. Dijkstra, S. G. van Duinen, A. G. Webb, A. J. Rozemuller, M. A. van Buchem and L. van der Weerd, *J. Alzheimers Dis.*, 2013, **34**, 1037.
156. M. D. Meadowcroft, J. R. Connor, M. B. Smith and Q. X. Yang, *J. Magn. Reson. Imaging*, 2009, **29**, 997.
157. M. D. Meadowcroft, S. T. Zhang, W. Z. Liu, B. S. Park, J. R. Connor, C. M. Collins, M. B. Smith and Q. X. Yang, *Magn. Reson. Med.*, 2007, **57**, 835.
158. J. T. Vaughan, C. J. Snyder, L. J. DelaBarre, P. J. Bolan, J. Tian, L. Bolinger, G. Adriany, P. Andersen, J. Strupp and K. Ugurbil, *Magn. Reson. Med.*, 2009, **61**, 244.
159. P. F. van de Moortele, C. Akgun, G. Adriany, S. Moeller, J. Ritter, C. M. Collins, M. B. Smith, J. T. Vaughan and K. Ugurbil, *Magn. Reson. Med.*, 2005, **54**, 1503.
160. G. J. Metzger, C. Snyder, C. Akgun, T. Vaughan, K. Ugurbil and P. F. van de Moortele, *Magn. Reson. Med.*, 2008, **59**, 396.

161. G. Adriany, P. F. van de Moortele, F. Wiesinger, S. Moeller, J. P. Strupp, P. Andersen, C. Snyder, X. Zhang, W. Chen, K. P. Pruessmann, P. Boesiger, T. Vaughan and K. Ugurbil, *Magn. Reson. Med.*, 2005, **53**, 434.
162. R. F. Lee, C. J. Hardy, D. K. Sodickson and P. A. Bottomley, *Magn. Reson. Med.*, 2004, **51**, 172.
163. R. F. Lee, C. R. Westgate, R. G. Weiss, D. C. Newman and P. A. Bottomley, *Magn.Reson.Med.*, 2001, **45**, 673.
164. X. Z. Zhang and A. Webb, *J. Magn. Reson.*, 2004, **170**, 149.
165. G. Adriany, P. F. van de Moortele, J. Ritter, S. Moeller, E. J. Auerbach, C. Akgun, C. J. Snyder, T. Vaughan and K. Ugurbil, *Magn. Reson. Med.*, 2008, **59**, 590.
166. C. J. Snyder, L. DelaBarre, G. J. Metzger, P. F. van de Moortele, C. Akgun, K. Ugurbil and J. T. Vaughan, *Magn. Reson. Med.*, 2009, **61**, 517.
167. J. Hoffmann, G. Shajan, K. Scheffler and R. Pohmann, *Magn. Reson. Mater. Phys.*, 2014, **27**, 373.
168. G. Shajan, M. Kozlov, J. Hoffmann, R. Turner, K. Scheffler and R. Pohmann, *Magn. Reson. Med.*, 2014, **71**, 870.
169. J. R. Porter, S. M. Wright and A. Reykowski, *Magn. Reson. Med.*, 1998, **40**, 272.
170. G. J. Metzger, E. J. Auerbach, C. Akgun, J. Simonson, X. M. Bi, K. Ugurbil and P. F. van de Moortele, *Magn. Reson. Med.*, 2013, **69**, 114.
171. D. Kajfez and P. Guillon, *Dielectric Resonators*, SciTech Publishing, London, UK, 1998.
172. S. A. Aussenhofer and A. G. Webb, *Magn. Reson. Med.*, 2012, **68**, 1325.
173. S. A. Aussenhofer and A. G. Webb, *NMR Biomed.*, 2013, **26**, 1555.
174. K. Haines, T. Neuberger, M. Lanagan, E. Semouchkina and A. G. Webb, *J. Magn. Reson.*, 2009, **200**, 349.
175. T. Neuberger, V. Tyagi, E. Semouchkina, M. Lanagan, A. Baker, K. Haines and A. G. Webb, *Concepts Magn. Reson., Part B*, 2008, **33B**, 109.
176. S. A. Aussenhofer and A. G. Webb, *J. Magn. Reson.*, 2014, **243**, 122.
177. D. O. Brunner, Z. N. De, J. Frohlich, J. Paska and K. P. Pruessmann, *Nature*, 2009, **457**, 994.
178. A. Webb, *Prog. Nucl. Magn. Reson. Spectrosc.*, 2014, **83**, 1.
179. B. Zhang, D. K. Sodickson, R. Lattanzi, Q. Duan, B. Stoeckel and G. C. Wiggins, *Magn. Reson. Med.*, 2012, **67**, 1183.
180. A. Andreychenko, J. J. Bluemink, A. J. Raaijmakers, J. J. Lagendijk, P. R. Luijten and C. A. van den Berg, *Magn. Reson. Med.*, 2013, **70**, 885.
181. A. Andreychenko, H. Kroeze, V. O. Boer, J. J. Lagendijk, P. R. Luijten and C. A. van den Berg, *Magn. Reson. Med.*, 2014, **71**, 1641.
182. Y. Pang, D. B. Vigneron and X. Zhang, *Magn. Reson. Med.*, 2012, **67**, 965.
183. A. Andreychenko, H. Kroeze, D. W. Klomp, J. J. Lagendijk, P. R. Luijten and C. A. van den Berg, *Magn. Reson. Med.*, 2012, **70**, 875.
184. F. H. Geschewski, D. Brenner, J. Felder and N. J. Shah, *Magn. Reson. Med.*, 2013, **69**, 1805.
185. J. Hoffmann, C. Mirkes, G. Shajan, K. Scheffler and R. Pohmann, *Magn. Reson. Med.*, 2016, **75**, 452.

186. S. Orzada, S. Maderwald, B. A. Poser, A. K. Bitz, H. H. Quick and M. E. Ladd, *Magn. Reson. Med.*, 2010, **64**, 327.
187. J. Murphy-Boesch, R. Srinivasan, L. Carvajal and T. R. Brown, *J. Magn. Reson., Ser. B*, 1994, **103**, 103.
188. O. G. Gruschke, N. Baxan, L. Clad, K. Kratt, D. von Elverfeldt, A. Peter, J. Hennig, V. Badilita, U. Wallrabe and J. G. Korvink, *Lab Chip*, 2012, **12**, 495.
189. W. M. Brink, V. Gulani and A. G. Webb, *J. Magn. Reson. Imaging.*, 2015, **42**, 855.

CHAPTER 4

B_0 *Shimming Technology*

ROBIN A. DE GRAAF*[a,b] AND CHRISTOPH JUCHEM[a,c]

[a]Magnetic Resonance Research Center, Yale University, School of Medicine, Department of Radiology and Biomedical Imaging, 300 Cedar Street, New Haven, CT 06520, USA; [b]Magnetic Resonance Research Center, Yale University, School of Medicine, Biomedical Engineering, 300 Cedar Street, New Haven, CT 06520, USA; [c]Magnetic Resonance Research Center, Yale University, School of Medicine, Neurology, 300 Cedar Street, New Haven, CT 06520, USA
*E-mail: robin.degraaf@yale.edu

4.1 Introduction

Spatial homogeneity of the main magnetic field B_0 is essential for the majority of MR applications. In high resolution NMR, spectral linewidths of small molecules in solution are less than 1 Hz, representing a magnetic field homogeneity within the sample of ~1 part-per-billion (ppb). Wider linewidths result in overlapping resonances and a lower SNR. In MRI, the presence of magnetic field inhomogeneity can lead to image distortion and signal loss, whereas spatial magnetic field variations in localized MR spectroscopy (MRS) lead to loss of sensitivity and spectral resolution. Besides manufacturing imperfections, such as minute variations in magnet coil windings, and environmental effects, such as the presence of large (metallic) objects (structural beams, laboratory equipment, gradient coils) close to the magnet, the majority of magnetic field imperfections are sample-induced. Variations in magnetic susceptibility between materials or tissue types lead to a disturbance of the surrounding and internal magnetic fields. In the human head,

New Developments in NMR No. 7
Magnetic Resonance Technology: Hardware and System Component Design
Edited by Andrew G Webb

Published by the Royal Society of Chemistry, www.rsc.org

for example, the largest differences in magnetic susceptibility occur between brain tissue and air in the nasal and auditory passages leading to strong magnetic field distortions in the frontal cortex and temporal lobes, respectively.

For more than 50 years, the standard approach to minimize magnetic field variations has been to superimpose secondary magnetic fields with a spatial variation governed by spherical harmonic (SH) functions (Chapter 1, Appendix B) in a process referred to as 'shimming'.[1–4] The term 'shimming' originates from the small pieces ('shims') of magnetic material that were used in the early days of NMR in order to improve the homogeneity of the magnetic field produced by pole magnets. Even though on modern day magnets the subject-specific magnetic field homogeneity is adjusted primarily using electromagnetic coils, the term 'shimming' remains. SH-based shimming has been the standard for so many decades because it performs admirably as a method of optimizing the magnetic field homogeneity across a 'bare' magnet devoid of a sample, or a vertical-bore magnet with a sample of uniform composition and geometry (*e.g.* a 5 mm NMR tube). However, the complex magnetic field distribution encountered throughout the human body challenges the capabilities of SH-based shimming. In addition, the limited availability of higher-order SH shim coils on most *in vivo* MR systems leads to compromised magnetic field homogeneity in many parts of the human body. As a result, the last decade has seen a resurgence in the development of shimming approaches more tailored towards *in vivo* systems and that are not necessarily based on SH functions.

This chapter will give an overview of shimming-related concepts and technology. Following a description of the origins of magnetic field inhomogeneity, the theory and practical implementation of SH shimming will be presented. Magnetic field mapping will be discussed as a crucial ingredient for optimal shimming. The flexibility of SH shimming can be substantially enhanced by sequentially employing optimal shim settings for different volumes, such as adjacent slices in multi-slice MRI, in a technique referred to as dynamic SH shimming. The theory and practical considerations of dynamic SH shimming are discussed in detail. The limitations of SH-based shimming have led to the development of a number of shimming approaches more tailored towards homogenizing the magnetic field across the human brain. Following an overview of these methods, the most successful techniques based on extended grids of uniform current loops will be discussed in more detail. The focus will be on the practical aspects of subject-specific *in vivo* shimming, with an emphasis on human brain applications.

4.2 The Origins of Magnetic Field Inhomogeneity

The homogeneity of the static B_0 magnetic field is, in addition to its strength, the most important parameter characterizing an MR magnet. It can be shown that a magnet designed as an infinitely long solenoid will produce a perfectly homogeneous magnetic field. Short truncated solenoidal coils with correction coils at either end, or optimized coil configurations or wire patterns, are the practical implementation of this theoretical solution, as described in Chapter 2. Figure 4.1A/B shows the magnetic field generated by a commonly employed 6-ring magnet. The magnetic field homogeneity is optimized over a diameter-of-spherical-volume (DSV), which is typically of the order of 40–50% of the magnet bore diameter. From Figure 4.1A/B it follows that a carefully designed magnet produced without manufacturing errors can generate a magnetic field with a homogeneity of better than 1 ppm over its DSV. However, minor manufacturing errors in the placement of tens of miles of superconducting wire for a typical magnet can have a devastating effect on the magnetic field homogeneity. An error of 0.01% in the current density of one of the outer rings will lead to a magnetic field inhomogeneity of >10 ppm over the DSV (Figure 4.1C). Similarly, large (metallic) objects placed close to the magnet (structural beams, laboratory equipment) or even inside the magnet (gradient and shim coils, patient bed) can greatly perturb the magnetic field homogeneity. The magnetic field inhomogeneity without a sample or subject is typically optimized once during MR system installation, as covered in Chapter 2. In contrast, magnetic field inhomogeneity generated by the sample is something that all MR users are affected by and will be the primary focus of this chapter.

In a perfectly homogeneous magnetic field B_0, the total magnetic field B_{total} inside a continuous and homogeneous material is given by

$$B_{total} = B_0 + \mu_0 M, \text{ with } M = (\chi/\mu_0)B_0 \tag{4.1}$$

where B_0 is the magnetic field in a vacuum and M represents the magnetization induced inside the material. The amount of magnetization that a material can acquire is proportional to the magnetic susceptibility, χ, a dimensionless parameter that describes how the magnetic permeability μ of the material deviates from the vacuum permeability μ_0 according to $\chi = (\mu/\mu_0) - 1$. From an MR point of view, materials can be classified based on the sign and magnitude of the magnetic susceptibility. Materials with a negative susceptibility are referred to as diamagnetic materials and decrease the magnetic field inside the material. In high resolution NMR, all of the commonly used deuterated solvents are diamagnetic. Copper, which is used to form the RF coil, is also diamagnetic. Since water is diamagnetic ($\chi = -9.05 \times 10^{-6} = -9.05$ ppm), most tissues have a negative magnetic susceptibility between −7 and −11 ppm.[3] Materials with a positive susceptibility are referred to as paramagnetic materials and increase the magnetic field inside the material. The most commonly encountered paramagnetic material is air ($\chi = +0.36 \times 10^{-6} = 0.36$ ppm), whereas other paramagnetic materials, such as titanium and chromium, may be encountered as part of metallic prostheses. Neither diamagnetic nor paramagnetic materials retain their magnetic properties when the external magnetic field is removed. In contrast, ferromagnetic materials have a very large, positive magnetic susceptibility and retain some of their magnetic properties even when the external magnetic field is removed. Although used for passive shimming the bare magnet,[4] ferromagnetic materials are generally incompatible with MRI applications.

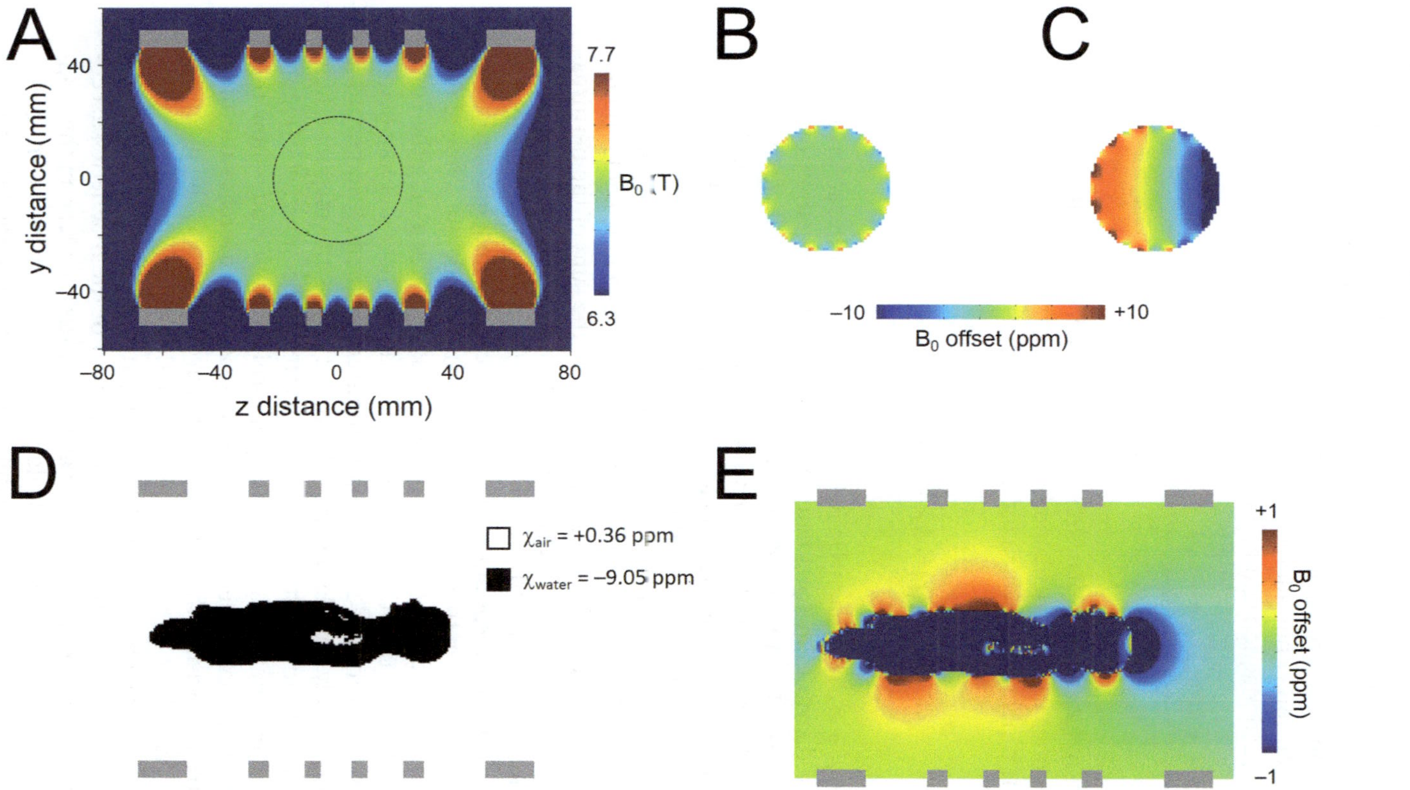

Figure 4.1 Origins of magnetic field inhomogeneity. (A) 7 T magnetic field (outer diameter 100 cm) generated by a commonly used 6-coil design. (B) Across the diameter spherical volume (DSV) of 45 cm the magnetic field homogeneity is better than 1 ppm. (C) When the current in the outermost right coil is 99.99% of its nominal value, the magnetic field homogeneity deteriorates to more than 10 ppm. (D) Simplified magnetic susceptibility distribution of the human body solely composed of water and air. (E) Magnetic field induced by the magnetic susceptibility distribution shown in (D). Note the reduced magnetic field inside the human body owing to the diamagnetic nature of water.

The presence of a homogeneous material will, according to eqn (4.1), slightly alter the magnitude of the magnetic field but will not lead to magnetic field inhomogeneity *per se*. Magnetic field inhomogeneity is created at magnetic susceptibility boundaries, for example between air and tissue. The change in induced magnetization, ΔM, is given by:

$$\Delta M = \left(\frac{\chi_{\text{tissue}}}{\mu_0}\right)B_0 - \left(\frac{\chi_{\text{air}}}{\mu_0}\right)B_0 = \left(\frac{\Delta\chi}{\mu_0}\right)B_0 \tag{4.2}$$

The spatial magnetic field distribution that is generated from this magnetization difference can be calculated by considering the *z*-component of the unit magnetic dipole field,[3] which for a volume element ΔV gives

$$\Delta B_z(r) = \Delta\chi \frac{B_0}{4\pi} \frac{3z^2 - r^2}{r^5} \Delta V \tag{4.3}$$

with $r^2 = x^2 + y^2 + z^2$. The magnetic field generated by an arbitrary magnetic susceptibility distribution $\Delta\chi(r)$ can be calculated by integration of $\Delta B_z(r)$ over all positions *r*. Figure 4.1D shows a magnetic susceptibility distribution map of a simplified human body composed only of air and water. Besides the general decrease in magnetic field strength owing to the diamagnetic nature of water, the magnetic field (Figure 4.1E) is characterized by strong, localized magnetic field inhomogeneity originating from the many magnetic susceptibility transitions visible in Figure 4.1D. The magnetic field inhomogeneity that the MR user ultimately encounters is the sum of the perturbations created by magnet imperfections and environment (Figure 4.1A–C) and the human body (Figure 4.1D/E). Magnetic field effects arising from motion, blood oxygenation and respiration also contribute to the final magnetic field inhomogeneity. However, these dynamic effects are outside the scope of this chapter and will only be mentioned in passing. Before discussing the complex magnetic field distribution of the human body and head in more detail, it is instructive to study the susceptibility-induced magnetic field perturbations of geometrically simple objects. Figure 4.2 shows the magnetic field lines and the magnetic field strength $B_z(r)$ generated by paramagnetic (Figure 4.2A/B) and diamagnetic (Figure 4.2C/D) ellipsoids. The results show that paramagnetic and diamagnetic ellipsoids increase and decrease the magnetic field strength within the material, respectively. Whereas the magnetic field in both cases is perfectly homogeneous within the ellipsoid, the magnetic field outside the object is highly inhomogeneous.

Besides the amplitude, the spatial distribution of the magnetic susceptibility (*i.e.* the geometry of the object) also has a strong effect on the magnetic field inhomogeneity. For example, Figure 4.2E shows the magnetic field distribution of a paramagnetic sphere. While the general features outside the sphere are similar to those displayed in Figure 4.2A for an ellipsoid, the magnetic field inside the sphere is identical to the applied magnetic

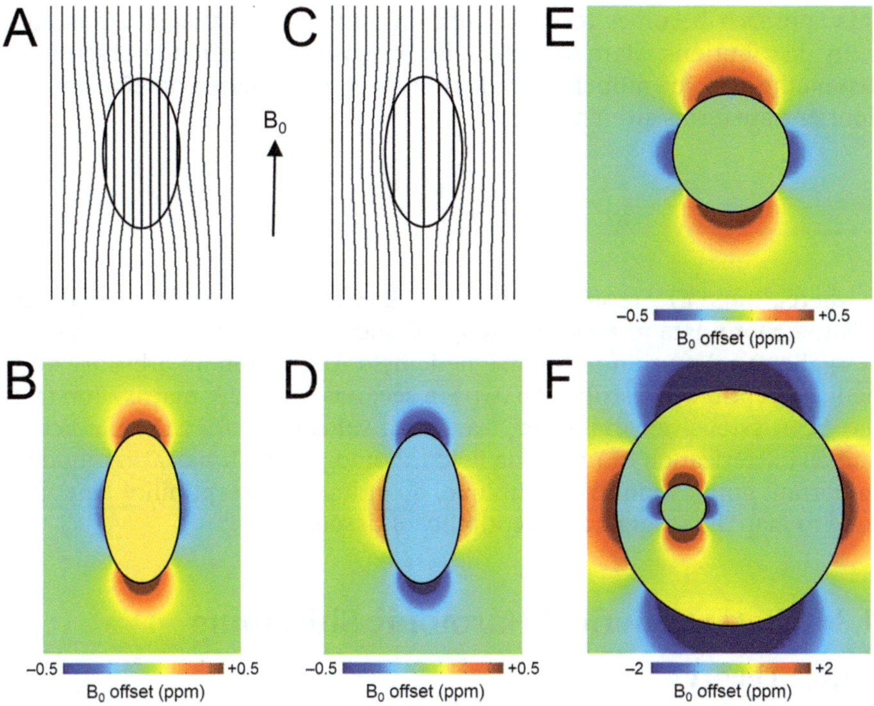

Figure 4.2 Magnetic field lines and strengths of geometrically simple objects. (A and C) Qualitative magnetic field lines and (B and D) quantitative induced magnetic field amplitudes of (A and B) paramagnetic (χ = +1.0 ppm) and (C and D) diamagnetic (χ = −1.0 ppm) ellipsoids in a vacuum. (E) Induced magnetic field of a paramagnetic sphere (χ = +1.0 ppm) in a vacuum. (F) Induced magnetic field of an air-filled sphere (χ = +0.36 ppm) inside a larger, water-filled sphere (χ = −9.05 ppm).

field despite the paramagnetic nature of the sphere. This demonstrates that the spatial geometry and the magnetic susceptibility of the object are both important in determining the final magnetic field perturbation. Small modifications, such as changing a flat bottom tube to a round bottom tube in an MR phantom, can have large consequences for the magnetic field homogeneity. Figure 4.2F shows the magnetic field distribution created by a small, air-filled sphere inside a large, water-filled sphere, showing that the smaller, paramagnetic sphere creates large magnetic field disturbances inside the larger, diamagnetic sphere. This simplified scenario represents actual conditions in MR applications studying the human brain. Small, air-filled cavities from the nasal passages and auditory tract surround the brain and cause strong, localized magnetic field distortions. More details can be found in Section 4.3.4.

For simple objects such as spheres and tubes, eqn (4.3) can be expanded to provide an analytical expression. For more complicated objects, eqn (4.3) can in principle be integrated numerically over all spatial positions. However, this approach becomes too time-consuming for larger magnetic susceptibility distributions. Marques[5] and Salomir[6] have independently described a fast method for the numerical evaluation of eqn (4.3) based on efficient fast Fourier transforms given by

$$\Delta B_z(r) = B_0 \mathrm{FT}^{-1}\left[\left(\frac{1}{3} - \frac{k_z^{\ 2}}{k^2}\right)\mathrm{FT}(\Delta\chi(r))\right] \tag{4.4}$$

where FT and FT^{-1} represent forward and inverse Fourier transformations, k is the coordinate in reciprocal k-space and $k^2 = k_x^{\ 2} + k_y^{\ 2} + k_z^{\ 2}$. Using eqn (4.4) allows calculation of the magnetic field $\Delta B_z(r)$ from an arbitrary magnetic susceptibility distribution $\Delta\chi(r)$ in seconds for a 128^3 matrix. Eqn (4.4) has been extensively evaluated[5–9] and is a valuable tool in the prediction and characterization of magnetic field inhomogeneity. It has also found an important application in a technique, quantitative susceptibility mapping, used to map the magnetic susceptibility of tissue *in vivo*.[10]

4.3 Static Spherical Harmonic Shimming

4.3.1 Theory

In the previous section, it has been demonstrated that magnetic field inhomogeneity has two primary origins, namely: (1) imperfections in the main magnetic field owing to manufacturing errors and environmental disturbances, and (2) magnetic susceptibility boundaries within or surrounding the sample or subject. Early papers on shimming were concerned primarily with the optimization of the homogeneity of the magnet.[1,2] As outlined in Appendix B of Chapter 1, the magnetic field inside an NMR magnet can be described mathematically by Laplace's equation $\nabla^2 B_z = 0$, which states that for a current-free region of interest magnetic field lines cannot form closed loops. In other words, Laplace's equation is valid when the magnetic field inhomogeneity inside the region of interest originates from perturbations outside the region of interest. For a bare NMR magnet without a sample or subject this assumption is perfectly valid and Laplace's equation describes the magnetic field to high accuracy. The solution to Laplace's equation for a spherical region around the magnet isocenter was given in Appendix B in Chapter 1 as an infinite series of spherical harmonic functions. It is very useful to visualize the different orders of spherical harmonic functions, and Figure 4.3 shows a graphical depiction for $n \leq 3$.

In many circumstances, such as magnetic field plotting of the bare magnet or shim coil design *via* analytical evaluation of SH expansions, it is advantageous to express the magnetic field and the SH fields in spherical coordinates. However, since modern MRI-based magnetic field B_0 mapping is almost always based on a Cartesian grid, it is often convenient to express the fields in terms of Cartesian coordinates. Using the relationships

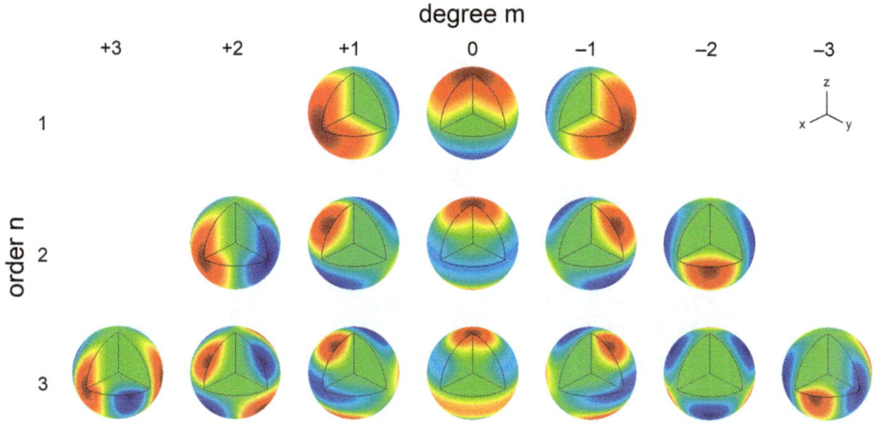

Figure 4.3 Graphical representation of all spherical harmonic terms m up to the third order n on the surface of a unit sphere. In order to evaluate the spherical harmonic functions internal to the spherical surface the positive octant has been removed. The trivial spherical harmonic function for $n = 0$, *i.e.* a pure offset, is not displayed.

$x = r\sin\theta\cos\phi$, $y = r\sin\theta\sin\phi$, $z = r\cos\theta$ and $r^2 = x^2 + y^2 + z^2$ the magnetic field can be expressed in Cartesian coordinates according to

$$B(x,y,z) = B_0 + \sum_{n=1}^{\infty}\sum_{m=-n}^{+n} C_{n,m}F_{n,m}(x,y,z) \tag{4.5}$$

The functions $F_{n,m}(x,y,z)$ are given in Table 4.1 up to $n = 5$. It should be noted that the common name of most spherical harmonics terms is an incomplete representation of the actual function in Cartesian coordinates. For example, a 'Z3' function ($n = 3$, $m = 0$) is not limited to the z axis but also has contributions in the x–y plane. Other terms, such as 'ZC2' ($n = 3$, $m = 2$) represent a mix between spherical and Cartesian coordinates and nomenclature.

The magnetic field inside an MR magnet can be quantitatively described by an SH expansion given by eqn (4.5). The next step towards optimizing the magnetic field homogeneity is to find a way to generate the SH-based magnetic field distributions in order to apply appropriate correction fields. In general, there are only two ways of modifying the magnetic field inside an MR magnet, namely by: (1) current-carrying wire or by (2) additional magnets. In the latter case, the magnets do not have to be permanently magnetized but could also be in the form of diamagnetic, paramagnetic or ferromagnetic materials magnetized by the main magnetic field. Figure 4.4A–C show the basic elements to create magnetic fields. The magnetic field distribution outside a sphere of radius R (Figure 4.4A) and magnetic susceptibility χ can be found by substituting $\Delta V = (4\pi/3)R^3$ into eqn (4.1). The magnetic field generated by a current loop (Figure 4.4B/C) of radius R, n number of turns and carrying current I placed in the magnet isocenter with the coil axis parallel to

Table 4.1 Spherical and Cartesian representation of low-order ($n \leq 5$) spherical harmonics functions.

Order n	Degree m	$P(\theta)$[a]	$F(x, y, z)$[b]	Common name
0	0	1	1	Z0
1	0	$\cos\theta$	z	Z
1	+1/−1	$\sin\theta$	x/y	X/Y
2	0	$1/2(3\cos^2\theta - 1)$	$z^2 - 1/2R^2$	Z2
2	+1/−1	$3\sin\theta\cos\theta$	$3zx/3zy$	ZX/ZY
2	+2/−2	$3\sin^2\theta$	$3(x^2 - y^2)/6xy$	X2Y2 (or C2)/ XY (or S2)
3	0	$1/2(5\cos^3\theta - 3\cos\theta)$	$z^3 - 3/2zR^2$	Z3
3	+1/−1	$3/2\sin\theta (5\cos^2\theta - 1)$	$6x(z^2 - 1/4R^2)/ 6y(z^2 - 1/4R^2)$	Z2X/Z2Y
3	+2/−2	$15\sin^2\theta\cos\theta$	$15z(x^2 - y^2)/30zxy$	ZX2Y2 (or ZC2)/ ZXY (or ZS2)
3	+3/−3	$15\sin^3\theta$	$15x(x^2 - 3y^2)/ 15y(3x^2 - y^2)$	X3 (or C3)/Y3 (or S3)
4	0	$1/8(35\cos^4\theta - 30\cos^2\theta + 3)$	$z^4 - 3z^2R^2 + 3/8R^4$	Z4
4	+1/−1	$5/2\sin\theta (7\cos^3\theta - 3\cos\theta)$	$10zx(z^2 - 3/4R^2)/ 10zy(z^2 - 3/4R^2)$	Z3X/Z3Y
4	+2/−2	$15/2\sin^2\theta (7\cos^2\theta - 1)$	$45(x^2 - y^2)(z^2 - 1/6R^2)/ 90xy(z^2 - 1/6R^2)$	Z2C2/Z2S2
4	+3/−3	$105\sin^3\theta\cos\theta$	$105zx(x^2 - 3y^2)/ 105zy(3x^2 - y^2)$	ZC3/ZS3
4	+4/−4	$105\sin^4\theta$	$105(x^2 - y^2)^2 - 420x^2y^2/ 420xy(x^2 - y^2)$	X4/Y4
5	0	$1/8(63\cos^5\theta - 70\cos^3\theta + 15\cos\theta)$	$z^5 - 5z^3R^2 + 15/8zR^4$	Z5
5	+1/−1	$15/8\sin\theta (21\cos^4\theta - 14\cos^2\theta + 1)$	$15z^2x(z^2 - 3/2R^2) + 15/8xR^4/15z^2y (z^2 - 3/2R^2) + 15/8yR^4$	Z4X/Z4Y
5	+2/−2	$105/2\sin^2\theta (3\cos^3\theta - \cos\theta)$	$105z(x^2 - y^2) (z^2 - 1/2R^2)/210zxy (z^2 - 1/2R^2)$	Z3C2/Z3S2
5	+3/−3	$105/2\sin^3\theta (9\cos^2\theta - 1)$	$420x(x^2 - 3y^2) (z^2 - 1/8R^2)/420y (3x^2 - y^2)(z^2 - 1/8R^2)$	Z2C3/Z2S3
5	+4/−4	$945\sin^4\theta\cos\theta$	$945z(x^2 - y^2)^2 - 3780 zx^2y^2/3780zxy(x^2 - y^2)$	ZC4/ZS4
5	+5/−5	$945\sin^5\theta$	$945(x^5 - 5xy^2 (2x^2 - y^2))/945(y^5 + 5x^2y(x^2 - 2y^2))$	X5/Y5

[a]Note that only $P(\theta)$ is given. The complete spherical harmonics function requires multiplication with $r^n\cos(m(\phi - \phi_{\mathrm{m}}))$.
[b]$R^2 = x^2 + y^2$.

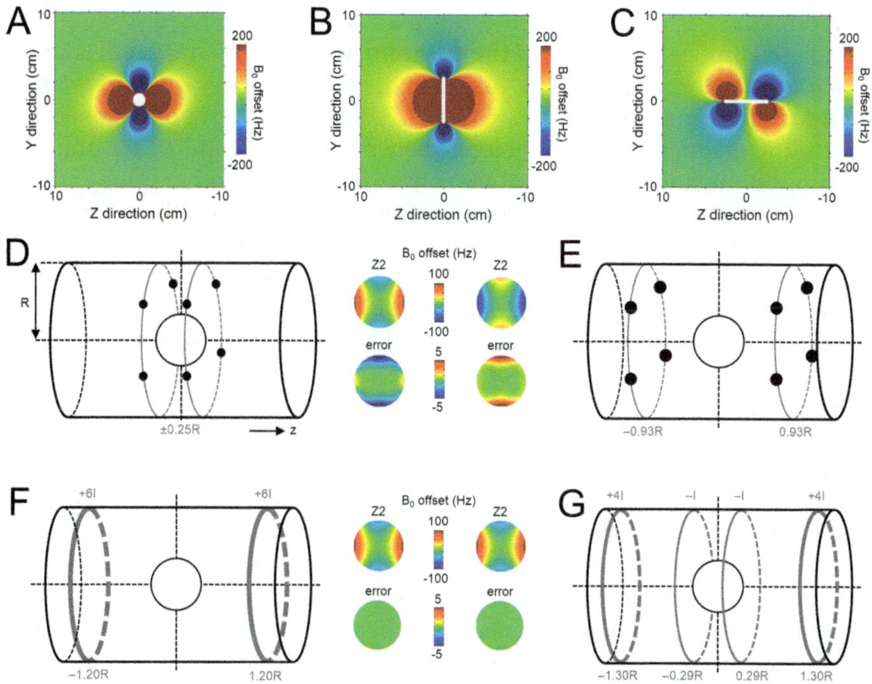

Figure 4.4 Elements for the generation of magnetic fields. (A) Magnetic field generated by a 1 mL spherical volume of paramagnetic niobium (χ = 237 ppm). (B and C) Magnetic field generated by a circular loop of wire (Ø 5 cm, current I = 1 A) with the coil axis (B) parallel and (C) perpendicular to B_0. (D) Arrangement of 8 pieces of steel to generate a +Z2 field across the DSV = (2/3)R. The generated +Z2 field is accompanied by a Z0 offset of −430 Hz. (E) Arrangement of 8 pieces of steel to generate a −Z2 field. *Circa* 50% more steel is required to generate the same Z2 field strength as in (D). The Z0 offset accompanying the −Z2 term was 131 Hz. (F) Arrangement of two current loops to generate a +Z2 field. The error (ideal field minus generated field) is greatly reduced when compared to the 8 piece design of (D). The Z0 offset is 834 Hz. (G) Arrangement of an improved four-coil design to generate a Z2 term. The Z0 term has been completely eliminated.

the z (or B_0) axis can be obtained through integration of the Biot–Savart law, as shown in Appendix A, Chapter 1, yielding:

$$
\begin{aligned}
B_x &= \frac{xzC}{r_{xy}^2\left(R^2+r^2+2Rr_{xy}\right)^{1/2}}\left[\frac{R^2+r^2}{R^2+r^2-2Rr_{xy}}E\left(\kappa^2\right)-K\left(\kappa^2\right)\right] \\
B_z &= \frac{C}{\left(R^2+r^2+2Rr_{xy}\right)^{1/2}}\left[\frac{R^2-r^2}{R^2+r^2-2Rr_{xy}}E\left(\kappa^2\right)+K\left(\kappa^2\right)\right]
\end{aligned}
\tag{4.6}
$$

where $r^2 = x^2 + y^2 + z^2$, $r_{xy}^2 = x^2 + y^2$ and $C = \mu_0 nI/2\pi$. $K(\kappa^2)$ and $E(\kappa^2)$ are the complete elliptical integrals of the first and second kind, respectively, with $\kappa^2 = 4Rr_{xy}/(R^2 + r^2 + 2Rr_{xy})$. The magnetic field along the y direction can be found from B_x, according to $B_y = (y/x)B_x$. Whereas NMR is only concerned with static magnetic fields (B_z) along the z direction, B_x and B_y are nonetheless important since B_x, B_y and B_z are converted into each other upon rotation of the shim coil. For example, rotating the shim coil about the x axis by an angle α leads to the transformation $B_y = B_y \cos(\alpha) + B_z \sin(\alpha)$ and $B_z = B_z \cos(\alpha) - B_y \sin(\alpha)$. Every rotation of the shim coil can be described by standard 3 × 3 rotation matrices. Early work on shim field design and construction analyzed the magnetic field produced by steel pieces, current loops and arcs in terms of analytical SH expansions.[1,2] The pieces or currents are then spatially arranged in such a way that undesired SH contributions are canceled, thereby ideally leaving a single, desired SH field $m = 0$, $n = 0$ which is perfectly homogeneous. The analytical evaluation can also be performed numerically and practical solutions can be obtained in seconds with modern day computer optimization routines.

As an example, Figure 4.4D shows the creation of a +Z2 ($n = 2$, $m = 0$) SH field using eight pieces of steel. Four pieces are placed equidistant on a circle spanning the magnet bore, whereas the other four pieces are placed symmetrically around the magnet isocenter. While the generated Z2 field is relatively pure, with up to 5% (primarily) fourth-order SH contributions, it should be realized that the elimination of a Z0 SH contribution is impossible with the configuration shown in Figure 4.4D. As the magnetic susceptibility of steel has a fixed sign, it is not immediately obvious how a −Z2 SH field can be created. Whereas the +Z2 SH field is primarily based on the negative lobe of a steel dipole field (Figure 4.4A), a −Z2 SH field would have to be based on the positive lobe. By placing the pieces further out along the z axis, the positive dipole lobe is emphasized leading to a −Z2 field, albeit in the presence of a positive Z0 offset. Note that the increased separation in Figure 4.4E to generate a −Z2 field demands the use of *circa* 50% more steel than the generation of a +Z2 field of equal magnitude. Figure 4.4D/E clearly shows that small pieces of steel can generate accurate SH fields. However, the amplitude of the generated fields is not readily adjustable to deal with subject-specific magnetic field variations. As a result, steel-based shimming is primarily used to homogenize the bare magnet. The required SH shim order is generally low ($n < 6$ to 8), although the inhomogeneity could be of relatively large magnitude (>100 ppm). Many gradient coils come with premade trays along the perimeter that can hold small pieces of steel or other metal. Following the generation of a magnet field plot of the bare magnet (see Section 4.3.2), a computer algorithm can determine the amount and position of metal that optimally opposes the magnetic field inhomogeneity. One of the challenges with steel shimming is that of saturation, in which the induced magnetic field no longer scales linearly with the applied external magnetic field. In addition, making the steel shims part of the gradient coil means that magnet shimming has to be redone upon replacement of the gradient coil. A final consideration is that the permeability of steel is temperature dependent, such that the magnetic field homogeneity could depend on, for example, heating induced by gradient switching.

Similarly to small pieces of steel, the placement of current loops to generate SH fields can be based on analysis of SH expansions or can be aided by computer minimization algorithms. Figure 4.4F shows the creation of a +Z2 field with a two-loop design (the minimum number of loops that can produce such a field). The obtained Z2 field is pure, but comes with the mandatory Z0 contribution. In the case of current loops the generation of a −Z2 field is trivial as it simply involves reversal of the current. Figure 4.4G shows the creation of a +Z2 field with a more commonly employed four-loop design. Whereas the strength and purity of the obtained Z2 field is similar to that of Figure 4.4F, the main advantage of the four-loop design is the absence of a Z0 contribution. Figure 4.5 shows shim coil designs for the first and second order SH functions largely based on analytical evaluation of SH expansions. Note that the tesseral SH functions (see Appendix B, Chapter 1) require the use of arcs placed with a periodicity $m\phi$.

The results shown in Figure 4.4 demonstrate the principles of magnetic field generation based on metal pieces, current loops and current arcs whereby their placement is guided primarily by arguments pertaining

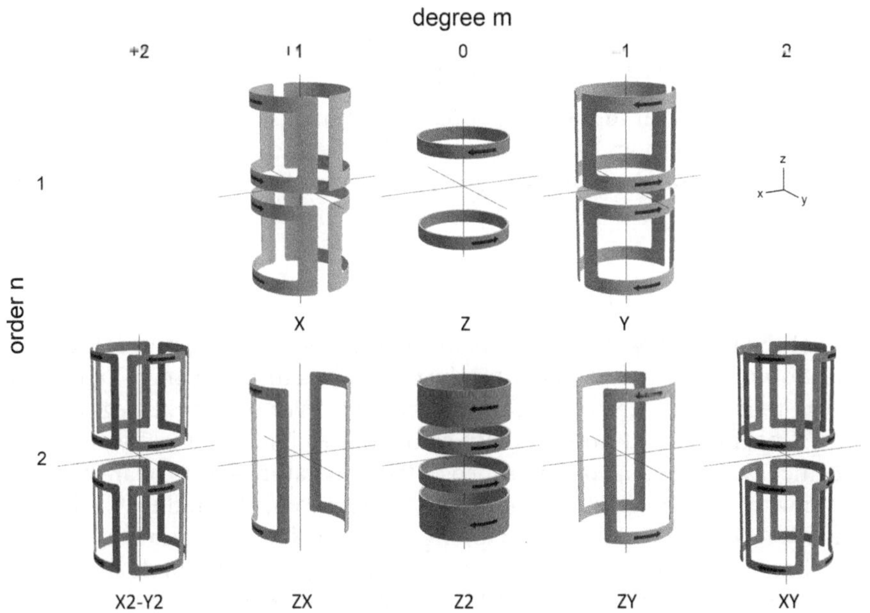

Figure 4.5 First and second order SH shim coils. The coils were designed based on current loops and arcs as described by Romeo and Hoult.[2] Note that wire parallel to B_0 does not contribute to the generated SH fields, but are essential in connecting various arcs. Black arrows indicate the current directions. (Courtesy of D. Green and S. Pittard.)

to symmetry, cancellation of specific spherical harmonics and construction feasibility. While elegant and instrumental in the advancement of MR methodology, these methods lack the flexibility to incorporate many additional design criteria. For example, many head gradient coils are of a short, asymmetric design in order to accommodate the subject's head and shoulders while still maintaining gradient performance and linearity. As a consequence, the shim coils also need to be designed asymmetrically.[11] As will be discussed in Section 4.4, dynamic SH shimming demands the rapid switching of shim coil currents, thereby making minimum-inductance shim coil designs desirable.[12] In addition, in order to minimize shim-induced eddy currents, actively shielded shim coils could be considered.[13–15] Other design criteria may be related to the desire for a larger homogeneous DSV, the presence of non-cylindrical mounting surfaces[16] or the need for minimum-power shim coil designs.[17] In all of these cases, a more general design method for shim coils is needed. The target field method first described by Turner[18,19] is a general method to design continuous current flows on cylindrical surfaces in order to generate high-performance linear gradients. The method allows optimization with respect to inductance, power and DSV. The target field method has been used by Forbes and Crozier[15,20,21] for the design of shielded zonal and tesseral shim coils on cylindrical and planar surfaces. Other methods for gradient coil design, such as the boundary element method,[22–24] can also be modified for shim coil design.[25]

4.3.2 Magnetic Field Mapping

With the theoretical foundation of SH shimming established, the next and arguably most important step in shimming is the characterization of the magnetic field inhomogeneity, a process commonly referred to as magnetic field mapping.

4.3.2.1 B_0 Field Mapping

One of the first steps following the installation of a new magnet is the optimization of the magnetic field produced by the empty magnet. Figure 4.1 shows that inevitable fabrication inaccuracies will lead to a degradation of the magnetic field homogeneity compared to its theoretical value. Site-specific issues, such as the presence of large metallic objects near the magnet or within the magnet (*e.g.* the gradient coil) can further reduce the homogeneity. As MRI-based B_0 field mapping is not feasible at the time of magnet installation, magnet manufacturers typically employ a magnetic field mapping approach based on a small NMR coil and sample inside a larger plotting rig (Figure 4.6A). The arm of the rig can be moved inside the magnet along spherical coordinates, whereby the NMR coil records the Larmor frequency. By plotting the magnetic field at regular polar and azimuthal angles the spatial magnetic field distribution of the bare magnet can be characterized. Figure 4.6B shows a 1D example ($\theta = 90°$) of the measured B_0 offsets as a function

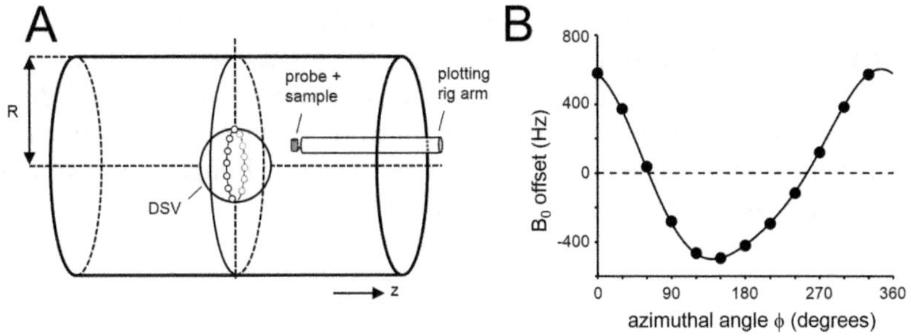

Figure 4.6 Principle of magnet field plotting. (A) The magnet field plotting rig is mounted on the magnet and is composed of a rigid arm containing a small NMR coil and sample. The arm is able to move in all three spatial dimensions within the magnet while maintaining a fixed orientation with respect to B_0. The resonance frequency is typically determined at discrete spatial locations sampling a 3D DSV sphere in the spherical coordinate system (shown are equiangular positions on the equatorial DSV circle, $\theta = 90°$). (B) Example data of measured B_0 offset *versus* azimuthal angle ϕ ($\theta = 90°$) at a radius of 10 cm. The sinusoidal variation implies magnetic field inhomogeneity according to *X* and *Y* linear gradients, whereas the distortion from a perfect sinusoid implies the presence of higher-order components. Fourier analysis quantitatively reveals contributions from $X = 50$ Hz cm^{-1}, $Y = -20$ Hz cm^{-1} and X2Y2 = 0.8 Hz cm^{-2}. Data were acquired on a 7 T MR system in which the room temperature shims were purposely adjusted to create the magnetic field distribution shown in (B). Magnetic field offsets were measured with a 1-channel 'field camera'[79] that could be rotated about the *z* axis.

of azimuthal angle at a radius of 10 cm from the isocentre. The sinusoidal field variation indicates the presence of first-order SH terms, whereas the distortion from a perfect sinusoid corresponds to the presence of higher-order SH contributions. Fourier analysis of the individual waveforms can be used to determine the exact SH contributions. For actual magnetic field mapping, the sampling of spatial positions is more extensive, typically sampling a 3D sphere along spherical coordinates throughout the inner diameter of the magnet bore.

4.3.2.2 *MRI-Based* B_0 *Field Mapping*

Having mapped the bare magnet, applied suitable current to the superconducting shim coils and placed small pieces of steel in the appropriate places within the magnet bore, the magnetic field inhomogeneity generated by the sample or subject becomes relevant when the MR system is fully operational. As such, the subject-specific magnetic field distribution is often obtained from the phase evolution during a period of free precession using an appropriate MRI sequence. While MRI-based B_0 mapping can essentially be based on any MRI pulse sequence, the fast gradient-echo (GE) method is commonly used owing to its speed, ease-of-use and inherent sensitivity to B_0

offsets. Besides the standard GE image acquired at an echo-time TE, an additional GE image is acquired at an echo-time TE + τ. As described in Chapter 1, Section 1.8.4, the phase difference $\Delta\phi$ between the two images is directly proportional to the magnetic field distribution, since $\Delta\phi(r) = 2\pi\Delta\nu(r)\tau$. The location-dependent phase difference between the two images is conveniently calculated by voxel-specific complex division:

$$\Delta\phi = \arctan\left(\frac{R_1 I_2 - I_1 R_2}{R_1 R_2 + I_1 I_2}\right) \tag{4.7}$$

where R and I refer to the real and imaginary components of the MRI signal, respectively, and 1 and 2 refer to the signals acquired with $\tau = 0$ and $\tau > 0$, respectively. The additional delay τ, typically set of the order of 1 ms, slightly decreases the signal intensity owing to T_2^* relaxation. However, since this only affects the signal intensity it has no consequence for the phase calculation given by eqn (4.7).

The most commonly encountered problem in B_0 mapping is so-called phase wrapping (Figure 4.7C). The arctan function in eqn (4.7) can only calculate the phase between $-\pi$ and $+\pi$ radians. When the real phase evolution is larger (*i.e.* for $\Delta\nu = 250$ Hz and $\tau = 3.0$ ms, $\Delta\varphi = 1.5\pi$), the calculated phase will wrap to -0.5π. In general, phase wrapping leads to an incorrect estimation of the local magnetic field (-0.5π would be equivalent to a -83 Hz offset rather than the correct +250 Hz). However, in the presence of a wide range of magnetic field inhomogeneity, phase wrapping will also lead to discrete phase jumps (*e.g.* two adjacent spatial points with real phases of 0.95π and 1.05π will be wrapped to 0.95π and -0.95π), making it impossible to accurately estimate spatial magnetic field distributions. The problem of discrete phase jumps can be solved through multidimensional spatial phase unwrapping algorithms.[26] However, these algorithms are typically not straightforward and are computationally intensive. A simpler solution is to select the evolution time τ small enough so that phase wrapping does not occur (*e.g.* the real phases are between $-\pi$ and $+\pi$, Figure 4.7B). Unfortunately, in the presence of noise the accuracy of the estimated offsets is not optimal when using a small τ delay. Accurate estimates of magnetic field offsets are obtained when the number of evolution delays, as well as the delay durations, are extended (see Figure 4.7D/E). While the acquired phase during the longer delays will likely be wrapped, the initial delay is chosen such that no phase wraps occur (Figure 4.7D). Then, based on the linear phase-time trend established by the first delay(s), the subsequent delays can be unwrapped by adding or subtracting an integer multiple of 2π to the calculated phase (Figure 4.7D). Once all phases have been temporally unwrapped, a linear least-squares fit of the phase-time curve will provide a best estimate of the magnetic field offset (Figure 4.7E). Typical computation times for phase calculation, phase unwrapping and linear least-squares fitting for a $128 \times 128 \times 64$ dataset is several seconds. Magnetic field B_0 mapping with multiple delays essentially trades increased data acquisition time for increased accuracy while

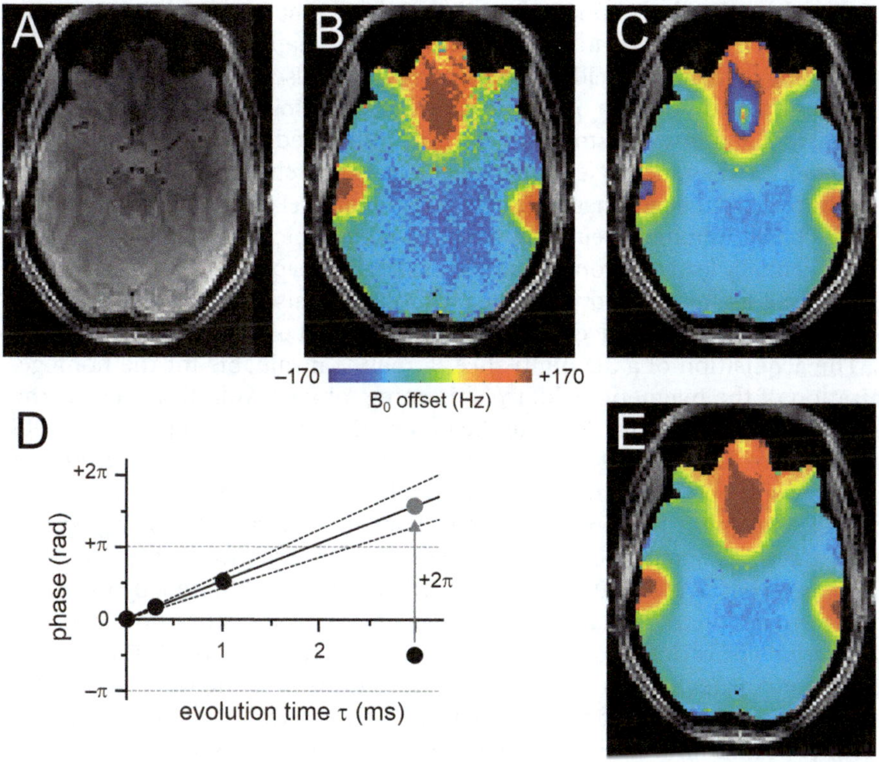

Figure 4.7 Principle of MRI-based magnetic field mapping. (A) Gradient-echo MR image and (B) calculated B_0 magnetic field map with $\tau = 0.33$ ms. While the B_0 map is devoid of phase-wrapping artifacts, the small evolution delay τ leads to low SNR. (C) B_0 map calculated with $\tau = 3.0$ ms. The SNR has greatly improved, but areas with a frequency offset $|\Delta\nu| > 167$ Hz display a phase wrapping artifact. (D) Temporal phase unwrapping based on multi-echo MR data with $\tau = 0.33$, 1.0 and 3.0 ms. For the pixel shown, the phase of the first two evolution delays (0.33 and 1.0 ms) is not wrapped and can establish a linear time-phase trend (solid linear line) together with confidence intervals (dotted linear lines). The phase of the final delay is wrapped and adding a multiple of 2π will bring it back to its proper, unwrapped value. (E) B_0 map following temporal phase unwrapping as shown in (D). The B_0 map is characterized by high SNR and the absence of phase wrapping artifacts.

also eliminating the need for multi-dimensional phase unwrapping. While this approach is strongly recommended for shim coil calibration (see Section 4.3.3), the increased accuracy may not be needed for routine *in vivo* shimming.

The calculated B_0 map is in principle independent of the pulse sequence used. The pulse sequence can nevertheless have a significant effect on the appearance of the B_0 map. Gradient-echo-based B_0 mapping can be fast and reliable, but the images may show areas of signal loss when strong magnetic field inhomogeneity is encountered owing to phase cancellation during the initial echo-time TE. In these cases, it is better to utilize spin-echo-based B_0 mapping, since phase evolution owing to magnetic field inhomogeneity is refocused at the top of the echo. B_0 maps based on EPI can be acquired very rapidly. However, EPI image quality (signal loss and image distortion) is heavily influenced by magnetic field inhomogeneity and as such is less than ideal to provide reliable B_0 maps. It should also be realized that any magnetic field variation not related to magnetic susceptibility based inhomogeneity that happens during the B_0 mapping acquisition can potentially lead to erroneous magnetic field B_0 maps. These effects include, but are not limited to, eddy currents owing to the pulsed magnetic field gradients, subject motion, subject breathing and field drift of the magnet.

The acquisition of a 3D multi-slice B_0 map is invaluable for the homogenization of the magnetic field, *via* shimming, of large volumes, such as the entire human head, for MRI applications. However, for single-voxel MRS applications the acquisition of a large 3D B_0 map dataset is redundant and possibly sub-optimal as only a small fraction of the B_0 map pixels describe the voxel magnetic field inhomogeneity. For applications such as single-voxel MRS that cover a small localized area, several fast alternatives to MRI-based B_0 mapping exist. FASTMAP (fast automated shimming technique by mapping along projections)[27,28] characterizes the magnetic field distribution along 1D projections of various orientations through the center of the MRS volume. For second-order SH shimming, six 1D projections with and without an extra evolution delay τ are sufficient to quantitatively determine all SH terms. As a result, optimal shim settings can be determined in seconds, rather than minutes. Several improvements of FASTMAP have been proposed for improved accuracy,[29] shorter acquisition times[30] and extension to 2D slices.[31]

4.3.3 Calibration of Shim Coil Efficiency

The final step in preparing the MR system for magnetic field homogenization is the calibration of the shim coil efficiency. In other words, the actual, as opposed to theoretical, magnetic field response of each shim coil to unit input current needs to be accurately known. The most straightforward approach is to spatially map the magnetic field distribution for a particular shim coil following a known input current using the previously described MRI-based B_0 mapping method. Figure 4.8A shows magnetic field B_0 maps following the application of different current settings to the Z2 ($n = 2$, $m = 0$) shim coil. When analyzed on a pixel-by-pixel basis it is clear that the magnetic field offset scales linearly with the applied current (Figure 4.8B). The slope of the line gives the efficiency of the Z2 shim coil in terms of generating a specified frequency offset at that particular spatial location. The frequency offset at zero current is indicative of the background magnetic field inhomogeneity and is discarded. Repeating the linear regression for all pixels produces a spatial efficiency map for the Z2 shim coil (Figure 4.8C). By visual inspection it is clear that the dominant field contribution is indeed a SH function describing a Z2 SH term. However, the clear asymmetry in the magnetic field reveals

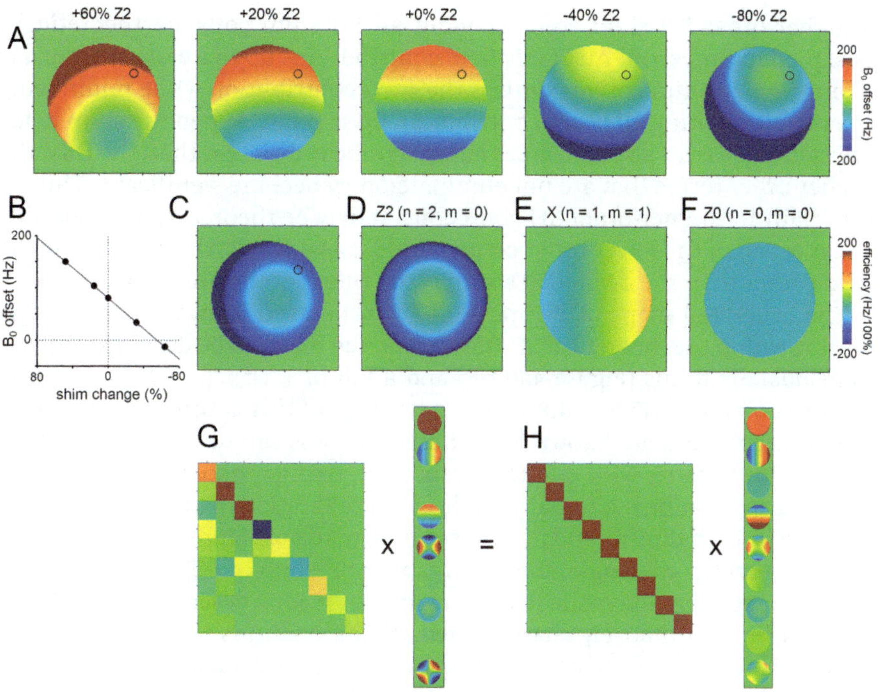

Figure 4.8 Principle of shim coil calibration. (A) 3D multi-slice B_0 maps acquired in the presence of different Z2 shim settings. (B) Magnetic field response for a single image pixel (black open circle in (A)) to the different Z2 shim settings. The magnetic field offset changes linearly with the shim change. The offset at 0% shim change reflects the background magnetic field inhomogeneity and is removed through linear regression (solid line). The slope of the linear is the efficiency of the shim coil to produce a B_0 offset for a 100% shim change. (C) Repeating the linear regression for every pixel leads to a calibrated shim efficiency map. (D–F) The shim calibration or efficiency map of (C) can be decomposed into contributions of pure SH functions, in this case (D) Z2, (E) X and (F) Z0 terms. (G and H) Repeating the steps outlined in (A)–(F) yields a $(n+1)^2 \times (n+1)^2$ matrix when calibrating shims up to order n. (G) When staying within the SH framework, the calibrated shims are decomposed into pure SH terms leading to a full rank calibration matrix. As the calibrated shim fields are shown for the axial plane ($z = 0$) the shim terms Z, ZX and ZY appear with zero-amplitude. (H) When the calibration map of (C) is used directly, the calibration matrix is essentially reduced to a $(n+1)^2 \times 1$ array.

that the Z2 shim coil also produces additional magnetic field components. A quantitative SH decomposition reveals smaller, but significant, low order SH field components (Figure 4.8D–F). This effect, in which a given SH coil produces a number of SH components, is very typical and can have a number of different origins. Firstly, incorrect shim coil placement or rotation relative to the gradient isocentre leads to low order SH contributions when driving a higher order SH shim coil. Secondly, all SH shim coils are theoretically designed to cancel out magnetic field contributions up to a certain order, *i.e.* a four-ring Z2 shim coil maximizes the $n = 2$ contribution while minimizing or even eliminating all other contribution up to $n = 4$.[2] When the shim fields are calibrated over a spatial area that is significantly larger than the DSV, the higher-order terms that are not eliminated may become significant. Thirdly, especially for higher-order SH shim coils, it may be theoretically impossible to cancel all of the lower order contributions. As a result, the presence of SH fields other than the desired SH term is the rule, not the exception.

The calibrated maps of Figure 4.8C–F can be used in two different ways. Staying within the theoretical SH framework, the map can be decomposed into individual SH terms (Figure 4.8D–F) and a full $(n + 1)^2 \times (n + 1)^2$ calibration matrix is obtained (Figure 4.8G). The advantage of this approach is that once the SH amplitudes are known the calibration maps can be readily resampled onto an MRI matrix with different resolution, orientation or field-of-view for different applications. Alternatively, the shim calibration map shown in Figure 4.8C can be used directly and the full-rank 2D matrix of Figure 4.8G reduces to a $(n+1)^2 \times 1$ array, *i.e.* a single efficiency map per SH shim coil. This approach has the advantage that magnetic fields that are not readily described by SH fields can still be quantitatively characterized. This scenario is encountered when venturing into the area of non-SH-based shimming as detailed in Section 4.5.

Since shim calibration is normally only performed once soon after magnet installation, it is imperative to achieve the highest possible accuracy in the calibrated shim efficiency maps. Any erroneous deviation will lead to systematically incorrect shim settings in *all* subsequent studies. The acquisition and processing pipeline shown in Figure 4.8 has several opportunities for determining quality control measures. Firstly, the MR images need to be of sufficient sensitivity and free of artifacts. Secondly, during the calculation of the B_0 maps, control can be enforced on the quality of the linear regression between phase and time (Figure 4.7D). Similarly, quality control can be enforced on the linear regression between B_0 offset and shim setting (Figure 4.8B). During all of these steps a given pixel can be discarded when certain thresholds are not met. Missing single pixels in the final calibrated efficiency map can be "filled in" through interpolation. In the case that many pixels need to be discarded one should consider re-measurement of the data under improved experimental conditions (which may simply require signal averaging and therefore more measurement time).

Once high-quality, calibrated efficiency maps ($F_{n,m,\text{cal}}$ for $(n + 1)^2$ SH fields up to order n or $F_{k,\text{cal}}$ for k non-SH shim coils) have been generated, the system is ready for subject-specific shimming. The shimming problem can be framed as a least squares minimization in which one needs to find the shim coil coefficients $C_{n,m}$ (or C_k) that for SH-based shimming minimizes:

$$\sum_{p=1}^{P}\left[\Delta B_0\left(x_p,y_p,z_p\right)+\sum_{n=0}^{\infty}\sum_{m=-n}^{+n}C_{n,m}F_{n,m,\text{cal}}\left(x_p,y_p,z_p\right)\right]^2 \tag{4.8}$$

Or for non-SH-based shimming minimizes:

$$\sum_{p=1}^{P}\left[\Delta B_0\left(x_p,y_p,z_p\right)+\sum_{k=1}^{K}C_kF_{k,\mathrm{cal}}\left(x_p,y_p,z_p\right)\right]^2 \tag{4.9}$$

ΔB_0 is the experimentally-measured, subject-specific magnetic field B_0 map and P is the total number of image pixels in the region-of-interest (ROI). Formulating the shimming process as a least squares minimization has the advantage that shim current constraints and weighting factors are easily accommodated.[32,33] Careful selection of the region-of-interest is critical for proper shimming performance. Any experimentally-acquired *in vivo* B_0 map will have pixels that are artifactual. For example, pixels in and around major blood vessels will often display large and variable B_0 offsets. Inclusion of these pixels into the ROI will negatively affect the shim optimization. For human head applications it is important to exclude extracranial tissues from the ROI. These tissues often have large lipid contributions which lead to artifactual B_0 offsets.

4.3.4 Static Spherical Harmonic Shimming of the Human Brain

The mathematical framework for SH shimming has been established for optimizing the magnetic field homogeneity of an empty magnet for which the assumption underlying Laplace's equation is completely satisfied. SH shimming has readily been adopted for optimizing the magnetic field homogeneity over cylindrical tubes[34,35] in high resolution NMR. In the situation in which the effects of the magnetic susceptibility transition at the fluid–air meniscus interface are minimized by susceptibility-matching plugs or ensuring that the transition is far outside the sensitive volume of the RF coil, Laplace's equation describes the magnetic field inside a cylindrical NMR tube to high accuracy. High-resolution, liquid-state NMR magnets are typically equipped with up to sixth-order ($n = 6$) shim coils, routinely providing a spectral resolution of 1 part-per-billion or better. The tremendous success of SH functions in magnet and NMR tube shimming led to a natural adaptation of SH-based shimming to *in vivo* samples for both animal and human MRI systems.

However, unlike homogeneous cylinders, the magnetic susceptibility distribution throughout the human body is heterogeneous with many abrupt changes between tissue and air. These changes in magnetic susceptibility lead to the formation of *local* magnetic field inhomogeneity, which is in direct violation of the conditions underlying Laplace's equation, which demand that magnetic field inhomogeneity observed in a given region-of-interest originates from effects *outside* the region-of-interest. In spite of these theoretical limitations, SH-based shimming has been the default shimming method for *in vivo* samples since the inception of *in vivo* MR imaging and spectroscopy. Figure 4.9 shows a typical SH shimming

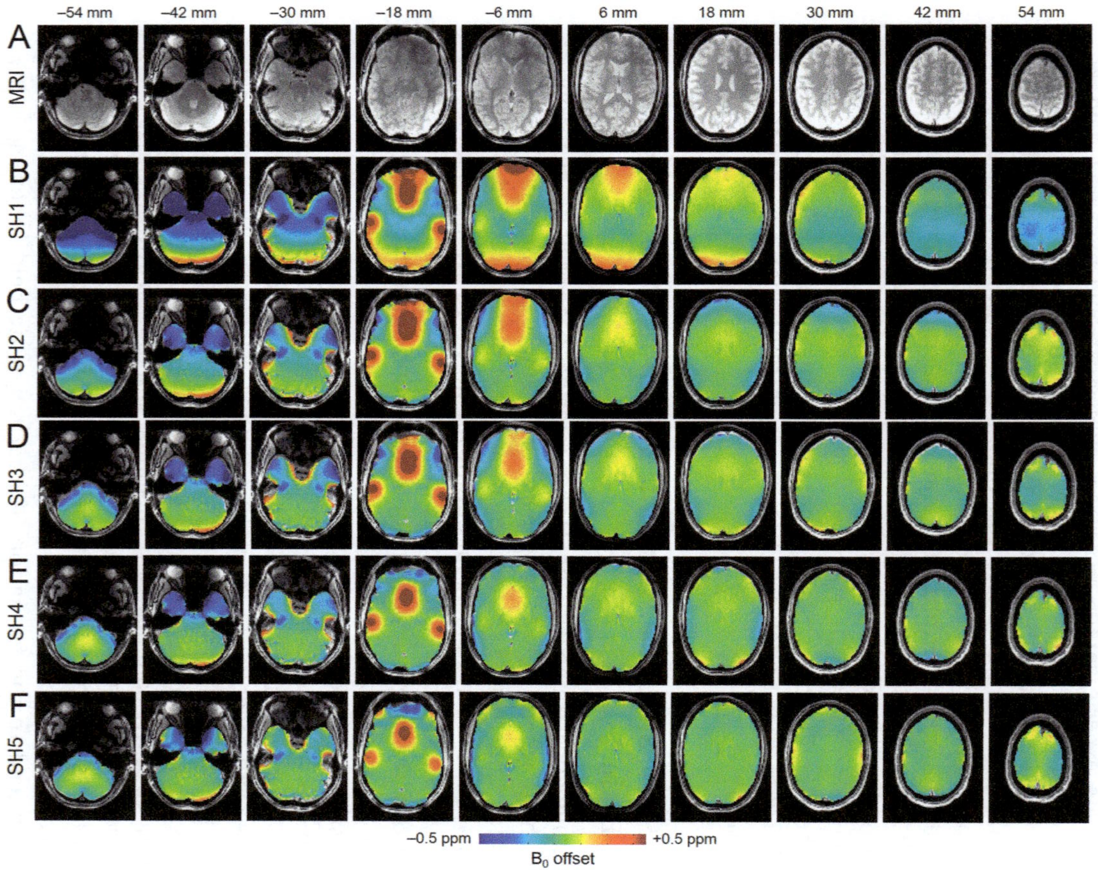

Figure 4.9 Static SH shimming of the human brain. (A) MR images and (B–F) B_0 maps selected from a multi-slice, whole brain dataset acquired at 4 T (66 slices of 2 mm, isotropic 2 × 2 mm in-plane resolution). The B_0 maps were reconstructed from four evolution delays, 0, 0.33, 1.0 and 3.0 ms, using 1D temporal phase unwrapping as detailed in Section 4.3.2.2. Residual B_0 maps were calculated following removal of SH terms corresponding to (B) $n \leq 1$, (C) $n \leq 2$, (D) $n \leq 3$, (E) $n \leq 4$ and (F) $n \leq 5$.

result for the human brain when including increasing numbers of SH orders. The figure shows that the magnetic field homogeneity across the majority of the brain is relatively high for $n \geq 2$. However, specific areas of the brain in close proximity to air–tissue susceptibility boundaries remain inhomogeneous even after the removal of up to fifth order SH terms. These areas typically occur in the frontal cortex, temporal lobes, hippocampus and auditory cortex, and correspond to areas that challenge the assumptions underlying Laplace's equation to the greatest extent. The magnetic field inhomogeneity observed in, for example, the frontal cortex originates from an air–tissue magnetic susceptibility boundary that is only millimeters below the region-of-interest. The steep and localized magnetic field gradients generated in the frontal cortex are well beyond the capabilities of SH shimming, even when including magnetic fields up to the fifth order. In Section 4.5 it is shown that localized magnetic field disturbances are better corrected with localized shim correction coils than with SH-based coils. Nevertheless, besides several specific areas of poor magnetic field homogeneity, SH-based shimming performs admirably across the majority of the human brain as well as many other organs.

Figure 4.10A provides a quantitative summary of SH-based shimming for static, global magnetic field homogenization of the human brain. The results are expressed in parts-per-million (ppm) since magnetic field inhomogeneity scales linearly with the applied magnetic field B_0. Therefore, even though the results were obtained at 4 T, the results when expressed in ppm are representative for any magnetic field strength. In addition to the commonly used standard deviation (SD), Figure 4.10A shows additional bars describing the frequency range in the B_0 map that covers 80, 90 and 95% of the pixels: these measures are used to account for the non-Gaussian character of the B_0 frequency distribution (Figure 4.10B). While the SD is a commonly used measure of magnetic field homogeneity, caution should be exercised when used on non-Gaussian frequency distributions. Whereas the frequency range of ±SD should cover *circa* 68% of the MR pixels if the distribution were Gaussian, Figure 4.10B shows that the actual range covering 68% is smaller for the chosen example. The use of frequency ranges that cover certain percentages of MR pixels is more generally applicable as it does not make assumptions about the normality of the frequency distribution. Figure 4.10 shows that the magnetic field homogeneity improves steadily with increasing SH orders. However, even after the removal of all fifth-order SH terms, 10% of the brain still has a magnetic field inhomogeneity of 0.3 ppm or worse. Nevertheless, Figure 4.10A quantitatively shows that inclusion of higher-order SH shims is beneficial in minimizing magnetic field inhomogeneity. This is also supported by experimental data with a fourth and partial fifth-order SH shim insert having demonstrated excellent homogeneity and high-quality MRSI in the human hippocampus.[36] Figure 4.10C shows a summary of the required SH shim strength needed to obtain the results shown in Figure 4.9 and Figure 4.10A. Similar to Figure 4.10A, the required shim strengths are expressed in ppm/cm^n thereby making the numbers relevant for any magnetic field

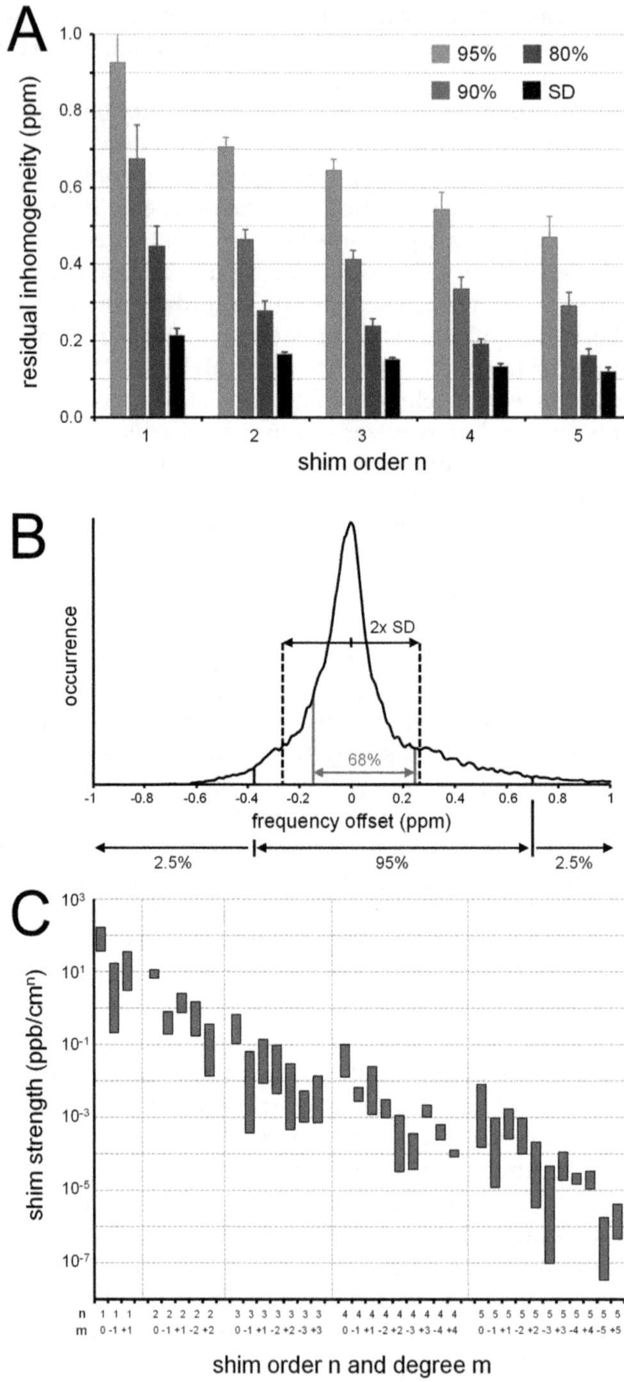

Figure 4.10 Quantitative summary of static, global SH shimming performance and shim requirements. (A) Residual magnetic field inhomogeneity following the removal of SH shim terms of increasing order up to

strength. While the magnetic field homogeneity across all subjects is very reproducible (typically less than 10% SD in Figure 4.10A), the required shim strengths to obtain that homogeneity can vary strongly between different human subjects with the range of many SH terms spanning 10–100 fold. This observation is in agreement with previous reports on inter-subject magnetic field distributions.[36,37] Nevertheless, the results in Figure 4.10C should provide a reasonable basis for the specification of SH shim sets for human brain applications.

4.4 Dynamic Spherical Harmonic Shimming

4.4.1 Principle of Dynamic Shimming

Dynamic SH shimming relies on the principle that a given magnetic field distribution of a large object can be better approximated across multiple, smaller and independent volumes, than by a single distribution across the entire object. This scenario is commonly encountered in multi-slice MRI and multi-voxel MRS. Figure 4.11 shows the principle of dynamic SH shimming for a 1D problem where the magnetic field inhomogeneity is (arbitrarily) represented by a third-order polynomial function. When analyzed over all spatial positions simultaneously (Figure 4.11A), as is done during static, global shimming, a linear magnetic field provides a poor approximation to the measured inhomogeneity. However, many volumetric MR studies are performed in a multi-slice manner, in which magnetization is excited and detected on a slice-by-slice basis. When the magnetic field analysis is restricted to the spatial positions of a single slice (Figure 4.11B), the linear magnetic field provides a much better approximation to the local magnetic field inhomogeneity. The magnetic homogeneity across the other slices is very poor, but since the magnetization of those slices lies along the longitudinal axis, this is not relevant. Once the signal of the desired slice is detected, the shims can be changed to a setting that is optimal for the next slice. In this manner, all slices can experience a higher magnetic field homogeneity sequentially

---

n = 5. Data is obtained on 7 healthy volunteers (2 women, 5 men) from 66-slice, whole brain B_0 maps (see Figure 4.9). Error bars represent the standard deviation across the 7 subjects. (B) To capture the non-Gaussian character of the residual B_0 offset distribution, residual magnetic field inhomogeneity is expressed as the frequency range that holds 80, 90 and 95% of all MR pixels in addition to the standard deviation (SD). The frequency distribution histogram was generated from a 20 mm thick slab centered at $z = -12$ mm (see Figure 4.9A) following second-order SH shimming. (C) SH shim strengths required to homogenize the human brain in a static, global fashion. The bar for each SH term represents the range between the minimum and maximum absolute-valued shim strengths needed to homogenize the magnetic field across 7 subjects. The approximate orthogonality of the SH terms across the human brain ensured that lower-order SH terms did not significantly change upon inclusion of higher-order SH terms.

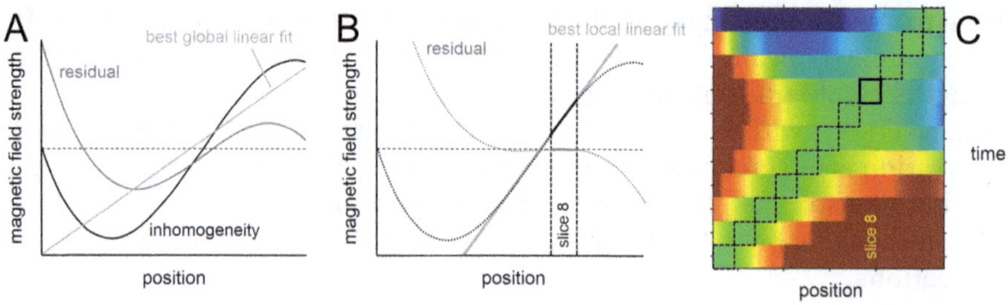

Figure 4.11 Principle of dynamic SH shimming. (A) The magnetic field inhomogeneity along a 1D spatial direction (solid black line) is poorly approximated by a first-order, linear SH shim field when analyzed over all positions (light gray line). (B) When the magnetic field homogeneity is improved on a slice-by-slice basis, the first-order, linear SH shim field provides a much better approximation to the local inhomogeneity (shown for slice 8 in an 11-slice experiment). (C) Repeating the analysis outlined in (B) for all other slices on a slice-by-slice basis shows that excellent magnetic field homogeneity can be obtained for all slice positions. The optimal shim settings for a given slice must be set when signal from that slice is about to be excited and detected. In other words, magnetic field homogeneity throughout the brain is obtained sequentially in time.

in time (Figure 4.11C). In general, it can be stated that for a given shim coil configuration, dynamic shimming will always provide an improved magnetic field homogeneity relative to static shimming. This is true for SH-based shimming as discussed here, but also for non-SH-based shimming, as discussed in Section 4.5.2.

4.4.2 Practical Considerations for Dynamic Shimming

One of the considerations involved with slice-specific SH shimming is that of shim degeneracy, a phenomenon for which shims of different order and degree appear identical when analyzed over a reduced ROI such as a 2D plane. For example, across an axial plane (constant z coordinate) the SH fields for $n = 1$, $|m| = 1$ (*i.e.* X and Y gradients) and $n = 2$, $|m| = 1$ (*i.e.* ZX and ZY shims) appear identical. Including all first and second order SH fields in the presence of significant shim degeneracy will lead to the failure of any standard form of least-squares shim optimization.

Two straightforward approaches to deal with shim degeneracy are shim degeneracy analysis[38,39] and multiple slice analysis.[40] In the first approach, the available SH shim fields are analyzed for degeneracy and only the unique terms are included. For example, to optimize the in-plane homogeneity for an axial plane (z = constant) only the shims Z0, X, Y, Z2, X2Y2 and XY are included for $n \leq 2$. With shim degeneracy analysis, the through-slice shims need to be estimated after the optimal in-plane shims have been applied. The second approach utilizes the fact that SH fields are always unique across a 3D volume. Therefore, rather than analyzing the magnetic field inhomogeneity across an infinitely narrow slice thickness, the approach includes slices on either side of the given slice. An attractive feature of multiple slice analysis is that through-plane inhomogeneity is automatically captured.

While dynamic SH shimming is theoretically a relatively straightforward concept, the experimental realization is complicated by shim-induced eddy currents and the need for modified hardware. Abruptly switching currents in a shim coil will lead to the generation of eddy currents in surrounding structures such as the magnet cryostat. Since shim coils are normally not actively shielded, the magnitude of such shim-induced eddy currents can be an order of magnitude larger than those observed for modern, actively-shielded gradient coils. It has been observed that pulsed shims of order n can generate an eddy current-induced artificial field term shaped similar to the pulsed shim itself, but also any shape (of degree m) of any order equal or smaller than $n - 1$. The eddy currents associated with linear imaging gradients ($n = 1$) are limited to B_0 magnetic field offsets ($n = 0$) and linear field gradients ($n = 1$), in which eddy current cross-terms between linear fields of different degree m are normally minimal. For higher order shims, however, a multitude of cross-terms can be generated. Figure 4.12A–C show the temporal magnetic field variations following a 100% change in the ZX ($n = 2$, $m = +1$) shim. From Figure 4.12C it is clear that the second order magnetic field takes almost 1000 ms to stabilize. Similarly to the linear gradients, a significant perturbations of the main magnetic field B_0 can also be measured for >2500 ms following the shim change (Figure 4.12A). A feature that is unique to higher order SH dynamic shimming is that of cross-term eddy currents wherein a change in the ZX shim lead to a perturbation in a lower-order Z shim (Figure 4.12B) that lasts for >1000 ms.

The characterization of shim-induced eddy currents as shown in Figure 4.12A–C can be a time-consuming operation as both the spatial and temporal dynamics of each SH shim need to be obtained. In order to minimize the experimental measurement time, some studies have used projection-based methods[38,40] to characterize the temporal dynamics of different shim orders and degrees sequentially. Other studies have used 3D MRI to spatially map all eddy currents orders and degrees simultaneously.[41] Most recently, the technology of MR field cameras has been used to characterize shim-induced eddy current with high speed and accuracy.[42,43] Once the temporal magnetic fields related to shim-induced eddy currents have been characterized, a method that can reduce or even eliminate them is needed. Eddy currents associated with pulsed linear gradient fields are classically dealt with *via* gradient pre-emphasis and B_0 correction.[44–46] In pre-emphasis, the beginning of the input gradient waveform is purposely driven higher than the nominal gradient pulse such that the overshoot exactly cancels the observed eddy current-induced gradient dampening. At the end of the gradient pulse, the waveform is purposely driven lower to achieve the reversed effect for the descending gradient ramp. B_0 correction works according to the same principle with the main difference that the compensation is applied to the Z0

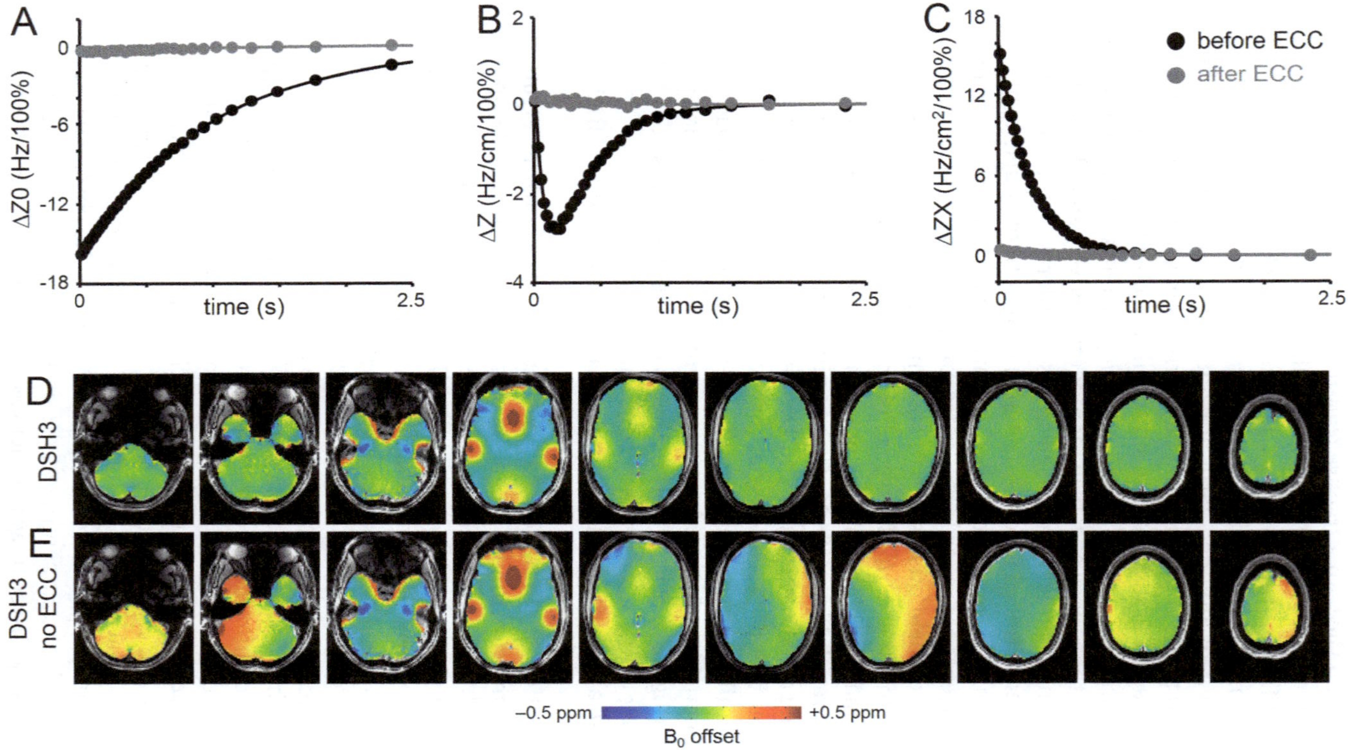

Figure 4.12 Measurement, compensation and effects of shim-induced magnetic field perturbation. (A–C) Magnetic field perturbations (black dots and line) following a 100% shim change in a ZX ($n = 2$, $m = +1$) SH shim. The magnetic field perturbations can be decomposed into temporal variations of (A) the main magnetic field, (B) a cross-term into the linear Z gradient and (C) a self-term of the ZX shim. Repeated measurements at different times following the shim change allows the determination of the eddy current amplitudes and time constants. Using shim pre-emphasis and B_0 correction, the shim-induced perturbations can be dramatically reduced (gray dots and line). (D and E) Residual B_0 maps following slice-by-slice dynamic, third-order SH shimming in the (D) absence and (E) presence of shim-induced eddy currents. Shim-induced eddy-current amplitudes and time constants were taken from ref. 40.

shim coil in order to cancel the B_0 eddy currents. The principle of pre-emphasis and B_0 correction can also be applied to dynamic SH shimming, albeit with several modifications.[38,40] Instead of having a 3 × 2 interaction matrix (linear-to-linear and linear-to-Z0 for each linear SH term), third-order SH shims could require a 16 × 16 interaction matrix. As the characterization of each interaction can potentially require the inclusion of several time constants (*e.g.* see Figure 4.12B), the number of compensation circuits can become unwieldy. Nevertheless, dynamic SH shimming with full pre-emphasis has been experimentally demonstrated up to the third order.[40] While not every theoretically possible interaction was observed, the pre-emphasis still covered 67 interactions and time constants. Figure 4.12A–C (gray line) show the performance of pre-emphasis for the ZX shim. Generally, the magnitude of temporal magnetic field variations is reduced by greater than 10-fold. Figure 4.12D and E show the effect of shim-induced eddy currents. With full pre-emphasis (Figure 4.12D), the magnetic field homogeneity obtained with third-order dynamic shimming is approximately 50% better than with static third-order SH shimming. However, without pre-emphasis (Figure 4.12E) the eddy current-related lower-order SH contributions degrade the magnetic field homogeneity to a level worse than that which can be achieved with static third-order SH shimming, thereby negating the advantage of dynamic SH shimming entirely. It should be noted that the degree of degradation depends on the history of shim changes from the preceding slices, *i.e.* it is subject- and ROI-specific. In addition, any change in slice order or repetition time will lead to a different eddy current field distribution.

4.4.3 Dynamic Spherical Harmonic Shimming of the Human Brain

Dynamic updating of linear shim terms was first presented in the multi-voxel spectroscopy work of Ernst and Hennig in 1991 [47], used in multi-slice imaging in 1996 by Blamire *et al.*,[48] followed shortly by Morrell and Spielman.[49] Higher-order dynamic SH shimming in multi-slice imaging was first demonstrated by de Graaf *et al.*[50] on rat brain and by Koch *et al.*[38] on human brain. The utility of higher-order dynamic SH shimming in human brain imaging was also established by Zhao *et al.*[51] through computer simulations. Application of higher-order dynamic SH shimming in multi-voxel MRS has been presented by Koch *et al.*[52] More recently, dynamic SH shimming has been applied at 7 T for MRI[40,53] and MRSI[54] applications.

Figure 4.13 shows typical results from the human brain of first- through fifth-order dynamic SH shimming, which can be directly compared to the results for static first-through-fifth order SH shimming shown in Figure 4.9. Figure 4.14 gives a quantitative summary of the performance of dynamic SH shimming that can be directly compared to Figure 4.10A. As a general rule-of-thumb it can be stated that, at the level of global statistics, dynamic SH shimming with order n gives comparable results to static SH shimming with order $n + 2$, in agreement with other studies.[40] However, at the individual

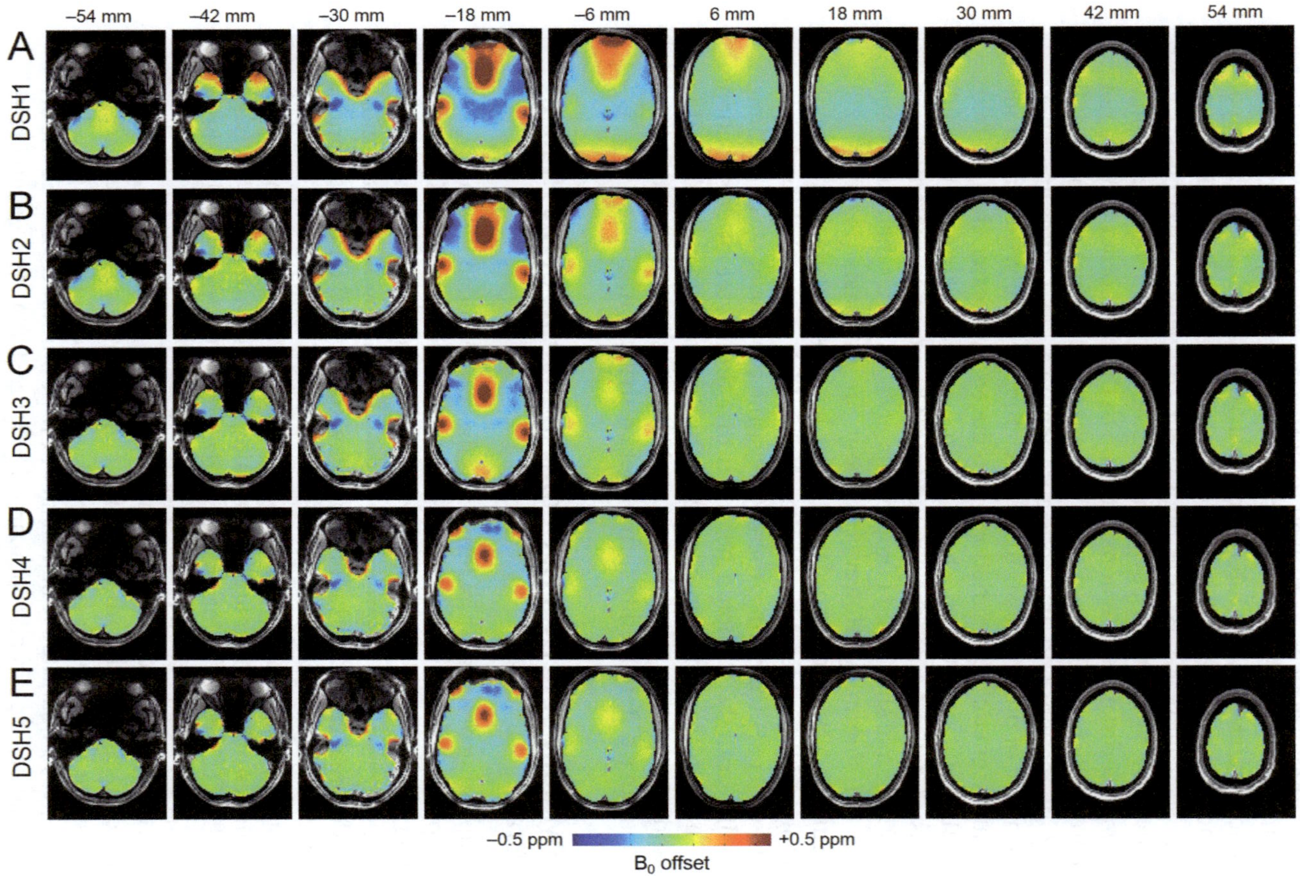

Figure 4.13 Dynamic SH shimming of the human brain. As the same subject is shown for static SH shimming, the results can be directly compared to Figure 4.9. Residual B_0 maps were calculated following removal of SH terms corresponding to (A) $n \leq 1$, (B) $n \leq 2$, (C) $n \leq 3$, (D) $n \leq 4$ and (E) $n \leq 5$. All calculations were performed over three-slice volumes composed of the target slice and the adjacent two slices. Maximum shim currents were limited to three times the maximum values shown in Figure 4.10, although unlimited shim currents gave essentially the same results.

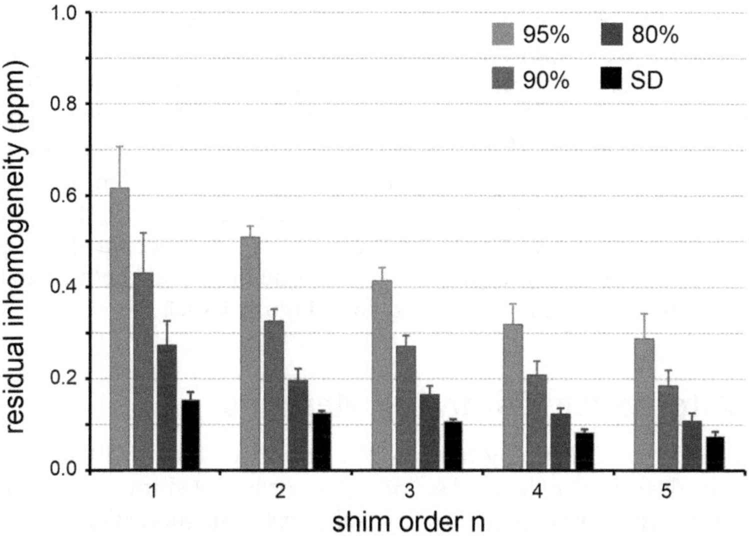

Figure 4.14 Quantitative summary of dynamic SH shimming performance. Residual magnetic field inhomogeneity following the removal of SH shim terms of increasing order up to $n = 5$. To capture the non-Gaussian character of the residual B_0 offset distribution, residual magnetic field inhomogeneity is expressed as the frequency range that contains 80, 90 and 95% of all MR pixels in addition to the standard deviation (compare Figure 4.10). Error bars represent the standard deviation across 7 subjects.

slice level, significant differences can still manifest themselves. Comparing static, fifth-order SH shimming (Figure 4.9) with dynamic, third-order SH shimming (Figure 4.13) reveals that static SH shimming has a number of areas of undesirable inhomogeneity (*e.g.* the slices positioned at −42 mm and −18 mm).

It should be noted that experimental dynamic SH shimming with full pre-emphasis has only been demonstrated up to third-order SH.[40] While the extension to fourth and fifth-order SH fields appears straightforward, the need for extensive pre-emphasis on all SH terms and their cross-interactions with lower-order SH terms may prevent an experimental realization. In addition to the increased experimental challenges, the inclusion of increasingly higher-order SH terms will also lead to diminishing returns in terms of absolute magnetic field homogeneity (see Figure 4.14 and ref. 40).

It should be realized that the shim current requirements for dynamic SH shimming can be many times greater than those summarized in Figure 4.10C. Firstly, global SH shimming is necessarily a compromise between the homogeneity in many different areas, which tends to lead to the compensation of an average inhomogeneity (see Figure 4.11A). During local shim optimization the number of contributing areas is greatly reduced, leading to the compensation of the actual, typically stronger, local inhomogeneity (see Figure 4.11B). Secondly, the shim current overshoot required to provide complete shim pre-emphasis can be in excess of 100%. For a given amplifier configuration this would translate to an effective reduction in available shim current range by 50%. The results in Figure 4.12 and 4.13 were obtained by limiting the available shim currents to three times the maximum values shown in Figure 4.10C. Repeating the minimization with unlimited shim currents gave essentially the same results. Taken together with the shim current overshoots required for pre-emphasis, a safe number for the required SH shim strengths to homogenize the magnetic field over the human brain is five times the maximum values shown in Figure 4.10C.

4.5 Alternative Shimming Methods

Besides the linear imaging gradients, the majority of (clinical) MR systems are equipped with a very limited set of higher-order SH shim coils, often only including a second-order SH set. As a result, severe magnetic field inhomogeneity in many parts of the human brain is the rule rather than the exception. The results in Figure 4.9 and 4.13 clearly demonstrate that it is beneficial to extend the MR system with higher order ($n > 2$) SH shim coils, but this comes at the cost of increased space requirements. Over the last decade, significant research efforts have focused on alternative approaches to homogenize the magnetic field across the human brain. The approaches fall in two categories, one employing subject-specific shimming with passive (*i.e.* no current carrying) materials and the other with active non-SH-based coils.

4.5.1 Passive Approaches

Passive shimming involves the placement of magnetic material at strategic locations so as to compensate the magnetic field inhomogeneity within an adjacent region-of-interest. As described previously, passive shimming is routinely employed in optimizing the magnetic field homogeneity of the bare magnet by placing small pieces of steel at the edge of the magnet. The same principle can be applied for the optimizing of subject-specific magnetic field homogeneity.

Probably the first example was shown by Jesmanowicz *et al.*,[55] who used copy-toner as a ferromagnetic material for passive shimming. Following the characterization of the required SH field terms, printed distributions of copy toner on sheets of paper were created which were subsequently rolled to form ferroshim inserts.

Juchem *et al.*[56] used a Ni–Fe alloy (permalloy) to create high strength second order SH fields to supplement the weaker, active SH shims in an animal MRI system. By using two permalloy strips per SH term, all five second order SH fields could be generated with good accuracy. Following magnetic field mapping to determine the required SH fields the subject-specific passive shim set was constructed to achieve a correction of the bulk magnetic field inhomogeneity. As a final step the active shims were fine-tuned to achieve the optimal magnetic field homogeneity.

As described earlier, the magnetic field inhomogeneity in the human frontal cortex is caused primarily by the susceptibility difference between air in the nasal cavity and water in the brain. As an air-filled sphere is a rough approximation of the nasal cavity, the magnetic field inhomogeneity is roughly dipolar in nature (Figures 4.2F and 4.9). While this is a great oversimplification of the problem, the positive lobe of a general paramagnetic dipole field can be seen in essentially all human subjects. Wilson *et al.*[57,58] proposed to reduce the magnitude of this roughly paramagnetic dipolar field distribution by placing a highly diamagnetic material in the mouth of subjects. The mouth cavity is typically located directly underneath the sinus cavity in the z direction, which is by definition the B_0 magnetic field direction. The negative dipole lobe of a diamagnetic material in the mouth can then greatly reduce the positive dipole lobe induced by the sinus cavity. Wilson *et al.*[57,58] used highly oriented, pyrolytic graphite (PG) as the diamagnetic material. The susceptibility of PG is highly anisotropic with $\chi \sim -450$ ppm perpendicular to and $\chi \sim -85$ ppm parallel to the graphite plane. In order to accommodate inter-subject magnetic field variations a range of PG-based mouth inserts were created. While the mouth inserts improved the magnetic field homogeneity in several brain areas, significant inhomogeneity remained in other parts. Ultimately, the use of PG-based diamagnetic mouth inserts has not been embraced by the larger MR community. Contributing factors might be subject discomfort and the inability to provide whole-brain magnetic field homogeneity.

The use of a single material with a fixed susceptibility (copier toner, permalloy, pyrolytic graphite) limits the flexibility in generating local magnetic field shapes. Koch *et al.*[59] extended the field shaping capability of passive shimming by using two materials with different magnetic susceptibility. Paramagnetic zirconium (χ = +92 ppm) and diamagnetic bismuth (χ = −164 ppm) provide a strong, local magnetic field perturbation, are easy to use and are readily available at reasonable cost. Using these materials in an animal MRI scanner the magnetic field homogeneity across the mouse brain was optimized using a cylindrical/conical array of 6 rows and 16 elements per row. The magnetic field maps were decomposed with the experimentally measured response of unit pieces of zirconium and bismuth to obtain the optimal metal distribution across the 96 elements. The magnetic field homogeneity improved across the entire mouse brain and outperformed standard second-order SH shimming. The two-material passive shimming approach has since been extended to human brain.[60,61]

The primary disadvantage of all passive shimming approaches is the limited ability to account for inter-subject variations during MR studies. The magnetic field homogeneity across the human brain varies strongly owing to anatomical differences between subjects, but also owing to positioning and angulation relative to the magnetic field direction within the magnet. This observation is supported by the large range of SH shim fields that exist for different subjects[37] (Figure 4.10C). A single, fixed passive shim can, therefore, not be expected to optimally improve the magnetic field homogeneity, and tailoring of the passive shim setup to the subject becomes mandatory. While this optimization leads to improved performance, it also greatly increases the time requirements to construct or adopt a subject-specific passive shim. The inability to rapidly and effectively deal with inter-subject magnetic field variations has led to a dismissal of passive shimming by the larger MR community. Nevertheless, the passive shimming approaches have demonstrated that non-SH-based shimming can be highly effective. This has naturally led to the development of active shimming approaches that do not rely on the SH framework.

4.5.2 Active Approaches

Hsu and Glover[62] were one of the first to show that *local* active shimming has the potential to greatly improve the *local* magnetic field homogeneity. By placing three independently adjustable current loops in the human mouth cavity, the total magnetic field can be sufficiently shaped to provide improved magnetic field homogeneity across the human frontal cortex in a subject-specific manner. While these studies provided proof-of-concept for local active shimming approaches, devices placed within the human body create patient discomfort and safety concerns that prevent acceptance within the MR community.

Another active approach is so-called single-channel shim coils.[63,64] A single, continuous wire can be routed in complicated ways to produce a magnetic field of almost arbitrary complexity. This has been used successfully to homogenize low-field, open MR magnets. The disadvantages for *in vivo* shimming are similar to that of passive approaches in that a single-channel coil will not have sufficient flexibility to capture the inevitable large inter-subject variations.

The single-channel shim coil concept has been extended by Harris *et al.*[65] with a technique referred to as the dynamically controlled adaptive current network. Instead of using a single, continuous wire Harris *et al.* used a cylindrical grid of wires in which each intersection is controlled by switches/transistors. By controlling the switches in several hundred intersections, the current flow can be forced along complicated trajectories similar to the single-channel approach. As the switches can be controlled dynamically, a wide range of current pathways can be achieved with the same setup. While the technique works well at a proof-of-principle level, a detailed assessment of the *in vivo* shimming performance is currently lacking.

Juchem *et al.*[66] extended the concept of local active shim coils to a setup that is entirely outside the human body. By using six independently adjustable current loops of varying sizes placed in specific, fixed locations a significant improvement in magnetic field homogeneity across the human frontal cortex was achieved without negatively affecting other parts of the brain. Whereas the design aim of this study was limited to the human frontal cortex, it proved that local active shimming using a multi-coil (MC) setup external to the human body can be highly successful. The generalization of the local, active shimming approach to a grid of 4 × 12 DC coils was presented in 2011 for the human brain,[67] based on earlier studies performed on mouse brain.[68,69] When applied in a dynamic fashion, MC-based shimming is referred to as DYNAMITE (*DYN*amic *M*ulti-co*I*l *TE*chnique).

4.5.2.1 *Principles and Considerations in Multicoil Shimming*

The implementation of MC shimming is to a large degree characterized by flexibility in the position, orientation and shape of the individual DC coils. Whereas the design philosophy of SH shim coils often hinges on achieving high purity of individual SH terms over a fixed DSV, MC setups essentially rely on the magnetic field generated by a simple circular current loop (Figure 4.4B and C). The accuracy or purity of an MC-generated shim field is therefore not determined beforehand, but is optimized for a given ROI. As a result, an MC setup can be tailored for specific experimental conditions, anatomical regions or species. For example, an MC setup for rat brain MRI has been described in which the placement of DC coils was restricted owing to the presence of a surface coil RF transceiver and electrophysiological equipment.[70] Despite the non-cylindrical distribution of DC coils, excellent magnetic field homogeneity was achieved across the rat brain. Using this flexibility it is not hard to see that specific organs, such as the human breast or spine, can greatly benefit from dedicated MC setups.

However, despite the flexibility there are a number of guiding principles that can maximize the performance of an MC setup. It is well-known that the magnetic field strength generated by a circular current loop decreases rapidly with increasing distance from the coil. For an axially oriented coil (Figure 4.4B) the magnetic field strength drops to 35% and 9% relative to that in the center of the loop at one and two radii away from the coil, respectively. Proximity of the DC coils to the subject under investigation is therefore important for maximum efficiency.

It is also well known that the presence of metal can have a significant effect of the RF field distribution of a loop coil.[71] In the first mouse MC prototype setup a dampening of the RF transmit field was observed when the MC setup was placed within the RF coil.[69] Placing the MC setup outside the RF shield in a second mouse prototype[68] eliminated all effects on the RF field distribution. Using this knowledge, the interaction between RF and DC coils in the MC setup designed for human brain studies was minimized through elimination of physical overlap between the two coil types.[67] Recent work has pursued the combination of RF and DC coils using the same physical wire.[72–74] While in these studies the RF eddy-current-induced dampening of the RF transmit and receive fields was minimal, the RF performance was nonetheless compromised, presumably owing to the increased complexity of the RF coil circuitry.

Shim coil efficiency, *i.e.* the ability to generate a certain shim field strength and shape for a given current and wire length, is an important parameter for any shim coil design. For classical SH shim coils the efficiency is a constant since the shim coil geometry and target DSV are fixed. For MC-based shim setups, efficiency is a dynamic parameter that depends on the ROI, MC geometry and desired shim field. Juchem *et al.*[75] performed a detailed analysis of the efficiency for MC setups to generate SH fields of various orders. It was found that a 6 × 8 MC setup, similar to the one used for MC shimming of the mouse brain,[68] could generate all first through third order SH fields over a spherical DSV with a similar efficiency as dedicated SH shim coils. When the ROI was reduced to a single slice, the MC efficiency increased several-fold, the amount depending on the exact SH term. Finally, when applying all SH fields together, the overall MC efficiency was found to be several-fold higher than the SH efficiency. The MC efficiency for specific SH terms could be enhanced by specific DC coil placement. However, this would reduce the flexibility of MC shimming to deal with any arbitrary magnetic field distribution. The MC accuracy and efficiency to generate lower-order SH fields can also be improved by replacing the simple circular DC coil shape with more complex, optimized coil shapes.[76] It remains to be proven if more accurate lower-order SH fields generated by an MC setup translate into an improved shimming performance. Currently, the MC efficiency is such that improved magnetic field homogeneity on mouse, rat and human brain has been achieved at high magnetic fields with only a modest ±1 A current on each DC coil.

A final consideration of MC-based shimming relates to potential interactions between the DC current loops and (1) other DC current loops, (2) the magnetic field gradients and (3) the magnet. Similar to dynamic SH shimming, the application of dynamic MC shim fields is inherently more powerful than the use of a single, fixed MC setting. Rapidly changing the current in one DC coil should, according to Faraday's law of magnetic induction, lead to induced currents in the surrounding coils. This effect has indeed been observed experimentally,[77] although the induced currents amount to less than 1% of the driving current when driven by amplifiers with a constant current topology. In addition, this effect is only visible during *changes* in DC current. Since these changes always occur prior to slice excitation, the effect is inconsequential for dynamic MC shimming. Similar to the interactions between DC coils, the effect of magnetic field gradients can also be observed as small (<1%) induced currents in the DC coils during magnetic field gradient transients.[77] As magnetic field gradients between signal excitation and reception are balanced in most MR pulse sequences, the total effect is largely self-canceling. Finally, DC current loops can, in principle, generate eddy currents in the magnet cryostat and other structures when pulsed. Fortunately, the close proximity of the MC setup to the subject automatically translates to a significant physical distance between the MC shims and the magnet bore. As a result, MC-induced eddy currents have not been observed experimentally.

4.5.2.2 Multi-Coil Shimming of the Human Brain

Figure 4.15 shows a typical MC shimming result on the human brain. Using a 48-coil setup similar to that described by Juchem *et al.*[67,78] the magnetic field homogeneity obtained after static, whole-brain shimming (Figure 4.15A) is comparable to static, fifth-order SH shimming or dynamic, third-order SH shimming. However, the magnetic field homogeneity following dynamic MC shimming (Figure 4.15B) is better than that achieved with any other shimming method. Figure 4.15C and G (left column) give the quantitative summary of the MC shimming performance which can be directly compared to Figures 4.10A and 4.14 for static and dynamic SH shimming, respectively.

Recently, the local DC current coils that are integral to the MC shimming approach have been combined with the high-frequency RF receive coils.[72–74] Using the same physical wire for the shim and RF currents has the advantage that both coil types can be close to the subject. The disadvantages are that (1) the shim coil is limited to a single turn, thereby demanding stronger shim amplifiers to provide the required magnetic field strength. (2) The RF coil design is substantially more complex with the need for multiple chokes and additional inductors and capacitors to allow the flow of DC current in an AC-optimized loop. The need for additional components per coil can lead to compromised RF sensitivity.[74] (3) The combined RF-shim setup is limited to studies that employ a receive coil array and finally (4) RF and MC do not necessarily share the same optimal geometry. Nevertheless, the combination of RF and shim coil into a single setup is attractive and Figure 4.15C and G (middle column) show the performance of a 48-coil ('helmet') RF-shim setup (Figure 4.15E). For global, static shimming the RF-shim setup is inferior to the four-ring, 48-coil set up shown in Figure 4.15D. However, for dynamic shimming the performance of the two setups is similar. This again underlines the flexibility of the MC shimming approach in that the performance is not critically dependent on the exact coil placement. The poor global shimming performance of the helmet RF-shim setup is due to the absence of DC coils around the nose and mouth areas, thereby preventing the proper shaping of the required global shim fields. This can be remedied by employing a hybrid setup consisting of 24 combined RF-shim coils and 24 DC coils in two rings of 12 (Figure 4.15F). The global shimming performance of this hybrid setup is markedly improved (Figure 4.15C and G, right column), while retaining excellent dynamic shimming capabilities. The results in Figure 4.15 demonstrate that superior shimming performance can be obtained with non-SH-based setups composed of local DC coils. After adhering to some general guidelines on reasonable proximity and coil distribution, the exact experimental implementation of the MC concept is highly flexible with many possibilities for RF and DC coil positioning and integration. It is expected that the future will bring study-specific MC setups, similar to the study-specific RF arrays commonly employed.

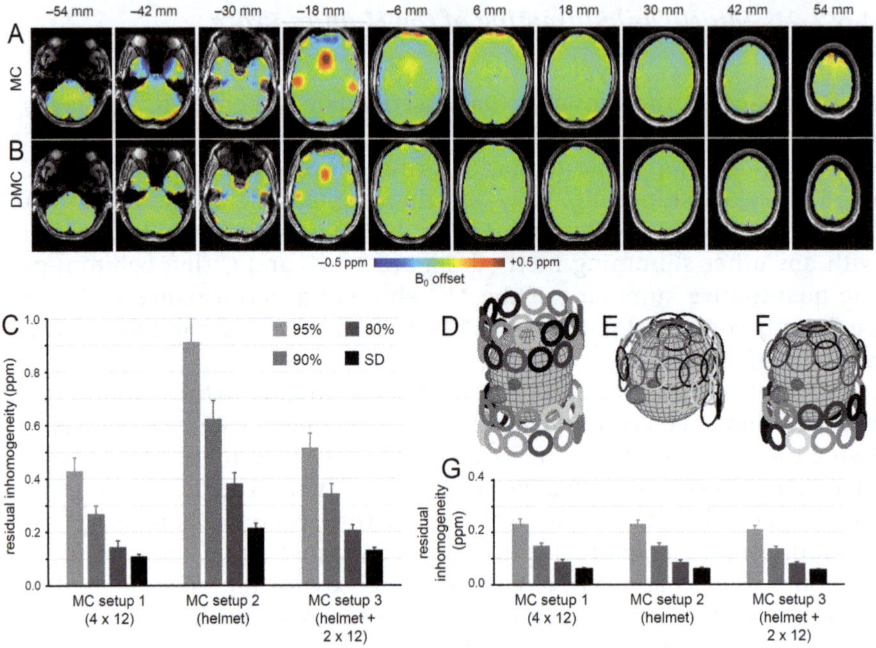

Figure 4.15 Multi-coil (MC) shimming of the human brain. (A and B) Residual B_0 maps following (A) static, global MC shimming and (B) dynamic, local MC shimming using a cylindrical 48-coil setup similar to that described in ref. 67. As the subject population is identical, the displayed B_0 maps can be directly and quantitatively compared to those obtained with static SH shimming (Figure 4.9) and dynamic SH shimming (Figure 4.13). (C) Quantitative evaluation of static, global MC shimming performance using the cylindrical 48-coil setup and two alternative setups, shown in (D–F). The combined 48-coil RF-shim setup (E) is referred to as a 'helmet', whereas the hybrid setup consisting of a 24-coil RF-shim helmet supplemented with two rings of 12 coils is shown in (F). The thicker coils in (D) and (F) represent 100-turn coils with a 1 A current limit, whereas the thinner coils in (E) and (F) represent single turn coil with a 3 A current limit. Note that the displayed coil thickness is not quantitatively linked to the actual coil thickness. The different coil gray scale colors reflect the fact that each coil has an independent current. (G) Quantitative evaluation of dynamic MC shimming performance for all three MC setups. In (C) and (G) the errors represent standard deviations over 7 subjects. The results of (C and G) can be directly compared with the quantitative performance measures shown in Figures 4.10 and 4.14 for static and dynamic SH shimming, respectively.

References

1. M. J. E. Golay, Field Homogenizing coils for nuclear spin resonance instrumentation, *Rev. Sci. Instrum.*, 1958, **29**, 313–315.
2. F. Romeo and D. I. Hoult, Magnet field profiling: analysis and correcting coil design, *Magn. Reson. Med.*, 1984, **1**, 44–65.

3. J. F. Schenck, The role of magnetic susceptibility in magnetic resonance imaging: MRI magnetic compatibility of the first and second kinds, *Med. Phys.*, 1996, **23**, 815–850.
4. D. I. Hoult and D. Lee, Shimming a superconducting nuclear-magnetic-resonance imaging magnet with steel, *Rev. Sci. Instrum.*, 1985, **56**, 131–135.
5. J. P. Marques and R. Bowtell, Application of a Fourier-based method for rapid calculation of field inhomogeneity due to spatial variation of magnetic susceptibility, *Concepts Magn. Reson.*, 2005, **25B**, 65–78.
6. R. Salomir, B. D. de Senneville and C. T. W. Moonen, A fast calculation method for magnetic field inhomogeneity due to an arbitrary distribution of bulk susceptibility, *Concepts Magn. Reson., Part B*, 2003, **19**, 26–34.
7. K. M. Koch, X. Papademetris, D. L. Rothman and R. A. de Graaf, Rapid calculations of susceptibility-induced magnetostatic field perturbations for *in vivo* magnetic resonance, *Phys. Med. Biol.*, 2006, **51**, 6381–6402.
8. S. K. Lee and I. Hancu, Patient-to-patient variation of susceptibility-induced B_0 field in bilateral breast MRI, *J. Magn. Reson. Imaging*, 2012, **36**, 873–880.
9. C. D. Jordan, B. L. Daniel, K. M. Koch, H. Yu, S. Conolly and B. A. Hargreaves, Subject-specific models of susceptibility-induced B0 field variations in breast MRI, *J. Magn. Reson. Imaging*, 2013, **37**, 227–232.
10. E. M. Haacke, S. Liu, S. Buch, W. Zheng, D. Wu and Y. Ye, Quantitative susceptibility mapping: current status and future directions, *Magn. Reson. Imaging*, 2015, **33**, 1–25.
11. L. K. Forbes and S. Crozier, Asymmetric zonal shim coils for magnetic resonance applications, *Med. Phys.*, 2001, **28**, 1644–1651.
12. P. Hudson, S. D. Hudson, W. B. Handler, T. J. Scholl and B. A. Chronik, Quantitative Comparison of Minimum Inductance and Minimum Power Algorithms for the Design of Shim Coils for Small Animal Imaging, *Concepts Magn. Reson., Part B*, 2010, **37b**, 65–74.
13. P. Mansfield and B. Chapman, Active magnetic screening of coils for static and time-dependent magnetic field generation in NMR imaging, *J. Phys. E: Sci. Instrum.*, 1986, **19**, 540–545.
14. M. A. Brideson, L. K. Forbes and S. Crozier, Winding patterns for actively shielded shim coils with asymmetric target-fields, *Meas. Sci. Technol.*, 2003, **14**, 484–493.
15. L. K. Forbes and S. Crozier, A novel target-field method for magnetic resonance shim coils: III. Shielded zonal and tesseral coils, *J. Phys. D: Appl. Phys.*, 2003, **36**, 68–80.
16. S. E. Ungersma, H. Xu, B. A. Chronik, G. C. Scott, A. Macovski and S. M. Conolly, Shim design using a linear programming algorithm, *Magn. Reson. Med.*, 2004, **52**, 619–627.
17. D. I. Hoult and R. Deslauriers, Accurate shim-coil design and magnet-field-profiling by a power-minimization-matrix method, *J. Magn. Reson., Ser. A*, 1994, **108**, 9–20.

18. R. Turner, A target field approach to optimal coil design, *J. Phys. D: Appl. Phys.*, 1986, **19**, L147–L151.
19. R. Turner, Minimum inductance coils, *J. Phys. E: Sci. Instrum.*, 1988, **21**, 948–952.
20. L. K. Forbes and S. Crozier, A novel target-field method for finite-length magnetic resonance shim coils: I. Zonal shims, *J. Phys. D: Appl. Phys.*, 2001, **34**, 3447–3455.
21. L. K. Forbes and S. Crozier, A novel target-field method for finite-length magnetic resonance shim coils: II. Tesseral shims, *J. Phys. D: Appl. Phys.*, 2002, **35**, 839–849.
22. S. Pissanetzky, Minimum energy MRI gradient coils of general geometry, *Meas. Sci. Technol.*, 1992, **3**, 667–673.
23. R. Lemdiasov and R. Ludwig, A stream function method for gradient coil design, *Concepts Magn. Reson.*, 2005, **26B**, 67–80.
24. M. Poole and R. Bowtell, Novel gradient coils designed using a boundary element method, *Concepts Magn. Reson.*, 2007, **31B**, 162–175.
25. M. S. Poole and N. J. Shah, Convex optimisation of gradient and shim coil winding patterns, *J. Magn. Reson.*, 2014, **244**, 36–45.
26. M. Jenkinson, Fast, automated, N-dimensional phase-unwrapping algorithm, *Magn. Reson. Med.*, 2003, **49**, 193–197.
27. R. Gruetter and C. Boesch, Fast, noniterative shimming of spatially localized signals, *in vivo* analysis of the magnetic field along axes, *J. Magn. Reson.*, 1992, **96**, 323–334.
28. R. Gruetter, Automatic, localized in vivo adjustment of all first- and second-order shim coils, *Magn. Reson. Med.*, 1993, **29**, 804–811.
29. J. Shen, R. E. Rycyna and D. L. Rothman, Improvements on an *in vivo* automatic shimming method [FASTERMAP], *Magn. Reson. Med.*, 1997, **38**, 834–839.
30. R. Gruetter and I. Tkac, Field mapping without reference scan using asymmetric echo-planar techniques, *Magn. Reson. Med.*, 2000, **43**, 319–323.
31. J. Shen, D. L. Rothman, H. P. Hetherington and J. W. Pan, Linear projection method for automatic slice shimming, *Magn. Reson. Med.*, 1999, **42**, 1082–1088.
32. H. Wen and F. A. Jaffer, An *in vivo* automated shimming method taking into account shim current constraints, *Magn. Reson. Med.*, 1995, **34**, 898–904.
33. D. H. Kim, E. Adalsteinsson, G. H. Glover and D. M. Spielman, Regularized higher-order *in vivo* shimming, *Magn. Reson. Med.*, 2002, **48**, 715–722.
34. G. N. Chmurny and D. I. Hoult, The ancient and honourable art of shimming, *Concepts Magn. Reson.*, 1990, **2**, 131–149.
35. L. F. Fuks, F. S. C. Huang, C. M. Carter, W. A. Edelstein and P. B. Roemer, Susceptibility, lineshape, and shimming in high-resolution NMR, *J. Magn. Reson.*, 1992, **100**, 229–242.

36. J. W. Pan, K. M. Lo and H. P. Hetherington, Role of very high order and degree B_0 shimming for spectroscopic imaging of the human brain at 7 Tesla, *Magn. Reson. Med.*, 2013, **68**, 1007–1017.
37. S. Clare, J. Evans and P. Jezzard, Requirements for room temperature shimming of the human brain, *Magn. Reson. Med.*, 2006, **55**, 210–214.
38. K. M. Koch, S. McIntyre, T. W. Nixon, D. L. Rothman and R. A. de Graaf, Dynamic shim updating on the human brain, *J. Magn. Reson.*, 2006, **180**, 286–296.
39. K. M. Koch, D. L. Rothman and R. A. de Graaf, Optimization of static magnetic field homogeneity in the human and animal brain *in vivo*, *Prog. Nucl. Magn. Reson. Spectrosc.*, 2009, **54**, 69–96.
40. C. Juchem, T. W. Nixon, P. Diduch, D. L. Rothman, P. Starewicz and R. A. de Graaf, Dynamic shimming of the human brain at 7 Tesla, *Concepts Magn. Reson., Part B*, 2010, **37**, 116–128.
41. A. Bhogal, M. Versluis, J. Koonen, J. C. W. Siero, V. O. Boer, D. W. Klomp, P. R. Luijten and H. Hoogduin, Image-based method to measure and characterize shim-induced eddy current fields, *Concepts Magn. Reson.*, 2013, **42A**, 245–260.
42. S. J. Vannesjo, M. Haeberlin, L. Kasper, M. Pavan, B. J. Wilm, C. Barmet and K. P. Pruessmann, Gradient system characterization by impulse response measurements with a dynamic field camera, *Magn. Reson. Med.*, 2013, **69**, 583–593.
43. S. J. Vannesjo, B. E. Dietrich, M. Pavan, D. O. Brunner, B. J. Wilm, C. Barmet and K. P. Pruessmann, Field camera measurements of gradient and shim impulse responses using frequency sweeps, *Magn. Reson. Med.*, 2014, **72**, 570–583.
44. P. Jehenson, M. Westphal and N. Schuff, Analytical method for the compensation of eddy-current effects induced by pulsed magnetic field gradients in NMR systems, *J. Magn. Reson.*, 1990, **90**, 264–278.
45. D. J. Jensen, W. W. Brey, J. L. Delayre and P. A. Narayana, Reduction of pulsed gradient settling time in the superconducting magnet of a magnetic resonance instrument, *Med. Phys.*, 1987, **14**, 859–862.
46. J. J. van Vaals and A. H. Bergman, Optimization of eddy-current compensation, *J. Magn. Reson.*, 1990, **90**, 52–70.
47. T. Ernst and J. Hennig, Double-volume ^{1}H spectroscopy with interleaved acquisitions using tilted gradients, *Magn. Reson. Med.*, 1991, **20**, 27–35.
48. A. M. Blamire, D. L. Rothman and T. Nixon, Dynamic shim updating: a new approach towards optimized whole brain shimming, *Magn. Reson. Med.*, 1996, **36**, 159–165.
49. G. Morrell and D. Spielman, Dynamic shimming for multi-slice magnetic resonance imaging, *Magn. Reson. Med.*, 1997, **38**, 477–483.
50. R. A. de Graaf, P. B. Brown, S. McIntyre, D. L. Rothman and T. W. Nixon, Dynamic shim updating (DSU) for multislice signal acquisition, *Magn. Reson. Med.*, 2003, **49**, 409–416.

51. Y. Zhao, A. W. Anderson and J. C. Gore, Computer simulation studies of the effects of dynamic shimming on susceptibility artifacts in EPI at high field, *J. Magn. Reson.*, 2005, **173**, 10–22.
52. K. M. Koch, L. I. Sacolick, T. W. Nixon, S. McIntyre, D. L. Rothman and R. A. de Graaf, Dynamically shimmed multivoxel ^{1}H magnetic resonance spectroscopy and multislice magnetic resonance spectroscopic imaging of the human brain, *Magn. Reson. Med.*, 2007, **57**, 587–591.
53. S. Sengupta, E. B. Welch, Y. Zhao, D. Foxall, P. Starewicz, A. W. Anderson, J. C. Gore and M. J. Avison, Dynamic B_0 shimming at 7 T, *Magn. Reson. Imaging*, 2011, **29**, 483–496.
54. V. O. Boer, D. W. Klomp, C. Juchem, P. R. Luijten and R. A. de Graaf, Multi-slice MRSI of the human brain at 7 Tesla using dynamic B_0 and B_1 shimming, *Proc. Int. Soc. Magn. Reson. Med.*, 2011, **19**, 142.
55. A. Jesmanowicz, V. Roopchansingh, R. W. Cox, P. Starewicz, W. F. B. Punchard and J. S. Hyde, Local ferroshims using office copier toner, *Proc. Int. Soc. Magn. Reson. Med.*, 2001, **9**, 617.
56. C. Juchem, B. Muller-Bierl, F. Schick, N. K. Logothetis and J. Pfeuffer, Combined passive and active shimming for *in vivo* MR spectroscopy at high magnetic fields, *J. Magn. Reson.*, 2006, **183**, 278–289.
57. J. L. Wilson, M. Jenkinson and P. Jezzard, Optimization of static field homogeneity in human brain using diamagnetic passive shims, *Magn. Reson. Med.*, 2002, **48**, 906–914.
58. J. L. Wilson, M. Jenkinson and P. Jezzard, Protocol to determine the optimal intraoral passive shim for minimisation of susceptibility artifact in human inferior frontal cortex, *NeuroImage*, 2003, **19**, 1802–1811.
59. K. M. Koch, P. B. Brown, D. L. Rothman and R. A. de Graaf, Sample-specific diamagnetic and paramagnetic passive shimming, *J. Magn. Reson.*, 2006, **182**, 66–74.
60. K. M. Koch, P. B. Brown, D. L. Rothman and R. A. de Graaf, External diamagnetic and paramagnetic passive shimming of the human brain, *Proc. Int. Soc. Magn. Reson. Med.*, 2007, 982.
61. S. Yang, H. Kim, M. O. Ghim, B. U. Lee and D. H. Kim, Local *in vivo* shimming using adaptive passive shim positioning, *Magn. Reson. Imaging*, 2011, **29**, 401–407.
62. J. J. Hsu and G. H. Glover, Mitigation of susceptibility-induced signal loss in neuroimaging using localized shim coils, *Magn. Reson. Med.*, 2005, **53**, 243.
63. D. Tamada, K. Kose and T. Haishi, A new Planar single-channel shim coil using multiple circular currents for magnetic resonance imaging, *Appl. Phys. Express*, 2012, **5**, 056701.
64. Y. Terada, S. Kono, K. Ishizawa, S. Inamura, T. Uchiumi, D. Tamada and K. Kose, Magnetic field shimming of a permanent magnet using a combination of pieces of permanent magnets and a single-channel shim coil for skeletal age assessment of children, *J. Magn. Reson.*, 2013, **230**, 125–133.
65. C. T. Harris, W. B. Handler and B. A. Chronik, A new approach to shimming: the dynamically controlled adaptive current network, *Magn. Reson. Med.*, 2014, **71**, 859–869.

66. C. Juchem, T. W. Nixon, S. McIntyre, D. L. Rothman and R. A. de Graaf, Magnetic field homogenization of the human prefrontal cortex with a set of localized electrical coils, *Magn. Reson. Med.*, 2010, **63**, 171–180.
67. C. Juchem, T. W. Nixon, S. McIntyre, V. O. Boer, D. L. Rothman and R. A. de Graaf, Dynamic multi-coil shimming of the human brain at 7T, *J. Magn. Reson.*, 2011, **212**, 280–288.
68. C. Juchem, P. B. Brown, T. W. Nixon, S. McIntyre, D. L. Rothman and R. A. de Graaf, Multicoil shimming of the mouse brain, *Magn. Reson. Med.*, 2011, **66**, 893–900.
69. C. Juchem, T. W. Nixon, S. McIntyre, D. L. Rothman and R. A. de Graaf, Magnetic field modeling with a set of individual localized coils, *J. Magn. Reson.*, 2010, **204**, 281–289.
70. C. Juchem, P. Herman, B. G. Sanganahalli, P. B. Brown, S. McIntyre, T. W. Nixon, D. Green, F. Hyder and R. A. de Graaf, DYNAmic Multi-coIl TEchnique (DYNAMITE) shimming of the rat brain at 11.7 T, *NMR Biomed.*, 2014, **27**, 897–906.
71. C. R. Camacho, D. B. Plewes and R. M. Henkelman, Nonsusceptibility artifacts due to metallic objects in MR imaging, *J. Magn. Reson. Imaging*, 1995, **5**, 75–88.
72. H. Han, A. W. Song and T. K. Truong, Integrated parallel reception, excitation, and shimming (iPRES), *Magn. Reson. Med.*, 2013, **70**, 241–247.
73. T. K. Truong, D. Darnell and A. W. Song, Integrated RF/shim coil array for parallel reception and localized B0 shimming in the human brain, *NeuroImage*, 2014, **103**, 235–240.
74. J. P. Stockmann, T. Witzel, B. Keil, J. R. Polimeni, A. Mareyam, C. LaPierre, K. Setsompop and L. L. Wald, A 32-channel combined RF and B shim array for 3T brain imaging, *Magn. Reson. Med.*, 2016, **75**, 441–451.
75. C. Juchem, D. Green and R. A. de Graaf, Multi-coil magnetic field modeling, *J. Magn. Reson.*, 2013, **236**, 95–104.
76. P. T. While and J. G. Korvink, Designing MR shim arrays with irregular coil geometry: theoretical considerations, *IEEE Trans. Biomed. Eng.*, 2014, **61**, 1614–1620.
77. T. W. Nixon, C. Juchem, S. McIntyre, D. L. Rothman and R. A. de Graaf, Design and implementation of a real time multi-coil amplifier system, *Proc. Int. Soc. Magn. Reson. Med.*, 2010, 1532.
78. C. Juchem, S. Umesh Rudrapatna, T. W. Nixon and R. A. de Graaf, Dynamic multi-coil technique (DYNAMITE) shimming for echo-planar imaging of the human brain at 7 Tesla, *NeuroImage*, 2015, **105**, 462–472.
79. N. De Zanche, C. Barmet, J. A. Nordmeyer-Massner and K. P. Pruessmann, NMR probes for measuring magnetic fields and field dynamics in MR systems, *Magn. Reson. Med.*, 2008, **60**, 176–186.

CHAPTER 5

Magnetic Field Gradients

RALPH KIMMLINGEN*[a]

[a]Wattstr.4, 90513, Zirndorf, Germany
*E-mail: ralph@dr-kimmlingen.de

5.1 Introduction

The acquisition of the first MR image was reported by Paul Lauterbur in 1973.[1] He used the first-order shim coils of his laboratory NMR system in order to measure spatially dependent magnetic resonance signals. These so-called ‘secondary fields’ from the shim coils were superimposed onto the much stronger B_0 static magnetic field. Rudimentary images were formed using a backprojection technique. Starting with these backprojection methods, together with later line scan[2] and echo planar imaging (EPI) methods[3] devised by Peter Mansfield, linear field gradient coils became an essential part of an MRI system. Incorporation of these gradient coils has also become standard in high resolution NMR systems, both for liquid and solid samples, as covered in Chapter 1.

5.1.1 Linear Magnetic Field Gradients

MRI requires spatially varying magnetic fields in all three dimensions (x,y,z), a requirement met most simply by three *orthogonal* fields $\vec{B}(\vec{r})$ in which the field amplitude varies linearly with the spatial coordinate. Current-carrying conductor geometries which generate these field types are named ‘constant

New Developments in NMR No. 7
Magnetic Resonance Technology: Hardware and System Component Design
Edited by Andrew G Webb

Published by the Royal Society of Chemistry, www.rsc.org

gradient coils' or simply 'gradient coils'. The Larmor frequency of a nucleus at position $\vec{r}$ is given by:

$$\omega = \gamma \left| \vec{B}_0 + \vec{B}(\vec{r}) \right| \tag{5.1}$$

The strength of the main magnetic field B_0 is typically two orders of magnitude higher than the maximum value imposed by the secondary gradient fields. The Larmor frequency is determined almost solely by the value of $B_z(r)$, with the effect of field components which are perpendicular to B_0 being almost negligible (see Figure 5.1):

$$\left| \vec{B}(\vec{r}_0) \right| = \sqrt{(B_0 + B_z(\vec{r}_0))^2 + B_x^{\,2}(\vec{r}_0) + B_y^{\,2}(\vec{r}_0)} \approx \left| B_0 + B_z(\vec{r}_0) \right| \tag{5.2}$$

In this chapter, the term '$B(r)$' will be used as an abbreviation for the magnetic field component $B_z(r)$. The gradient fields for the three dimensions x,y,z will be referred to as $G_x(r)$, $G_y(r)$ and $G_z(r)$, respectively.

5.1.2 Spatial Encoding and Geometric Distortion

As covered in Section 1.11, the ideal gradient coil generates a magnetic field whose spatial derivative is constant over the entire spatial domain:

$$G_x(\vec{r}) = \frac{\partial B_z}{\partial x} = \text{constant},\; G_y(\vec{r}) = \frac{\partial B_z}{\partial y} = \text{constant},\; G_z(\vec{r}) = \frac{\partial B_z}{\partial z} = \text{constant} \tag{5.3}$$

Maxwell's laws, however, dictate that this requirement can only be met in a *finite* region of space, the so-called linearity volume (LV), which is usually expressed as a diameter-of-spherical-volume (DSV) with a certain linearity radius r_{LV} (see Figure 5.2). If part of the sample lies outside the LV, its spin density information will be projected non-linearly onto the frequency axis. The resulting images show geometric distortions and foldback artifacts. Deviations of the spatial derivative of the gradient from the nominal value (without a change in sign) cause local compression or elongation of the spin density projection on the frequency axis. In the spatial domain, this is manifested as local image distortions and deviations from the true signal intensity. Since the characteristics of the non-linearity are determined by the particular gradient coil design, the image distortions can be corrected by interpolation

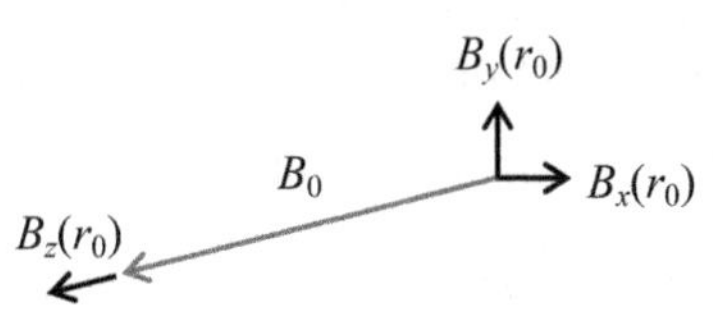

Figure 5.1 Vector components of the magnetic field at position r_0.

('distortion correction') after measurement. However, this method cannot correct the deviation from the nominal spatial resolution within the interpolated region, resulting in image blurring and loss of spatial resolution.

An aliasing artifact is generated if part of the excited sample lies beyond the peak of $B(r)$, *i.e.* in a region in which the sign of the first derivative of the gradient field has changed from positive to negative, or *vice versa*, as depicted as region C in Figure 5.2. In this case the spin density is projected onto a frequency range which is exactly the same as that encoded within part of object A, and therefore an artificial signal overlap is seen in the reconstructed image. A key element in gradient design is, therefore, to obtain as large a volume as possible over which the gradients are linear, and to avoid any regions that can cause aliasing.

Another factor that is very important to consider in gradient coil design is the generation of "eddy currents" in other conducting structures within the magnet system. By Faraday's law, the time-dependent switching of the magnetic field will induce currents in any continuous conductor surface, and Lenz's law states that the direction of induced current flow will be such that its magnetic field opposes the change of magnetic field that caused the current flow. These eddy currents will have several spatially dependent terms, which can decay with significantly different time constants. The presence of these time-varying eddy currents during imaging and/or spectroscopy sequences can have serious effects on the data quality, and so minimization of eddy currents is a major design criterion, as will be covered in detail later in this chapter. The basic approach is to use a secondary shielding coil, placed outside the main gradient coil, in an "actively shielded gradient set".

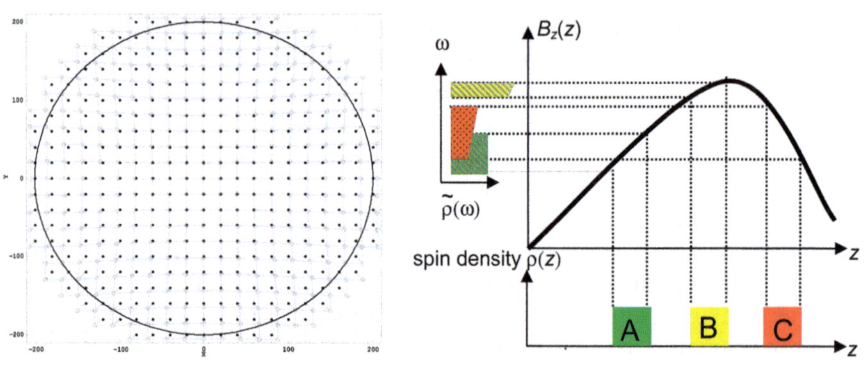

Figure 5.2 (Left) Simulated distortion plot of two identical transverse gradient axes (X,Y) in the $Z = 0$ plane. Within a circle with a diameter of 400 mm, the pixel misregistration is less than 10 mm. (Right) Imaging characteristics of a typical gradient field G_z. The magnetic field amplitude B_z is plotted against the spatial coordinate z. The imaging sequence projects the spin density $\rho(z)$ of the three objects A, B, and C on to the frequency axis, ω. A represents a linear projection as desired, B displays geometric distortion at the boundary of the linearity volume, and C exhibits back-folding/aliasing from outside the linearity volume.

5.1.3 Classification of Design Methods

According to Turner,[4] gradient coil design methods can be divided into two categories (see also Figure 5.3). The earliest methods, which were the mainstay until the mid-1980s[5,7] are based on optimization of the geometry (size/position) of independent current loops ('discrete windings'). More advanced methods introduced later are based on the design of a *continuous* current density surface. This can be performed using methods such as finite element,[8] as covered in more detail in Chapter 8. The continuous current density can be realized in its calculated form, or can be discretized to map on to distinct current loops in a second step, which produces 'distributed windings'. These two categories of methods are referred to as *Biot–Savart* ('discrete windings') and *current density* ('distributed windings') methods, respectively, and are outlined in the next two short sections.

5.1.4 Biot–Savart Methods

As shown in Chapter 1, Appendix A, the magnetic field $\vec{B}$ of an arbitrarily chosen conductor path driven with a current I (and divided into a finite number of pieces $d\vec{l}$, each located at position $\vec{r}'$) can be calculated using Biot–Savart's law (which can be derived directly from Maxwell's equations in the magnetostatic case):

$$d\vec{B}(r) = \frac{\mu I d\vec{l}}{4\pi} \times \frac{(\vec{r} - \vec{r}')}{|\vec{r} - \vec{r}'|^3} \tag{5.4}$$

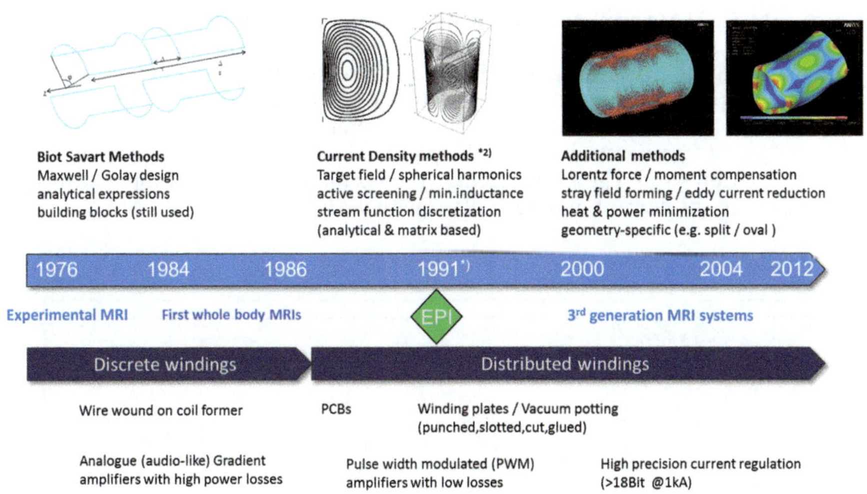

Figure 5.3 History of gradient coil design methods. *) In 1991, PNS (peripheral nerve stimulation) was shown to limit gradient performance, *2) R. Turner, Gradient Coil Design, A Review of Methods, *Magn. Reson. Imaging*, 1993, **11**, 903.

The magnetic field at the position in space $\vec{r}$ can be computed by numerical integration of the contributions $d\vec{B}$ of all conductor pieces $d\vec{l}$.

5.1.4.1 Examples

As covered in Appendix B in Chapter 1, a linear gradient in the axial (z) direction can be produced by two suitably spaced windings, with opposite currents, on a cylindrical surface (the so-called 'Maxwell coil' shown in Figure 5.4a). A linear gradient in the transverse (x,y) direction requires a slightly more complicated conductor geometry, and can be produced by four symmetric arcs, spaced appropriately on a cylindrical surface, as depicted in Figure 5.4b. Both coil types generate a linear field gradient within a sphere with a diameter of about half that of the coil cylinder (see Figures 5.5 and 5.6).

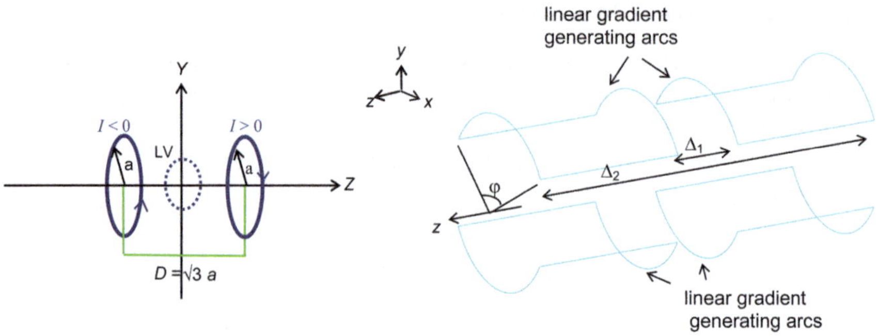

Figure 5.4 (left) Maxwell coil: two windings on a cylindrical surface with radius a and spacing $d = a\sqrt{3}$ generate an approximately linear field gradient in the z direction. The deviation from an ideal field gradient within a sphere of $r_{LV} = 0.5a$ is less than 5%. (right) Golay coil: two pairs of arcs with radius a and geometry parameters $\varphi = 120°$, $\Delta_1 = 0.78a$ and $\Delta_2 = 5.13a$ generate a linear field gradient in the y direction ($r_{LV} = 0.4a$, deviation <5%). The wire current in the x–z plane with $y > 0$ is opposite in sign to the current in the x–z plane with $y < 0$.

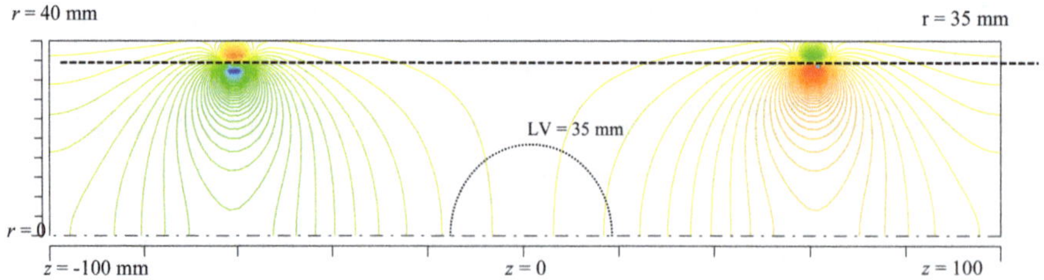

Figure 5.5 Contour plots of the magnetic field (B_z) of a Maxwell coil in the rz-plane ($I = 20$ A, $r = 35$ mm, contour steps 0.23 mT, red and green contours indicate opposite signs). The deviation from an ideal field gradient within a 35 mm sphere (LV) is less than 5%.

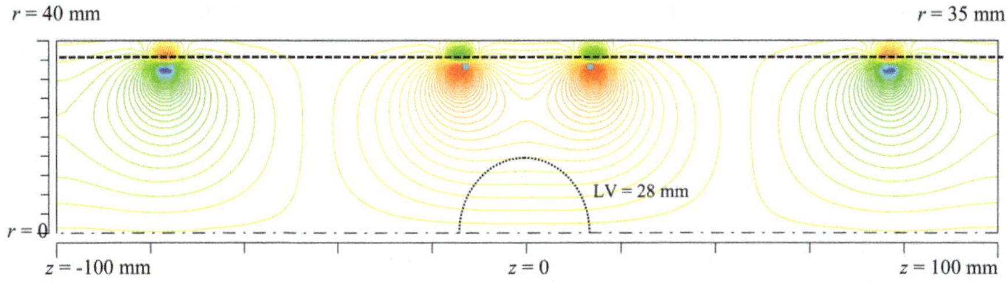

Figure 5.6 Contour plots of the magnetic field (B_z) of a Golay coil in the *rz*-plane (I = 20 A, r = 35 mm, contour steps 0.29 mT, red and green contours indicate opposite signs). The deviation from an ideal field gradient within a 28 mm sphere (LV) is less than 5%.

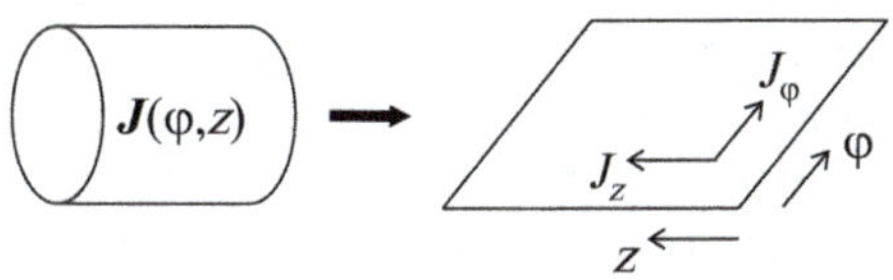

Figure 5.7 Schematic of the current density $\mathbf{J}(\varphi,z)$ of a cylindrical coil with vector components $\mathbf{J}_z$ and $\mathbf{J}_\varphi$. The two vector components are linked *via* the magnetostatic continuity equation.

So-called *building block* methods use several parameterized conductor loops in order to increase the size of the linearity volume. An axial gradient coil could, for example, be formed from three pairs of circular loops on a cylindrical surface. The axial positions of the loops could be used as an optimization parameter for an iterative algorithm. After each step (*i.e.* modification of the positions of the loops), the net magnetic field is calculated with the Biot–Savart law and compared with the target specification, with an optimization algorithm used to minimize the difference between the calculated and target fields. Examples of such optimization algorithms that have been used in gradient coil design include gradient descent, the least-square method, and simulated annealing.[5–7]

5.1.5 Current Density Methods

The magnetic field of a gradient coil is characterized by the direction and strength of the current flow in the conductor loops. It is advantageous to generalize this description to a continuous current density on the coil surface. In general, the resulting coil design will show higher spectral purity and less inductance in comparison to Biot–Savart methods.[4]

The current density (J) on the surface of a cylindrical coil can be written as (see also Figure 5.7):

$$\vec{J}(\phi,z) = J_\phi(\phi,z)\,\vec{e}_\phi + J_z(\phi,z)\,\vec{e}_z \tag{5.5}$$

where ϕ is the azimuthal angle. Both vector components are linked to each other *via* the magnetostatic continuity equation. Hence, the time derivative of the electric charge density ρ becomes zero (*i.e.* in this case of a constant current):

$$\nabla\vec{J} = -\frac{\mathrm{d}\rho}{\mathrm{d}t} = 0 \Rightarrow \frac{\partial\vec{J}}{\partial z} = -\frac{\partial\vec{J}}{\partial\phi} \tag{5.6}$$

According to eqn (5.6), all elementary currents on the current density surface form closed conductor loops. This property can be exploited by numerical optimization methods, which divide the current density surface into elementary rectangular conductor loops.

Several analytical[9] and numerical[10] optimization methods have been developed in order to determine the current density function $J(\phi,z)$.

The analytical target field method developed by Turner[11] calculates the current density distribution on a cylindrical surface, which generates a desired target field variation inside the coil cylinder. If there is no restriction to the coil length, the current density for each gradient axis (x,y,z) can be calculated on two separate (concentric) cylinder surfaces, as introduced by Bowtell and Mansfield.[12] The innermost or primary surface generates the desired gradient field within the target field volume. The outer or secondary surface cancels the stray field outside of the gradient coil and prevents induction of eddy currents on conductive surfaces of the main magnet (Figure 5.8).

One of the earliest finite element approaches divides the coil surface into small current density elements.[10] The magnetic field contribution of each element to a set of points within the region of interest is then used to formulate a matrix equation. The equation is solved in such a way that the resulting discrete current density distribution produces a minimum deviation from

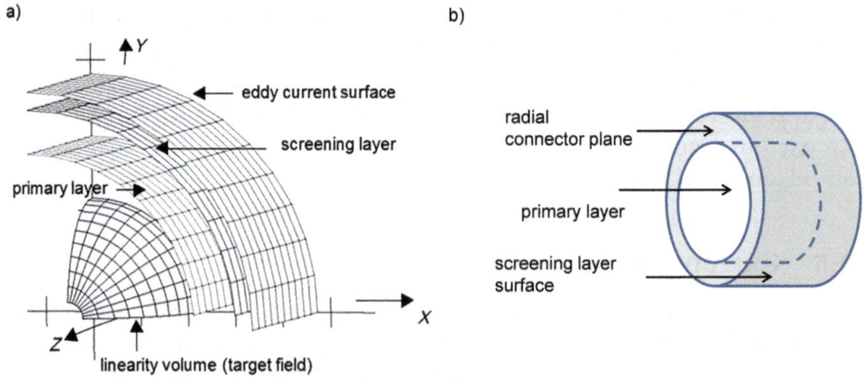

Figure 5.8 (a) Isometric view of the finite-element mesh data of a typical cylindrical actively screened coil. A numerical optimization algorithm computes both direct and indirect (induced eddy current) field contributions of each mesh element and stores them in a matrix. (b) Current density surfaces of an actively screened gradient coil. In some cases, as shown in the next two figures, it is beneficial to connect the primary and secondary surfaces by introducing an additional radial (connector) plane.

a given target field in the region of interest. Wire patterns for both optimization approaches may be derived from contours of the current density's integral ('stream function').[13] These types of methods have been extended by adding optimization criteria, such as minimum inductance, minimum power, or minimum torque with respect to the main magnet.[14,15]

Eddy current effects of a given gradient coil design can be influenced and optimized on the design level. Mansfield and Chapman[12] minimized the stray field of a cylindrical gradient coil by connecting a concentric screening layer in series with the primary layer. In order to fully screen the field of the primary layer, the screening layer needs to be longer (at least by the distance between primary and screen).

For short whole-body gradient sets with a length/diameter ratio of <2 and DSV > 40 cm, the resistance of the gradient coil becomes an important issue. With conventional shielded coil designs, current returns consume nearly half of the surface. To achieve high gradient efficiency and allow very high currents (>400 A), the optimum utilization of space in the radial dimension is essential. The so-called '3D coil design' is one approach to solve this problem.[16] In 3D design, the optimization process takes place on two radially connected current density surfaces (see Figures 5.9 and 5.10). As a result, primary and secondary half-loops form a common '3D-loop', which saves up to 50% in space compared to a conventional wire pattern. In addition, the undesired field contributions produced by current returns are reduced.

The following section describes the practical implementation steps in designing a gradient coil using the current density surface approach, with

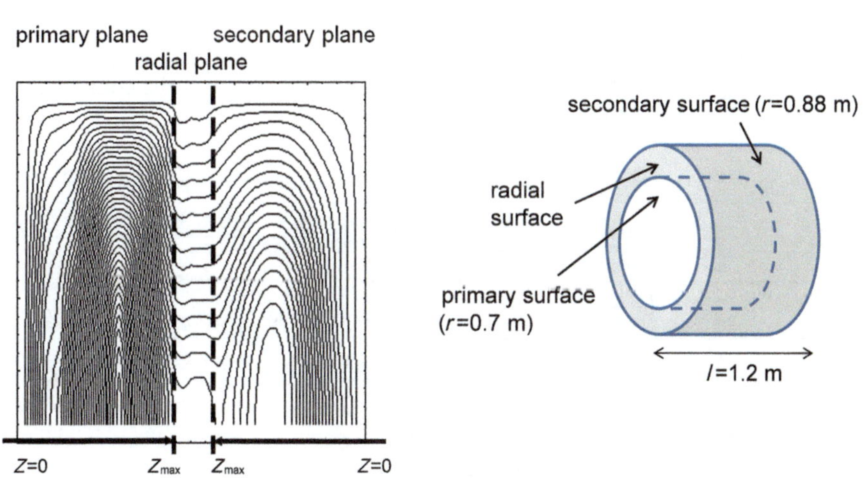

Figure 5.9 Example of a transverse cylindrical gradient coil incorporating the concept of 3D coil design. (left) Contour lines of the current density (1/8 of the cylindrical surface is shown for clarity: the remainder can be reconstructed by symmetry). (right) The particular dimensions of the coil geometry. Reproduced from ref. 39 with permission from John Wiley and Sons. Copyright © 2002 Wiley-Liss, Inc.

the particular example of a saddle-shaped coil. First, the current density surface of a cylindrical coil is divided into N adjacent elementary areas, each with an individual mesh current I_j, where $j = 1...N$. Each current forms a closed loop at the boundary of its elementary area. The target field $B_{\text{target},i}$ is a vector of magnetic field values with $i = \{1...M\}$. The target field coordinates are denoted as z_i (*e.g.* M points on a sphere, the outer surface of which defines the linearity volume). The magnetic field $B_i(z_i)$ at a target field point can be computed by adding up the field contributions of all N elementary areas, as shown in Figure 5.11. This contribution consists of a geometry-dependent

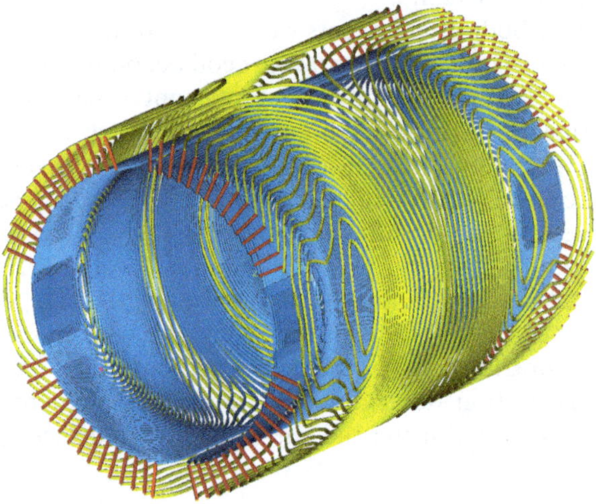

Figure 5.10 Wire patterns of an actively shielded transverse gradient coil with 3D design. The primary layer is shown in blue, the radial connectors in red and the shielding layer in yellow.

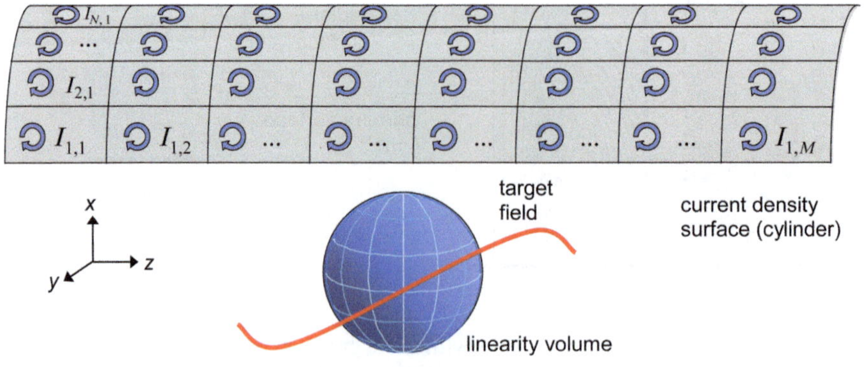

Figure 5.11 The current density plane of a cylindrical gradient coil is divided into $N \times M$ mesh elements. The resulting magnetic field on the surface of the target field sphere can be calculated by adding the contributions of each mesh element with individual mesh currents $I_{n,m}$.

matrix element f_{ij} (numerically calculated with the Biot–Savart law) and the individual mesh current I_j:

$$B_i(z_i) = \sum_{j=1}^{N} f_{ij} I_j \tag{5.7}$$

Using eqn (5.7), one can determine the magnetic field vector at the target field coordinates z_i for a given current vector I.

In order to prepare the input for a quadratic optimization algorithm, one needs to define a target function Q. This function contains at least a quadratic weighting term of the deviation from the target field ($B_i - B_{\text{target},i}$) and in most cases also a weighting term for the field energy (denoted as w_e) and the inductance (L) of the coil.

$$Q = \sum_{i=1}^{M} \left(B_i - B_{\text{target},i}\right)^2 + w_e \sum_{j=1}^{N} \sum_{k=1}^{N} \left(L_{jk} \cdot I_j \cdot I_k\right) \tag{5.8}$$

The matrix elements L_{jk} contain the self and mutual inductance of every mesh element. In order to exclude invalid (non-physical) solutions, further boundary conditions need to be defined. The maximum/minimum mesh currents and differences between adjacent mesh currents must be chosen according to the maximum allowable wire current and wire cross-section.

With all this information properly transformed into vector and matrix form, a quadratic simplex algorithm, for example, can be used to find the optimum current vector I_{opt}. Each $I_{\text{opt},j}$ corresponds to a particular rectangular mesh element, representing the optimized current density. Hence it is possible to obtain the stream function *via* numerical integration of the two dimensional representation of the current density. Finally, the wire pattern for the coil is extracted from the contour lines of the stream function (see Figure 5.12).

However, this is only the first step towards the development of a fully functional gradient system. The wire pattern needs to be adapted to the available wire cross-sections and the particular manufacturing methods for the coil layer. The physical and electromagnetic interactions of the gradient coil with other components of the MR system are also important aspects of gradient design. One major factor is the coupling with the main magnet. Interactions with large surfaces with high conductivity, transferred vibrations, rigid body forces and energy deposition into the helium vessel all need to be analyzed and may be included in the optimization algorithms. One example is the induction of eddy currents on the cryoshield (see Figure 5.13). These can be introduced into the target function Q as a linear boundary condition or even as a quadratically weighted term. The surface of the cryoshield (*eddy current surface*) is modeled as a mesh similar to the primary and secondary current density plane of the gradient coil ($j = \{1\ldots N_{\text{eddy}}\}$). The calculation of the magnetic field vector of the eddy currents on this surface following a gradient pulse can be performed in two steps. First, it is assumed that the eddy current surface behaves as a superconductor and induction is always

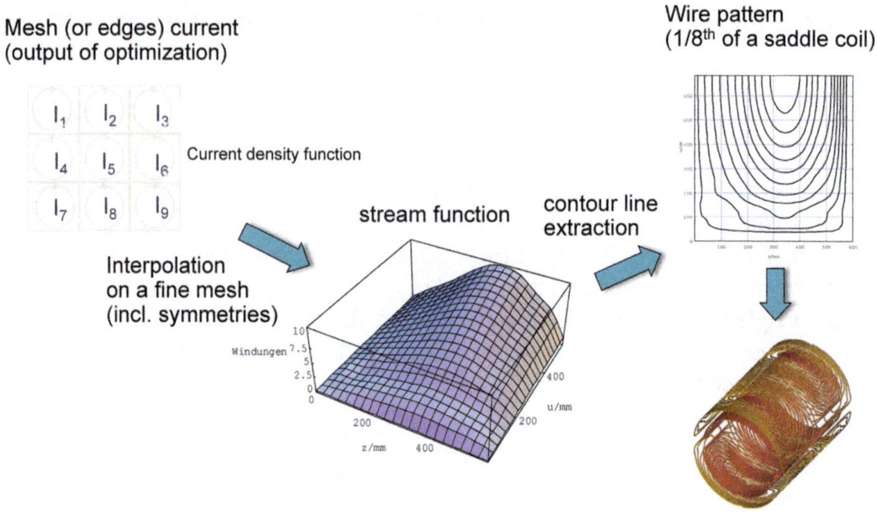

Figure 5.12 Example of a gradient coil design workflow with a discrete current density as the optimization output (showing 1/8th of a saddle coil to produce either an *x*- or *y*-gradient). The stream function is calculated by numerical integration on an interpolated fine mesh. The wire pattern is generated from the contour lines of the stream function.

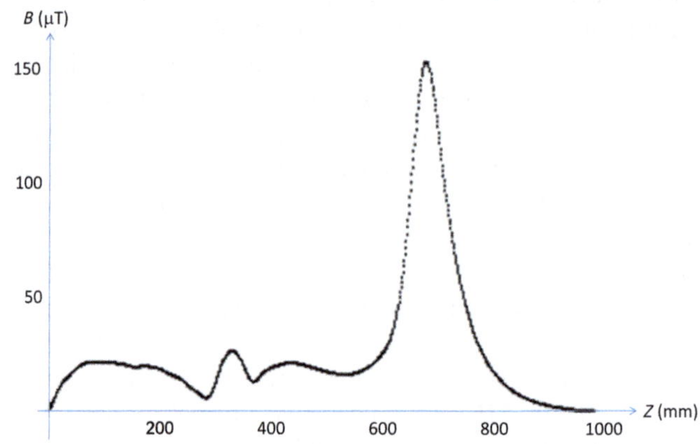

Figure 5.13 Typical stray field plot of a shielded gradient axis. The absolute value of the magnetic field is calculated outside the gradient coil body (which has a length of 1.2 m and a reference gradient strength of 1 mT m^{-1}) at the position of the cryoshield of the main magnet.

normal to the surface. Hence, the relationship between the stray magnetic field $B_{\text{stray},i}$ of the gradient coil and the induced current immediately after the gradient pulse in each mesh element of the eddy current surface $I_{\text{eddy},j}$ can be described by the matrix element m_{ij}:

$$B_{\text{stray},i} = \sum_{j=1}^{N_{\text{eddy}}} m_{ij} I_{\text{eddy},j} \tag{5.9}$$

In the second step, the effect of all eddy currents in the mesh elements on the target field points is calculated. Similar to eqn (5.7), the magnetic field contribution of each mesh element of the eddy current surface is stored in a vector $B_{\mathrm{eddy},i}(z_i)$. Using the geometrically defined matrix element element n_{ij} one can write:

$$B_{\mathrm{eddy},i}\left(z_i\right) = \sum_{j=1}^{N_{\mathrm{eddy}}} n_{ij} I_{\mathrm{eddy},j} \tag{5.10}$$

Combining eqn (5.9) and (5.10), the magnetic field vector of the eddy currents acting on the target field point z_i can be written with the matrix elements mn_{ij} (and may be added to the target function Q):

$$B_{\mathrm{eddy},i}\left(z_i\right) = \sum_{j=1}^{N_{\mathrm{eddy}}} mn_{ij} B_{\mathrm{stray},j} \tag{5.11}$$

5.1.6 Methods Using Spherical Harmonics

As covered in Appendix B in Chapter 1, it is also possible to use techniques involving cancellation of particular spherical harmonics in order to design gradient coils. For gradient coil design it is usually sufficient to consider harmonic expansions up to the 3rd order. Since examples have been given in both Chapters 1 and 4, these techniques are not described in detail here.

5.1.7 Definition of Gradient Performance Parameters

Conventional clinical MR scanners are capable of switching a gradient amplitude of 30–80 mT m^{-1} within 200–400 µs. The highest performance demands are associated with diffusion-weighted sequences, which require fast gradient switching for the EPI readout pulses, at least a 30 mT m^{-1} peak amplitude for the diffusion pulses, high duty cycle, and excellent shielding of eddy fields. Gradient system specifications and performance consist of many measures, including peak gradient amplitude, slew rate, linearity volume and free bore diameter. Optimizing the design necessarily involves compromises since, for example, increasing the free bore diameter reduces the maximum gradient amplitude that can be achieved. Of course, one must also consider that gradient coils optimized for whole-body applications with a large linearity volume and inductance also require high-voltage/-current power supplies to drive them, and there are limits on the performance (power/response time *etc.*) of these supplies that affect the characteristics of the overall gradient system (covered further in Section 5.2.2).

The inductance L and ohmic resistance R_{L} define the maximum *slew rate* (SR) of the gradient coil current $I(t)$. An electric circuit consisting of a coil and a voltage supply $V(t)$ can be described by the differential equation:

$$V(t) = L\frac{\mathrm{d}I(t)}{\mathrm{d}t} + R_{\mathrm{L}} I(t) \tag{5.12}$$

If one assumes that the maximum supply voltage V_{max} is constant during the switching time corresponding to the transition from $I = 0$ to $I = I_{max}$, $V(t)$ can be replaced by V_{max} in eqn (5.12). If the contribution of R_L is very small, a good approximation of the gradient *rise time*, t_{rise}, is given by:

$$t_{rise} \approx L \frac{I_{max}}{V_{max}} \tag{5.13}$$

Assuming a peak gradient amplitude of G_{max} at maximum current I_{max} and maximum voltage V_{max}, the value of the SR of the gradient system can be written as:

$$\mathrm{SR} = \frac{G_{max} V_{max}}{L V_{max}} \tag{5.14}$$

One needs to keep in mind that this approximation is only valid if the parasitic voltage drop, given by $R_L \times I(t)$, across the coil resistance is approximately two orders of magnitude smaller than the available voltage of the power supply. This assumption is usually fulfilled for whole-body gradient systems: indeed, otherwise the power requirements of these large gradient coils would exceed the capability of commercially-available gradient power supplies and cooling systems.

To a first approximation, the MRI spatial resolution improves with higher gradient amplitude (*i.e.* the spatial derivative of the gradient field, see Figure 5.2), because the bandwidth of the MR system can be projected onto smaller physical dimensions within the sample. A high SR, or equivalently a short t_{rise}, and fall time (which can be considered to be equal to the rise time) are desirable, since the total time required to play out a gradient pulse, $t_{rise} + t_{flattop} + t_{rise}$ is reduced and so, for example, shorter TE values within an imaging sequence are possible giving a higher SNR for tissues in which $T_2 \leq$ TE.

Neglecting the interaction of the gradient system with other components of the MR system at this point, the requirements on a gradient coil can be summarized by three design criteria: (i) the highest possible linear field gradient or gradient strength, (ii) the lowest inductance or fastest slew rate, and (iii) the maximum extent of the linear field gradient inside the linearity volume.

It is convenient to use standardized terms for the comparison of gradient coil designs of different nature. The most common terms are given below.

5.1.7.1 Gradient Efficiency, η, in Units of [$mT\,A^{-1}\,m^{-1}$]

The gradient efficiency is given by:

$$\eta = \frac{G}{I} \text{ for a given number of windings } N \tag{5.15}$$

Typical values vary from ~10 mT A^{-1} m^{-1} (coil radius ~30 mm) for MR microscopy gradients to around 0.1 mT A^{-1} m^{-1} (coil radius ~350 mm) for whole-body gradients.

5.1.7.2 *Linearity Radius* r_{LV} *[m]*

This is defined as the radius of a sphere where the deviation of the field gradient from the reference value is less than a defined limit (*e.g.* 5%). Typical linearity volume radii for MR microscopy gradients are 25 mm and for whole-body gradients 250 mm.

5.1.7.3 *Slew Rate in Units of* $[mT\,m^{-1}\,ms^{-1}]$

The SR is defined as the achievable gradient strength (mT) after 1 ms (or a lesser time if the maximum gradient strength is reached earlier). Typical slew rate values of MR microscopy systems with low inductance are 1000 mT m^{-1} ms^{-1}, whereas whole-body gradients typically have values one order of magnitude less (100–200 mT m^{-1} ms^{-1}).

5.1.7.4 *Performance Index PI in Units of* $[mT^2\,m^{-2}\,ms^{-1}]$

The performance index is defined as the slew rate multiplied by the maximum gradient strength.

In terms of the design trade-offs associated with the different performance indices, the following relationships are useful:

(a) The gradient efficiency scales as $1/r^2$, where r is the gradient coil radius.
(b) The size of the linearity volume increases linearly with the gradient coil radius.
(c) The gradient coil inductance increases quadratically with the number of windings.
(d) The gradient coil inductance scales as r^5 (for a constant gradient efficiency).

The last two rules in particular show the importance of efficient use of the radial space inside the magnet bore. The best gradient performance is achieved by using the minimum necessary number of windings driven by the highest available current. In practice, the maximum current is mainly limited by power amplifier technology, the physical forces on gradient connectors, the wire cross-section and cooling efficiency (the latter two parameters being related to the operating temperature of the gradient coil conductors). High performance targets in terms of shielding efficiency, linearity volume and a low degree of non-linearities in the gradient field result in an increased required number of windings, forming a significant challenge to the coil designer.

5.1.8 Developments in "Conventional" Gradient Designs

In the early 1980s, discrete coil winding methods and "audio" gradient amplifiers were sufficient to achieve target specifications for gradient performance, with imaging sequences typically taking tens of minutes. However, starting with the first applications of the ultrafast EPI imaging method in the early 1990s,[3] gradient slew rate in particular became a overarching challenge for gradient system design.

The first attempts to reduce the required rise time of EPI readout gradients involved additional resonant circuits, so called 'EPI boosters':[18,19] although successful in research environments, this approach was not adopted commercially. The principle is very simple, namely resonating the inductance of the gradients with an external capacitance. Although this arrangement could provide very fast smooth sinusoidal waveforms, it had several disadvantages. A more sophisticated approach, which could also be used to produce gradients with a more desirable trapezoidal shape, is shown in Figure 5.14, in which three different resonant frequencies are combined together.

A more conventional approach to increase slew rate performance involved partitioning of the coil windings and parallel driving with two or more amplifiers.[20] With conventional gradient design methods, an increase in slew rate (other parameters being constant) can be achieved by either increasing the voltage of the gradient power amplifier (GPA) or reducing the inductance of the coil design (see Figure 5.15). Increases in the GPA voltage have become available over the intervening years, with current systems standardly providing 2 kV: in terms of gradient coil design, one needs to ensure that the insulation properties of the gradient coil system can withstand these very high voltages. The alternative, but also complementary, approach of reducing the inductance of the gradient coil by reducing the number of windings has a negative influence on the gradient linearity and maximum gradient strength. The decrease in G_{max} can be counteracted by an increase in the maximum GPA current I_{max}. Combined with improved gradient design (*i.e.* new

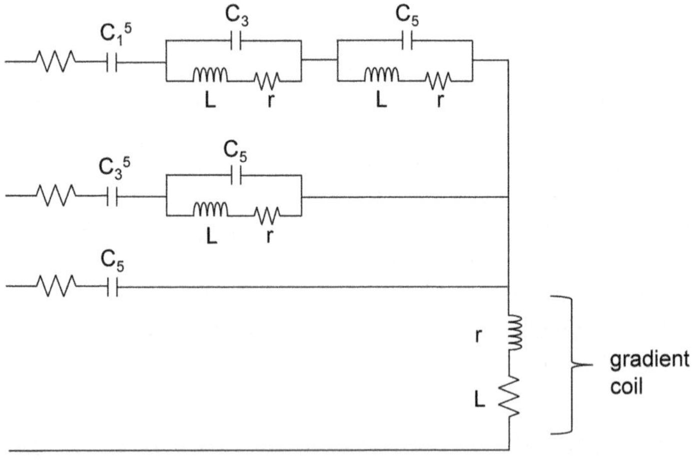

Figure 5.14 Circuitry showing a resonant gradient design, in which a number of different resonant frequencies can be added together to give a pseudo-trapezoidal gradient waveform.

optimized wire patterns for the linear axes), newer GPAs with 50% increased I_{max} can easily achieve a doubling of the gradient slew rate.

The third basic approach to achieve higher slew rates is to reduce the inner radius of the coil system. As the slew rate increases inversely to the 5th order of the coil radius, even small absolute decreases in the diameter can have significant effects, and this parameter is a crucial part of the MR system definition. As an example, assume a patient bore of 60 cm and a gradient coil inner radius of 35 cm. Reducing the radial space for the RF transmit coil and patient bore tube from 5 cm to 4 cm leads to a factor of $(35/34)^5 = 1.156$ slew rate increase. The 1 cm reduction can either be used to increase the gradient coil thickness (further increasing the slew rate advantage and reducing power losses owing to an increase in the wire cross-section that can now be used), or to reduce the inner bore of the main magnet and thus saving significant system cost and weight.

Historically, the availability of new GPA technology in the late 1990s (see Section 5.2.2) led to the development of gradient coils capable of withstanding currents and voltages of up to 1 kA/2 kV. Water cooling technology for the GPA cabinet and the coil is mandatory to remove the large amounts of heat generated (up to ~40 kW average power). Using these new GPAs, gradient amplitudes of ~45 mT m^{-1} and slew rates of ~200 T m^{-1} s^{-1} could be achieved.

However, the high technological effort, cost and power consumption of this somewhat brute-force approach meant that gradient performance did not change much in the late 1990s. In the early 2000s it was recognized that significant advances would be most easily achieved by considering the magnet/gradient/RF coil system as a whole, rather than simply improving each element separately. Progress in magnet, gradient and RF technology, simulation tools and system integration enabled joint concepts to be developed for all major MRI components. As a result, a new generation of MRI scanners appeared. Interestingly, the gradient system concepts focused not only on higher efficiency and gradient strength, but also on aspects of patient

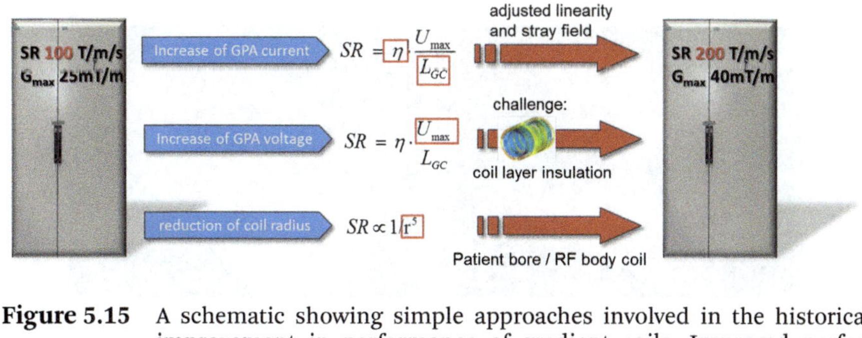

Figure 5.15 A schematic showing simple approaches involved in the historical improvement in performance of gradient coils. Improved performance parameters can be achieved by increasing GPA power and adjusting the coil design, or by a reduction of the (inner) coil radius and adjustments to the RF body coil and/or the patient bore diameter.

comfort (see next subsection) and environmental "friendliness". As a consequence, power consumption, magnet weight and helium usage became the main focuses.

In order to minimize the economic and ecological footprint of a MR system, the magnet bore size and gradient system power consumption should be reduced. The most established magnet bore diameter is approximately 90 cm, combined with a patient bore diameter of 60 cm. The radial space between the patient bore and magnet bore is usually divided between the gradient coil, RF transmit coil and magnet cover, which includes noise damping. Moving the design target of an MR system towards lower magnet weight and smaller size, minimum power consumption and a defined extent of clinical usage, resulted in significant consequences for gradient coil design. As mentioned above, owing to the inverse fifth order dependence of the slew rate on coil radius, even a slightly reduced coil size results in strongly reduced power requirements of the GPA. Reduction of the shielding coil radius has a direct effect on the magnet bore size (and thus cost and weight). If combined with a reduced primary coil radius, the slew rate advantage can still outweigh the disadvantage of the smaller shield radius. The challenge is to find the best compromise for all functions within the magnet bore (RF, gradient, noise damping, patient table support) and the desired clinical applications (*i.e.* maximum gradient strength, slew rate, linearity and magnet homogeneity).

As an example, dedicated system design could potentially decrease the power requirements by an order of magnitude. The most efficient way to achieve this is to reduce the maximum GPA current I_{max}. Owing to its quadratic relationship to power loss, all electrical components in the gradient power circuit would benefit from less complexity and lower cost. Although such a reduction of I_{max} to ~100 A greatly simplifies GPA design, it does not directly affect the current density within the gradient coil (see Figure 5.16). The current density is independent of I_{max} as long as the required number of windings in the coil layer can be driven with the specified slew rate (limited by the available GPA voltage).

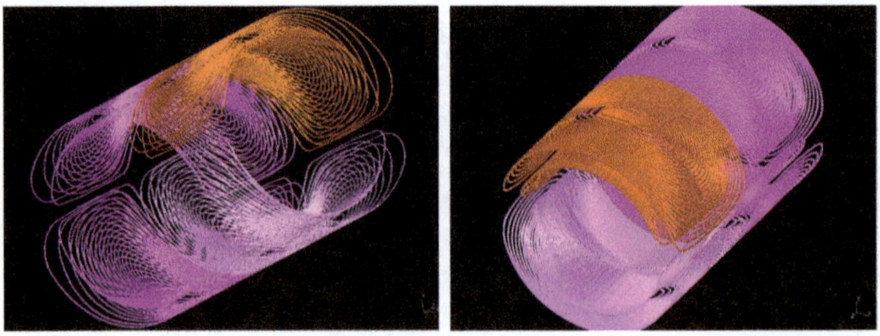

Figure 5.16 Design studies of different geometries of transverse gradient coils. Left: high-current design (I_{max} 1 kA, with thick wires gives a value of η = 50 μT A^{-1}), right: a low-current design (I_{max} 100 A, with a much larger number of thinner wires gives η = 200 μT A^{-1}).

5.1.9 Integrated Gradient and RF Designs

The RF transmit coil of a clinical whole-body 1.5 T or 3 T MR scanner is integrated within the patient bore tube. Therefore, there is always a discussion from a design point-of-view about the best compromise for the relative amounts of radial space to be assigned to the gradient coil and the RF coil. Today's gradient coils are actively shielded, and as a consequence there is a lower limit to the total "thickness" of the gradient coil assembly that still results in efficient field generation. The efficiency decreases strongly if the distance between primary and shield layer becomes smaller than a geometrically defined limit, meaning a severely increased power to produce a certain gradient amplitude. For whole-body gradients, this limit is in the range of ~70 mm, defined by a shield coil that needs to cancel more than 50% of the primary field amplitude (at the radius of the shield coil layer). Similar considerations are true for the RF transmit coil: as covered in Chapter 3, currents on the RF shield of an RF coil reduce the maximum B_1^+ field per unit input power, and the closer the shield to the RF coil the greater the reduction in efficiency. Efficient implementations use the patient bore tube as a support structure for the RF transmit coil, and the RF return flux is guided by an RF screen (usually a thin, slotted copper cylinder) positioned on the inside of the gradient coil. This configuration helps to prevent excessive RF absorption around the gradient conductors, which show a broad resonance spectrum in the MHz range.

Several attempts have been made to integrate the functions of RF and gradient coils into a single component. For example, the concept of a barrel-shaped magnet supporting both gradient and RF return flux with minimal radial space requirements was published in 2005.[21] The concept requires about a 10 cm wide gap into the centre of the gradient coil. This can be realized by splitting the gradient in two separate coils or by simply separating the primary windings in the center of a conventional gradient coil (see Figure 5.17). The extended RF-flux volume beyond the gap needs to be decoupled from the gradient conductors with additional screening layers.

The RF resonator can either be an integral part of the gradient coil body or can be attached independently to the patient bore tube (see Figure 5.18). The advantage of saving radial space is counterbalanced by design restrictions for both the gradient- and RF-coil. The required gap in the center of the gradient coil limits the freedom for shaping the RF-transmit field, and also limits the gradient linearity and efficiency. Figure 5.19 compares the performance of such an integrated design with a conventional separate gradient and RF coil.

5.1.10 Increased Bore-Size Systems

With superconducting magnet technology becoming more efficient and allowing shorter magnet designs, the amount of space available for larger patients, or for increased patient comfort, in the MR scanner has become a very important design aspect, and this clearly has implications for

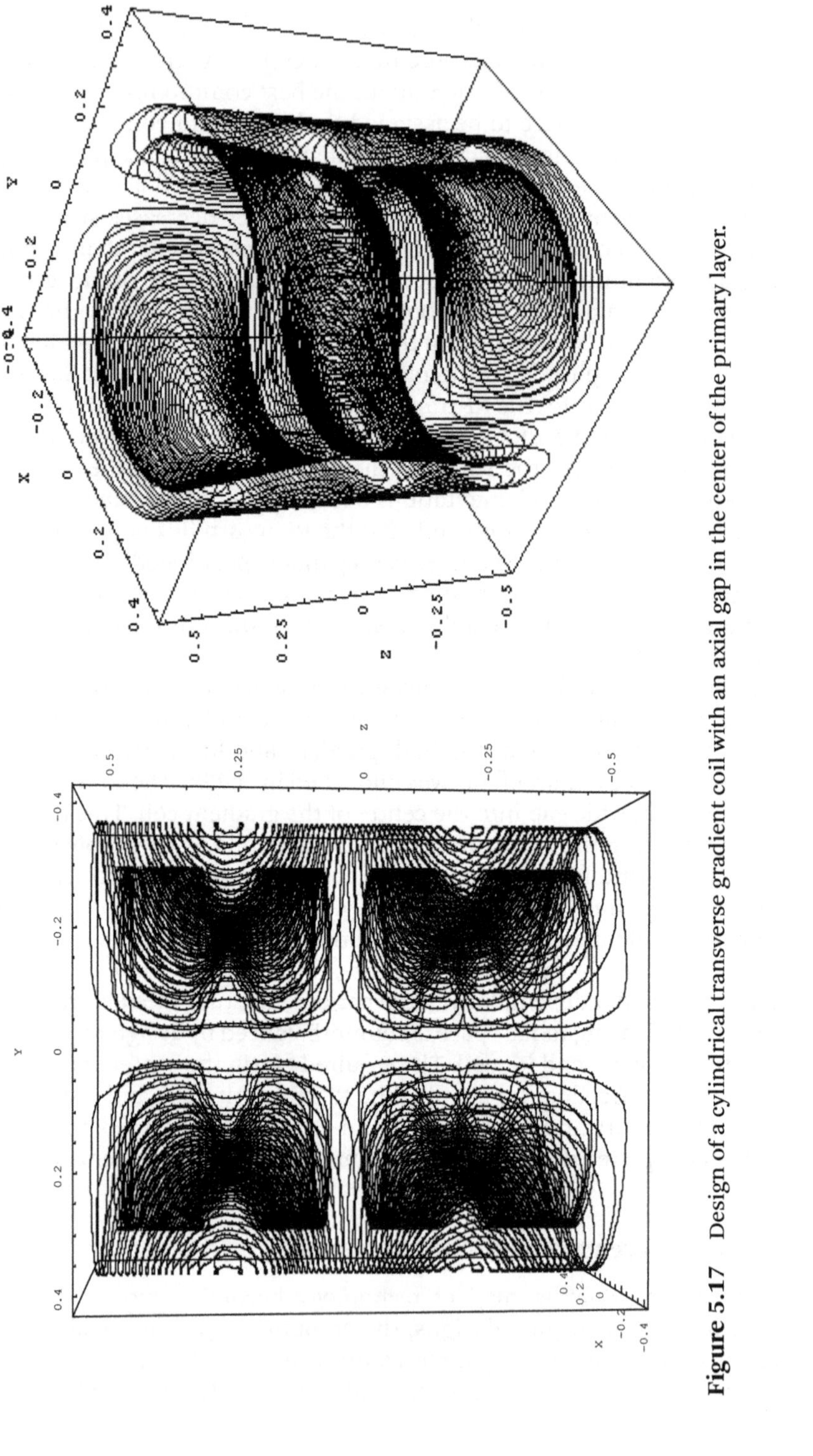

Figure 5.17 Design of a cylindrical transverse gradient coil with an axial gap in the center of the primary layer.

gradient design. Historically, the issue of having a large bore magnet was first addressed by planar MR systems (*e.g.* C-shape magnets) with a low magnetic field strength, much less than 1 T. However, the image quality for large field-of-view imaging, in particular, was very low. Initially, commercial vendors concentrated on increasing the field strength of these types of magnets to at least 1 T: however, at this point the cost of the required superconducting magnet technology made it commercially unviable, and now less than 5% of the world's MRI market consists of such systems.

The second, and current, strategy for clinical MRI systems combines cost-efficient cylindrical designs of the main magnet, gradient coil and RF transmit coil, aimed towards a short patient bore (1.2–1.8 m) with a clear bore opening up to 70 cm. RF integrated design (see previous section) is an aspect of

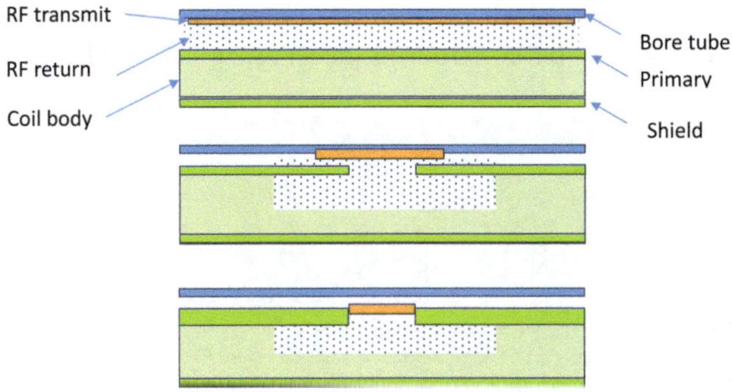

Figure 5.18 Example of an integrated RF design geometry. (top) Conventional, (centre) hybrid, and (bottom) integrated.

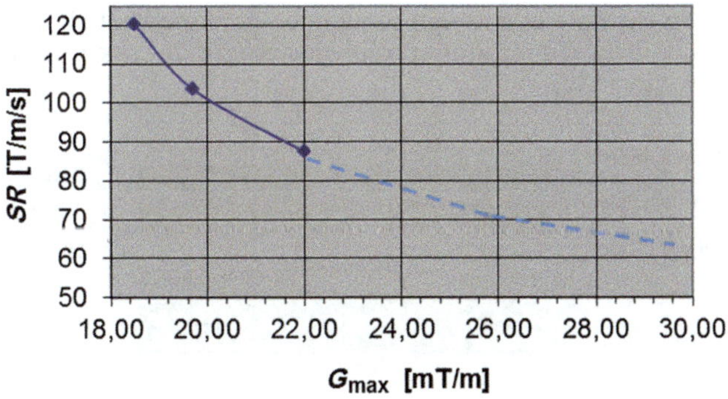

Figure 5.19 Plots of SR *vs.* G_{max} for of a cylindrical, transverse gradient axis design with constant wire thickness and radii of the primary and shield layers. The blue line denotes a design with a central gap of 10 cm in the primary layer. It shows a slew rate decrease with increasing maximum gradient strength G_{max}. Valid design results are not achievable above G_{max} = 22 mT m^{-1}. The dashed line represents a conventional design without central gap, which allows a wider range of G_{max} or gradient efficiency.

system design that helps to save radial space. Another factor is the improvement in imaging sequences and distortion correction algorithms (for gradient nonlinearities), which have allowed the DSV over which the gradients must be linear to be adjusted to the actual needs of a clinical MRI product, rather than the more difficult-to-achieve specifications of previous generations of MR systems. System optimization plays an important role during the system definition phase, as parameters such as magnet cost, RF efficiency or gradient slew rate are influenced significantly by the choice of magnet bore, as illustrated in Figure 5.20, which shows the trade-off between G_{max}, SR and LV.

Today's MRI scanners are usually designed as multi-purpose machines, *i.e.* capable of imaging any part of the body. Hence, the general approach of optimizing radial space between patient bore, RF transmit coil and gradient coil has become the important design criterion for industry. As one example, in order to achieve maximum openness of a cylindrical bore, a conical geometry may be considered (see Figure 5.21). The design of the

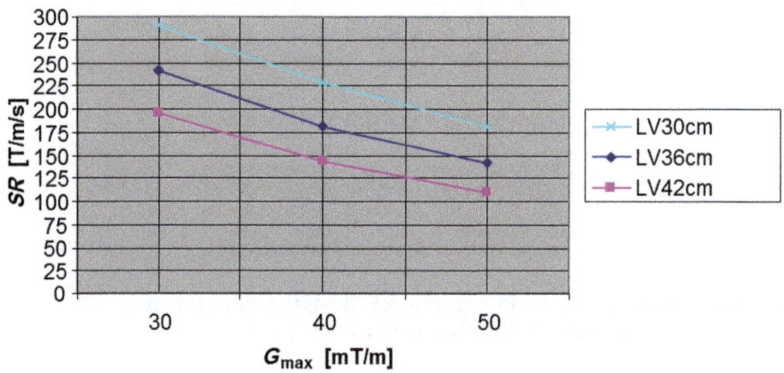

Figure 5.20 An example of a design of an open-bore gradient showing the interplay between maximum gradient strength G_{max}, slew rate and linearity volume LV.

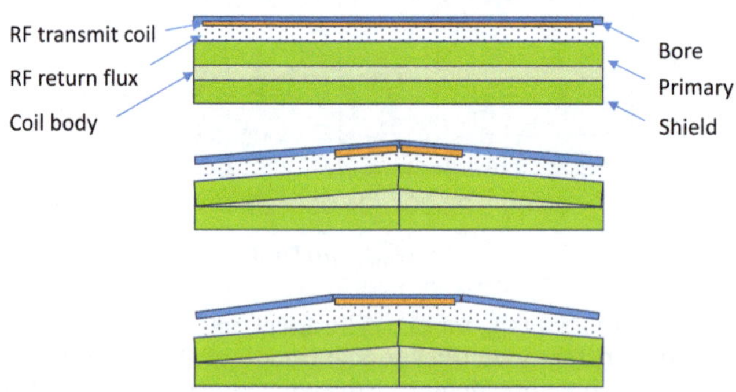

Figure 5.21 Example of a conical open-bore geometry. (top) Conventional, (centre) gradient and RF coils with conical symmetry, and (bottom) conventional RF coil and conical gradient.

RF transmit coil may either follow the same conical form (which is not favourable from a purely RF design perspective owing to the break in symmetry) or may maintain a cylindrical shape along the resonator's length. Owing to the conical shape of the primary gradient layer, the gradient efficiency is slightly reduced (in comparison to a cylindrical shape with the same minimum inner radius). However, this is more than compensated by the reduced inductance of the coil (resulting from the return conductors at both ends of the coil), thus giving a net advantage of 10–20% inductance. In practice, the tight requirements for fabrication tolerance of the conical layer structure reduce the advantages of this approach somewhat over the theoretical value.

5.1.11 Non-Cylindrical Designs

Though cylindrical designs dominate modern gradient coils, several other geometries have been considered and implemented. The coil geometry obviously is determined mostly by the boundary conditions set by the main magnet design, *e.g.* planar magnets are usually equipped with planar gradient sets, and low-field (≤1 T) MR systems use pole shoe magnets with round pole plates and similarly shaped gradient sets (see Figure 5.22).

Insert gradients can technically neglect the basic symmetry of the main magnet, but rarely do so. Indeed, owing to the interactions between the magnet and gradients (*e.g.* concomitant fields, forces and torques), it is always advantageous to keep as much symmetry as possible in the basic design. Dedicated insert gradients for different body parts are known (*e.g.* spine or knee[22]), but have not been commercially successful in replacing their whole-body counterparts: dome-shaped head-only gradient inserts have been the only ones to find use, and even these are not widely distributed. The electrical and mechanical design aspects of non-cylindrical geometries are not

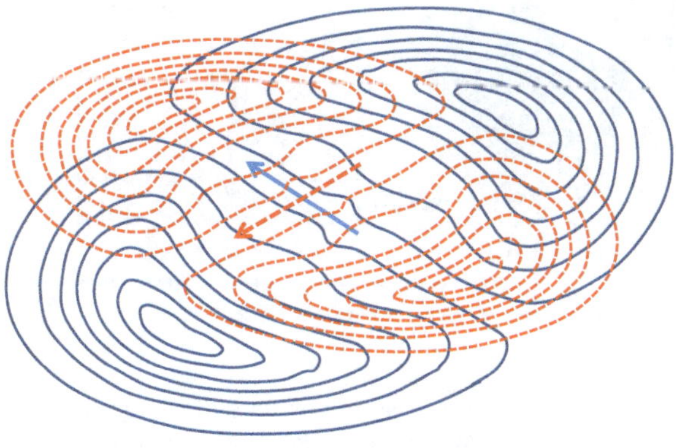

Figure 5.22 Example of x- and y-planar gradients used in an open MRI system.

very different from the whole body gradients, *i.e.* the target field method can be applied to calculate the primary and shield current density on a circular plane. The wire pattern and axial connections are then generated by extracting the contour lines of their stream function.

5.2 Gradient System

5.2.1 Overview

The overall gradient system consists of a number of sub-components, each forming a subsystem of its own (see Figure 5.23).

(i) *The gradient coil system* comprises the three linear field coils (*X*,*Y*,*Z*) including their respective shielding coils, higher-order shim coils, cooling layers, temperature sensors, RF screen, current and water connectors, as well as the coil support and suspension structure.

(ii) *The gradient cooling system* consists of the hospital's or university's primary water supply, the vendor's secondary water cooling (*i.e.* heat exchanger) and the cooling circuit inside the magnet room (*i.e.* the connection to the cooling circuit within the gradient coil).

(iii) *The GPA system usually* consists of the hospitals' main power supply, the power distributor for the gradient circuit, the amplifier cabinet (small-signal unit, final stages, transformer and supervision circuits), cables, filters and chokes which connect the GPA with the gradient coil and the MR control system.

(iv) *The gradient control system* comprises the computer, which calculates and plays out the digital gradient envelopes, the real time controller,

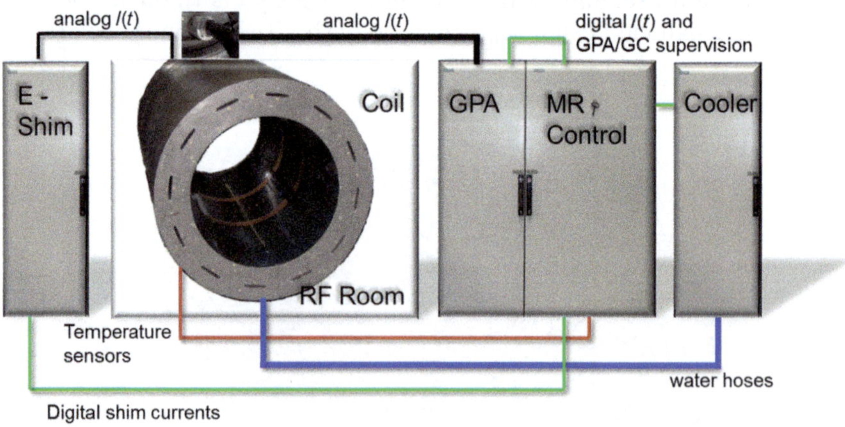

Figure 5.23 The overall gradient system consists of the gradient coil system, the gradient cooling system, the GPA, the shim amplifier and the gradient control system, which is integrated into the MR control system.

which transmits the digital signal and supervises the analogue current signal to-and-from the GPA, the gradient safety watchdog, and service loops for system diagnosis. Gradient distortion correction algorithms are implemented on the image reconstruction computer and rely on gradient-coil-specific pre-calculated or measured magnetic field distributions of the three gradient axes. A history of MR gradient system evolution is shown in Figure 5.24.

In the following sections, each of the above sub-components is explained in more detail.

5.2.2 Gradient Coil System

Generation of magnetic fields with high amplitudes over large volumes is a difficult task. Historically, in one embodiment, wooden structures were used to support the race-track-like windings (see Figure 5.25) of some very early gradient coils! Rapid gradient switching leads to strong vibrations owing to the Lorentz forces on the conductors within the main magnetic field. These vibrations can be counteracted by a high stiffness of the support structures (*e.g.* with glass fiber reinforced plastic) and by linking all of the layers to a single body (*e.g.* with epoxy resin). In addition, a high dielectric strength of all of the gradient layers is required so that electrical breakdown does not occur at the minimum (<2 mm) radial distance between the conductors. Hence, gradient coils today mainly consist of epoxy resin, which has excellent dielectric strength and high geometric flexibility at low cost. The challenge in the fabrication process is to ensure that all layers and subcomponents of the coil system are completely impregnated. With epoxy, this can only be achieved with a complex vacuum potting procedure. Every step of this procedure needs to be defined precisely and controlled thoroughly.[23] Process and material parameter, such as the temperature profile, filling material, filling percentage and curing time, need to be adjusted to the geometry and inner structure of the gradient coil body. A successful potting procedure is usually validated with a high voltage or electrical discharge test of the coil. This test step ensures that the dielectric strength is high enough to withstand amplifier voltages of up to 2 kV for the lifetime (10–20 years) of the MR system.

Cylindrical gradient coils are usually assembled on a roll shaft (see Figure 5.26). The innermost layer may consist of a thin slotted copper foil, which helps to minimize coupling of the coil system with the adjacent RF transmit system. The gradient conductors can be prefabricated (*e.g.* copper pipes with integrated cooling circuits) or directly wound on the layer below (typical for longitudinal axes). Care needs to be taken that adjacent layers are fully impregnated while maintaining the minimum radial dimension. The wire cross-section is usually chosen to be as large as possible to minimize power loss. If slotted or water-cut copper plates are used, additional measures need

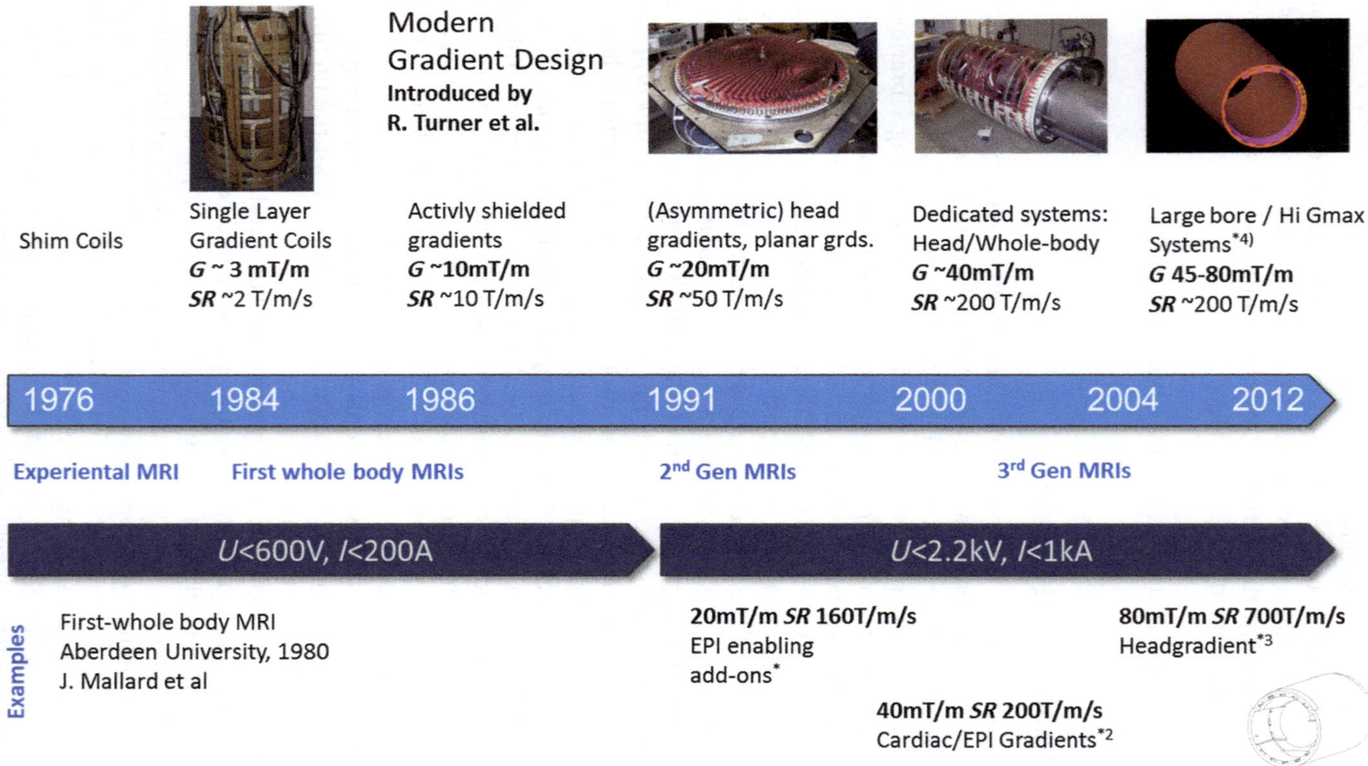

Figure 5.24 History of MRI system development. *1) M. S. Cohen and R. M. Weisskoff, *Magn. Reson. Imaging*, 1991. *2) F. Schmitt and E. Eberlein *et al.*, *ISMRM*, 1999. *3) A. Vom Endt *et al.*, *ISMRM*, 2000. *4) Equals an increase of gradient power (slew rate × amplitude) of three magnitudes over 20 years.

to be taken to minimize eddy currents within the coil layer itself. Typical wire cross-sections for whole-body scanners are several tens of square millimeters. Animal systems or NMR scanners with a magnet bore size of 30 cm and less do not allow more than a few square millimeters per wire. Temperature monitoring and simulations are essential to prevent damage to such small conductive structures with their associated limited heat capacity.

The radial space between the primary and shielding coils can be used for various purposes. Magnets without superconducting shims can benefit from an iron shim system, which can be inserted into slots in the gradient coil body. Electrical shim coils for patient shimming can also be located within this space: the five second-order shim coils usually fit. Although these second-order shims are sufficient for clinical whole-body applications, third-order and higher-order shims (see Chapter 4) are important for high field human MRI systems as well as high resolution NMR systems. These shim

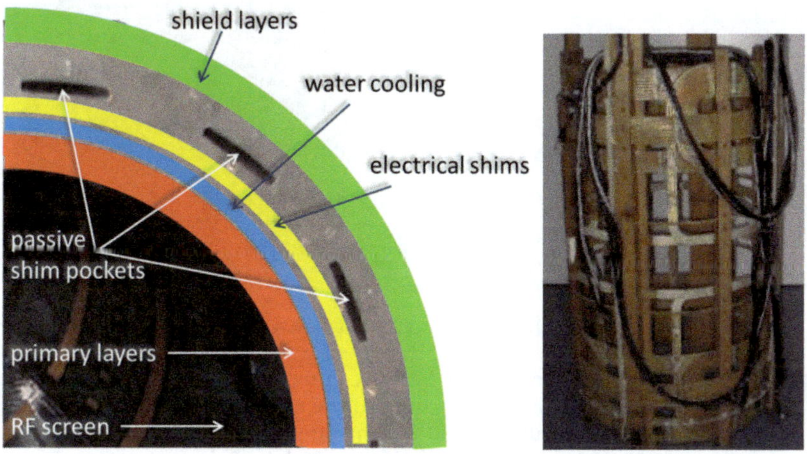

Figure 5.25 Left: schematic of the different layers of a cylindrical coil system. Right: a historic 'race track' gradient coil with wooden support frame, capable of a G_{max} of 3 mT m^{-1} and slew rate of 10 T m^{-1} s^{-1}!

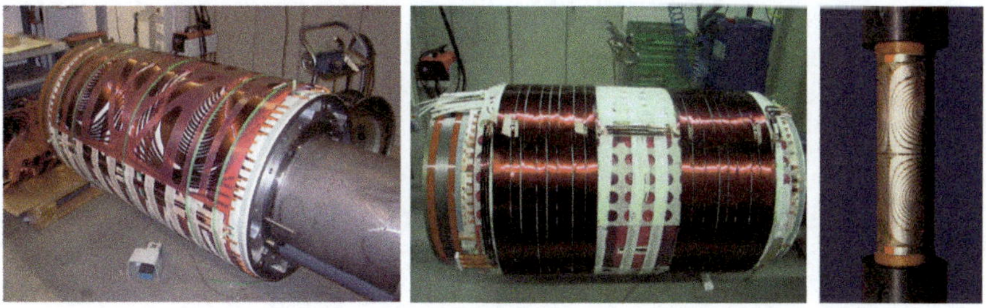

Figure 5.26 Assembly of cylindrical gradient coils. Left: transverse whole body gradient on a roll shaft. Middle: longitudinal whole body gradient on a roll shaft. Right: transverse microscopy gradient on a roll shaft.

coils (*e.g.* Figures 5.27) are more challenging to integrate, as their field energy is considerably higher. Hence, only a small selection of these higher order shim coils may be accommodated within the gradient coil body.

5.2.3 Gradient Power Amplifier (GPA) and Connectors

Until the late 1980s, gradient coil and GPA technology was limited to operating parameter values of ~600 V and ~200 A using analogue amplifier technology from the HiFi audio market. In the early 1990s, the pulse width modulation (PWM) technique was implemented for MRI gradient amplifiers: the principle is shown in Figure 5.28. High current transistors with low

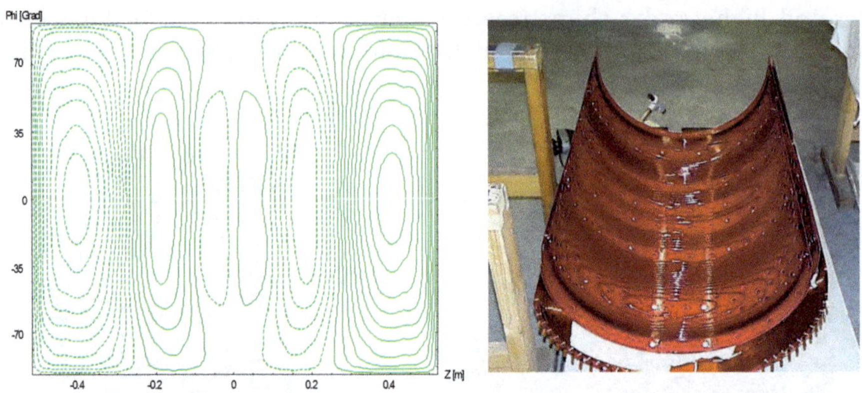

Figure 5.27 Left: contour lines of the current density of a third-order transverse shim coil (one-quarter of the cylinder surface is shown), with $r = 0.381$ m, dashed lines denote negative current density. With the maximum current of 65 A passing through the coil, the A_{31}/B_{31} field term shows a field amplitude of 130 μT on the surface of an iso-centric sphere of 400 mm diameter. Right: fabricated A_{31}/B_{31} coil segment. The wire pattern is glued on to a thin suspension plate and pre-adjusted to the target radius.

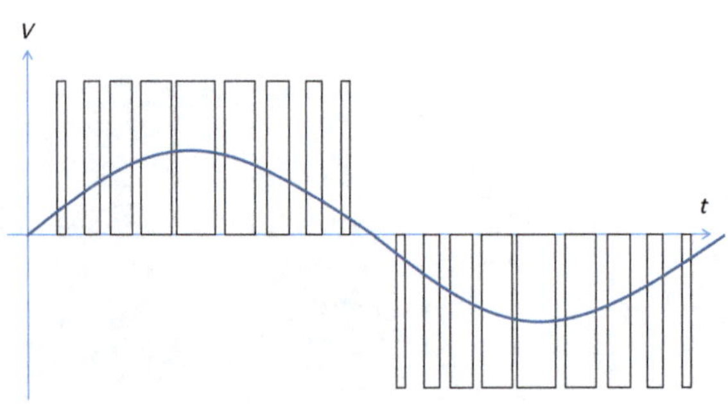

Figure 5.28 Principle of a pulse width modulation (PWM) amplifier in which the output function (sinusoidal) can be produced by a series of rapidly switched pulses.

switching losses (Insulated Gate Bipolar Transistor *IGBT* with a drive current of 50–200 A) can be used to apply ultra-short voltage packages (10–20 μs at 400–500 V). The required voltage and current for a gradient envelope can be generated by modifying the on/off ratio of the transistors with a regulation circuit. In order to achieve ppm precision for MRI, high-precision current sensors (closed-loop Hall effect sensors) are built into the circuit.

Today's GPAs are capable of simultaneously generating a driving current of 1 kA and voltage of 2 kV. These values are possible to achieve by cascading the outputs of several PWM stages[24,25] and requires a transformer capable of taking several tens of kVAs. With such 'CASCADE' type amplifiers (V_{max} = 2000 V, I_{max} = 500 A, three channels per amplifier) the current signal is controlled by a proportional-integral-derivative (PID) regulator.[26] As shown in the schematic in Figure 5.29, the amplifier output voltage $v(t)$ at time t is given by the weighted sum of the proportional (P), integral (I) and derivative (D) error, $e(t)$, of the current from its target value:

$$v(t) = Pe(t) + I\int e(t)\,\mathrm{d}t + D\frac{\mathrm{d}e}{\mathrm{d}t} \tag{5.16}$$

As the characteristics of the feedback circuit of the amplifier are sensitive to any changes in inductive load, it is important to find a set of parameters (P,I,D) that ensures a stable transient response. In the ideal case, the transient response of the amplifier current shows the behaviour of a critically damped oscillation. In practice, however, a single overshoot cycle of the controlled variable is acceptable. The stability of a parameter set can thus be described by the overshoot amplitude and its decay time (see Figure 5.30).

In the case of an error occurring, owing for example to a bad electrical connection or faulty components within the gradient circuit, oscillations are detected and supervised by a dedicated small-signal unit. The same is true for gradient load conditions that exceed the specification (*e.g.* excessive power loss within the gradient coil, final stages and chokes).

Even when the gradient system is running within specifications, the current-carrying connectors of the coil experience high strain. These connectors are exposed to dynamic currents and voltages up to the full scale values of 2 kV/1 kA. In most cases, these special cables are located within the main magnet stray field gradients. The sequence-dependent frequency spectrum of the

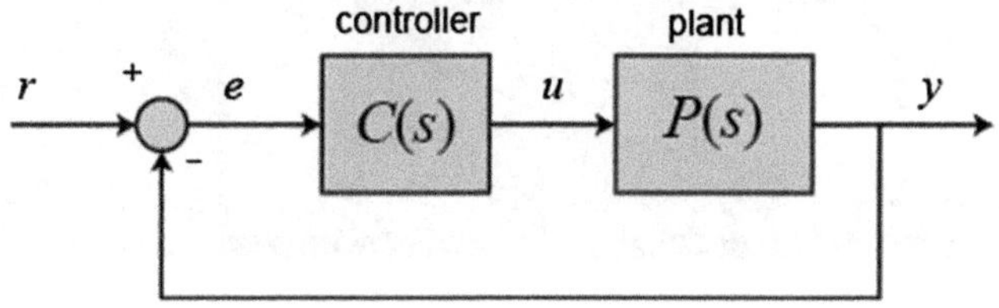

Figure 5.29 Schematic of a proportional-integral-derivative circuit used to control the output current of the gradient amplifier.

MR protocol/experiment may cause the connectors to vibrate significantly owing to the Lorentz forces. Over the years, the design of these cables has changed from clamped single wires over partial coaxial cables to unscrewable, actively cooled coax tubes used in ultra-high-power gradients sets (see Figure 5.31).

5.2.4 Gradient Cooling System and Temperature Supervision

The gradient layers (X,Y,Z) are usually cooled with one or more layers of water cooling. The preferred position of these circuits is the space between the primary and secondary layers. Owing to the inverse fifth-order dependence of the slew rate on the coil radius, any other positioning within the coil body would degrade gradient performance considerably.

In practice, however, this placement can rarely be achieved. As covered later in this chapter, one exception is small-bore MR microscopy gradients,

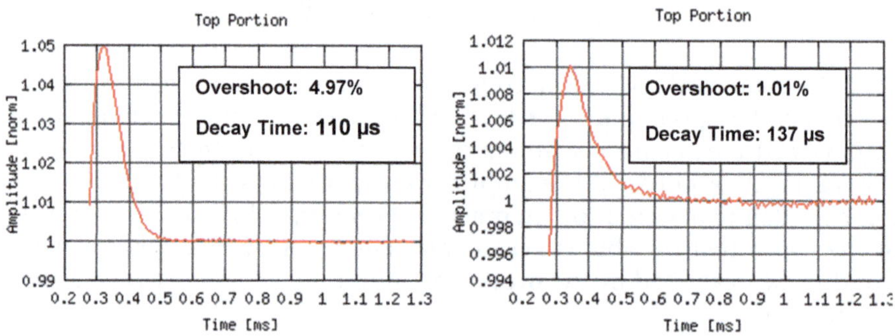

Figure 5.30 Current transient response to a 200 μs ramp under two different load conditions (*i.e.* coil inductance), measured with the internal current probe of a CASCADE gradient amplifier. Reproduced from ref. 39 with permission from John Wiley and Sons. Copyright © 2002 Wiley-Liss, Inc.

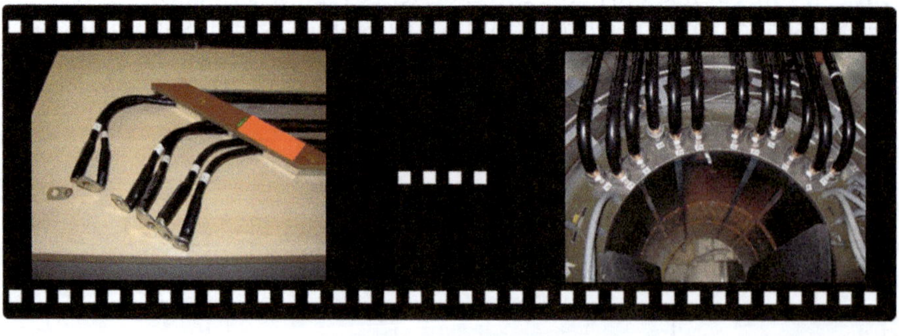

Figure 5.31 Development of gradient connector cables. (left) Early implementations used single-wire connectors and were later reinforced with clamps. More recent implementations use coaxial cables with reinforced connector bridges. (right) The latest developments use water-cooled coax connectors.

which can be fabricated with ~1 mm thin copper layers per axis. When glued together with thermally conducting epoxy and insulated by a ~100 μm layer, it may be sufficient to add a single cooling layer on top of the primary axes (see Figure 5.32).

As soon as the cross sections of the wire and the support structures become larger, heat removal of the adjacent layers becomes more difficult. The thermal resistance across three layers may become so high that the temperature hot-spots of the innermost layer (with the largest distance to the cooling layer) cause safety shutdown of the gradient system while the temperature of the outermost layer is actually several tens of degrees lower. As a result, whole-body gradient sets are either optimized for a thin package of coil layers (*e.g.* slotted copper plates) or equipped with a thin cooling layer between all of the coil layers (see Figure 5.25). The advantages of the latter approach are the balanced heat flow between the axes and the additional degree of freedom for the choice of the wire cross-section. Another approach combines both electrical conductor and cooling into an actively cooled coil design. Copper tubes have a very low thermal resistance and are the preferred material for the construction of a cooling circuit. However, owing to their high electrical conductivity, heating owing to eddy currents needs to be considered. This is especially problematic if the gradient wires are made of hollow rectangular copper tubes. In this case, the dielectric strength of the cooling liquid and copper tube is of concern. This can be solved by using additional insulation layers, with the drawback of increased thermal resistance. An alternative to copper-based cooling is the use of a material with good insulation properties and low thermal resistance. To this end, special epoxy resins with the required properties have been developed for the commercial market.

Another important aspect of cooling design is the cross-section of the tubes and the resulting flow behavior and flow resistance. Turbulent water flow is required for efficient heat removal from the walls of the tubes and needs to be implemented by geometrical design. This is usually investigated with finite-element analysis tools (*e.g.* ANSYS), which can simulate heat transfer and temperature distributions within a given gradient and cooling

Figure 5.32 Water-cooled microscopy gradient prototype. The water cooling tubes are wound in bifilar fashion directly on the primary gradient coils (X,Y,Z).

design. The design result can most easily be validated experimentally with an infrared camera (see Figure 5.33).

The maximum allowable temperature of a gradient coil is defined by the material with the lowest limit for heat exposure. Most materials can be chosen to withstand more than 100 °C. However, the particular epoxy resin may be limited to less than 100 °C. A specialized epoxy has to simultaneously satisfy multiple specifications, such as dielectric and mechanical strength, as well as moisture and/or thermal resistance. If the gradient coil temperature exceeds the given limit, the epoxy resin begins to degrade and both mechanical and electrical properties become much worse. In order to avoid these adverse temperature conditions, a temperature supervision circuit is usually integrated into the gradient coil body. The supervision circuit may consist of optical or resistive (negative temperature coefficient, NTC, or positive temperature coefficient, PTC) sensors. An important aspect of these sensors is their robustness over the lifetime of the gradient coil. All sensors have to survive tens of thousands of thermal cycles under high-amplitude vibrations and high-voltage ramps. Usually, about ten sensors are installed at the measured hot-spot positions or near adjacent structures, such as the RF screen, water cooling or gradient connectors.

In order to simulate and test the working conditions of an MRI scanner, all manufacturers define a so-called "duty cycle" for their gradient systems. A simple formula to describe the heat load duty cycle (DC) of the gradient system without any knowledge of the actual coil or amplifier hardware is given by eqn (5.17).

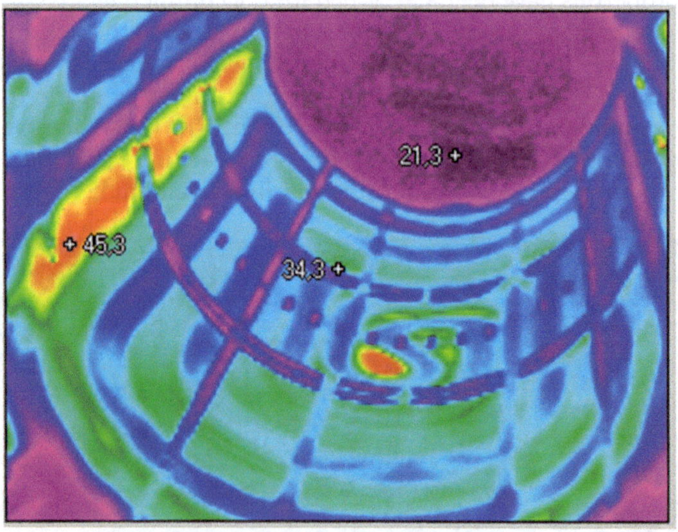

Figure 5.33 Thermal hot spots of a cylindrical gradient coil prototype measured with an infrared camera on the inside of the cylinder bore: temperatures color coded from blue (cold) to red (hot). The measurements show that, in this particular case, the water cooling efficiency needs to be improved in specific local regions.

$$\mathrm{DC}(t_1,t_2) = \frac{\int_{t_1}^{t_2} G(t)^2\,\mathrm{d}t}{G_{\max}{}^2(t_2 - t_1)} \cdot 100\% \tag{5.17}$$

Here $G(t)$ denotes the gradient amplitude *vs.* time graph and G_{max} the highest constant (or bipolar) amplitude that the gradient system can provide continuously. There is a quadratic dependence of the electrical losses on the gradient amplitude, *i.e.* reducing the G_{max} by 50% corresponds to a 25% reduction in DC. This simple approach is useful for pulse sequence designers to understand the load conditions of a specific sequence, but in order to design a gradient hardware according to system requirements, a more sophisticated approach is necessary, as described below.

The power dissipation within the components of the gradient system (gradient coil, GPA, cables, filters) needs to be evaluated according to the frequency- and amplitude-specific behavior of each individual subcomponent. Simplified to considering only the gradient coil and GPA, the average power within an imaging sequence interval TR can be written as:

$$\overline{P}_{\mathrm{TR}} = \frac{1}{\mathrm{TR}}\left(\int_{\mathrm{TR}} I(t)\cdot V_{L,\mathrm{GPA}}\mathrm{d}t + \int_{\Delta\upsilon}^{\sim 2} I(\upsilon)\cdot R(\upsilon)_{\mathrm{GC}}\mathrm{d}\upsilon\right) + P_0 \tag{5.18}$$

Here $I(t)$ denotes the gradient current *vs.* time graph, $V_{L,GPA}$ the voltage drop of the semiconductors in the final stage of the GPA, $I(v)$ the spectral distribution of the gradient current, $R(v)_{GC}$ the frequency-dependent resistance of the gradient coil (see Figure 5.34) and P_0 the standby power losses of the gradient system.

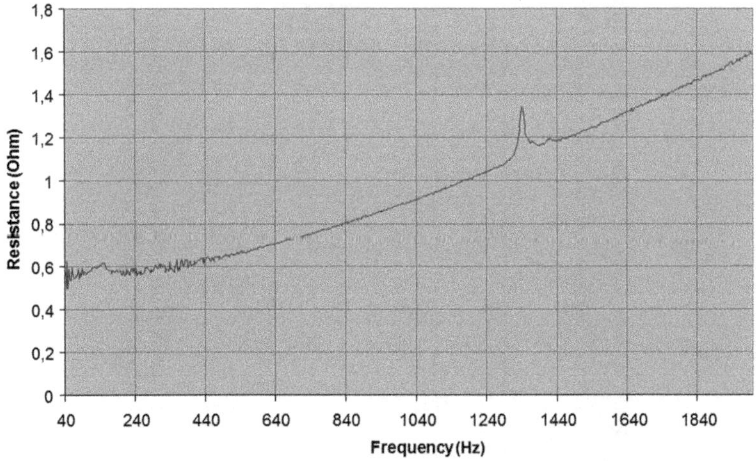

Figure 5.34 Frequency-dependent resistance of the *z*-axis of a cylindrical gradient coil prototype. The measured ohmic resistance for gradient waveforms with high frequency (*e.g.* 1400 Hz) components is twice that for those with low frequency (*e.g.* 400 Hz) components. The resistance shows a peak at ~1300 Hz where one of the major mechanical resonances of the gradient coil body is located.

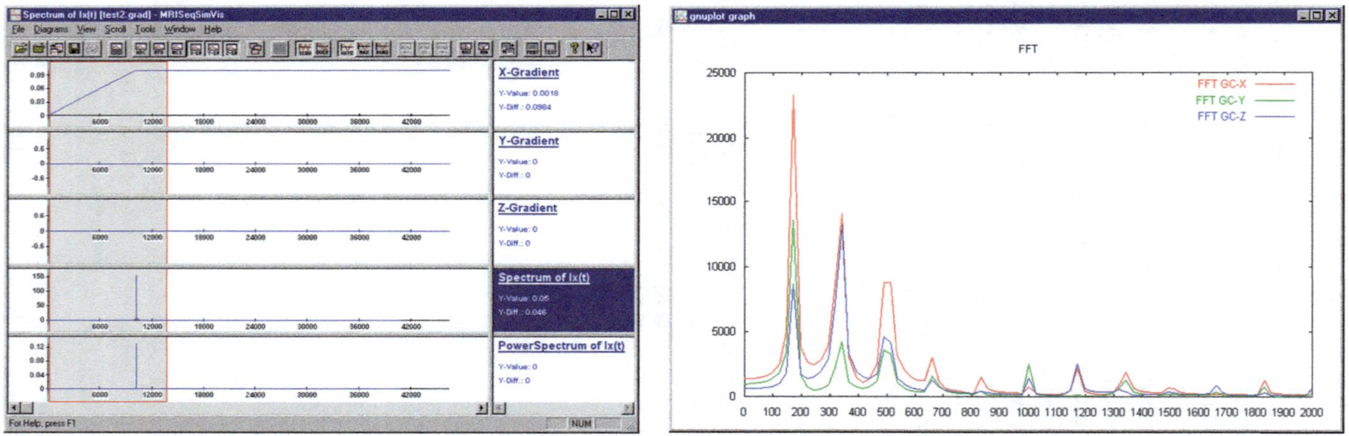

Figure 5.35 Left: simulation of power dissipation within a sequence development tool. Right: Fourier spectrum of a TrueFISP sequence with dominant spectral components in the 150–500 Hz range for all three of the gradients.

Eqn (5.18) can be used for continuous simulation of the duty cycle of all three gradient axes and amplifier power stages (see Figures 5.35 and 5.36). Automatic simulation of all desired sequences of a "to-be-defined" MR scanner gives the design engineer the required input to estimate the average power. Validation of the system design is performed either by running the most demanding sequences on the prototype hardware or by performing an equivalent 'duty cycle sequence' test.

5.2.5 Gradient Control System and "Safety Watchdog"

The control system of an MR scanner synchronizes and drives the activities of all the hardware components (see Figure 5.37). The interface of the MR controller to a logical hardware component usually requires a sub-control system of its own. The gradient controller and the PNS safety watchdog are fed by logical gradient shapes (*read*, *phase*, *slice*) and gradient-coil-specific limits for PNS, respectively. Static correction data acquired during the first startup phase of the MR scanner ('tune-up data') are transferred to the gradient controller and GPA before the start of the measurement sequence.

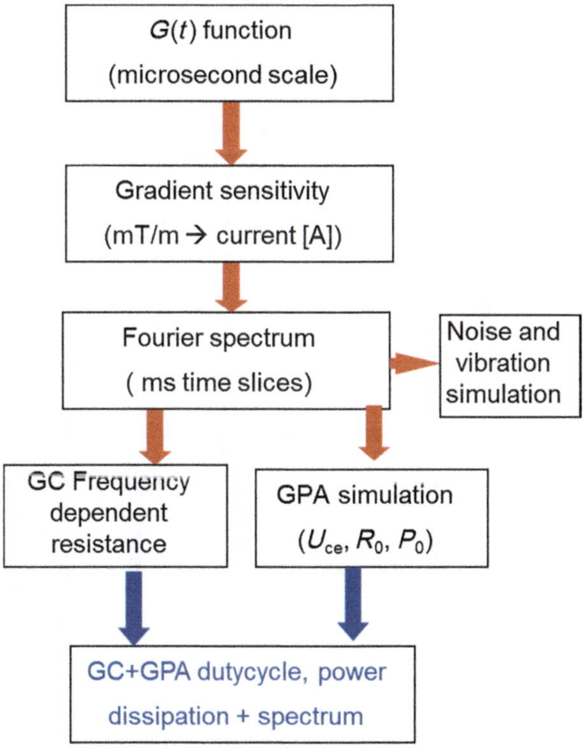

Figure 5.36 Flow chart of a simulation program for gradient system power dissipation. The calculated Fourier spectrum could also serve as an input for acoustic noise and vibration simulations.

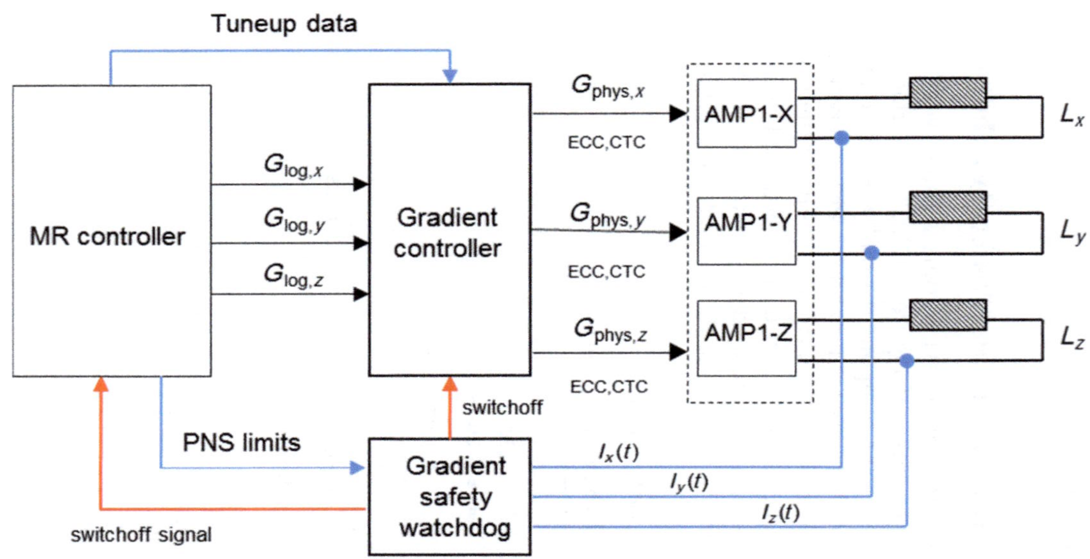

Figure 5.37 Control system of a typical MR scanner and its interface to the gradient controller. The gradient controller and the PNS safety watchdog are fed with logical gradient shapes (G_{log}) and gradient-coil-specific limits for peripheral nerve stimulation. The logical shapes are converted to physical shapes (G_{phys}) and modified by filter functions for eddy current and cross-term compensation (ECC, CTC) before being fed to the gradient amplifier's final stages (Amp $X/Y/Z$). The current in the gradient circuit with coil impedance L is measured by high precision current probes and fed to the gradient safety watchdog unit. In case the PNS limit is exceeded, a safety switch-off signal is sent to the gradient and MR controller.

After the start of a pulse sequence, the logical gradient shapes are converted into physical gradient shapes. This procedure consists of several steps. The first step is the discretization of the shape with a sampling time in the microsecond range. The resulting three-dimensional vectors can be rotated numerically to the desired target orientation (*e.g.* the imaging plane). Using the gradient sensitivity data from the tune-up phase, the required magnetic-field value is converted to units of gradient amplifier current $I(t)$. In order to minimize eddy current effects, additional filters based on tune-up data (*e.g.* eddy current pre-emphasis) may be applied to $I(t)$ before transfer to the gradient amplifier component. The filter parameters can be determined, for example, with an MR sequence that measures the spatial and temporal dynamics of the eddy currents while scanning a phantom.[27]

The output of the gradient amplifier circuit is monitored by high-precision current sensors. These output signals are used for control and supervision of the gradient amplifier cabinet. In addition, these values can be used for the stimulation supervision ('gradient safety watchdog', see Figure 5.38). The limits for human exposure to time-varying magnetic fields are defined by the International Electrotechnical Committee (IEC). In order to exploit the maximum performance of a new gradient system, an appropriately approved volunteer study typically results in a factor-of-two increase in PNS levels compared to the general IEC limitation.

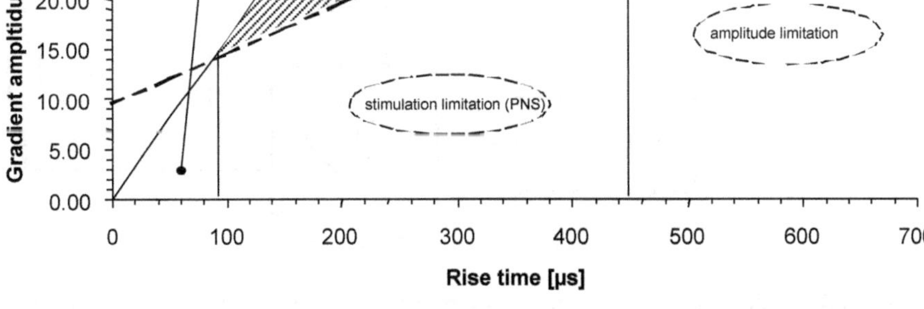

Figure 5.38 Limitations of gradient system performance for human use. The main limitation is given by the peripheral nerve stimulation threshold (PNS limit). If the technical performance of the gradient system is not adapted to this limit, a significant part of the parameter space is not available for imaging human subjects. Reproduced from ref. 39 with permission from John Wiley and Sons. Copyright © 2002 Wiley-Liss, Inc.

5.3 Examples of Specific Gradient Coil Designs

5.3.1 High Strength Gradients for a 7 T Horizontal Bore Animal Magnet

A 7 T/300 MHz horizontal bore magnet was the platform for the development of a microscopy gradient coil insert.[29] The magnet has a warm-bore diameter of 210 mm and a pre-installed commercial gradient coil with a free bore diameter of 120 mm. The overall gradient system comprises three water-cooled linear axes as well as high-order shims. The gradient controller can drive three linear and 15 higher-order shim channels. The gradient channels are driven with three COPLEY amplifiers with a maximum current/voltage of 100 A/150 V. The regulation circuit of the amplifiers can be adjusted to different impedance gradient coils by individually chosen RC filter banks (see Figure 5.39).

Although the commercial system was equipped with an unshielded microscopy insert with a free bore diameter of 60 mm (which fits inside the 120 mm inner diameter larger gradient set), the linearity volume is relatively small

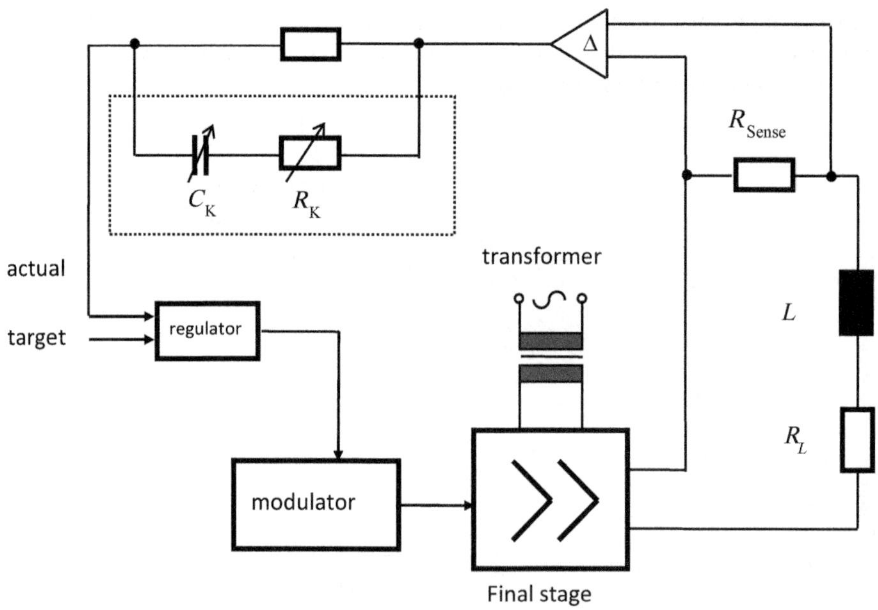

Figure 5.39 Block diagram of a switching-mode current amplifier with feedback regulation loop. The inductive load of the circuit is denoted by L and R_L, respectively. The actual value of the current in the gradient circuit is measured *via* the voltage drop over R_{sense} with the differential amplifier Δ. This signal is send to the regulator circuit after correction of the frequency response with the adjustable filter C_K, R_K. The regulator compares the actual with the target value and adjusts the modulator circuit accordingly.

(40 mm) and hence the design target of the new gradient set was an extended region of improved linearity. The commercial gradients had a very low inductance in the range of 100 μH and thus an extremely high (theoretical) slew rate: however, this was restricted by the frequency response of the gradient amplifier and its regulation circuit (see Table 5.1).

The prototype gradient inserts were designed with an analytical design method based on Bessel Functions.[28] The central expression of the design method is the Fourier transform of the current density $J(\phi,r)$ on a cylinder surface, expressed by Fourier coefficients $j_{\phi}(u)$. For the transverse gradient coils, the current density can be derived as:

$$J_{\mathrm{tra}}\left(\phi,\tfrac{z}{a}\right)=\frac{2}{\pi a}\begin{pmatrix}\cos\phi\int_0^{\infty}du\cos\left(u\tfrac{z}{a}\right)j_{\phi}(u)\\ \sin\phi\int_0^{\infty}du\sin\left(u\tfrac{z}{a}\right)\dfrac{j_{\phi}(u)}{u}\end{pmatrix} \tag{5.19}$$

where z/a and φ are the normalized cylinder coordinates, a = coil radius, u and m represent conjugate variables in the frequency domain, and $\mathrm{j}_{\phi}^{m}\left(\tfrac{u}{a}\right),\mathrm{j}_{z}^{m}\left(\tfrac{u}{a}\right)$ are the Fourier coefficients of the current density.

For longitudinal (z) coils, the current density is given by:

$$J_{\mathrm{long}}\left(\phi,\tfrac{z}{a}\right)=\frac{1}{\pi a}\begin{pmatrix}\int_0^{\infty}du\sin\left(u\tfrac{z}{a}\right)\mathrm{j}_{\phi}(u)\\ 0\end{pmatrix} \tag{5.20}$$

The Fourier coefficients were obtained by calculus of variations of a Lagrange function. The latter includes expressions for coil inductance, linearity and power dissipation. The linearity is controlled by the number of

Table 5.1 Properties of the commercial NMR gradient system (data sheet values) and the newly designed microscopy prototypes for a horizontal 7 T animal system. Gradient field properties for the prototypes were calculated with the program package MATHEMATICA (based on a continuous current density function derived with an analytic design method). The table lists the gradient axis with the highest inductance.

Microscopy gradients	BGA-12	BG-60	Prototype A	Prototype B
Inner diameter [mm]	120	60	60	60
Slew rate @ 150 V, 100 A [T m^{-1} s^{-1}]	2500	20 000	12 808	6543
Realized slew rate [T m^{-1} s^{-1}]	714	4166	3255	3407
Theor./realized rise time [μs]	80/280	50/240	61/240	125/240
Coil radius [mm]	>60	>30	34.5	39
Gradient efficiency [μT Am^{-1}]	2000	10 000	7770	8200
Linearity volume diam. [mm] (<5% deviation in sphere)	80	40	53.2	60
Inductance [μH]	120	75	91	188

Lagrange multipliers used. An example of the function $f_\varphi(u)$ and its stream function is shown in Figure 5.40.

Starting with about ten contour lines, a good approximation of the continuous current density can be achieved. The maximum number of contours is limited by the cross-section of the wire or metal plate which is used for fabrication of the coil layer (see Figure 5.41). Owing to the limited rise time of the amplifier circuit, it was possible to accept a higher inductance without significant degradation of slew rate performance. In comparison to a whole-body gradient system, the slew rate of NMR microscopy gradients is about one order of magnitude higher. The same holds true for gradient efficiency.

The magnetic field distribution was calculated using the general-purpose electromagnetic simulation software package "MAFIA" (CST—Computer Simulation Technology AG). The conductor structures (*i.e.* contour lines of the stream function including modifications for internal connectors) were

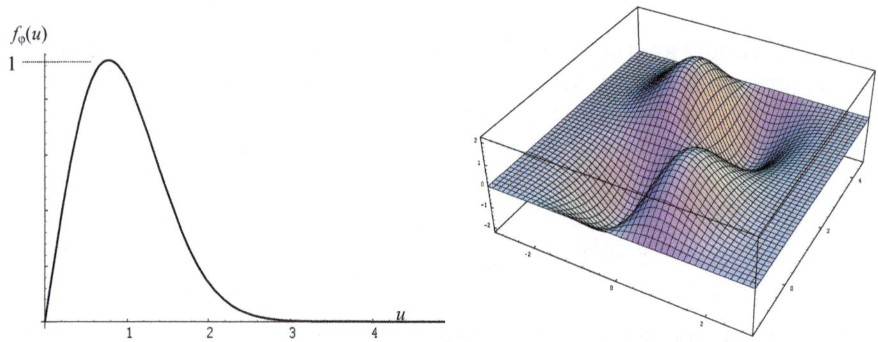

Figure 5.40 Left: normalized function $f_\varphi(u)$ of a transverse linear field gradient. Right: stream function (four quadrants) of a transverse linear field gradient.

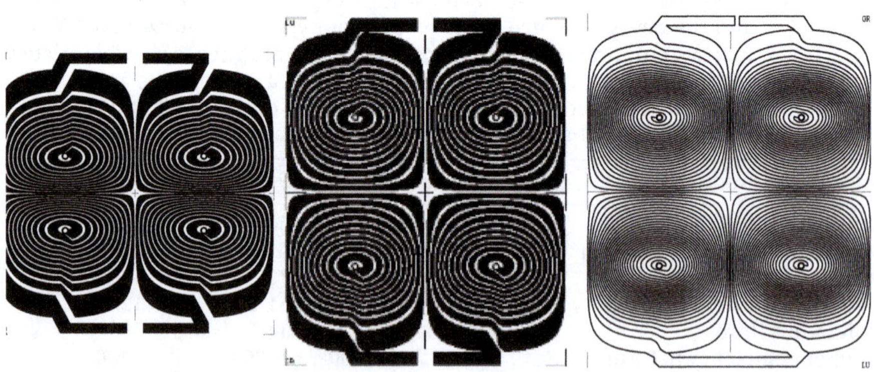

Figure 5.41 Left: transverse gradient layout (1/2 coil) calculated with single Lagrange multiplier (1st order) and discretized with $N = 10$ windings (etching pattern). Middle: 3rd order, $N = 10$. Right: 3rd order $N = 27$ (milling map).

imported from the calculus package 'Mathematica' (Wolfram Research). Hence, it was possible to determine and check the linearity volume and the gradient strength. In Figures 5.42 and 5.43, a comparison between the two prototype designs with identical linearity target (*i.e.*, equal number of Lagrange multipliers) but different discretization precision is shown.

Similar calculations were performed for the longitudinal gradient axis (see Figures 5.44 and 5.45). The resulting winding positions on the cylinder former are shown in Figure 5.46.

The first prototype transverse coil design was built with 2 × 100 μm double-sided etched copper foil. With these dimensions, the number of windings was limited to $N = 10$ (and 3rd-order Lagrange multipliers, see Figure 5.43) allowing for a minimum conductor width of 4 mm and a cross-section of 0.8 mm^2. The resistance of the complete transverse axis (*i.e.* two etching patterns

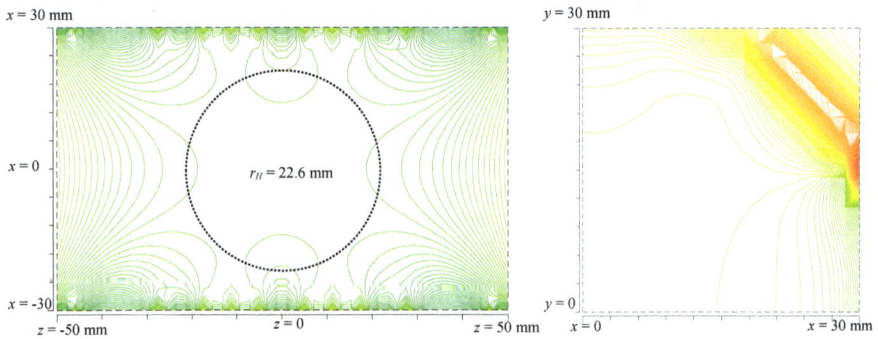

Figure 5.42 Contour lines of the *x* gradient axis (prototype A, $N = 10$ windings) in the (left) *xz*-plane and (right) *xy*-plane. The contour lines include areas where the deviation of the calculated magnetic field gradient is less than 2.5% (7.5%, 12.5%,...) from the ideal value.

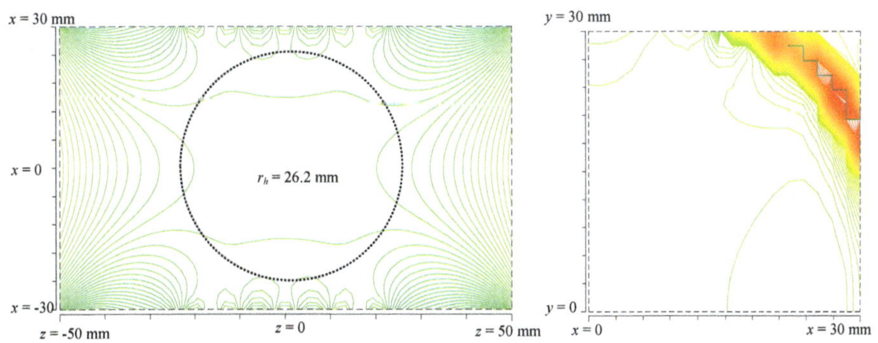

Figure 5.43 Contour lines of the *x* gradient axis (Prototype B, $N = 23$ windings) in the (left) *xz*-plane and (right) *xy*-plane. The contour lines include areas where the deviation of the calculated magnetic field gradient is less than 2.5% (7.5%, 12.5%, ...) from the ideal value.

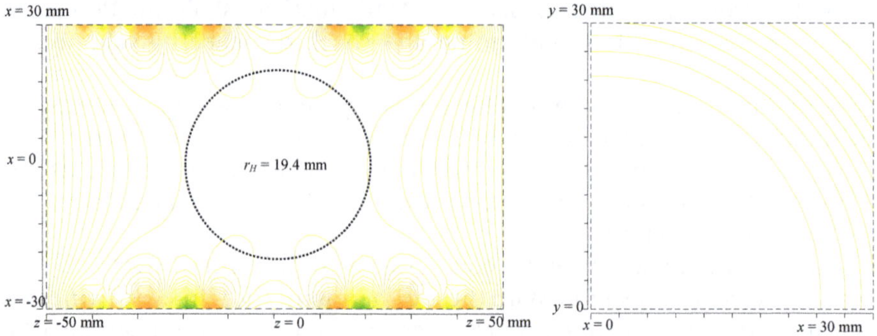

Figure 5.44 Contour lines of the z gradient axis (prototype A, $N = 10$ windings) in the (left) xz-plane and (right) xy-plane. The contour lines include areas where the deviation of the calculated magnetic field gradient is less than 2.5% (7.5%, 12.5%, ...) from the ideal value.

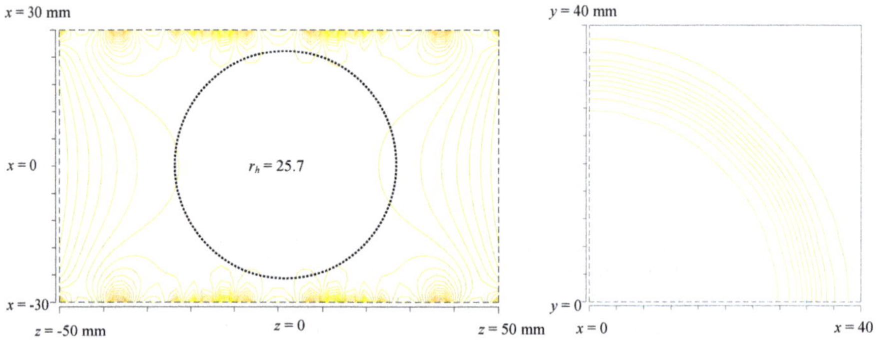

Figure 5.45 Contour lines of the z gradient axis (prototype A, $N = 30$ windings) in the (left) xz-plane and (right) xy-plane. The contour lines include areas where the deviation of the calculated magnetic field gradient is less than 2.5% (7.5%, 12.5%, ...) from the ideal value.

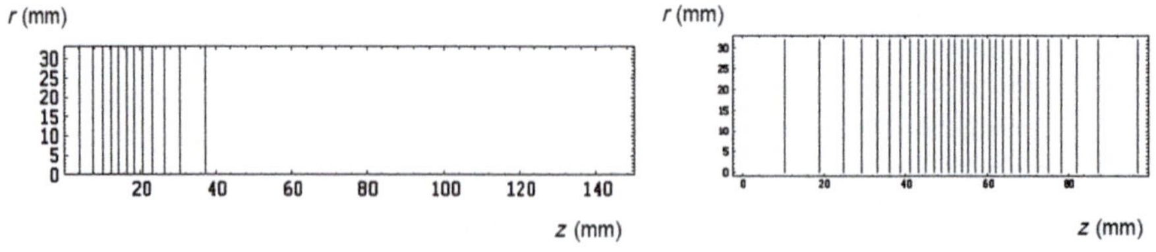

Figure 5.46 Winding positions of longitudinal gradients. Left: 1st-order Lagrange multipliers, current density discretized with $N = 2 \times 10$ windings. Right: 3rd-order Lagrange multipliers, current density discretized with $N = 2 \times 30$ windings.

covering 2 × 180°) was calculated to be 750 mΩ compared to the 200 mΩ of the commercial BG-60 gradient (which has a power dissipation of ~500 W at 50% G_{max}) and would thus limit the usable gradient strength. Increasing the copper foil thickness by electroplating was tried, but resulted in increased stiffness and insulation problems between the windings. Therefore, a water cooling circuit was applied directly onto the transverse coil layers. Four PTC resistors were placed at positions corresponding to the minimum conductor cross section (see Figure 5.47). These probes show a step function in resistance if the temperature exceeds 60 °C, forcing the NMR system into a safety shutdown. The longitudinal gradient was wound on machined grooves within the glass fiber reinforced plastic (GRP) former, see Figure 5.48. In order to minimize radial space, two adjacent wires per calculated conductor

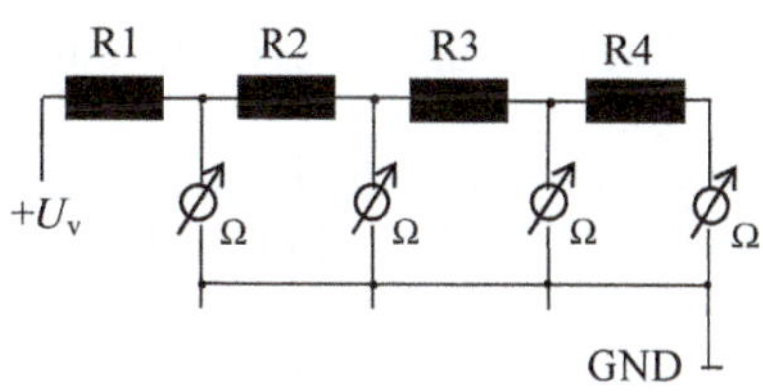

Figure 5.47 Left: temperature supervision circuit with PTC resistors. Right: water cooling made of neoprene hoses wound on milled GRP combs.

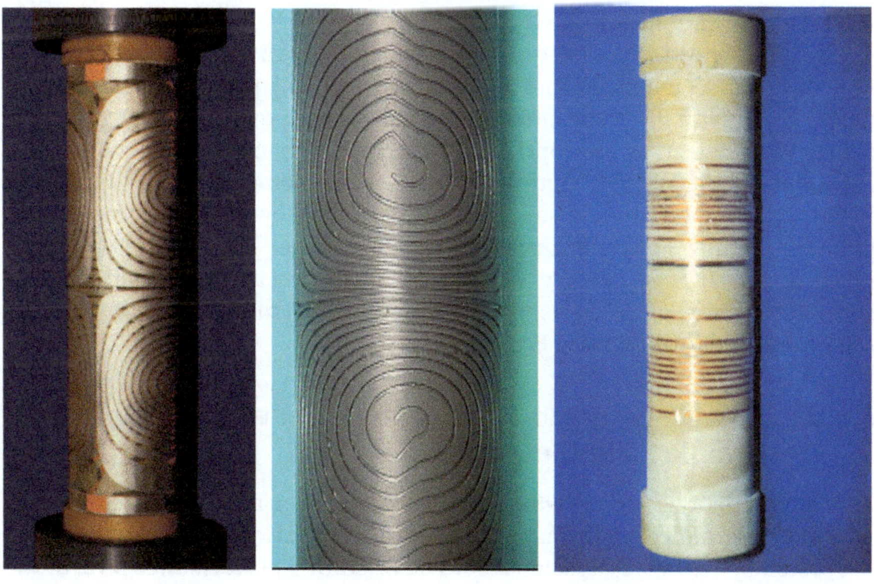

Figure 5.48 Left: prototype A gradient coil for the 7 T horizontal bore animal magnet with transverse gradients mounted. Centre: prototype B transverse gradient pattern milled onto a PVC cylinder. Right: prototype A coil body (machined GRP) with *z* coil windings mounted.

position were used. The net resistance including interconnection wires was measured to be 208 mΩ, which is about 50% greater than the 148 mΩ value of the commercial gradient.

Alternative construction methods for the transverse gradients were considered for the second prototype, *e.g.* laser cutting of copper plates (suitable for MR microscopy gradients with a coil diameter of ~20 mm), water cutting (suitable for larger gradients with a coil diameter of ~200 mm), and mechanical punching (suitable for whole-body gradients with a coil diameter of ~600 mm). Finally, a sandwich-like fabrication method was chosen. The coil consists of a thin copper cylinder that was joined with a thin GRP cylinder on a smaller radius. The wire pattern was generated using a 3D milling machine with a precision of less than 0.1 mm. In a second step, the wall thickness of the GRP cylinder was reduced to less than 1 mm on the inside. The advantage of this method is the increase in conductor cross-section and the high precision of the conductor positions. Hence, the discretization of the current density could be realized with $N = 30$ windings.

Gradient inductance, rise time, sensitivity and (non-) linearity were measured with an impedance meter, oscilloscope and fast gradient echo imaging. The measured values show on average a roughly 8% lower sensitivity and linearity with respect to the theoretical calculations. The major contribution to this deviation can be explained by the difference between the continuous current density and the discrete windings of the contours of the stream function. The remaining deviation originates from tolerances of the chosen fabrication method. These are higher for transverse gradients owing to their higher construction complexity.

As a pre-condition for MR measurements, the gradient amplifier circuit was impedance matched to the new gradient coil prototype: impedance matching essentially corresponding to maximum power transfer. This was achieved by selecting appropriate resistor and capacitor values within the regulator circuit. Without optimized settings, the achievable rise time would be far from the theoretical value (which is limited by the inductance of the gradient circuit). For the transverse gradients, it was possible to reduce the rise time to 120 μs (see Figure 5.49). After completing the optimization of the regulator circuit, gradient echo imaging was used to characterize gradient sensitivity and linearity (see Figure 5.50).

5.3.2 A Whole-Body Modular Gradient Set with Continuously Variable Field Characteristics

As described previously, PNS strongly depends on the LV and limits the usable performance for human imaging: on conventional whole-body gradient systems, the LV has a fixed value. The operating performance level of the particular gradient coil may be chosen as a general-purpose compromise, or optimized for dedicated applications (*e.g.* cardiac MRI). Since the efficiency and PNS limits of a gradient coil can be improved by reducing the coil radius, dedicated whole-body insert coils with small diameters have been built,[30]

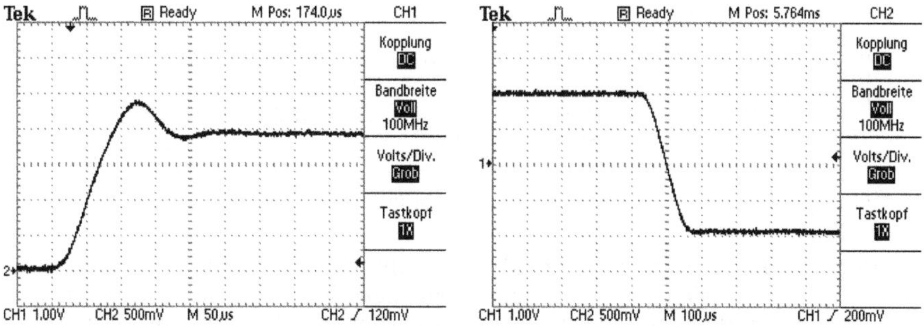

Figure 5.49 Electrical measurements performed on the microimaging gradient coil prototype A for the optimization of the current regulator in the GPA. (Left) The overshoot after gradient ramp up ($I = 0$ A $\rightarrow I = 20$ A) limits the minimum (stable) rise time to 250 µs. (Right) The overshoot amplitude with optimized RC values for the regulation circuit is close to zero, showing that a rise time of about 120 µs is possible.

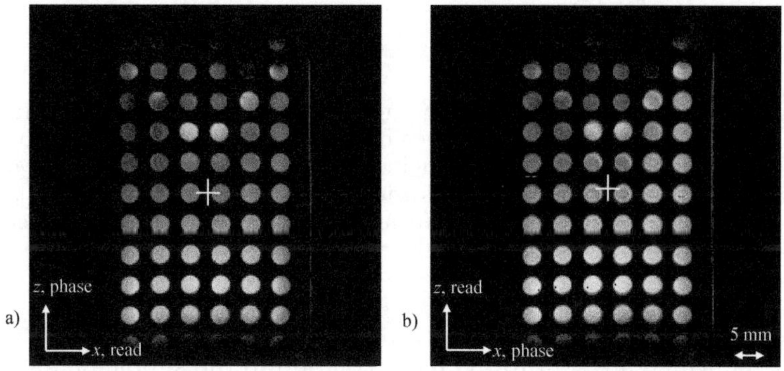

Figure 5.50 Proton density images of a water-filled phantom acquired with a fast gradient echo sequence using gradient microscopy coil prototype A. The selected slice lies in the *xz*-plane at $y = 0$. Imaging parameters are field-of-view (FOV) 30×30 mm^2, data matrix 256×256, bandwidth 83 Hz, TR 4.39 ms. (a) Using the *z*-gradient for phase-encoding, (b) using the *x*-gradient for phase-encoding.

although these have not been widely implemented. In 1996, a new concept of switchable gradients was introduced, which allows switching between elliptical LVs of ~20 cm and ~50 cm (see Figure 5.51).[31,32] Three years later, the first gradient system with continuously variable field characteristics was constructed.[33] This section describes the design of the latter type of gradient coil setup, which consists of two physical coils, a small field-of-view *basic coil*, and a field-of-view extending supplementary coil see Figure 5.51.

The first step was to develop a new gradient controller concept, which allows instantaneous switching between different gradient field characteristics within a measurement sequence. An arbitrary *field type* can be assigned to each single gradient pulse within the sequence. The controller

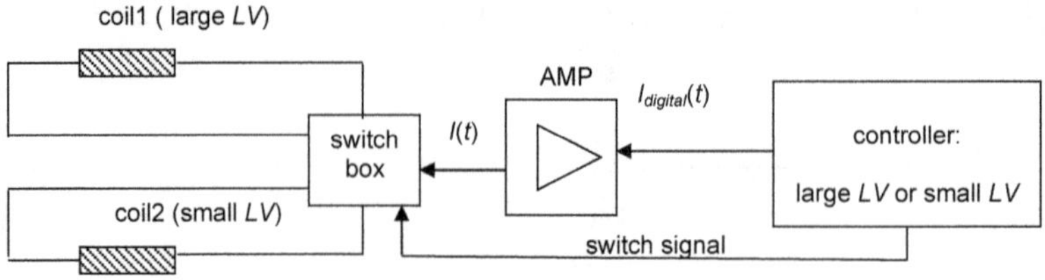

Figure 5.51 Block diagram of a gradient system with switchable field characteristics using two independent gradient coils (coil1, coil2) and a switching mechanism (switch box) for the current path from the amplifier (AMP).

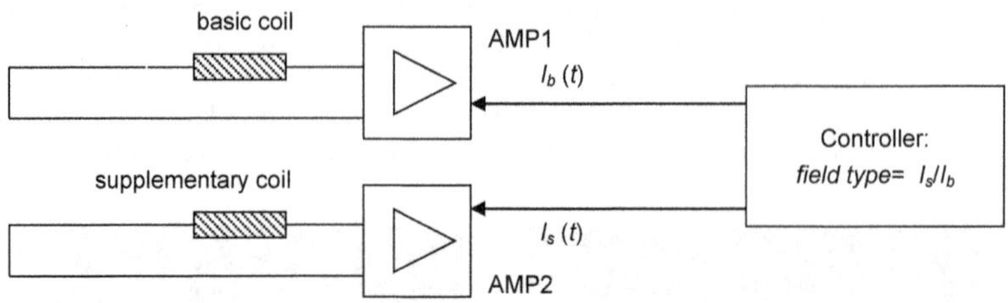

Figure 5.52 Gradient controller providing continuously variable field characteristics. The basic coil is driven by the first amplifier (AMP1), which generates a strong, rapidly changing gradient field in a small imaging volume. The supplementary coil is driven by the second amplifier (AMP2), which extends (or reduces) the linearity volume of the basic coil. The field type is defined by the ratio of the two currents and the sign of I_s.

then generates two differently scaled gradient pulses based on the *field type* parameter, which are sent to the basic coil and to the supplementary coil, respectively (see Figure 5.52).

The field characteristics of each axis can be described as a linear superposition of the basic and the supplementary coils' fields. The magnetic field $B(r,\theta,\phi)$ can be expressed using the coefficients $C_{n,m}$ of the spherical harmonics Y_{nm} (as described in Appendix B, Chapter 1).

The relative weights of these terms, and thus the field properties, depend on the ratio α of the basic coil's maximum current, $I_{\mathrm{b,max}}$, and that of the supplementary coil, $I_{\mathrm{s,max}}$, and its polarity, sign (I_s):

$$\alpha = \left|\frac{I_{\mathrm{s,max}}}{I_{\mathrm{b,max}}}\right| \cdot \mathrm{sign}(I_s) \tag{5.21}$$

The magnetic field produced by the gradient system can be written as a function of the coefficients of the basic coil $A_b(n,m)$ and supplementary coil $A_s(n,m)$ and the current ratio α:

$$B_{\mathrm{b+s}}(r,\theta,\phi) = \sum_{n,m}\left((C_b)_{n,m} + \alpha (C_s)_{n,m}\right) r^n Y_{nm}(\theta,\phi) \tag{5.22}$$

The supplementary coil generates spherical harmonics that increase the linearity of the basic coil at a greater radius. Scaling of I_s relative to I_b results in a proportional change in the homogeneity volume (see Figure 5.53). As the linearity volume of each physical gradient axis results from linear superposition of independent coil sets, multiple linearity volumes may be mixed on the same physical axis. This additional degree-of-freedom allows assigning arbitrary field types to *logical* gradient axes, in contrast to a conventional switchable system.

Two different design strategies for modular gradient coils can be used: first, stacking of independent, shielded coil sets,[32] which results in a reduced patient bore or lower duty cycle (owing to reduced conductor cross-section) in comparison to a conventional coil. The second approach involves surface partitioning, in which two coil sets share a common surface. The partitioning approach (see Figure 5.54) was used to calculate the current loops for a gradient coil of 1.2 m length and 0.68 m inner bore diameter. The chosen optimization algorithm includes boundary conditions for inductance, resistance and eddy current effects within the target field volume. Rigid body movement of the longitudinal coil owing to spatial variations of the radial component of the main magnetic field $B_{0,r}$ was minimized by balancing the forces of the primary and secondary current surfaces. Optimal gradient efficiency and shielding ratio was achieved by calculating the current density on

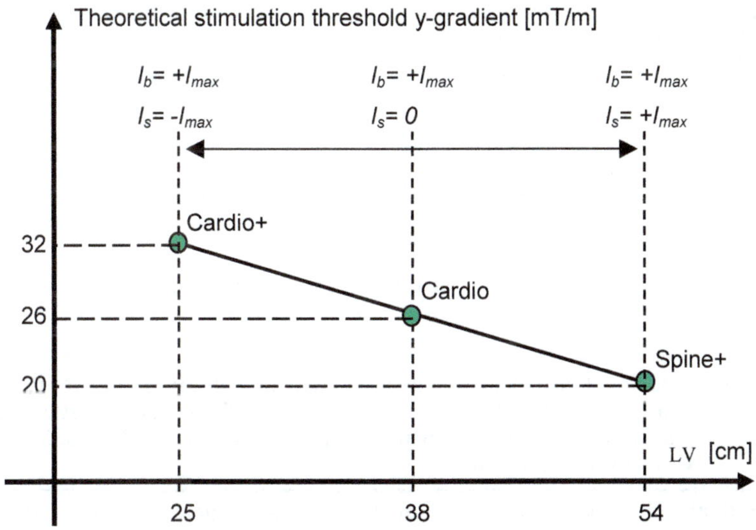

Figure 5.53 Dependence of the linearity volume and stimulation threshold of the *y* gradient axis from supplementary coil current I_s for a constant basic coil current I_b. The stimulation threshold of the DSV 54 system was determined in a clinical study and extrapolated to smaller DSVs using the Irnich model. The three major linearity volumes are depicted as 'Cardio+', 'Cardio' and 'Spine+', according to their preferred applications. Reproduced from ref. 39 with permission from John Wiley and Sons. Copyright © 2002 Wiley-Liss, Inc.

a three-dimensional surface including both primary and secondary planes. In the first step, calculations for a large DSV of 54 cm were made. In a second step, a suitable combination of conductors for both basic and supplementary coil was determined (Table 5.2).

A modular 6-channel gradient coil was built for a 1.5 T scanner and tested with standard system tune-up procedures. Two gradient amplifiers (2 kV/500 A) were connected in parallel to drive the basic and supplementary coils. System performance was measured for three representative field modes (see Table 5.2). Geometric distortions over three representative DSVs of the 6-channel gradient coil were verified with phantom images and TSE head scans (Figures 5.55 and 5.56).

For the experimental determination of the stimulation thresholds of the gradient system, a trapezoidal bipolar gradient wave form (64 periodic cycles, rise times 200 μs, 300 μs, and 400 μs, flat top 500 μs) was applied in eleven healthy adult volunteers (9 male, 2 female, average age 38 years). Each

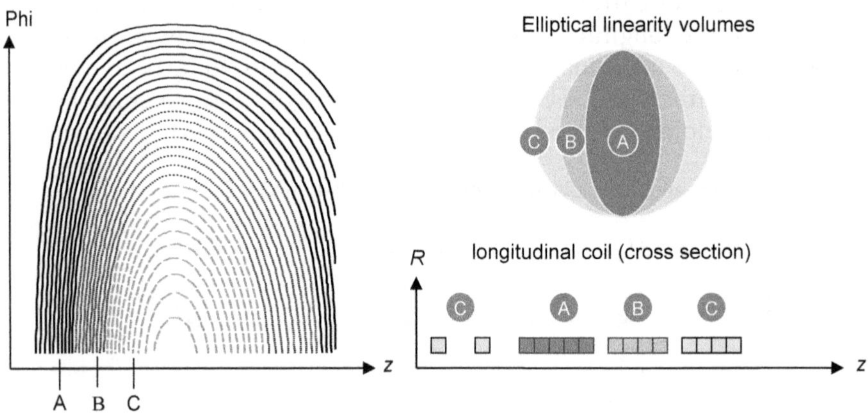

Figure 5.54 Principle of surface partitioning for transverse gradient coils (left) and longitudinal coils (right).

Table 5.2 Calculated properties of the individual and combined coils for the modular gradient set with continuously variable field characteristics. Reproduced from ref. 39 with permission from John Wiley and Sons. Copyright © 2002 Wiley-Liss, Inc.

Longitudinal coil	Sensitivity [μT A^{-1} m^{-1}]	Inductance [μH]	DSV (±10%) [cm]
Basic	81.2	660	42
Supplementary	n/a	327	n/a
Combined (B + S)	95.3	1272	56
Transverse coil	Sensitivity [μT A^{-1} m^{-1}]	Inductance [μH]	DSV (±10%) [cm]
Basic	85.2	666	38
Supplementary	n/a	370	n/a
Combined (B + S)	95.6	1316	54

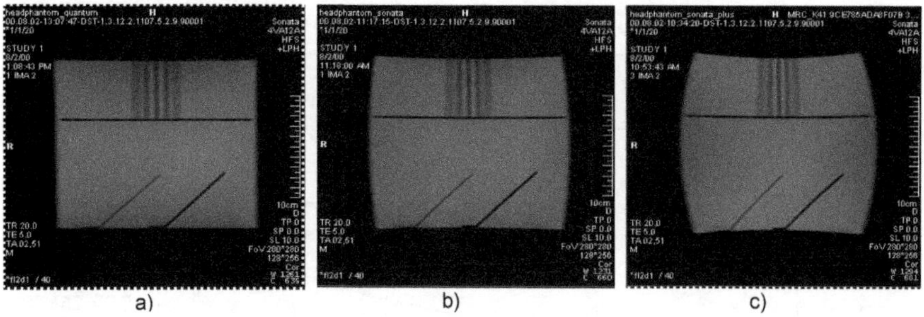

Figure 5.55 Examples of geometric distortion of MR images at three different linearity volumes for a coronal slice with FOV 28 × 28 cm. (a) DSV 54 cm, (b) DSV 38 cm, (c) DSV 25 cm. Reproduced from ref. 39 with permission from John Wiley and Sons. Copyright © 2002 Wiley-Liss, Inc.

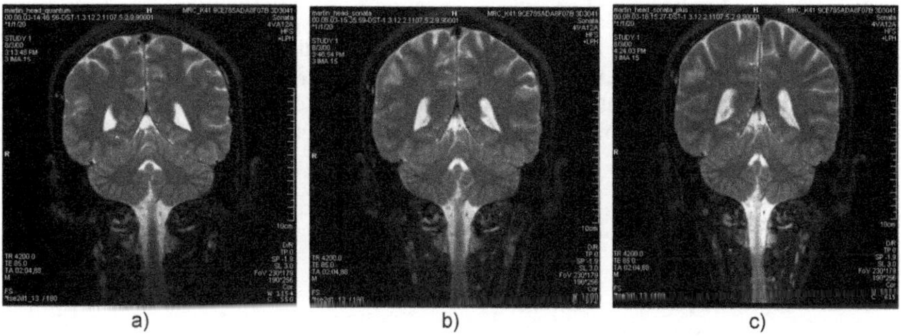

Figure 5.56 Coronal TSE head scans (FOV 23 × 18 cm) at three different linearity volumes (a) DSV 54 cm, (b) DSV 38 cm, (c) DSV 25 cm. Reproduced from ref. 39 with permission from John Wiley and Sons. Copyright © 2002 Wiley-Liss, Inc.

volunteer was positioned with their nose in the isocenter of the gradient coil. The gradient amplitude was increased in steps of 1 mT m^{-1} until subjects perceived a slight stimulation. This experiment was repeated for DSVs of 25 cm, 38 cm and 54 cm (see Figure 5.57). For a rise time of 200 μs, the stimulation threshold for the 25 cm DSV was increased by a factor of 1.5 to 1.7 in comparison to the 54 cm DSV. The experimental results agree well with theoretical predictions and support the assumption of a linear relationship between stimulation threshold and B_{max}.

5.3.3 A Head Gradient Coil Insert

As mentioned earlier in this chapter, the limits in using a whole-body gradient coil with maximum gradient amplitude in combination with high slew rates are primarily defined by PNS.[34] Suitable "countermeasures" include the reduction of the LV for a conventional gradient system or implementation

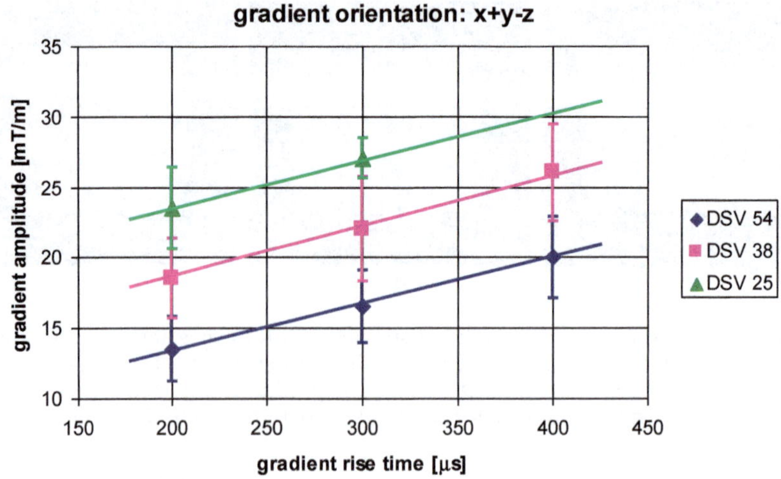

Figure 5.57 Comparison of stimulation experimental thresholds for the gradient axis combination $(x + y - z)$. In case of a 400 µs rise time and 25 cm DSV, stimulation was not perceived. Reproduced from ref. 39 with permission from John Wiley and Sons. Copyright © 2002 Wiley-Liss, Inc.

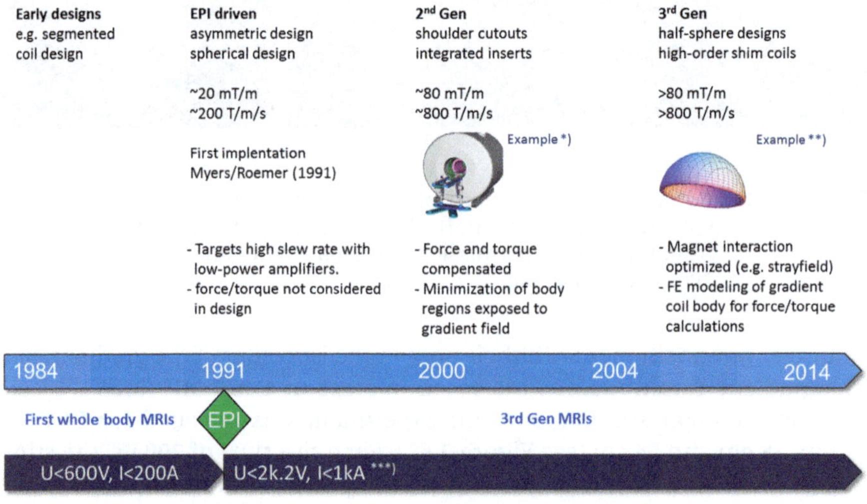

Figure 5.58 History of head gradient design. *) R. Kimmlingen, M. Gebhardt and E. Eberlein, *et al.*, *ISMRM*, 2004. **) M. Poole and R. Bowtell, Dome gradients, *ISMRM*, 2007. ***) Not fully applicable owing to current density/heat removal limitations (*e.g.* gradient inserts).

of a variable LV as covered in the previous section. These approaches may help to keep the usable slew rate of a whole-body MR scanner in the region of 200 T m^{-1} s^{-1}, but cannot enable ultra-high slew rates of 400 T m^{-1} s^{-1} and above. These values can only be reached by using smaller gradient coils, such as head-sized coils (see Figure 5.58). The benefits of reduced-diameter gradients are twofold. First, PNS limits become less restrictive owing to the fact that only a small part of the human body is exposed to the time-varying fields. Second, coil inductance is reduced by the fifth order of the radius, thus enabling higher slew rates and maximum gradient amplitudes at the same time. The disadvantage of an MR scanner with gradient systems for dedicated body regions is obvious: it can no longer be used as a general-purpose machine. For head imaging, this limitation can be overcome by designing the gradient coil as an *insert* to the whole-body system. This section focuses on the gradient coil specifics of such an approach.

A high-performance head gradient insert coil was developed specifically for use inside a clinical whole-body magnet with a patient bore diameter of 60 cm.[35] The total length of the insert gradient was 70 cm with an inner diameter of 36 cm and an outer diameter of 52 cm. The coil combines transverse asymmetric saddle coils together with longitudinal windings, which leave space for shoulder cut-outs (see Figure 5.59 and Tables 5.3 and 5.4). These

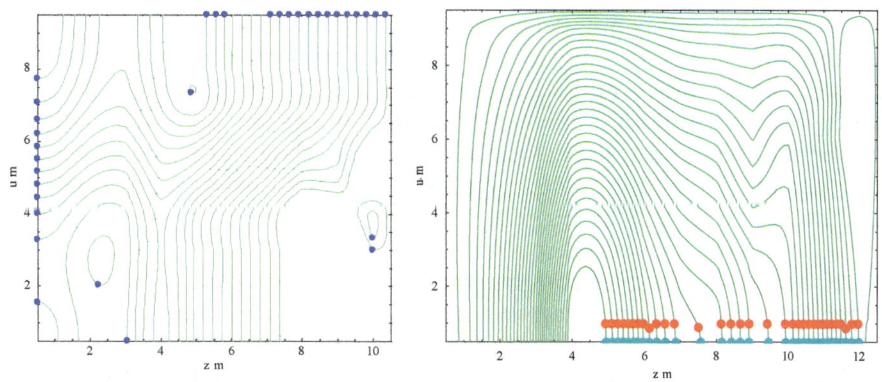

Figure 5.59 Contour lines of the stream function of an asymmetric head gradient coil (design example). Left: (symmetric) *z*-axis octant with shoulder cutouts. Right: asymmetric transverse axis quadrant.

Table 5.3 System performance as measured on the weakest gradient axis (*z*) of the head gradient insert coil. The performance index (PI) is given as product of peak gradient amplitude and slew rate. The DSV relates to the least linear gradient axis (*y*). Reproduced from ref. 39 with permission from John Wiley and Sons. Copyright © 2002 Wiley-Liss, Inc.

Fieldmode (transverse)	Peak gradient @ 475 A [$mT\ m^{-1}$]	Slew rate [$T\ m^{-1}\ s^{-1}$]	PI [$T^2\ m^{-2}\ s$]	Ramp time @ 2 kV [μs]	DSV (±10%) [cm]
Spine+	40	190	7.6	210	54
Cardio	36	200	7.2	180	38
Cardio+	32	210	6.7	150	25

Table 5.4 Technical data of the head gradient coil insert 'AC88'. The theoretical slew rate values are calculated for a peak current of 450 A without considering filters, chokes and the performance of the amplifier's regulation circuit.

Gradient axis	Sensitivity [μT A^{-1} m^{-1}]	Inductance [μH]	SR @ 2 kV [T m^{-1} s^{-1}]	LV 22 cm Dev [%]
X	206	396	936	6.7
Y	199	390	904	6.6
Z	193	258	877	7.5

allow proper positioning of the subject's head within the imaging volume. The gradient coil is shielded and torque-balanced, and is vacuum-potted with a filled epoxy resin. Special skids allow easy longitudinal movement into and out of the patient bore. The design was integrated into a 3 T whole-body MRI system, since the majority of clinical neuroimaging studies are performed at 3 T. The gradient coil was used in combination with a special mechanical handling device that allows efficient switching of the operational mode of the system between whole-body and head insert gradients. This switching procedure includes moving the coil manually by a crank and switching the gradient power cables and water cooling to the desired gradient coils with a single Y-switch. Both gradient operational modes use the same gradient power amplifier and system electronics. Close to the magnet, the cables for the gradient insert are guided by rolls to constrain movement caused by Lorentz forces. The switching procedure requires about 15 minutes and can be performed by a single person.

Acoustic noise produced by the gradient coil is an issue to be considered carefully. Some sequences can reach noise levels above the legal limits if driven at a mechanical resonance frequency. The reason for the potentially high vibration and noise amplitude of a head insert is the light weight and slim structure. In order to allow for quick exchange, the gradient coil including noise damping cover needs to fit into the whole-body patient bore. This limits the wall thickness of the gradient coil body and makes it difficult to provide a vibration-optimized suspension inside the bore. Wider use in a clinical environment could be enabled with a slightly modified electrical design (*e.g.* enabling more radial space for noise dampening measures by sacrificing some slew rate performance). The gradient suspension could be improved by increasing the wall thickness of the coil body and fixed installation in the magnet bore.

5.3.4 Ultra-Strong Whole-Body Gradients

Implementation of ultra-high gradient amplitudes with head-sized gradients inside a whole-body magnet bore is one approach to tackle the problem of limited gradient performance. However, the advantage of simultaneous ultra-high slew rate and gradient amplitude is counteracted by the limited space inside the coil. It is difficult to design a combined transmit/receive RF coil with 32 receive channels, good intra-channel decoupling and high SNR within the available space of a head insert gradient. Reducing the interactions between the gradient and the transmit coil is also extremely challenging. In addition, patient handling is difficult. It is not easy to find volunteers who are willing to expose themselves to the near-claustrophobic conditions of such an experimental device. Even with asymmetric gradient design, the limited size of the shoulder cutouts excludes a significant proportion of the normal population from being scanned. This makes the question of whether ultra-high gradient strength (>100 mT m^{-1}) can be achieved with a *whole-body gradient design* highly relevant. This section focuses on the gradient coil specifics of this latter approach.

As part of the National Institutes of Health Blueprint initiative in North America, the Human Connectome Project (HCP) was set up to explore the connections in the brain. The major MRI applications in the HCP include resting-state fMRI, diffusion-weighted MRI, and task-related fMRI. The NIH funded two MRI scanners dedicated to this effort, one at the Center for Magnetic Resonance Research (CMRR) in Minneapolis, later moving to Washington University, St. Louis, the other at the Massachusetts General Hospital (MHG)'s Martinos Center, Boston, in cooperation with University of California Los Angeles (UCLA). Two different gradient systems were specially designed for the NIH blueprint initiative.[17]

The design targets for a whole-body gradient were very high gradient amplitudes and an LV limited to the brain (thus entailing less restrictive PNS thresholds). Two versions of this gradient design were designed and built. Version 1 (SC72) supports a G_{max} of 100 mT m^{-1} with an SR of 200 T m^{-1} s^{-1} powered by a single GPA cabinet (2 kV, 1 kA). It was designed to match the forces and stray field of a 3 T magnet and provide space for passive iron shims. It had a length of 158 cm, an inner diameter of 64 cm and outer diameter of 81 cm. This yielded a robust, easy-to-use "engine" for diffusion-weighted imaging.

Version 2 (AS302) represented a quantum leap in whole-body gradient performance, *i.e.* a G_{max} of 300 mT m^{-1} at an SR of 200 T m^{-1} s^{-1}. Design studies with reduced linearity constraints showed that a G_{max} of 150 mT m^{-1} could be reached with a single gradient power amplifier. Owing to the large volume and usable length of the coil body, wire cross-section and cooling did not impose a duty-cycle limit at this amplitude. The amplitude was increased to 300 mT m^{-1} by doubling the number of current density layers within the coil body. To drive this high inductance at the SR = 200 T m^{-1} s^{-1} needed for EPI readout, a new gradient system concept involving multiple gradient amplifiers was developed. In order to achieve a slew rate of >200 T m^{-1} s^{-1}, each of the three axes X,Y,Z was split into four independently-driven segments (see Figure 5.60). The transverse axes gradients were split into four saddle coil stacks

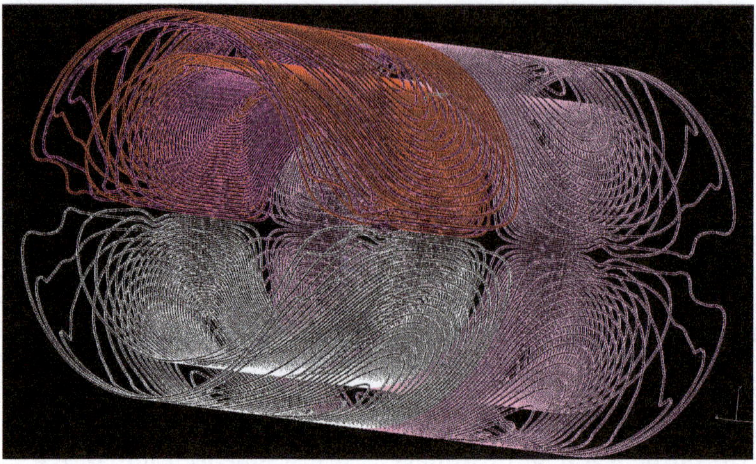

Figure 5.60 Design study of a transverse gradient axis capable of generating 150 mT m^{-1}. Each of the four actively shielded saddle coils can be driven independently. A G_{max} of 300 mT m^{-1} is achieved by stacking two saddle coils for each segment and connecting them in series.

which show only moderate mutual coupling. The longitudinal gradients were also split into four sections which had similarly low mutual coupling. Stray field and forces were matched to the 3 T magnet used. The MR control system was extended to drive four sets of gradient amplifiers independently. The new architecture requires storing the calibration data for each of the 12 final stages driving the gradient coil segments.

The gradient waveform is logically split and fed to four individual gradient controllers (see Figure 5.61). This architecture allows arbitrary field characteristics to be generated for each gradient coil axis, which can be used to optimize eddy current compensation. Mutual coupling of the 12 gradient coil segments poses a challenge for the GPA regulator (PID control) and thus image quality for fast imaging sequences such as EPI. The GPA regulator architecture was extended to account for the dynamic differential control (D) of the driving signal. This allows the induced voltages in each segment coil owing to mutual coupling to be counteracted. A new RF body coil was developed which uses the existing clinical patient table with minor mechanical modifications and supports a patient bore of 560 mm. The magnet covers were modified accordingly. Two additional cooling cabinets are needed for thermal management of the gradient system, all installed in the technical room.

In diffusion imaging of the brain, typical TEs for b = 1000 s mm^{-2} on a whole-body scanner (G_{max} ~ 40 mT m^{-1}) are around 70–80 ms when Stejskal–Tanner encoding is applied without additional parallel imaging techniques.[36] Higher b values can only be achieved with a penalty in SNR, as

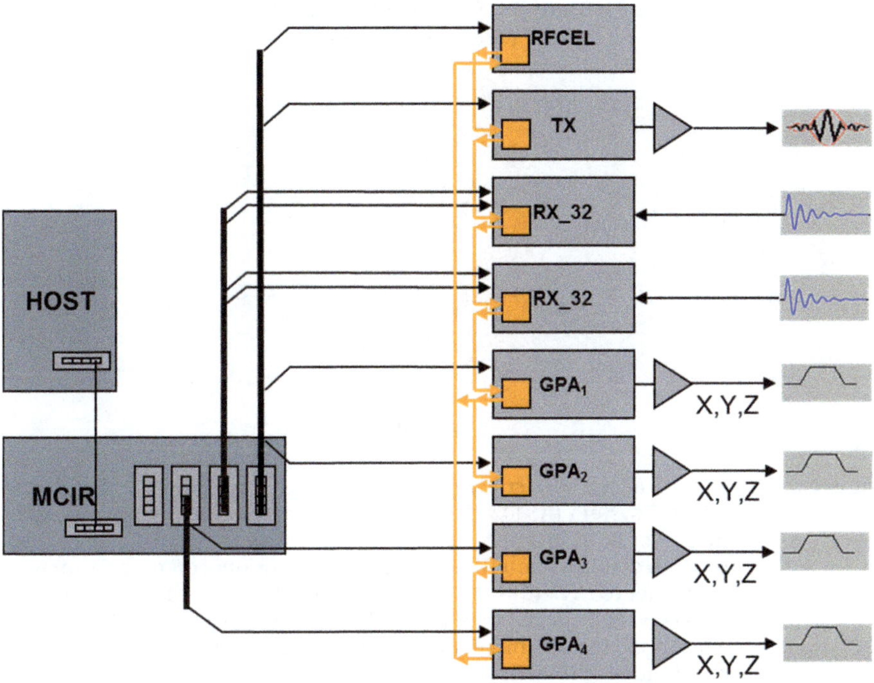

Figure 5.61 Modular control system architecture for the 300 mT m^{-1} gradient, driven by four gradient amplifiers.

TE increases with the duration of the diffusion lobe. Given the quadratic scaling of b with G_{max} and the exponential loss of SNR at longer TE, increasing the maximum gradient amplitude is highly desirable. Results using the connectome project gradients included diffusion spectral imaging (DSI) with b = 10 000, 512 directions and TE = 71 ms performed successfully at MGH. In addition, resting-state fMRI using 6-fold slice acceleration with the multi-band approach[37] was demonstrated with the SC72-based system at CMRR. PNS studies performed on the AS302 and SC72 coils showed that it is possible to use EPI readout amplitudes of 40 mT m^{-1} at SR 200 T m^{-1} s^{-1} without PNS. Gradient pulses with long rise times at high-amplitude are limited by the regulatory required cardiac monitor, which was implemented in hardware. While performing the PNS study, visual effects, resembling those described in the literature decades ago as phosphenes[38] were experienced by the first few subjects. Data are shown in Figure 5.62, taken with a trapezoidal EPI readout pulse on the AS302 gradient system, located below the PNS and cardiac thresholds. The PNS study was therefore only performed with the rise time range <800 μs (below the threshold for visual effects) and is extrapolated for higher rise times.

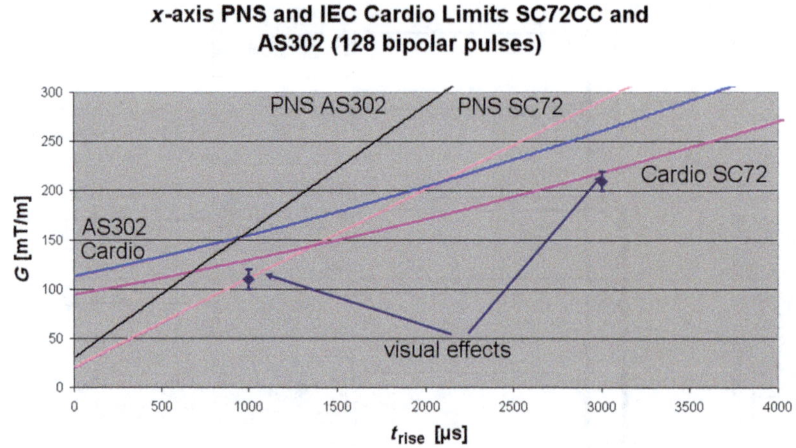

Figure 5.62 Results of the PNS study for the 100 mT m^{-1} system (SC72) and 300 mT m^{-1} system (AS302). The linearity volume of both coils was optimized for head imaging, hence the usable slew rate is 200 T m^{-1} s^{-1}. In order to conform with IEC standards, an additional d*B*/d*t* monitor was implemented (Cardio SC72, Cardio AS302). Additional stimulation effects (visual effects) were notified in the study of the AS302 gradient, while applying 128 bipolar pulses with a ramp duration of ≥1 ms and amplitudes >100 mT m^{-1}.

References

1. P. Lauterbur, *Nature*, 1973, **242**, 190.
2. P. Mansfield, *et al.*, *Br. J. Radiol.*, 1977, **50**, 188–194.
3. P. Mansfield, *J. Phys. C: Solid State Phys.*, 1977, **10**, 155.
4. R. Turner, *Magn. Reson. Imaging*, 1993, **11**, 903.
5. F. Romeo and D. I. Hoult, *Magn. Reson. Med.*, 1984, **1**, 44.
6. S. Crozier and D. M. Doddrell, *Magn. Reson. Imaging*, 1995, **13**, 615.
7. H. Siebold, *IEEE Trans. Magn.*, 1990, **26**, 897.
8. R. C. Compton, US Patent 4,456,881, 1982.
9. R. Turner and R. M. Bowley, *J. Phys. E*, 1986, **19**, 876.
10. S. Pissanetzky, *Meas. Sci. Technol.*, 1992, **3**, 667.
11. R. Turner, *J. Phys. D*, 1986, **19**, L147.
12. P. Mansfield and B. L. W. Chapman, *J. Magn. Reson.*, 1986, **66**, 573.
13. R. Turner, *J. Phys. E: Sci. Instrum.*, 1988, **21**, 948.
14. R. Bowtell and P. Mansfield, *Proc. SMRM*, 1989, 977.
15. G. Pausch, US Patent 5,309,107, 1992.
16. W. Arz, M. Gebhardt, F. Schmitt and J. Schuster, British Patent Application 2,347,505, 1999.
17. R. Kimmlingen, *et al.*, *Proc. ISMRM*, 2012, 696.
18. M. S. Cohen and R. M. Weisskoff, *Magn. Reson. Imaging*, 1991, **9**, 1.
19. F. Schmitt, *et al.*, *Proc. ISMRM.*, 1991, 116.
20. E. Yamamoto and H. Kohno, US Patent 4,959,613, filed 1989.

21. M. Vester, *et al.*, *Proc. ISMRM*, 2005, 13.
22. S. Patz, *et al.*, *Int. J. Imaging Syst. Technol.*, 1999, **10**(**3**), 216.
23. A. Kaindl, *et al.*, *Proc. Coil Winding Conf.*, 2001, 551.
24. O. M. Mueller, P. Roemer, J. N. Park and S. P. Souza, *Proc. ISMRM*, 1991, 130.
25. K. H. Ideler, S. Nowak, G. Borth, U. Hagen, R. Hausmann and F. Schmitt, *Proc. SMRM*, 1992, 1044.
26. H. Lenz, German Patent 1, 985, 6800, filed 1998.
27. V. Weissenberger, German Patent 1, 985, 9501, filed 1998.
28. H. Adolf, PhD thesis, University of Würzburg, 1996.
29. R. Kimmlingen, PhD thesis, University of Würzburg, 1996.
30. F. Schmitt, M. Gebhardt, P. Wielopolski, W. Renz, H. Siebold, H. Wuerch, K. Meier, H. Fischer and J. W. Goldfarb, *Proc. ISMRM*, 1996, 499.
31. P. R. Harvey and E. Katznelson, *Magn. Reson. Med.*, 1999, **42**, 561.
32. P. Harvey, *MAGMA*, 1999, **8**, 43.
33. R. Kimmlingen, *et al.*, *Magn. Reson. Med.*, 2002, **47**, 800.
34. W. Irnich, *MAGMA*, 1994, **2**, 43.
35. R. Kimmlingen, *et al.*, *Proc. ISMRM*, 2004, 1630.
36. P. J. Basser and C. Pierpaoli, *Mag. Res. Med.*, 1998, (**6**), 928.
37. D. A. Feinberg, *et al.*, *PLoS One*, 2010, **5**(**12**), e15710.
38. T. F. Budinger, *IEEE Trans. Nucl. Sci.*, April 1979, **26**(**2**), 2821.
39. R. Kimmlingen, M. Gebhardt, J. Schuster, M. Brand, F. Schmitt and A. Haase, *Magn. Reson. Med.*, 2002, **47**, 800.

CHAPTER 6

Radiofrequency Amplifiers for NMR/MRI

NEAL A. HOLLINGSWORTH[a], KRISHNA KURPAD[b], AND STEVEN M. WRIGHT*[a,c]

[a]Department of Electrical, and Computer Engineering, Texas A&M University, College Station, TX 77843 USA; [b]Biotronik, Inc., Lake Oswego, OR, 97035 USA; [c]Biomedical Engineering, Texas A&M University, College Station, TX 77843 USA
*E-mail: smwright@tamu.edu

6.1 Introduction

The RF sub-system of an MR scanner consists of the transmit chain and the receive chain, separated by the transmit/receive switch. Figure 6.1 shows a simplified block diagram of the RF transmit chain, which consists of a frequency synthesizer that produces a sinusoidal carrier at the Larmor frequency and a waveform generator that amplitude-modulates the carrier. The amplitude-modulated carrier is then fed into the RF power amplifier (RFPA). The output of the RFPA is connected to the transmit RF coil *via* the transmit/receive switch. If a quadrature transmit coil is used, then a quadrature hybrid circuit is added in-line in order to provide two inputs (I and Q) with an appropriate 90° phase difference. Circuits for transmit/receive switches and quadrature hybrids are covered in Chapters 7 and 3, respectively.

The RFPA is a device that produces a high power pulsed RF output from a low power, amplitude-modulated RF drive. The primary objective of RFPA design is

New Developments in NMR No. 7
Magnetic Resonance Technology: Hardware and System Component Design
Edited by Andrew G Webb

Published by the Royal Society of Chemistry, www.rsc.org

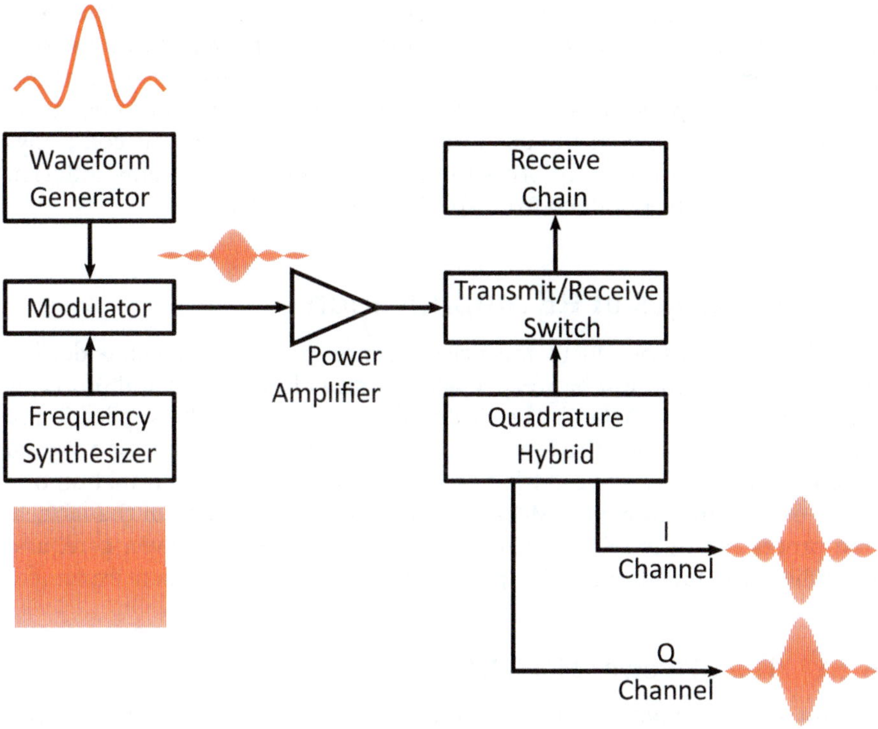

Figure 6.1 Simplified block diagram of the transmit chain of the MRI RF subsystem.

to ensure that the output waveform is an exact amplitude-scaled replica of the input waveform. The peak output power of MR RFPAs is typically of the order of 1–2 kilowatts for high resolution NMR, and up to tens of kilowatts for whole-body MRI. RFPAs may be also configured as continuous wave (CW) amplifiers. MRI RFPAs are high gain systems, where the gain is typically close to 60 dB. Such high gain is usually achieved using multiple stages of amplification. Most MRI RFPAs use solid state active devices such as bipolar junction transistors (BJTs) and metal–oxide–semiconductor field-effect transistors (MOSFETs) owing to their low cost, high reliability, small size and ease of cooling. BJTs and MOSFETs exhibit similar behaviour with the main difference between the two being that the former is a *current-controlled* device whereas the latter is a *voltage-controlled* device. A detailed treatment of the differences between the BJT and MOSFET may be found in, for example, Grant and Gowar.[1]

The principles of RF power amplification are independent of the configuration (CW or pulsed) or the type of device (BJT or MOSFET), although individual circuits may differ widely depending on the device type and design goals. In this chapter, the MOSFET is chosen as the active device to explain the principles of RF power amplification. First, a brief summary of MOSFET fabrication is given, followed by an explanation of the principles of RF power amplification based on the DC characteristics of the active device. The chapter then discusses the various types of distortions introduced into the output waveform owing to non-linear device behaviour. Finally, two relatively new types of amplifier designs, current source and low output impedance, which are finding increasing use in combination with transmit coil arrays are described. Additional information on RF amplifiers for many different applications can be obtained from specialized textbooks.[2,3]

6.2 Principles of RF Amplification

Field Effect Transistors (FETs) are a class of three-terminal solid state devices in which the current flow between a pair of terminals, the drain and source, is regulated by an electric field. The electric field is created by the application of a control voltage to the third terminal called the gate. The MOSFET is a special class of FET in which the electric field is created owing to the formation of a metal-oxide-semiconductor (MOS) capacitor. A brief summary of the fabrication process and the DC characteristics of the MOSFET are given here, and from there the operational principles of MOSFET RF amplifiers are explained.

6.2.1 The RF Power MOSFET

The RF power MOSFET is fabricated on a semiconductor (most often silicon) substrate called a die. MOSFET die fabrication involves several steps of solid state processes, which can be summarized as four general steps. These are illustrated by the schematics in Figure 6.2 showing the development of an N-channel enhancement mode MOSFET.

The fabrication process starts with a p-type substrate, which is a semiconductor where holes (positively charged virtual particles) are the majority carriers and the electrons (negatively charged particles) are the minority carriers. This is shown in Figure 6.2(a). Two low resistivity, n-type regions are then diffused into the substrate as shown in Figure 6.2(b). These form the drain and source regions of the MOSFET. The region between the drain and the source is where the channel is formed. The process of channel formation is explained in the following section. Following this a thin layer of silicon dioxide is deposited on the surface of the die as shown in Figure 6.2(c). The characteristics of the MOSFET are very sensitive to the purity and thickness of the oxide layer. Hence, these two parameters are carefully controlled during the deposition of the oxide. The oxide layer is prone to contamination by sodium ions and moisture present in the atmosphere, so the oxide is passivated with a layer of silicon nitride, which is impervious to both. The final fabrication step is metallization to allow electrical connections to be made to the die. Metal is overlaid on the insulator, covering the entire channel region, to form the gate terminal. Additionally, holes are etched into the oxide and nitride layers to allow for metallic contact with the drain and source regions that form the remaining terminals. Figure 6.2(d) shows the final structure of the MOSFET die.

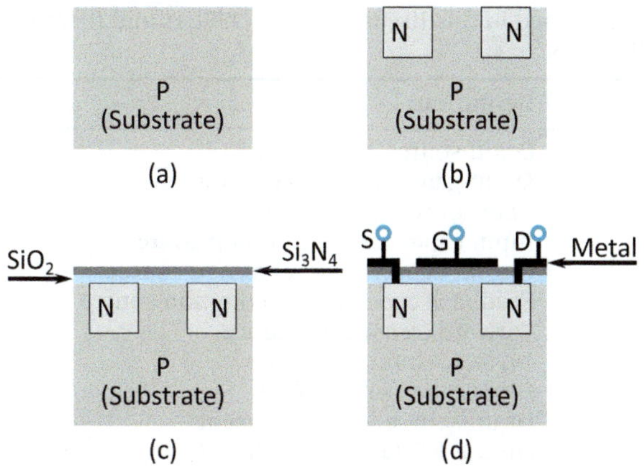

Figure 6.2 Summary of the steps involved in the fabrication of a MOSFET. Starting with the P-type substrate (a), the source and drain regions are formed by the diffusion of N-type wells into the substrate (b). The third step involves the deposition of the insulating oxide layer and the passivating nitride layer (c), and the final step involves metallization in order to develop the gate, drain and source contacts (d).

The N-Channel Enhancement mode MOSFET is characterized by the conduction of current from the drain to source, being facilitated by the formation of an n-type channel in a p-type substrate. The formation of the conduction channel requires that both the drain and gate are at a positive potential with respect to the source, in which case a drain-source current can flow. Depletion mode devices and P-channel devices also exist, but the N-Channel Enhancement mode MOSFET is the device most commonly used in the design of modern solid state amplifiers. The term "MOSFET" will be used to indicate an "N-Channel Enhancement mode MOSFET" unless otherwise stated. Table 6.1 shows a list of useful symbols and their definitions commonly used in MOSFET and RF amplifiers.

6.2.2 DC Characteristics

A curve-tracing circuit, *i.e.* one that measures the current and voltage characteristics of a particular circuit, shown in Figure 6.3, can be used to measure the DC characteristics of a MOSFET. The MOSFET is connected in a "common source" configuration, with the source terminal grounded. Variable DC voltage sources, V_{GG} and V_{DD}, are connected to the gate and drain terminals of the MOSFET, respectively. The ammeter A_D and voltmeters V_G and V_D are used to measure the drain current (I_D), gate voltage (V_{GS}), and drain voltage (V_{DS}), respectively.

This test circuit can be used to generate the "transfer curve" of the MOSFET, which shows how the drain current changes as the gate voltage is varied (these measurements are defined in the saturation region of the I–V curve as discussed below). The measurement is made by setting V_{DD} to be a constant

Table 6.1 Symbols used in the discussion of MOSFETs and RF amplifiers and their definitions.

Symbol	Definition
C_{DS}	Drain-source capacitance
C_{DG}	Drain-gate capacitance (same as C_{rs})
C_{GS}	Gate-source capacitance
C_{is}	Input capacitance, common source
C_{os}	Output capacitance, common source
C_{rs}	Feedback capacitance, common source
g_{fs}	Forward transconductance
I_D	On state drain current
I_{DQ}	Drain quiescent current
I_{DSS}	Drain-source leakage current
I_{om}	Output RF current amplitude
$R_{DS(on)}$	On-state drain-source resistance
R_L	Load resistance
θ	Angular time in radians; $\theta = \omega t$
$V_{BR(DSS)}$	Drain-source reverse bias breakdown voltage
V_{DSat}	Drain-source saturation voltage
V_{DD}	Drain DC supply voltage
V_{DS}	Drain-source voltage
$V_{DS}(\theta)$	Drain-source instantaneous RF voltage
V_{GG}	Gate DC supply voltage
V_{GS}	Gate-source voltage
$V_{GS}(\theta)$	Gate-source instantaneous RF voltage
V_{im}	Input RF voltage amplitude
V_{om}	Output RF voltage amplitude

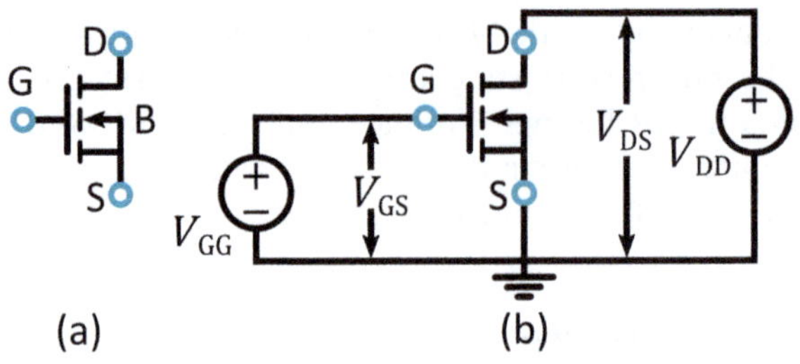

Figure 6.3 Symbolic representation of a MOSFET. (a) The symbolic representation shows the gate (G), drain (D), source (S) and bulk (B) contacts. (b) A schematic of the curve tracer circuit that is used to determine the DC characteristics of the MOSFET.

voltage that is relatively large (on the order of tens of volts). V_{GG} is then varied from zero to some positive value of a few volts while recording V_{GS} and I_D. A plot of I_D against V_{GS} results in the characteristic curve of Figure 6.4. The transfer curve can be divided into four major regions based on the relationship between I_D and V_{GS}. In the first region, where $V_{GS} < V_T$, I_D is zero. In the second region, a small quadratic increase in I_D relative to V_{GS} is seen when $V_T < V_{GS} < V_L$. This is followed by the third region in which I_D increases linearly when $V_L < V_{GS} < V_U$. Finally, I_D begins to saturate when $V_{GS} > V_U$.

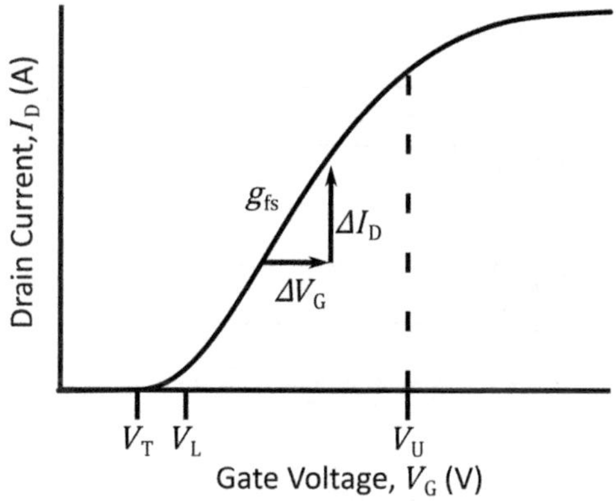

Figure 6.4 Transfer curve of a MOSFET shows the drain current as a function of applied gate voltage.

The forward transconductance (g_{fs}) of the MOSFET, which governs its ability to amplify, is commonly defined for the linear region of the transfer curve as:

$$g_{fs} = \frac{\Delta I_D}{\Delta V_{GS}} \tag{6.1}$$

The RF behaviour of the MOSFET is based on the periodic traversal of the transfer curve corresponding to the periodic swing of the applied gate voltage. The transfer curve is used to define the quiescent (or operating) point of the RF amplifier based on this periodic traversal. The same basic test system can be used to generate a set of *I–V* curves that relate the drain current to the drain voltage at fixed gate voltages. In this case, V_G is fixed at some voltage and V_D varied from zero to V_{DD}, while measuring V_{DS} and I_D. This is repeated for different values of V_G between zero and V_{GS}. The resulting characteristic curves are shown in Figure 6.5.

Three distinct regions of operation, marked in Figure 6.5, are defined as follows:

- In the "cutoff" region, when $V_{GS} < V_T$, I_D is zero. This region can be represented in circuit form by Figure 6.6(a) which contains back-to-back diodes.
- In the "active" region, when $V_{GS} > V_T$ and $V_{DS} < V_{DSat}$, I_D increases linearly with V_{DS}. The projection of OA′ on the abscissa in Figure 6.5 represents the range of V_{DSat} over the range of V_G. This is symbolically represented as a series resistor, $r_{DS(on)}$, in the drain-source path seen in Figure 6.6(b).
- In the "saturation" region, when $V_{GS} > V_T$ and $V_{DS} > V_{DSat}$, I_D is effectively independent of V_D. The MOSFET behaves as a true current source in this region, shown symbolically in Figure 6.6(c).

These measurements, and the resulting curves, define the basic operation of a MOSFET. They are key to understanding how a MOSFET functions as an

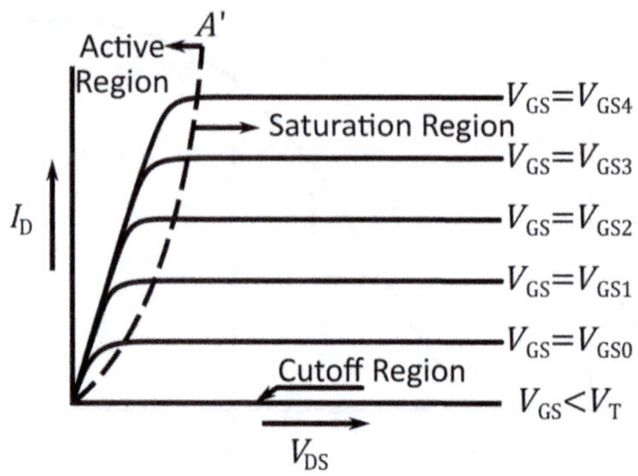

Figure 6.5 *I–V* curves of a MOSFET. Each curve corresponds to a specific setting of V_{GS}. The three regions of operation (active, saturation and cutoff) are shown.

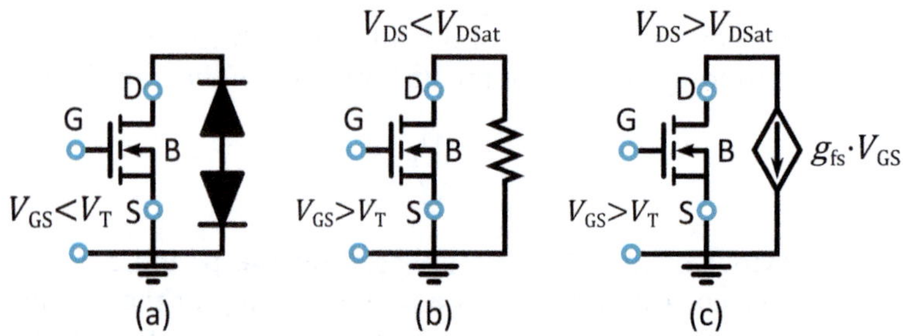

Figure 6.6 Symbolic representation of the three MOSFET states showing the cut off (a), active (b) and saturation (c) states of the MOSFET.

amplifier, but are not sufficient to understand its behavior at RF frequencies, so the following sections take an expanded view of the high frequency characteristics of a MOSFET.

6.2.3 RF Characteristics

The physical structure of the MOSFET results in the formation of parasitic capacitances, *i.e.* capacitance that occurs owing to the proximity of different physical structures within the device, that are not significant at low frequencies but which do have a significant effect on the MOSFET response at RF. These capacitances occur between the various nodes that are isolated by insulating regions, shown diagrammatically in Figure 6.7. The interface of the drain and substrate forms a diode that is reverse biased, resulting in a junction capacitance C_{DS}. The value of C_{DS} depends on the applied reverse-voltage across the diode. The gate metallization partially overlaps the diffused N-type regions of the gate and drain, giving rise to C_{GS} and C_{GD}. These are not voltage-dependent because they are not junction capacitances.

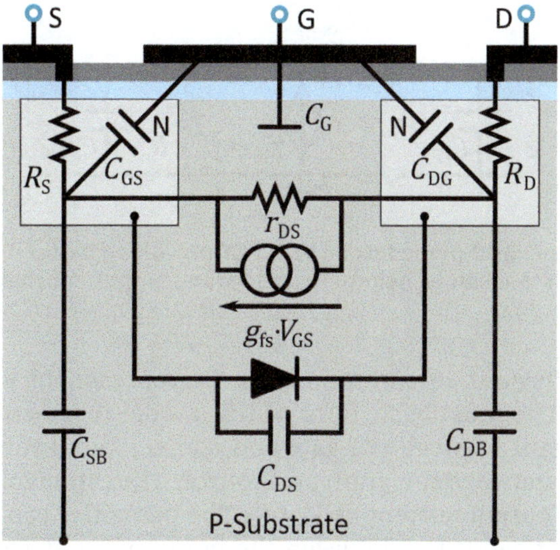

Figure 6.7 Parasitic capacitances of a MOSFET. C_{GS} and C_{DG} are the parasitic capacitances formed owing to overlap of the gate metallization over the source and drain regions, respectively. C_{DS} is the drain-source capacitance formed owing to the reverse-biased diode at the drain-substrate interface.

Typically, a MOSFET datasheet does not specify C_{DS}, C_{GS}, and C_{DG}. Instead, values of the output capacitance, C_{os}, input capacitance, C_{is}, and feedback capacitance, C_{rs}, are reported, values which represent hybrids of the previously mentioned capacitances. These two sets of values are related as follows:

- C_{os} is measured with V_{GG} set to ground potential and is given by:

$$C_{os} = C_{DS} + C_{GD} \tag{6.2}$$

  C_{os} appears as a capacitance across the output terminals of the MOSFET. Since the internal resistance of the MOSFET current source is very high, the output impedance of the MOSFET essentially comprises the output capacitance.
- C_{is} is measured with the drain shorted to the sourced terminal, and appears across the input terminals of the MOSFET. The value of C_{is} is given by:

$$C_{is} = C_{GS} + C_{GD} \tag{6.3}$$

  C_{is} does not vary significantly with the applied gate voltage. The effect of C_{is} is to reduce the input impedance of the MOSFET. This makes matching to a 50 Ω signal source more difficult.
- C_{rs} is the feedback capacitance and is the same as C_{GD}. This capacitance can be minimized through proper layout of the MOSFET.

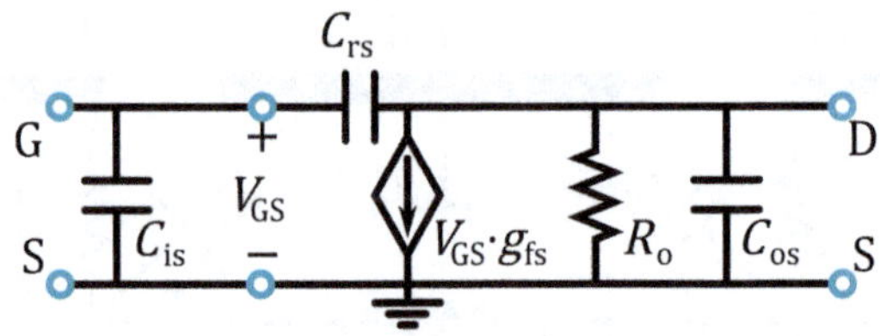

Figure 6.8 The hybrid-pi model of a MOSFET provides a basic circuit level description that can be helpful to understand amplifier behaviour.

MOSFET behaviour in saturation can be represented with a *hybrid-pi circuit model*.[4] The most basic form of this model represents the MOSFET as an open circuit between the gate and source, and a voltage-controlled current source between the drain and source. The input voltage node (V_{GS}) modulates the output current as $V_{GS}\cdot g_{fs}$. The parasitics can be included as lumped element capacitors between the nodes of the device, as shown in Figure 6.8. The increased complexity of these additional elements improve the model's accuracy, but the inherently non-linear operation of the MOSFET in large signal operation makes it difficult to model its behaviour completely.

It is important to remember that most elements of this model are modulated by the bias point, and the values of V_{DS} and I_{DQ} can have especially strong influences on C_{os} and g_{fs}, respectively. Furthermore, C_{OS} is a function of the drain-source voltage. So for a given operating point (*i.e.* a pair of gate and drain DC bias voltages) one can model the device with a fairly simple circuit.

One of the first major design questions is how the quiescent point of the amplifier should be selected, and what impact it has on performance. This is, essentially, a question of what "class" an amplifier should be.

6.2.4 Amplifier Classes

There are a number of different classes of RFPA. An amplifier's class of operation is determined by the percentage of the sinusoidal cycle of an oscillating input that results in conduction at the output of the amplifier. This is controlled by the quiescent point of the amplifier, which is a function of the DC gate and drain voltages. Generally, a higher bias voltage at the gate leads to a more linear amplifier. The basic linear classes of operation for amplifiers are referred to as A, AB, B, and C, in order of the highest to lowest bias voltage, as shown in Figure 6.9. Each of these classes is considered briefly here to illustrate how they are biased, and some of the ramifications of selecting a given class of amplifier.

Class A operation is obtained when the gate bias is set to roughly the middle of the linear region of the transfer curve of the MOSFET. This corresponds to $V_{GS} = V_{GS(A)}$ in Figure 6.9, and results in a quiescent current

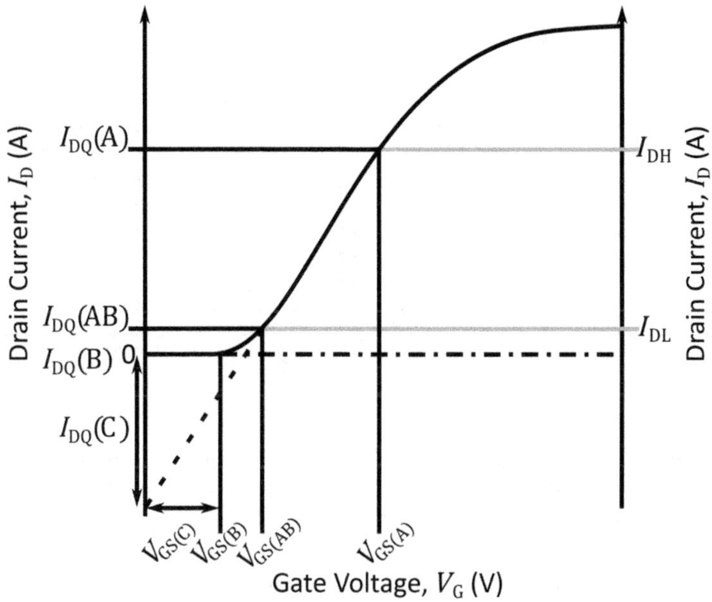

Figure 6.9 Transfer curve with gate bias points for linear amplification.

$I_{DQ} = I_{DQ(A)}$, which constitutes a large constant current of the order of a few amps for a high power device. This translates to a low efficiency for the amplifier (approximately 25%) and high power dissipation even when no signal is being amplified. However, application of an RF input voltage (Figure 6.10(a)) whose amplitude is less than $(V_{GS(A)} - V_T)$ ensures that the MOSFET never goes into the cut-off region. One can also characterize different classes of amplifiers in terms of a conduction angle, *i.e.* the fraction of the wave period during which the device is acting as an amplifier: the conduction angle is measured in degrees. For a class A amplifier, the conduction angle is 360°, with a very high degree of linearity.

Class B operation occurs at a lower bias point than class A, $V_{GS} = V_{GS(B)} = V_T$. Since the DC gate voltage is the same as the turn-on voltage it follows that $I_{DQ} = 0$. For this configuration, V_{GS} is less than V_T for half the sinusoidal cycle as shown in Figure 6.10(b), so the MOSFET is ON for only the positive half of the input voltage sinusoid, *i.e.* a class B amplifier has a conduction angle of 180°. The complete sinusoidal waveform at the output is obtained either by the use of a push–pull configuration in the case of a broadband amplifier or by a tuned output network in the case of a narrowband amplifier. The result is lower linearity than a class A amplifier, but a much higher efficiency of around 78%.

Class AB amplifiers represent a compromise between classes A and B, and are characterized by $V_{GS(B)} < (V_{GS} = V_{GS(AB)}) < V_{GS(A)}$. The result is a non-zero quiescent current, $I_{DQ} = I_{DQ(AB)}$, that is lower than a class A amplifier. $V_{GS(AB)}$

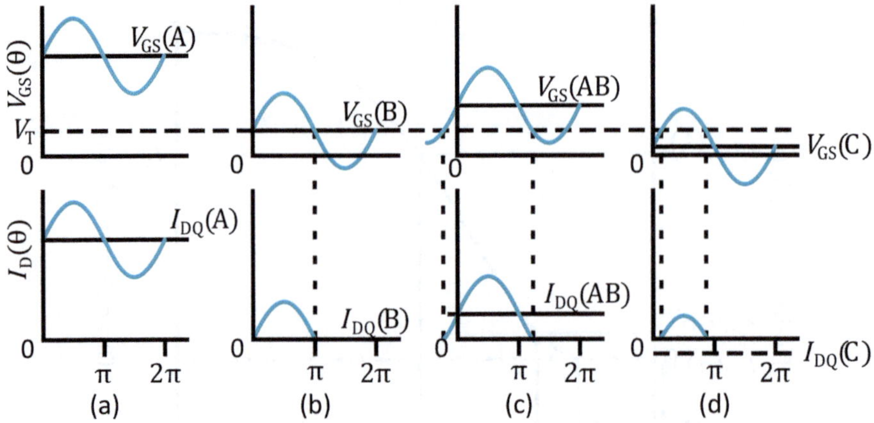

Figure 6.10 The upper half of the figure represents the input RF voltage and the lower half represents the output current. The bias points of class A (a); class B (b); class AB (c) and class C (d) amplifiers are shown.

is usually set to a value at the beginning of the linear portion of the transfer curve, so the resulting output is more linear than class C (see below). This gives a conduction angle of between 180° and 360°, as shown in Figure 6.10(c) and an efficiency that is typically around 50%.

Class C amplifiers have the lowest bias point of the basic classes at $0 < V_{GS(C)} < V_T$ and $I_{DQ} = I_{DQ(C)}$, which is shown to be negative in both Figures 6.9 and 6.10(d). However, in practice, $I_{DQ(C)}$ is zero because it lies below the threshold voltage of the device. The conduction angle for a class C amplifier is usually between 90° and 180°. The theoretical efficiency approaches 100%, though this can never be realized and a realistic efficiency is typically slightly better than class B. In exchange for the higher efficiency, the low conduction angle contributes to a loss of linearity.

In addition to the conventional A, B, AB and C class amplifiers, there are also G and H class amplifiers that modulate the drain voltage V_{DS}. Class G amplifiers use multiple supply rails of different voltages that are switched, based on the amplitude of the signal. This reduces the amount of power dissipated in the device by using a lower V_{DS} when the envelope allows. Class H amplifiers extend this idea by modulating V_{DS} as well as V_{GS}, further improving the efficiency of the design. In general, the gate bias point could be set to match any of the previous classes of operation, though class AB is perhaps the most commonly used. The gains in efficiency seen with class G and H amplifiers come at the cost of higher design complexity and some loss of linearity. Additionally, some of the device parasitics vary as a function of the signal level, potentially adding complexity to the matching network design, discussed later in this chapter.

Linear class amplifiers represent the classical amplifier design, but are by no means the only option. Non-linear switching amplifiers have been developed, and have also started to see applications in RF amplifier designs.

6.2.5 Switch-Mode Amplifiers

Switch-mode amplifiers are designed to operate with the output MOSFET in either the ON or OFF state, representing a different design from the linear devices discussed previously. Switch-mode significantly reduces the amount of power dissipated in the MOSFET since there is either no current flowing through it, or else only a small voltage drop occurs across the internal resistance. The final output is obtained by using a tuned output circuit that acts as a reconstruction filter, removing the higher order harmonics caused by switching. A diagram of the basic operation of such a class D amplifier is shown in Figure 6.11. This group of switch-mode amplifiers, which also includes classes E, F, and S, is highly efficient compared to linear class amplifiers, but this increased efficiency is achieved at a cost of linearity and complexity. Only relatively recently have these designs become viable for RF amplifiers, and their design and use are an area of active research.[5–7] Further details about the circuit and specific characteristics of each class of switching mode amplifiers may be found in the literature.[8] In terms of MR applications, linear class amplifiers are used almost exclusively, and so further analysis is restricted to this type of amplifier.

6.2.6 Mechanism of RF Power Amplification

The principle of operation of a MOSFET RF amplifier can be analyzed on the basis of the DC characteristics of the MOSFET. A class A amplifier will be analyzed in this section for simplicity, but other linear class amplifiers can be analyzed similarly. Switch-mode class amplifiers operate on the same fundamental principles, but the analysis proceeds somewhat differently because of their highly non-linear operation.

The analysis proceeds from a basic common source RF amplifier, as shown in Figure 6.12, with symbols defined in Table 6.2.

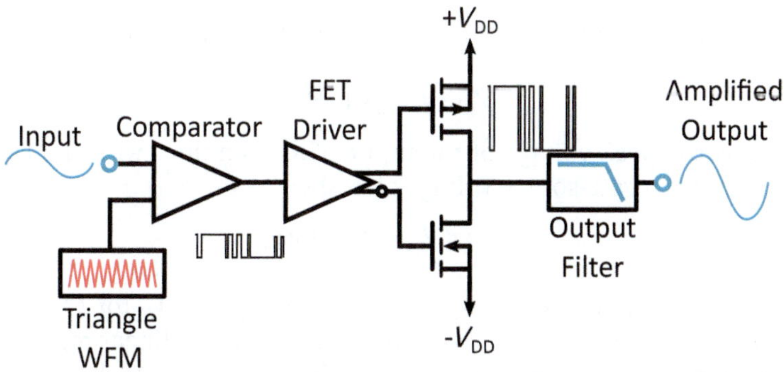

Figure 6.11 The basic class-D amplifier operates in a switch mode. The input is converted into pulses that are used to drive the output FETs. This signal is then filtered to obtain an amplified version of the input.

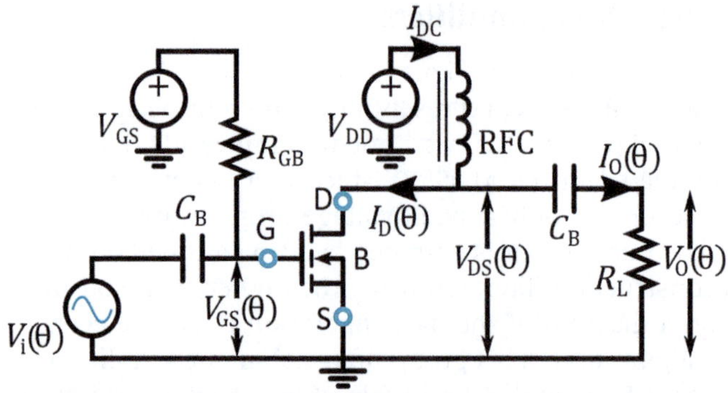

Figure 6.12 Schematic of the basic RF amplifier. The parasitic output capacitances described in Section 6.2.3 are ignored here. In practice the parasitic capacitances are compensated for and the load resistance represented by R_L is a virtual load resistance.

Table 6.2 Symbols used in RF amplifier design and their definitions.

Symbol	Definition
C_b	DC block capacitor
I_{DC}	DC quiescent current drawn by MOSFET
$I_d(\theta)$	Time varying drain current
$I_o(\theta)$	Time varying load current
R_{GB}	Gate bias resistor
R_L	Load resistor
RFC	RF choke
V_{GG}	Gate DC supply voltage
V_{DD}	Drain DC supply voltage
$V_{GS}(\theta)$	Time varying gate-source voltage
$V_{DS}(\theta)$	Time varying drain-source voltage
$V_i(\theta)$	Time varying input voltage
$V_o(\theta)$	Time varying output voltage

The RF source produces a sinusoidal voltage $V_i(\theta)$ such that:

$$V_i(\theta) = V_{im}\sin(\theta) \tag{6.4}$$

where V_{im} is the peak voltage of the input. This combines with the DC gate bias resulting in a gate-source voltage given by:

$$V_{GS}(\theta) = V_{GG} - V_{im}\sin(\theta) \tag{6.5}$$

For class A operation, $V_{im} < (V_{GS} - V_T)$, hence $V_{GS}(\theta) > V_T$ for $t > 0$, so the entire voltage waveform presented to the gate lies in the linear region of the transfer curve. The amplified output of the device is found as the multiplication of eqn (6.5) with the transconductance g_{fs} (defined in eqn (6.1):)

$$I_{DS}(\theta) = I_{DQ} - I_{om}\sin(\theta) \tag{6.6}$$

The quiescent current I_{DQ} flows through the RF choke and the RF current $I_{om}\sin(\theta)$ flows through the blocking capacitor into the load, R_L. Here, I_{om} is the peak current of the output RF wave. This creates a voltage drop of $V_o(\theta)$ across the load, given by:

$$V_o(\theta) = R_L I_{om}\sin(\theta) = V_{om}\sin(\theta) \tag{6.7}$$

where V_{om} defines the peak voltage of the output RF wave. This results in the power dissipated in R_L, defined as the output power P_o, being given by:

$$P_o = \frac{I_{om}^2}{2R_L} = \frac{V_{om}^2}{2R_L} \tag{6.8}$$

From eqn (6.7), it is clear that instantaneous drain-source voltage $V_{DS}(\theta)$ must be:

$$V_{DS}(\theta) = V_{DD} + V_{om}\sin(\theta) \tag{6.9}$$

The amplification processes described by these equations is illustrated in Figure 6.13. From Figure 6.13(c) and (f), it can be seen that the drain current is minimum when the voltage across the drain-source terminals is a maximum, and *vice versa*.

Eqn (6.8) implies that the output power varies linearly as the amplitude of the output voltage swing. Furthermore, V_{om} itself varies linearly with R_L so eqn (6.8) implies that the MOSFET behaves as a current source, with a larger

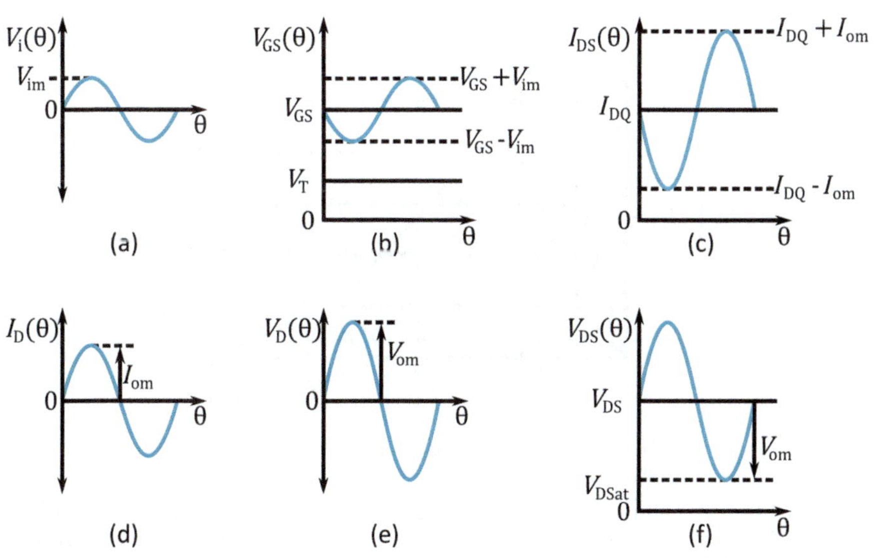

Figure 6.13 Waveform diagram of a class A amplifier. Figures (a) through (f) are pictorial representations of eqn (6.4) through (6.9), omitting eqn (6.8).

load resistance resulting in greater output power. This behavior is bounded by the fact that the drain-source voltage must be larger than the drain saturation voltage at all times, that is $V_{DS}(\theta) > V_{DSat}$. From this consideration, eqn (6.9) imposes an upper bound on the peak output voltage of:

$$V_{om} \leq V_{DD} - V_{DSat} \tag{6.10}$$

Initial assessment of eqn (6.9) would seem to indicate that V_{DD} could be increased arbitrarily to accommodate the condition imposed by eqn (6.10). However, the avalanche breakdown voltage, $V_{BR(DSS)}$, of the reverse-biased diode formed by the drain-substrate interface imposes an upper bound on V_{DD} as:

$$V_{DD} \leq V_{BR(DSS)} - V_{om} \tag{6.11}$$

From eqn (6.8) and the conditions set by eqn (6.10) and (6.11), the optimum value of R_L can be calculated as:

$$R_L = \frac{(V_{DD} - V_{DSat})^2}{2P_o} \tag{6.12}$$

This analysis indicates that the MOSFET may be described as a *high resistance current driver*. The RF operation of the amplifier is shown overlaid on the DC characteristic curves of the MOSFET in Figure 6.14. The line DD′ represents the variation of the drain current and voltage, I_D and V_{DS}, with the

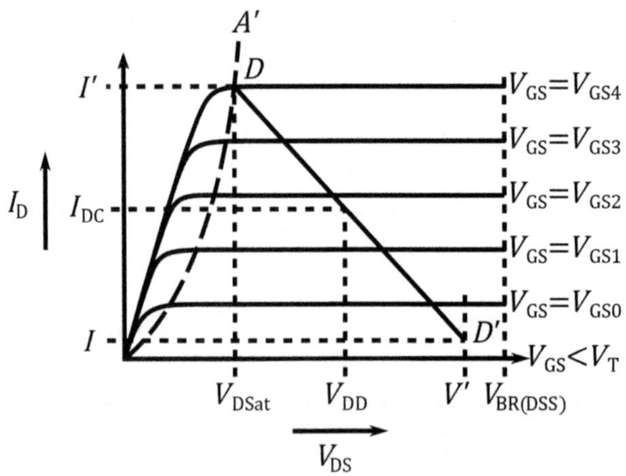

Figure 6.14 RF operation of a class A amplifier shown overlaid on the DC characteristic curves of a MOSFET. DD′ is the locus of V_{DS} and I_{DS} during RF operation of the MOSFET. V_{DS} swing is represented by VV'. The I_{DS} swing is represented by II'. The lower limit for the V_{DS} swing is V_{DSat}. I_D swings about I_{DQ}, the quiescent current.

sinusoidal variation of the gate voltage above the threshold voltage. DD′ is a straight line because I_D and V_{DS} are 180° out-of-phase for a purely resistive load. A purely reactive load would cause DD′ to be a perfect circle, while for the mixed loads found in most practical situations DD′ would become an ellipse. This is caused by the phase shift in the current-to-voltage relationship owing to reactive impedances.

6.3 Matching Networks for Amplifiers

The principles of impedance matching for RF coils were covered in Chapter 3 and very similar principles apply for RF amplifiers. Matching networks are used to minimize power loss and reflection. Amplifiers will not achieve their desired performance if they are not properly impedance matched. This suboptimal performance could manifest itself as low output power, or oscillation in some cases. As already mentioned in Chapter 3, there are a large number of possible designs of matching networks,[9–13] so some of the most important features for RF amplifiers are summarized here.

6.3.1 Basics of Matching Networks

The fundamental purpose of a matching network is to transform the impedance of a device or system from one value to another. This may be to minimize reflected power, obtain the minimum noise figure (*e.g.* for a preamplifier as covered in Chapter 7), or obtain the maximum power output from an amplifier. The principle of conjugate matching for maximum power transfer between a source impedance Z_s and a load impedance Z_L was covered in Chapter 3. Figure 6.15 shows impedance matching networks placed both at the input to the amplifier (between the frequency source and the active MOSFET) and also at the output of the amplifier, in front of the transmit/receive switch.

One can consider matching networks to be passive and lossless, as this is a reasonable approximation of their behaviour. While the passivity of the

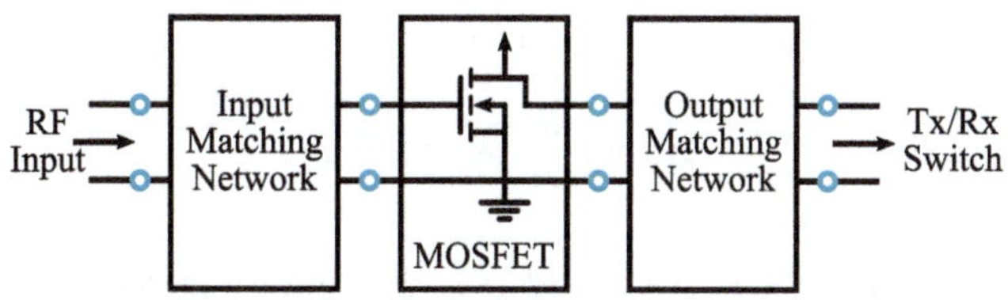

Figure 6.15 Two impedance matching networks are required at the input and output of the MOSFET-based power amplifier: one is placed between the output of the frequency synthesizer and the input to the MOSFET, and the other between the output of the MOSFET and the input of the transmit/receive switch.

network is absolute (they do not generate power, only transform it), real networks are lossy as all components dissipate some amount of power. Good quality, high quality (Q)-factor reactive components limit this problem, and the use of resistive elements should generally be avoided.

The use of reactive elements in the matching network means that it exhibits a frequency-dependent response. The matching network is designed to operate at the particular Larmor frequency, and the network will have a finite bandwidth over which an acceptable match is obtained. A common "acceptable match" is to have a Voltage Standing Wave Ratio (VSWR) of 1.5 : 1, which translates to a Return Loss (RL) of 14 dB: based on the value of the bandwidth, matching networks can be divided into either narrowband (10% fractional bandwidth or less) or broadband (above 10% fractional bandwidth). Narrowband designs can be used, for example, for clinical MRI systems in which transmission at the proton frequency is almost exclusively used. If multinuclear experiments, for example ^{23}Na, ^{31}P or ^{13}C, are desired then a lower frequency broadband RF amplifier is typically used. Similar arguments hold for NMR in which the proton amplifier is typically narrowband and the X-nucleus amplifier is broadband.

6.3.2 Narrowband Matching

In designing an amplifier for a clinical 3 tesla MRI system (operating at a Larmor frequency of ~127.8 MHz), the bandwidth of an excitation pulse would not be expected to exceed roughly 200 kHz, which corresponds to a fractional bandwidth of approximately 0.15%. This is well below the 10% fractional bandwidth that defines the use of narrowband matching techniques. The simplest versions of such impedance matching networks use two or three lumped element capacitors and/or inductors. These are arranged in one of the networks shown in Figure 6.16, which are identical to the networks covered in Chapter 3 for impedance matching RF coils to 50 Ω. In the L-network shown in Figure 6.16(a) the design equations are given by:

$$Q = \sqrt{\frac{R_1}{R_2} - 1} \tag{6.13}$$

$$X_p = \pm\frac{R_1}{Q}, X_s = \mp R_2 Q \tag{6.14}$$

with the condition that $R_2 > R_1$, and Q is the quality factor as defined previously. If $R_1 < R_2$ the series and parallel components are swapped, *i.e.* the series arm of the L-network should be connected to the smaller of the two resistances. The bandwidth of this network is given by ω/Q, so the higher the Q value the smaller the bandwidth. Note that with this type of two-element

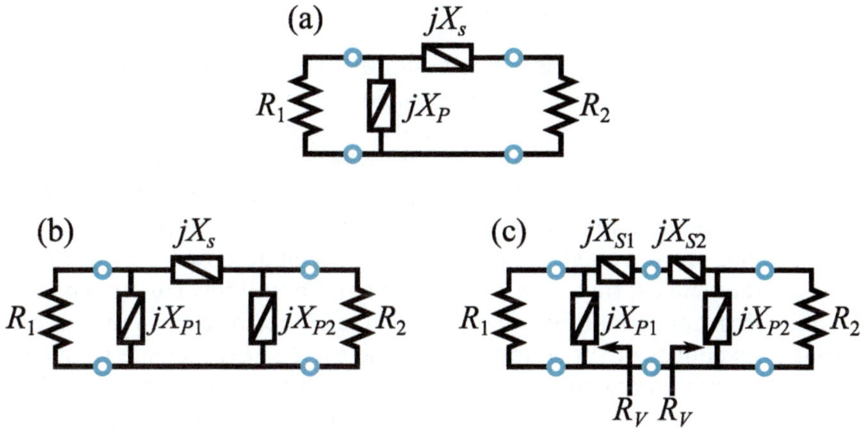

Figure 6.16 Two-element and three-element matching networks which can be used for narrowband impedance matching. (a) An L-network in which the bandwidth is fixed, (b) a pi-network in which the Q of the network can be specified as a design parameter, (c) equivalent circuit to that shown in (b) defining a virtual resistance, R_v, used to derive the design equations for the three elements of the network.

matching network, it is not possible to change the bandwidth of the network. In order to do this, one must use additional elements. A simple example is the three element pi-network, shown in Figure 6.16(b). Analysis is most easily performed by splitting the central lumped element into two, with a virtual ground between them, as shown in Figure 6.16(c), and evaluating the virtual resistance R_v. The design equations are given by:

$$Q \approx \sqrt{\frac{R_2}{R_v} - 1} \tag{6.15}$$

$$X_{C2} = -\frac{R_2}{Q},\; X_{C1} = -\sqrt{\frac{R_1 R_2}{\left(Q^2+1\right) - \frac{R_2}{R_1}}},\; X_L = \frac{R_2 Q + R_2\sqrt{\frac{R_1}{R_2}\left(Q^2+1\right) - 1}}{Q^2+1} \tag{6.16}$$

In addition to lumped elements, transmission lines can be used for impedance matching by connecting appropriately sized stubs, either short-circuited or open-circuited, as covered in Chapter 3.[14,15] One advantage to this latter method is that well-designed microstrip stubs can mitigate problems with coupling between the input and output of an amplifier. This can occur owing to flux coupling between inductors in the respective impedance matching networks, which may lead to oscillation of the amplifiers.

6.3.3 Broadband Matching

RF amplifiers for multi-nuclear NMR and MRI must be able to operate efficiently over a broad range of frequencies. For example, on a 9.4 tesla system (proton frequency 400 MHz), then RF pulses may have to be transmitted for nuclei ranging from ^{15}N (40.5 MHz) to ^{31}P (162 MHz), which equates to a fractional bandwidth of 200%. While using different amplifiers tuned for each frequency is possible, this quickly becomes expensive and impractical. A better solution is to use a single amplifier that has been designed to cover all the relevant frequencies, that is to say a broadband amplifier. This means that the matching networks must also be broadband, converting the input and output impedances of the amplifier to close to 50 Ω over a wide frequency range. Although broadband matching is a very involved subject, an intuitive approach would be to cascade a number of sub-circuits that progressively produce a 50 Ω match. The result is a relatively low Q network where the frequency response will be broadened. Such a network is shown schematically in Figure 6.17. This type of circuit is very similar to a broadband filter, and indeed many of the design criteria for such a network are identical to those used in filter design, so concepts such as the Butterworth or Chebyshev filters can easily be incorporated into the determination of the exact L and C values for the matching networks. These types of multi-segment filter can also be implemented using transmission lines or microstrips. Many papers give detailed descriptions of these and alternative approaches, such as tapered transmission lines,[10,16] multi-section matching techniques,[17,18] numerical design techniques,[19] as well as the fundamental limits of broadband systems.[20,21]

6.4 Amplifier Performance Considerations

The ideal amplifier described in Section 6.2.6 exactly reproduces the input signal at a higher amplitude, with no other changes. Real world amplifiers cannot fully achieve this, and they introduce some degree of distortion into the output signal. A number of different mechanisms can introduce distortions, and understanding these is helpful for evaluating a transmit amplifier for NMR or MRI use.[22] Additionally, it is important to consider performance parameters, such as efficiency, as these can limit the operation of the overall system.

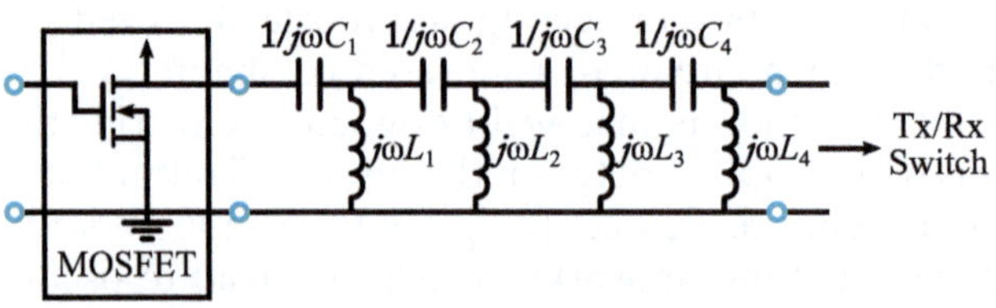

Figure 6.17 Schematic of a broadband matching network.

6.4.1 Linearity

An ideal amplifier would have the same gain at all input levels, which is to say that the gain is completely linear. Practical amplifiers are limited in this respect, and have a finite range of input powers over which they are approximately linear. The maximum input/output power for linear operation is often denoted as the P_{1dB} point, or the 1 dB gain compression point. This is the input/output power level at which the gain has decreased by 1 dB from the nominal value, as shown in Figure 6.18. It is worth noting that the less linear classes of amplifiers (*e.g.* class C) also exhibit non-linear gain at low input levels as well as at high input levels. While the 1 dB compression point is useful for comparing different amplifiers the actual linearity required will vary depending on the particular application.

Compression can introduce two primary problems: (1) harmonic distortion, and (2) calibration error. Harmonic distortion occurs when the variation in gain causes harmonics to be generated at the output of the amplifier, which may result in increased power dissipation in the filter components of the system. Calibration error is where an expected output power level is not achieved. Typically, RF calibration on MRI systems is performed *in situ* for each patient, and this calibration may be performed using relatively low peak amplitude pulses. If there is a significant error owing to non-linearity in the amplifier performance, then the tip angles proscribed in the particular imaging protocol may not be produced, *e.g.*, under-tipping of the 180° pulse in a spin echo sequence. In some cases, such as using complicated RF

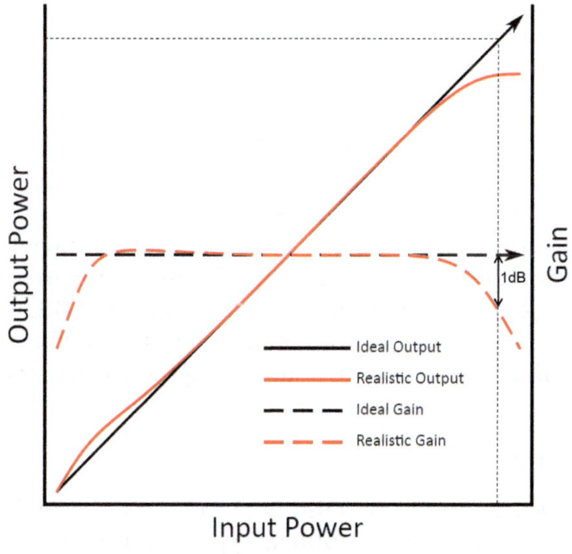

Figure 6.18 The ideal gain of an amplifier is completely flat (constant over all input powers). In reality, this is unobtainable. A typical measure of the linearity is the 1 dB compression point, that is the input power level at which the gain has decreased by 1 dB from nominal.

waveforms with a very high dynamic range for spatially selective excitations, a very high linearity may be needed to obtain accurate excitation profiles, with more complicated imaging errors being produced if the RF pulse shape is not faithfully reproduced.[23] There are a number of techniques that can be used to improve the linearity of amplifiers, including feedback,[24–27] feed-forward,[26] and predistortion.[23,28]

6.4.2 Noise Gating

RF amplifiers typically operate at output power levels many orders of magnitude higher than the MR signal that enters the receive system, and so can easily inject significant amounts of noise into the receive chain even with no RF pulses being transmitted. A noise gating, or blanking, scheme is used to prevent his problem. This process consists of two steps, first removing the gate bias from the device at the input and second, introducing a solid-state RF switch at the output of the amplifier.[29,30] Removing the bias from the amplifier, except during transmit, significantly reduces the gain of the device. This, in turn, reduces the amount of noise that is injected into the MR system. The RF switch at the output of the amplifier further attenuates any noise. A solid-state switch (*e.g.* a PIN diode based switch) is used because the amplifier output may be switched multiple times during a pulse sequence. Switching times in the order of 1 μs are needed, which makes electro-mechanical switches impractical. Using these two methods in concert ideally results in 80 dB or better of isolation between the input of the amplifier and the output of the switch when the output is blanked.

6.4.3 Dynamic Range

Dynamic range is a measure of the usable output range of a system. It is typically bounded at the high end by the maximum linear power that can be produced and at the low end by the noise floor. In general terms, RF amplifiers are designed to have a dynamic range of between 40 dB and 60 dB. However, it is worth noting that 60 dB is a high dynamic range for high power transmit systems, and would not normally be needed for NMR or MRI. Careful design would be needed to achieve this amount of dynamic range, as moderate and high power amplifiers tend to have relatively high noise figures.

6.4.4 Efficiency

Efficiency is a measure of the amount of power delivered to the load *vs.* the total power consumed by the amplifier. Typically, amplifiers for NMR and MRI have not needed to be very highly efficient in the past, and commercial amplifiers essentially designed for other purposes have been modified for MR use. However, there are increasingly challenging demands imposed both by wider bore (owing to the increasing physical size of the population), and higher field (7 tesla and above) MRI systems. With that said, as the output power of the amplifier increases, the efficiency becomes more of a concern. The primary problem is that energy not delivered to the load will be dissipated as heat. For example a 60 kW amplifier that delivers 10 kW to the load but is only 20% efficient would need to dissipate 50 kW in heat. This quickly becomes difficult to manage. A secondary concern is operating cost: the heat generated translates to wasted power both in generation and in removal, which drives up the operating cost of the system. Typically, high power amplifiers would be expected to be between 50% and 75% efficient.

6.4.5 Stability

Stability can be divided into two separate concepts: the propensity of the amplifier to oscillate under varying load conditions; and the change in performance metrics over time. Ideally, an amplifier would be "unconditionally stable", which is to say it would not oscillate with any load, where here the load corresponds to the impedance of the loaded RF coil. However, this is not theoretically possible for an amplifier with gain greater than unity. Instead, stability within the VSWR 3:1 circle is a more realistic design criterion. This should account for any mistuning of RF coils and/or for RF coils under very different loading conditions. The MR system should provide safety checks to prevent operation for extreme loads (*e.g.* an open-circuit owing to a coil being disconnected).

Temporal stability is caused by drift in the value or performance of different components. This may manifest itself in changes in gain, maximum output power, or output phase. Large shifts in performance can be caused by heating of the amplifier, so it is important to have adequate cooling. Some drift is inevitable, but would be expected to be kept within approximately 0.2 dB gain variation and 5° phase variation over the duration of the scan. Again, these values may have to be more stringent for extremely taxing MR applications.

6.4.6 Technical Specifications for Commercial RFPAs for MRI

Having covered the design and performance of RFPAs for magnetic resonance applications, Table 6.3 summarizes the technical specifications of such an amplifier used for a 3 tesla MRI system. In practice, several of these units may be used in parallel, since this is much cheaper than using a single very high power amplifier, if indeed such units even exist for the required frequency.

6.5 Amplifiers for Multi-Channel Transmission

As outlined in Sections 3.11 and 3.12 in Chapter 3, there are significant challenges in obtaining a homogeneous B_1^+ field at very high magnetic fields where the wavelength in tissue is a substantial fraction of, or greater than, the dimensions of the body part being imaged. This has motivated the use of multiple transmit channels, with independent control over the magnitude and phase

Table 6.3 Characteristics of a commercial RF amplifier for 3 tesla MRI.

Frequency range	10–130 MHz
Pulse power (50 Ω load)	8000 watts
CW power (50 Ω load)	100 watts
Gain (1 mW input power)	60 dB
Gain linearity (±1 dB)	40 dB dynamic range
Phase linearity (over 40 dB dynamic range)	5°
Harmonic content	−20 dB (2nd harmonic), −12 dB (3rd harmonic)
Input impedance	50 Ω
Output impedance	50 Ω
Input VSWR	<2 : 1
Amplitude rise/fall time	750 ns
Output noise (blanked)	20 dB above thermal noise
Blanking delay	~1 μs

of the input to each channel. A second reason to use multiple transmit channels is the use of parallel transmit techniques. RF excitation pulses that are selective in multiple dimensions, often known as spectral-spatial or simply 2D pulses, enable the tailored excitation of desired regions from within a larger FOV, therefore increasing the imaging efficiency. These multi-dimensional RF pulses are more complicated that regular 1D slice selective RF pulses, generally requiring significantly longer time periods, therefore extending the minimum TE of a particular imaging sequence, among other considerations. Drawing from parallel receive concepts, which use multiple receive coils to enable reduced sampling of *k*-space, Zhu and Katscher independently introduced the concept of parallel transmit MRI, often called Transmit SENSE.[31,32] This uses multiple transmit coils, each with independent RF channels, to produce multi-dimensional RF pulses with significantly reduced durations compared to those using conventional single channel MR systems.

However, the transmit array coils that are needed to implement these methods do not come without their own difficulties, particularly in terms of the presence of mutual coupling between individual transmit elements. One method to mitigate this problem of mutual coupling is to use a different type of amplifier than discussed previously in this chapter. A key element in this different type of amplifier design is the need to control the *current* through each of the elements for the successful application of Transmit SENSE,[31] or B_1-shimming. Most systems perform RF pulse tip angle calibration in terms of a power factor sent to the coil, but in fact it is not the power that generates the rotation of the nuclear spins, *i.e.* the particular tip angle, but instead it is the current on the coil that generates the tip angle, and the current is a function of both the power delivered to the coil and the coil impedance. If the impedance of individual elements were to change for some reason, for example a change in temperature or patient movement causing changes in loading, then the current needed for a given tip angle would remain the same, but the absorbed power and terminal voltage would vary with the change in impedance. This will be considered further in the next section.

6.5.1 Mutual Coupling in Transmit Arrays

As mentioned above, one of the key concepts in the design of transmit arrays is to control the relative currents on the elements, as these are what generate the B_1^+ fields. Driving the arrays with current sources would achieve this control, as shown schematically in Figure 6.19(a). This "forced current excitation", if achievable, would ensure a specified current at the terminals of each element. Under these circumstances, a specific RF waveform could be input to each element *regardless* of the interaction, or coupling, between elements of the transmit array.

A more realistic model is that of each element being driven by a voltage source with an internal generator resistance, shown in Figure 6.19(b). The "free voltage excitation" configuration specifies the voltage internal to the generator, so that the coil terminal voltage and current are functions of the element impedance. The difficulty lies in the fact that the element impedance may vary depending on the coupling between elements. Coupling between elements is dependent on a number of factors, including element spacing and the environment. This means that the element impedance is sensitive to the movement of the coils and environment, rendering specification of the currents very difficult in practice.

For a single antenna, as illustrated in Figure 6.20(a), the current delivered by the amplifier is a fixed value. Depending upon the mismatch between the

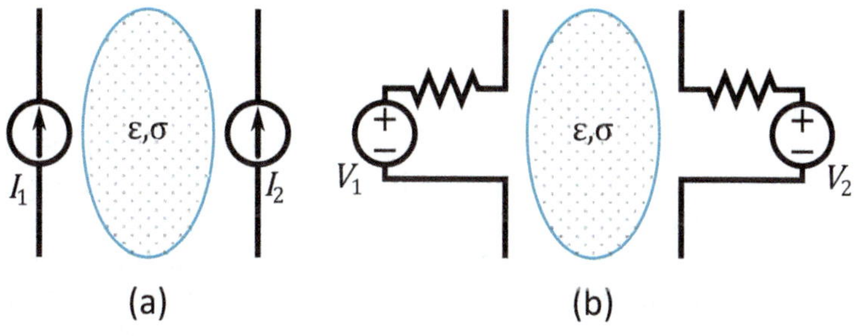

Figure 6.19 (a) Forced current excitation of a transmit array, in which the input current to each element of the array is driven by a current source or some other means of applying known currents. (b) Free voltage excitation of an array, the more common implementation. In this case, the applied voltage at the transmitter is known, but the actual voltage and current at the elements are a function of the input impedance of the antenna. This input impedance can vary owing to changes in loading or relative currents on the elements, so the currents on the array elements may not follow the applied drive voltages. The variables ε and σ represent the permittivity and conductivity of a loading sample.

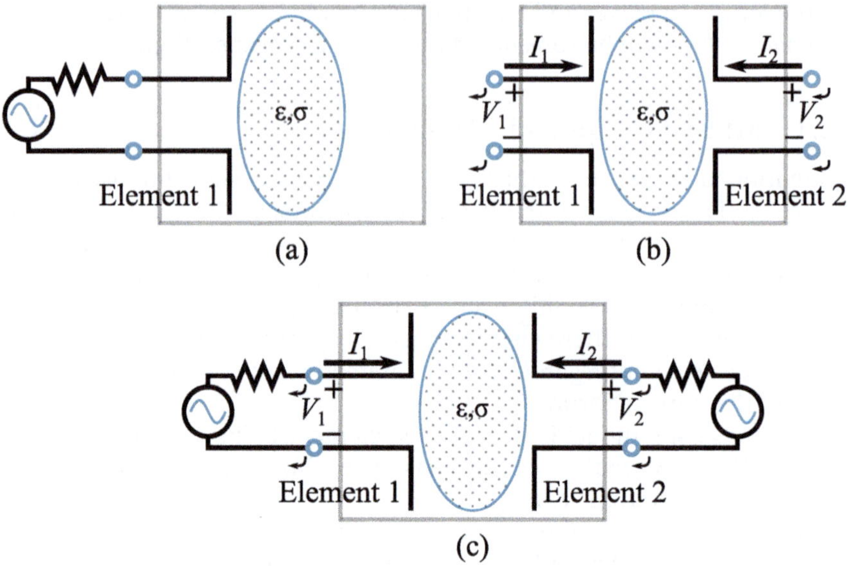

Figure 6.20 (a) Network representation of a single element defined by the terminal voltage and current, and the generator voltage and resistance. The interaction with the load is contained in the element input impedance, *i.e.* the ratio of the element voltage and current. (b) Two-port representation of a two-element array defined by the element terminal voltages and currents. The interactions with the load and between the elements are contained in the port-open circuit impedance matrix, *i.e.* the relationships between the port voltages and currents. (c) Network representation of a two element array and sources. The element currents, I_1 and I_2, in general will not directly follow the respective generator voltages owing to coupling between the elements.

amplifier impedance and the antenna element impedance there may be a degree of reflected power which reduces the current delivered to the antenna. However, the fidelity of the waveform will not be impacted, assuming that the amplifier is properly protected.

The situation is potentially much different when using arrays of elements. Forming an array by introducing another element means that the presence of a second can affect the input impedance of the first. Specifically, the fields produced by element 1 induce a current, I_2, in element 2, which in turn induces a voltage back across the first element. The net result is that the currents on the individual elements are disturbed by the mutual coupling between the two elements (as described previously in Section 3.5.2). This behavior can be analyzed as a two port system, illustrated in Figure 6.20(b). The mesh equations for this two-port are given by:

$$\begin{aligned} V_1 &= Z_{11}I_1 + Z_{12}I_2 \\ V_2 &= Z_{21}I_1 + Z_{22}I_2 \end{aligned} \tag{6.17}$$

or in matrix form:

$$\begin{bmatrix} V_1 \\ V_2 \end{bmatrix} = \begin{bmatrix} Z_{11} & Z_{12} \\ Z_{21} & Z_{22} \end{bmatrix} \begin{bmatrix} I_1 \\ I_2 \end{bmatrix} \tag{6.18}$$

where the impedance terms are:

$$\begin{aligned} Z_{11} &= \left.\frac{V_1}{I_1}\right|_{I_2=0} \quad & Z_{12} &= \left.\frac{V_1}{I_2}\right|_{I_1=0} \\ Z_{21} &= \left.\frac{V_2}{I_1}\right|_{I_2=0} \quad & Z_{22} &= \left.\frac{V_2}{I_2}\right|_{I_1=0} \end{aligned} \tag{6.19}$$

These are the "open-circuit" port impedance matrix elements and can be determined by measurement, or in principle using commercially available electromagnetic modeling software covered in Chapter 8. One needs to be able to specify the element currents because the current produces the B_1^+ field. This can be achieved by using current source amplifiers which are described later in Section 6.6. However, most RF amplifiers are described best as a voltage source with a series internal resistance, usually 50 Ω. The equivalent circuit in this case is illustrated in Figure 6.20(c) in which the RF amplifier voltage is controlled rather than specifying terminal voltages. This is accomplished by controlling the input voltage to the amplifier and knowledge of the amplifier gain. In this simplified model, the generator voltages can be found by adding the effect of the generator impedances to eqn (6.18):

$$\begin{bmatrix} V_{g1} \\ V_{g2} \end{bmatrix} = \begin{bmatrix} Z_{11} & Z_{12} \\ Z_{21} & Z_{22} \end{bmatrix} + \begin{bmatrix} R_g & 0 \\ 0 & R_g \end{bmatrix} \begin{bmatrix} I_1 \\ I_2 \end{bmatrix} \tag{6.20}$$

or simply $V_g = (Z_{oc} + Z_g)I$, where $Z_{oc} = \begin{bmatrix} Z_{11} & Z_{12} \\ Z_{21} & Z_{22} \end{bmatrix}$ is the open circuit port impedance matrix.

Dividing equation 6.17 by the current I_1 at port 1, one obtains:

$$\begin{aligned} Z_{1,\text{in}} &= \frac{V_1}{I_1} = Z_{11} + Z_{12}\left(\frac{I_2}{I_1}\right) \\ Z_{2,\text{in}} &= \frac{V_2}{I_2} = Z_{22} + Z_{12}\left(\frac{I_1}{I_2}\right) \end{aligned} \tag{6.21}$$

i.e. the driven port impedances depend on the mutual impedance and the port current ratio. The "active impedance" of each port can be solved for in a similar manner, and it can be quite different from the self-impedance of the port. For port n, the input impedance is given by:

$$Z_{n,\text{in}} \sum_{m=1}^{N} Z_{mn}\left(\frac{I_M}{I_n}\right) \tag{6.22}$$

Note that the active impedance is a function of both the mutual impedances between elements and also the currents on the ports. So if the currents on the elements vary, whether in the process of B_1 shimming or applying different pulses to different elements as in Transmit SENSE, the impedances presented by each element to its transmitter also change. There are a variety of approaches to mitigating this problem. The most straightforward approach for arrays with just a few elements may simply be to eliminate coupling (as far as possible) between the ports of the arrays, as discussed for receive arrays in Section 3.7.[33,34]

Alternative approaches use RF amplifier architectures that behave as current sources to ensure that prescribed currents are delivered to the coil terminals. These amplifiers effectively duplicate the technique almost universally used in receive arrays, that of using a low-impedance RF preamplifier to present a high impedance to the receiving element, as originally introduced by Roemer.[33] This current source design is discussed in detail in the following sections. It should be noted that there are other techniques available, including the use of feedback-based decoupling, which involves measuring the current on each element, comparing it against the desired waveform, and correcting the input. While this principle has been shown to be feasible, it does require a very high bandwidth feedback system.[35] Combined with resonant systems it also tends to lead to instability. An iterative approach can reduce some of the problems associated with this approach, but may still drive the output of the amplifiers to a minimum in some cases.

6.6 Current Source Amplifiers

One approach to decoupling individual elements of a transmit array is to use an RF current source amplifier,[36,37] which is a modified version of the classical RF amplifier. The current source amplifier takes advantage of the fact that FETs behave as controlled current sources. Their drain-to-source resistance appears very high, in the order of kilo-ohms, when they are operated in the saturation region (see Section 6.2.2). The current source amplifier, in contrast to the standard power amplifier, is designed to present a high impedance to any induced voltage at the expense of not being power-matched.[37,38] In essence, a single element matching network is used to remove any reactance owing to parasitic output capacitance, leaving only the resistance of the MOSFET drain. The coil is series-resonated and connected directly to the amplifier so that there are no other current paths except through the amplifier output. Effectively, the amplifier has a low impedance load and the coil has a high impedance load, shown in Figure 6.21(a). The result is that coupling to the coil produces a voltage but the induced current is minimized, reducing the effective coupling. The downside is that the matching networks used do not result in a power-match for the amplifier, reducing the maximum current on the coil to the rating of the active device in the amplifier.

The RF current source concept is similar to the principle of the antennafier element[39] and active integrated antennas.[40–43] Using these devices, studies have shown that integrating the antenna with the active device results in the feed network of antenna arrays becoming less sensitive to the effects of mutual impedance than equivalent passive arrays driven by classical RF amplifiers, owing to the unidirectional nature of the active devices.[44] The design of this network directly trades an optimal power-match for higher isolation.

6.6.1 Matching Networks for Current Source Amplifiers

A simplified circuit schematic of the RF current source is shown in Figure 6.22. As mentioned before, the RF power MOSFET is an RF voltage-controlled current source. Comparing Figures 6.22 and 6.11, the RF current source is implemented by replacing the optimal load resistance of the classical RF

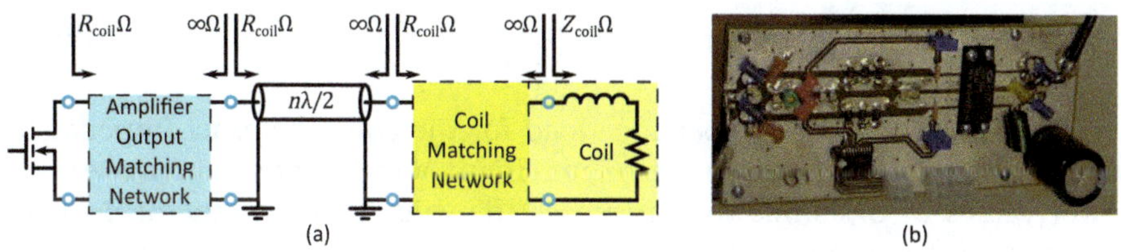

Figure 6.21 (a) The matching network for a current source amplifier is characterized by presenting a high impedance to the coil. The active device does not see an optimal match, so the output is limited to the current rating of the device. (b) An example amplifier module.

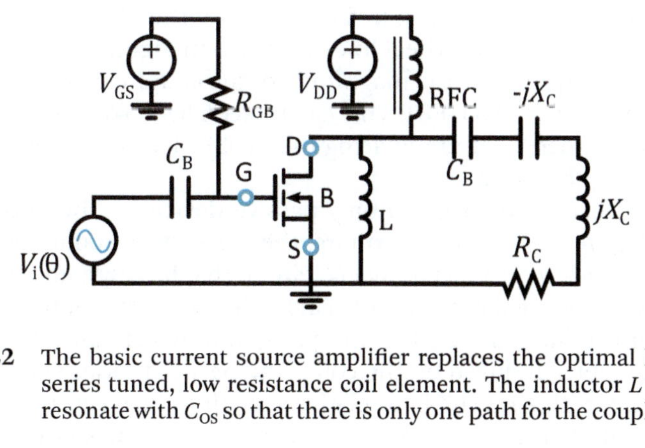

Figure 6.22 The basic current source amplifier replaces the optimal load with a series tuned, low resistance coil element. The inductor L is tuned to resonate with C_{OS} so that there is only one path for the coupled current.

amplifier by an RF coil that is tuned to series resonance at the Larmor frequency such that it presents a low resistance load to the MOSFET. If the output capacitance (C_{os}) is not compensated for by a conjugate match, it forms a parallel path for the current, reducing the amplitude of the current driven through the RF coil. For improved RF current source performance, C_{os} is tuned out using a parallel inductor such that the RF current driven through the RF coil is maximized.

This introduces some difficulty if the RF coil is to be used as a transmit/receive coil. The coil matching network used with current source amplifiers does not lend itself to receive systems. It is possible to integrate a matching network with the transmit/receive switch, so that the coil is appropriately tuned for the two cases, but a simpler approach is to use this as a transmit-only coil and have a separate receive array.

6.6.2 Analysis of the Current Source Network

The amplifier matching network is made up of a single inductive element resonated with the output capacitance C_{OS} such that:

$$L = \frac{1}{\omega^2 C_{OS}} \tag{6.23}$$

This results in a high impedance load for the coil. The exact impedance will be affected by the Q of the resonant circuit, which tends to increase with lower values of C_{OS} since this increases the value of the inductor, L. This design does not attempt to transform the low impedance of the tuned coil, so the MOSFET will not have an optimal load resistance. Therefore, the same RF current amplitude, I_0, driven into the RF coil will result in a voltage drop, V_0, that is significantly reduced compared to the optimal load resistance in a classical amplifier design. This reduces the V_{DS} swing, which results in a load line that no longer spans the saturation region of the MOSFET DC characteristic curve shown in Figure 6.14. In fact, the lower the RF coil resistance, the closer the load line is to a vertical line, as shown in Figure 6.23.

The induced current (*i.e.* the magnitude of the coupling) is reduced by the high impedance of the MOSFET. This can be seen by looking at the amplifier with the coil connected, and replacing the MOSFET with a hybrid pi model, as shown in Figure 6.24(a).

The circuit can be reduced to only passive devices at the output of the amplifier (Figure 6.24(b)). Furthermore, there are two resonant circuits: the MOSFET output has a tuned parallel LC circuit that has high impedance, and the coil has a series resonant circuit. This reduces the circuit to R_o and R_C in series, as shown in Figure 6.24(c). Coupling into this circuit would be manifested as a voltage induced on the coil element, *i.e.* a series voltage source. Since R_o is large, this results in a relatively small induced current. This represents the fundamental mechanism by which current source amplifiers reduce coupling.

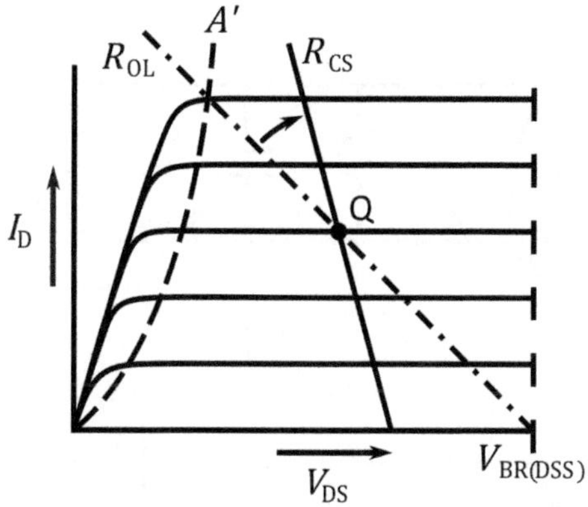

Figure 6.23 The current source amplifier has a lower load resistance than is optimal. This corresponds to a steeper load-line, meaning that for a given swing in output current there is a smaller swing in drain-source voltage.

In the case of the standard amplifier the RF coil is conjugate-load matched to a 50 Ω system. So based on this, an induced voltage (V_{EMF}) produces a current on the coil given by:

$$I_{EMF} = \frac{V_{EMF}}{Z_C + Z_C^*} = \frac{V_{EMF}}{2R_C} \tag{6.24}$$

Similarly, the current source has an induced current I_{EMF} of:

$$I_{EMF} = \frac{V_{EMF}}{R_o + R_C} \tag{6.25}$$

Combining eqn (6.24) and (6.25) allows the calculation of the ratio of currents for the two cases.

$$\frac{I_{CS}}{I_{50}} = \frac{\frac{V_{EMF}}{2R_C}}{\frac{V_{EMF}}{R_o + R_C}} = \frac{2R_C}{R_o + R_C} \tag{6.26}$$

Since $R_o \gg R_C$, then $I_{CS}/I_{50} \ll 1$, *i.e.* the current source amplifier significantly reduces the current coupled into a coil.

Considering the active case and not holding the output of the MOSFET at zero, then the total voltage across the MOSFET terminals is the vector sum of $V_{DS}(\theta)$ and $V_{EMF}(\theta + \phi)$. Examining the characteristic curves in Figure 6.23 shows that any variation in the drain-source voltage has a negligible impact on the output current, I_O. This means that I_O is completely controlled by the input voltage, so the coil is decoupled from all other coils.

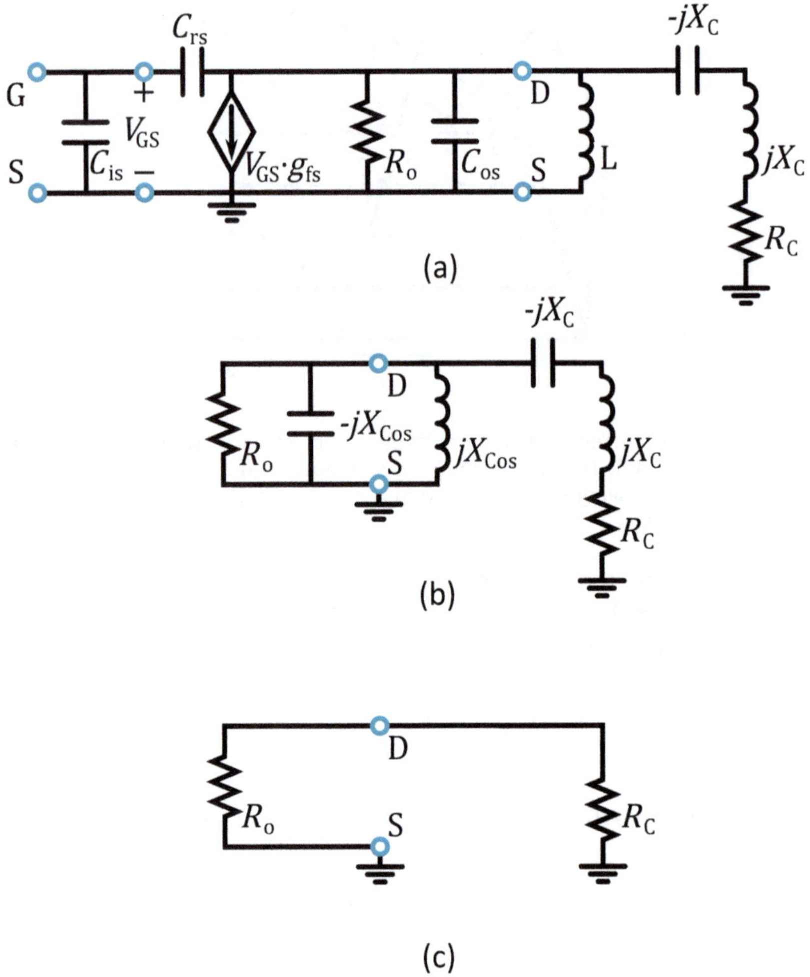

Figure 6.24 (a) The MOSFET is replaced by the hybrid pi model, and the coil is represented by a lumped element RLC circuit. The coil is tuned so that the L and C elements resonate. Only the output of the MOSFET is relevant here, so the gate and associated components are neglected. (b) The current source can be replaced with an open circuit, and the inductor L is tuned to resonate with the output capacitance C_{OS}. (c) Removing the resonant sections results in two series resistances.

6.7 Low Output Impedance Amplifiers

The low output impedance (LOI) amplifier design[45] attempts to obtain both a high output power and high inter-coil isolation. This is achieved in a manner analogous to the low input impedance pre-amplifier for receive coil arrays, where the amplifier matching network optimizes performance and also presents a low impedance to the coil.[33] The LOI design is such that the MOSFET sees an optimal power-load, and the coil port sees a low impedance, shown

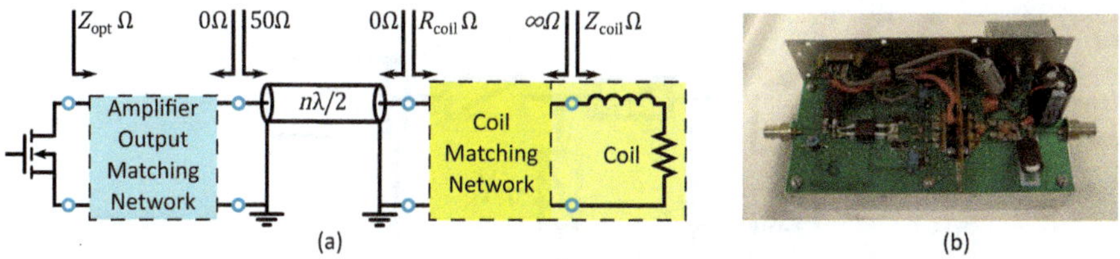

Figure 6.25 The matching network for a low output impedance power amplifier is characterized by presenting the optimal load to the active device and a low impedance to the coil, as shown in (a). An example amplifier module is shown in (b).

in Figure 6.25. The coil's matching network is designed to form a tuned tank circuit across the coil element, preventing induced currents. The design of the coil matching network is identical to those used with low input impedance preamplifiers.[46]

There are two simultaneous requirements for the amplifier output matching network: (1) that the matching network transforms a 50 Ω load into the optimal impedance for the active device, and (2) that the network transforms the drain-source impedance of the device into a small, purely real impedance. Using these two conditions a matching network for the amplifier can be designed.

Using the hybrid-pi model from Section 6.2.2, the output is modelled as a voltage-controlled current source paralleled by a resistance (R_O) and capacitance (C_{OS}). Varying the V_{DS} changes the value of this capacitance, but a class A/B linear amplifier operates at a fixed V_{DS} so one can treat it as a simple fixed value of approximately 100 pF. The output resistance of a MOSFET is typically on the order of a few kilo-ohms. This leads to an output impedance of the device that typically has a very small real component with some amount of reactance.

The basic matching network is built of four components, as shown in Figure 6.26: a shunt inductor (L_1) that cancels C_{OS}, then a reactive Tee-network (labeled X_S and $-X_S$) with the shunt and series elements being series-resonant. This combination allows both power matching and a low impedance load for the coil. Selecting the value of L_1 is reasonably straightforward as it simply resonates with C_{OS}: this removes the reactance from the output impedance. The Tee-network can be either high-pass (series capacitor, shunt inductor) or low-pass (series inductor, shunt capacitor). Both configurations are valid, but one may result in component values that are impractical in the real world, such as a very small inductance. The values of the series elements are found based on the optimal load resistance (R_{OL}) and the system impedance (50 Ω):

$$R_{OL} = \frac{1}{50\omega^2 C^2} = 50\omega^2 L^2 \tag{6.27}$$

The shunt element is then selected to resonate with the series element. These conditions produce both a power match and a low impedance load for the coil.

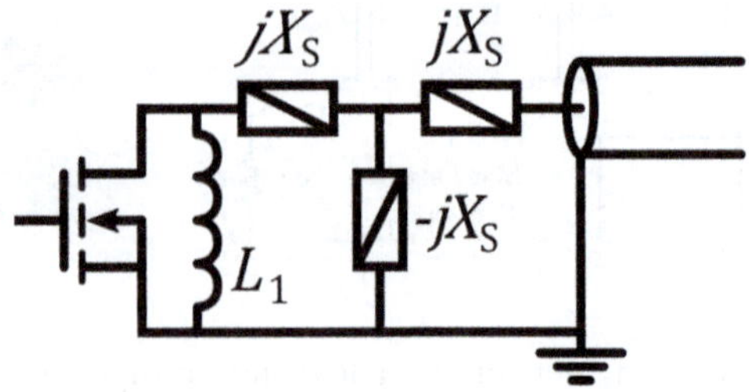

Figure 6.26 A simplified output matching network for an LOI amplifier.

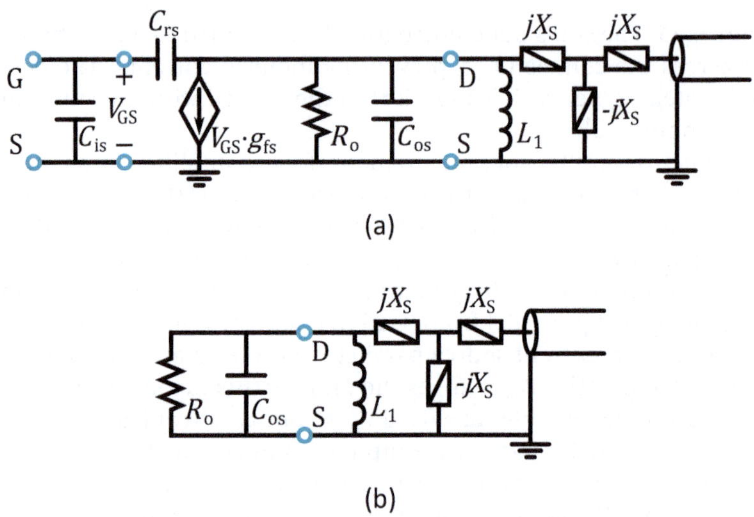

Figure 6.27 (a) Combination of the matching network for an LOI amplifier with the device replaced by the hybrid-pi model. (b) Equivalent circuit represented by only passive devices.

Combining the hybrid-pi model with the output matching network gives the circuit shown in Figure 6.27(a). Initially considering the output only, the input matching/biasing networks are not shown and the gate-side components can be removed. The feedback capacitance C_{rs} should be negligible in a well-designed MOSFET, so can also be ignored. Furthermore, a current source has infinite impedance, so can be replaced with an open-circuit when evaluating a circuit. Therefore, analysis can proceed using a simple circuit made only of passive devices, as shown in Figure 6.27(b). L_1 is designed to resonate with C_{OS} so a high impedance tank circuit is formed. This allows C_{OS} and L_1 to be replaced with an open circuit, so the load (Z_{OUT}) seen by the coil is given by:

$$Z_{OUT} = -\mathrm{j}X_S \| \left(R_o + \mathrm{j}X_S\right) + \mathrm{j}X_S = \frac{X_S^2}{R_o} \tag{6.28}$$

This shows that large values of R_o will result in a low impedance being presented to the coil. Similarly the load (Z_L) seen by the MOSFET for a 50 Ω system can be found as:

$$Z_L = \mathrm{j}\omega L_1 \| \left(\mathrm{j}X_S - \mathrm{j}X_S \| \left(\mathrm{j}X_S + 50\right)\right) = \left(\frac{50}{X_S^2} + \frac{1}{\mathrm{j}\omega L_1}\right)^{1} \tag{6.29}$$

The optimum load impedance for the MOSFET (Z_{OL}) may be considered to be the optimum load resistance in parallel with parasitic output capacitance (C_{OS}) so that:

$$Z_{OL} = R_{OL} \| C_{OS} = \frac{1}{R_{OL} - \mathrm{j}\omega C_{OS}} \tag{6.30}$$

In order to achieve the highest output power possible for the MOSFET the load impedance must be equal to the optimal load for the MOSFET ($Z_L = Z_{OL}$). Substituting eqn (6.29) and (6.30), and resonating the output capacitance (*i.e.* $\omega L_1 = 1/\omega C_{os}$) gives:

$$\frac{1}{\frac{1}{R_{OL}} - \mathrm{j}\omega C_{OS}} = \frac{1}{\frac{50}{X_S^2} - \frac{\mathrm{j}}{\omega L_1}}, R_{OL} = \frac{X_S^2}{50} \tag{6.31}$$

Maintaining these conditions results in an amplifier capable of producing the maximum rated power of the active device, and also a low impedance load for the coil. It is worth seeing how this is implemented in a real world configuration, as there are practical concerns to be managed.

6.7.1 Coil Matching Network for an LOI amplifier

Proper operation of the amplifier depends on a suitably designed coil, so we will briefly examine the coil matching network. The coil needs to satisfy two primary conditions: (1) the coil must be matched to 50 Ω and (2) the coil's matching network must form a trap when connected to the amplifier. The first condition is necessary to prevent reflected power at the coil port as well as presenting the optimal load. The second condition is the mechanism by which the isolation is actually achieved. A three element matching network can be used which both matches the coil to 50 Ω and forms a high impedance tank, shown in Figure 6.28. In practice, this network sometimes produces component values that cannot be realized (as discussed previously and in ref. 46), but other networks that achieve the same goals have been described.[47]

An example design using the ARF475FL power MOSFET (Microsemi PPG) for 3 T will serve to demonstrate these calculations. This device is rated for 1 kW output power at 128 MHz as a pulse amplifier, so is suitable for use in MRI.

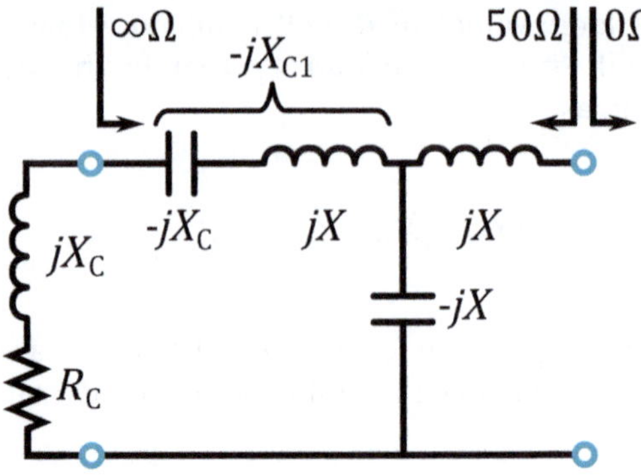

Figure 6.28 The basic matching network for a coil designed to work with LOI amplifiers. This is identical to the network used with low input impedance preamplifiers. The matching network forms a high impedance tank across the coil, preventing current from being induced.

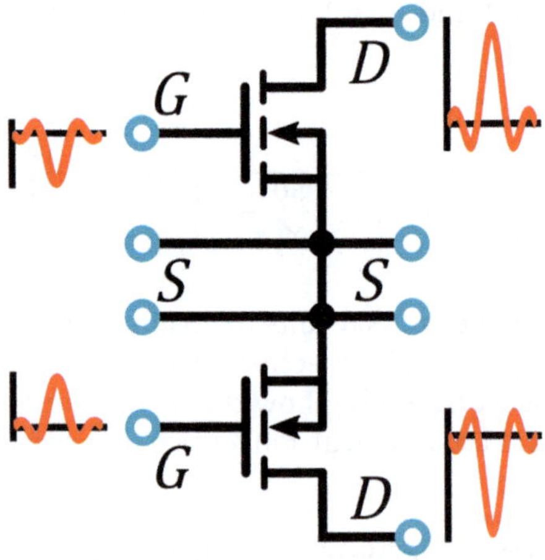

Figure 6.29 The ARF475FL is built as a push–pull device, with two separate NMOS devices in the package. The source pins are bonded together, so a common-source amplifier architecture is needed for this device.

The ARF475FL is a common source push–pull pair with two N-type metal-oxide-semiconductor (NMOS) devices in a single package, as shown in Figure 6.29. The design equations that have been formulated are for a single ended device, and push–pull, so they cannot be directly applied. Fortunately, a straightforward conversion can be used to transform a single-ended network design into a balanced design.[17] This allows the design of one-half of the matching network with the device matched to 25 Ω instead of 50 Ω. This difference accounts for the conversion from balanced to unbalanced in the matching network.

The data sheet for the MOSFET specifies the optimum load impedance at 120 MHz (a negligible difference from 128 MHz) as 10 − j20 Ω measured between the drains of the two devices, so it has a single-ended impedance of 5 − j10 Ω. The value of L_1 that resonates with C_{os} is 15.5 nH, and the impedance becomes 25 + j0 Ω. This means that R_{OL} is 25 Ω, which is needed for calculating the component values in the T-network. Substituting R_{OL} and the 25 Ω system impedance into eqn (6.27) gives values for C and L as:

$$R_{OL} = \frac{1}{25\omega^2 C^2} = 25\omega^2 L^2 \tag{6.32}$$

$$C = \sqrt{\frac{1}{25\omega^2 R_{OL}}} = \sqrt{\frac{1}{25\left(2\pi 128\cdot 10^6\right)^2}} = 49.7\text{pF} \tag{6.33}$$

$$L = \sqrt{\frac{R_{OL}}{25\omega^2}} = \sqrt{\frac{25}{25\left(2\pi 128\cdot 10^6\right)^2}} = 1.25\text{nH} \tag{6.34}$$

Of the two choices of series element, the more practical one is the 49.7 pF capacitor since an inductor of 1.25 nH is difficult to obtain. The shunt element must now be an inductor that resonates with the capacitor:

$$L = \frac{1}{\omega^2 C} = \frac{1}{\left(2\pi 128\cdot 10^6\right)^2 49.7\cdot 10^{-12}} = 31\text{ nH} \tag{6.35}$$

The design can be converted to a balanced network by doubling the value of the shunt inductors, and replicating the series elements in the other branch. The conversion takes advantage of the virtual ground between the two halves of the circuit, which is why the inductor value is doubled (shunt capacitances would be halved). The full matching network is shown in Figure 6.30. The final component values are L_1 = 31 nH, C = 50 pF and L = 31 nH.

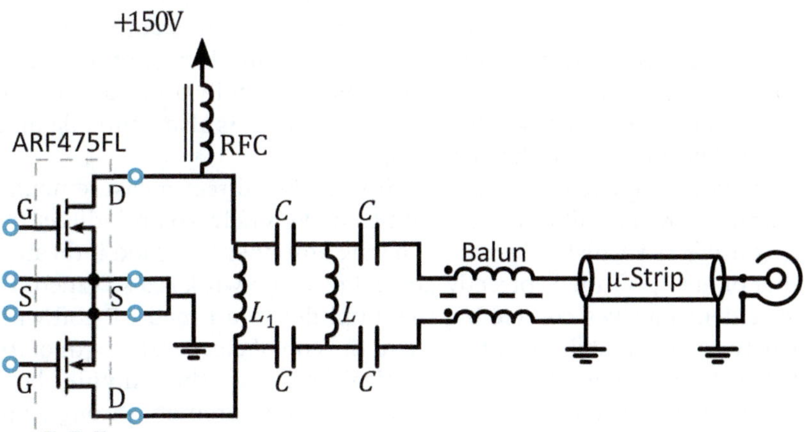

Figure 6.30 Matching network for the ARF475FL device.

The computed values for the matching network will be close to what is needed, but manufacturing tolerances mean that fine-tuning the amplifier will result in improved performance. Additionally, there will be some amount of parasitics and phase delay owing to the circuit board that needs to be compensated. There are too many degrees of freedom in the network to simply vary the values of all of the components, so a systematic method of measuring and tuning the amplifier is needed.

Figure 6.30 shows the matching network as well as a balun and a section of transmission line. The balun converts the 25 Ω balanced impedance to 50 Ω unbalanced, while the transmission line comprises a section of microstrip on a PC board and an output connector. This delay line is not needed from a matching network perspective, but is needed for a physical connection to the amplifier. The phase delay introduced must be accounted for when tuning the network. Tuning involves measuring the phase delay, adjusting the Tee network to have a minimum impedance, and finally connecting the amplifier to a test RF coil and setting L_1 to maximize isolation. The four series capacitors need to be exactly the same value, so it would be inconvenient to have to tune by adjusting all four elements. The tuning process is greatly simplified, and still achieves good results, when limiting tuning to the shunt elements.

With this basic strategy in mind, the first step is to measure the delay of the microstrip and output connector because any phase delay will affect the measured values of "later"components and propagate errors into the tuning procedure. This is done by taking a S_{11} measurement in the form of $|\Gamma| - \theta$ looking into the board with the balun detached from the microstrip. The value of the phase delay will vary depending on the exact board layout and construction: for the board shown in Figure 6.30 the phase delay was measured to be 23°.

Next, the Tee network is installed. The calculated value of the shunt inductor is 31 nH, however a tunable inductor of approximately that value should be used. The balun should also be fully connected at this point. The S_{11} measurement is repeated and L adjusted to obtain a $|\Gamma| - \theta$ that corresponds to a short circuit phase, shifted by the previously measured amount (in our case $|\Gamma| = 0.995$ and $\theta = 157°$). The Tee network should be fully tuned at this point, so the MOSFET and L_1 can be added to the circuit. L_1 is also a tunable inductor that will be adjusted in the next step.

L_1 should be adjusted to resonate with C_{os}, but direct measurements of the output of the amplifier are difficult to make safely, so an indirect technique is used. The isolation provided by the amplifier is a good indicator of whether or not L_1 is tuned correctly, since the Tee-network is also tuned. This measurement involves connecting the amplifier to a tuned RF coil with a cable between the matching network and the coil of electrical length of $n\lambda/2$, shortened to compensate for the electrical length on the amplifier board. The isolation measurement is made as an $|S_{21}|$ measurement through a pair of decoupled pickup loops that are positioned near to the RF coil. The amplifier drain is biased and L_1 is adjusted until maximum decoupling (minimum

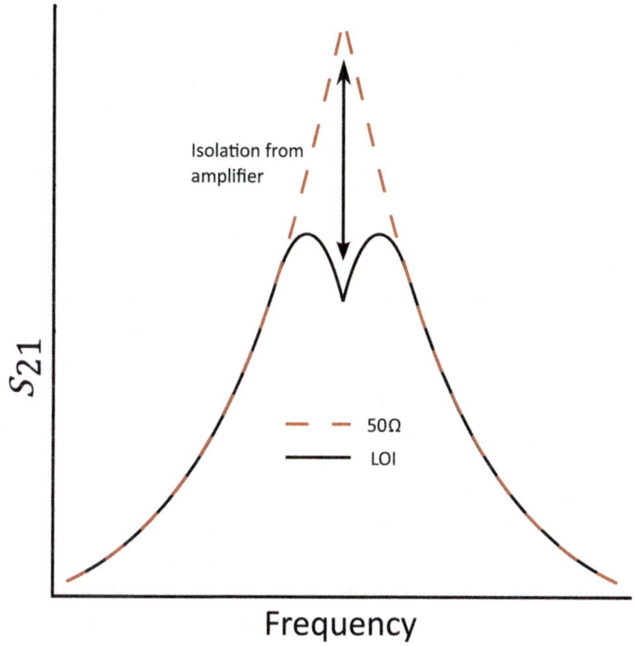

Figure 6.31 The $|S_{21}|$ measurement between the probes shows a distinct dip, or "dog-ear", with the isolating amplifier attached. The difference between the $|S_{21}|$ with the amplifier and a 50 Ω termination is the isolation that the amplifier provides.

$|S_{21}|$) is seen. This is indicated by a "dog-ear" in the $|S_{21}|$ curve at the frequency of interest, as shown in Figure 6.31.

The matching network should be fully tuned after this process. The power output, gain, and isolation of the amplifier can now be measured.

6.8 Testing and Comparison of Amplifiers Architectures

Evaluation of the RF amplifier designs is as important as understanding their operation. By measuring the performance both in test fixtures and by performing magnetic resonance experiments one can better understand their relative merits. An understanding of the test fixtures is important prior to discussing the actual results.

6.8.1 Amplifier Bench Measurements

The primary bench measurements that can be performed are the power delivered to the load and, in the case of multiple transmit elements, the intrinsic isolation provided by the amplifier. Conceptually, both of the measurements are fairly straightforward.

For the standard amplifier and LOI amplifier, the power delivered to the load can be measured using a through-line power meter, with a tuned coil acting as the load. It should be noted that it is important to measure the power while a series of RF pulses are being transmitted. Measuring the output of the current source amplifier requires a calibrated pickup loop. A basic calibration can be accomplished by positioning the pickup loop over an RF coil that is excited by a standard amplifier. The pickup loop reading can be taken with an oscilloscope, while power measurements are made with a through-line power meter, as shown in Figure 6.32(a). A lookup table is constructed by taking measurements at a number of different power levels. The pickup loop and calibration table can then be used for power measurements of the current source amplifier. This requires that the calibration fixture and the test fixture be as close to identical as possible.

Directly measuring the isolation provided by an amplifier is difficult, so instead it is measured *via* the $|S_{21}|$ between two pickup loops positioned over a tuned RF coil, as shown in Figure 6.32(b). Two measurements are needed in order to extract the degree of isolation. The first measurement corresponds to the case of the coil being connected to the amplifier. The amplifier is set up for normal operation, but the input is terminated in 50 Ω and the gate bias line grounded to prevent any possibility of high power being transmitted. The second measurement is performed with the coil connected to a 50 Ω termination. In both cases, the coil must be tuned and connected as it would be in normal operation. Two separate pickup loops are positioned near the coil so that they are isolated from each other, and are connected to the network analyzer for the $|S_{21}|$ measurement. It is important that the pickup loops not be moved between these two measurements. The difference between these two values (in dB) is the amount of decoupling provided by the amplifier.

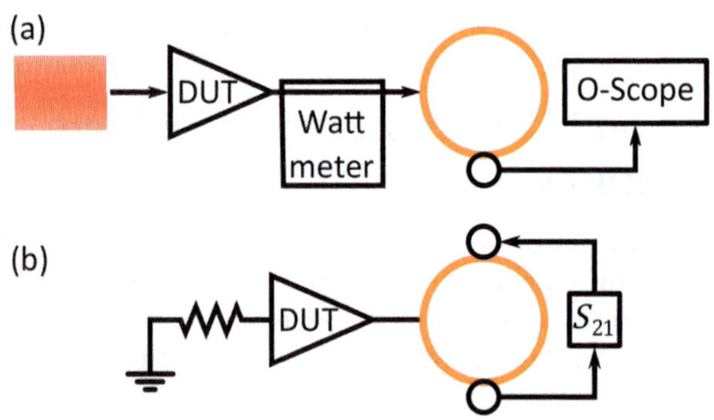

Figure 6.32 (a) Power measurements can be performed using a through-line watt meter for the standard and LOI amplifiers. For the current source amplifier, a calibrated pickup loop is needed. The basic calibration can be accomplished using this setup to generate a lookup table. Care must be taken in the location of the pickup loop to achieve accurate results. (b) The basic isolation measurement setup involves a S_{21} measurement *via* isolated pickup loops placed near the coil.

6.8.2 Amplifier Testing Using MRI

In addition to testing the delivered power and inter-element isolation on the bench, these performance metrics can also be gauged in the magnet using MR measurements. This has the advantage of including other effects such as potential coupling of the RF coil elements to the RF shield and gradients, and represents of course the ultimate performance test. One example of this testing is shown in Figure 6.33 with different designs of eight channel transmit array coils.[48] For the standard amplifier, a single channel of the coil was excited and all other channels were connected to 50 Ω terminations. The resulting image shows that there is significant coupling between the single excited element of the array and many of the other elements, particularly the one geometrically directly opposite. For the LOI amplifier test, one channel was excited, with each channel connected to its respective LOI. The images show that the maximum power delivered is very similar to a standard amplifier, but that the coupling to other elements is significantly reduced. Finally, the experiment was repeated for the current source amplifiers. In this case, a different transmit array structure was used because the maximum power that could be delivered is much less than for the other two forms of amplifier. The images show that there is an even higher isolation between the elements.

An additional performance test is to use all eight amplifiers at once, with a B_1-shimming routine using the transmit array to synthesize a uniform excitation profile. Each of the eight channels of a transmit system using the low output impedance amplifiers was used, with a set of complex weights applied

	Standard Amp	CS Amp	LOI Amp
Max Power	1 kW (device Rating)	150 W (device current rating)	1 kW (device Rating)
Isolation	0 dB	~30+ dB	~14 dB
Single Channel Excitation	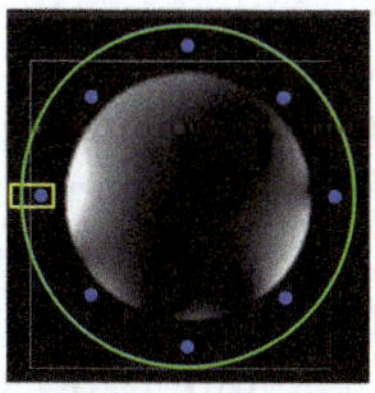	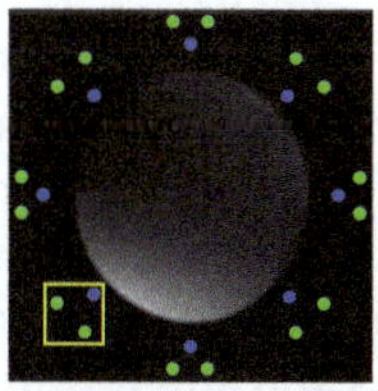	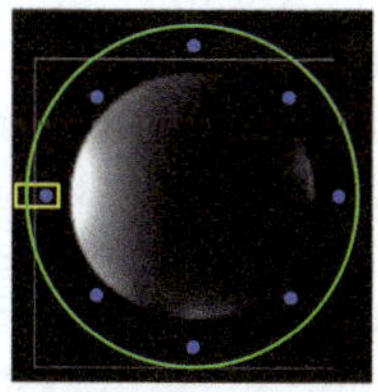

Figure 6.33 MRI-based performance tests of the different types of amplifiers. The isolation of the standard amplifier is defined as zero as the reference point. The power output of the current source amplifier is limited by the device current rating. Each image is overlaid with a basic display of the coil used. Blue represents the current path of the imaging element, and green the return/shield path. The active element of the array is shown in a yellow box.

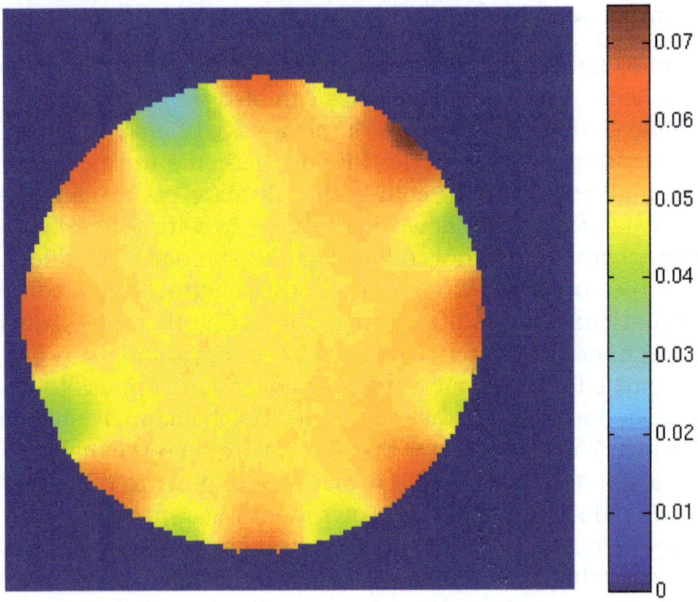

Figure 6.34 The signal from the channels of a transmit array were combined using B_1 shimming in order to generate a uniform excitation.

to the phase and magnitude of the driving signal applied to each individual channel. The high degree of inter-channel isolation and intrinsic load-independence allow the signals to be combined in a simple, linear fashion and thereby to synthesize a nearly-uniform excitation pattern shown in Figure 6.34.

6.9 Selection of Amplifier Architecture

Three different types of amplifier have been described in detail in this chapter: the "standard" power amplifier, the current source amplifier and the low output impedance amplifier. Each of these architectures has different advantages and disadvantages. There is no single clear choice that is best for all situations. Instead, different amplifiers might be considered for different applications. The selection between architectures is summarized in Figure 6.35.

Standard power amplifiers are still the best option when broadband or very high powers are needed. This is the relevant operating mode for high resolution NMR systems and single and dual channel MRI systems. The standard amplifier becomes less desirable as the channel count, *i.e.* the number of independent transmit elements, increases.

Current source amplifiers are limited to a few hundred watts in the amount of power they can produce, but can isolate elements of a transmit array to −30 dB or better. This makes them useful for arrays with a large number of elements that are spaced relatively close together (*e.g.* a 32-channel transmit head array). These types of arrays have high intrinsic coupling between

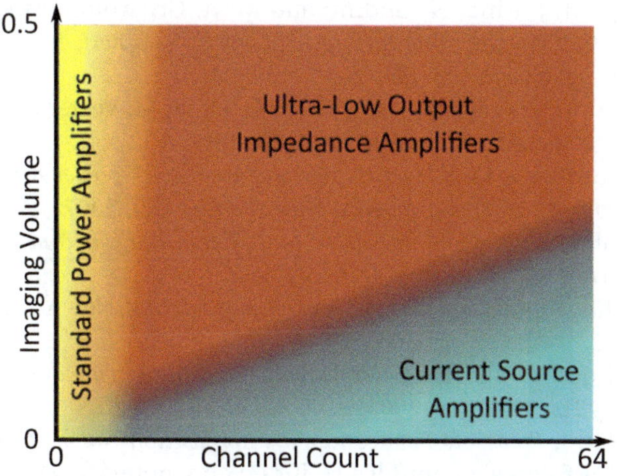

Figure 6.35 Schematic to reflect the behaviour of different amplifier architectures. Generally, in low transmit channel count situations a standard amplifier is a good choice. When the number of channels *vs.* imaging volume is low (*e.g.* an eight channel body transmit array) then low output impedance amplifiers have a number of advantages. Current source amplifiers are well suited to situations where there are a large number of channels in a small region, *e.g.* a 32 channel head transmit array.

elements that is hard to compensate for in other ways, and each individual channel does not require large amounts of power.

Low output impedance amplifiers can produce powers at the kilowatt level (the rating of the device used), and also provide around 14 dB of isolation for the array. This makes the LOI amp well suited to cases where a moderate number of channels are needed, with the elements spaced well apart (*e.g.* an eight channel transmit head array). Higher density arrays tend to have higher coupling between elements that the LOI cannot adequately mitigate, as well as lower power requirements.

References

1. D. A. Grant and J. Gowar, *Power MOSFETS: theory and applications*, New York, Wiley, 1989.
2. S. C. Cripps, *RF power amplifiers for wireless communications*, Artech House, Inc, Boston, 2002.
3. H. L. Krauss, C. W. Bostian and F. H. Raab, *Solid state radio engineering*, New York, Wiley, 1980.
4. A. S. Sedra and K. C. Smith, *Microelectronic circuits*, New York, Oxford University Press, 2004.
5. N. Gudino, M. J. Riffe, L. Bauer, J. A. Heilman and M. A. Griswold, *Proceedings of the 18th Scientific Meeting International Society for Magnetic Resonance in Medicine*, Stockholm, Sweden, 2010, vol. 2, p. 43.

6. M. Twieg, M. J. Riffe, N. Gudino and M. A. Griswold, *Proceedings of the 21st Scientific Meeting International Society for Magnetic Resonance in Medicine*, Salt Lake City, Ut, USA, 2013, p. 0725.
7. N. Gudino, J. A. Heilman, M. J. Riffe, O. Heid, M. Vester and M. A. Griswold, *Magn. Reson. Med.*, 2013, **70**, 276–289.
8. A. Grebennikov and N. O. Sokal, *Switchmode RF power amplifiers. [electronic resource]*, Burlington, MA, Newnes/Elsevier, 2007.
9. G. Gonzalez, *Microwave transistor amplifiers: analysis and design*, Upper Saddle River, N.J., Prentice Hall, 1997.
10. R. E. Collin, *Foundations for microwave engineering*, New York, IEEE Press, 2001.
11. I. J. Bahl and P. Bhartia, *Microwave solid state circuit design*, Hoboken, N.J., Wiley-Interscience, 2003.
12. B. Becciolini, *Freescale Semiconductor Application Notes*, 2005, pp. 1–16, http://www.freescale.com/files/rf_if/doc/app_note/AN721.pdf.
13. D. Pozar, *Microwave Engineering*, 2005.
14. K. Feng, A 64 channel transmit system for single echo acquisition MRI, Ph.D. thesis, Texas A&M Universtiy, 2011.
15. S. M. Wright, M. P. McDougall, K. Feng, N. A. Hollingsworth, J. C. Bosshard and C.-W. Chang, *IEEE Engineering in Medicine and Biology Society*, 2009, vol. 2009, pp. 4053–4056.
16. G. N. French, E. H. Fooks, H. E. Green, J. R. Pyle, S. B. Cohn and S. B. Cohn, *IEEE Trans. Microwave Theory Tech.*, 1968, **16**, 885–886.
17. L. Besser and R. Gilmore, *Practical RF Circuit Design for Modern Wireless Systems*, Artech House, Inc, Boston, 2003.
18. P. L. D. Abrie, *Design of RF and microwave amplifers and oscillators*, Boston, Artech House, 1999.
19. G. D. Vendelin, A. M. Pavio and U. L. Rohde, *Microwave circuit design using linear and nonlinear techniques*, Hoboken, N.J., Wiley-Interscience, 2005.
20. H. W. Bode, *Bell Teleph. Lab. Ser.*, 1945, 551.
21. R. M. Fano, *J. Franklin Inst.*, 1950, **249**, 139–154.
22. G. C. Scott, *Proceedings of the 21st Scientific Meeting International Society for Magnetic Resonance in Medicine*, Salt Lake City, Ut, USA, 2013, vol. 21, p. 2013.
23. F. Chan, J. Pauly and A. Macovski, *Magn. Reson. Med.*, 1992, **23**, 224–238.
24. G. T. Chico, RF Power Amplifier Linearity Compensation for MRI Systems, MSc Thesis, Massachussetts Institute of Technology, 2010.
25. D. I. Hoult, H. G. Kolansky, D. L. Kripiakevich and S. B. King, *J. Magn. Reson.*, 2004, **171**, 64–70.
26. S. C. Cripps, *Advanced Techniques in RF Power Amplifier Design*, Artech House, Inc, Boston, 2002.
27. M. G. Zanchi, P. Stang, A. B. Kerr, J. M. Pauly and G. C. Scott, *IEEE Trans. Med. Imaging*, 2011, **30**, 512–522.
28. P. Stang, A. B. Kerr, W. Grissom, J. M. Pauly and G. C. Scott, *Proceedings of the 17th Scientific Meeting International Society for Magnetic Resonance in Medicine*, 2009, vol. 17, p. 395.

29. J. F. White, *Microwave Semiconductor Engineering*, Van Nostrand Reinhold Company, New York, NY, 1982.
30. W. E. Doherty, Jr. and R. D. Joos, *The PIN Diode Circuit Designers' Handbook*, Microsemi, Watertown, Massachusetts, 1998.
31. U. Katscher, P. Boernert, C. Leussler, J. S. Van Den Brink, P. Börnert, C. Leussler and J. S. Van Den Brink, *Magn. Reson. Med.*, 2003, **49**, 144–150.
32. Y. Zhu, *Magn. Reson. Med.*, 2004, **51**, 775–784.
33. P. B. Roemer, W. A. Edelstein, C. E. Hayes, S. P. Souza and O. M. Mueller, *Magn. Reson. Imaging*, 1990, **16**, 192–225.
34. S. M. Wright and L. L. Wald, *NMR Biomed.*, 1997, **10**, 394–410.
35. D. I. Hoult, G. Kolansky and D. Kripiakevich, *J. Magn. Reson.*, 2004, **171**, 57–63.
36. K. N. Kurpad, E. B. Boskamp and S. M. Wright, *Proceedings of the 13th Scientific Meeting International Society for Magnetic Resonance in Medicine*, Miami Beach, Florida, USA, 2005, vol. C, p. 16.
37. W. Lee, E. B. Boskamp, T. M. Grist and K. N. Kurpad, *Magn. Reson. Med.*, 2009, **62**, 218–228.
38. K. N. Kurpad, S. M. Wright and E. B. Boskamp, *Concepts Magn. Reson., Part B*, 2006, **29(B)**, 75–83.
39. K. Chang, K. A. Hummer and G. K. Gopalakrishnan, *Electron. Lett.*, 1988, **24**, 1347.
40. K. Chang, R. A. York, P. S. Hall and T. Itoh, *IEEE Trans. Microwave Theory Tech.*, 2002, **50**, 937–944.
41. C. G. Parini, *Symposium on EDMO*, 1999, pp. 53–58B.
42. C. W. Pobanz and T. Itoh, *IEEE Potentials*, 1997, **16**, 6–10.
43. J. Lin and T. Itoh, *IEEE Trans. Microwave Theory Tech.*, 1992, **42**, 2186–2194.
44. A. P. Anderson, W. S. Davies, M. M. Dawoud and D. E. Galanakis, *IEEE Trans. Antennas Propag.*, 1970, **19**, 1970–1972.
45. X. Chu, X. Yang, Y. Liu, J. Sabate and Y. Zhu, *Magn. Reson. Med.*, 2009, **61**, 952–961.
46. A. Reykowski, S. M. Wright and J. R. Porter, *Magn. Reson. Med.*, 1995, **33**, 848–852.
47. A. Reykowski, Ph.D. thesis, Theory and Application of Synthesis Arrays in Magnetic Resonance Imaging, Texas A&M University, 1996.
48. K. L. Moody, N. A. Hollingsworth, F. Zhao, J.-F. Nielsen, D. C. Noll, S. M. Wright and M. P. McDougall, *J. Magn. Reson.*, 2014, **246C**, 62–68.

CHAPTER 7

The MR Receiver Chain

DENNIS W. J. KLOMP*[a] AND ANDREW WEBB[b]

[a]Center for Image Sciences, University Medical Center Utrecht, Utrecht, The Netherlands; [b]C.J.Gorter Center for High Field MRI, Department of Radiology, Leiden University Medical Center, Leiden, The Netherlands
*E-mail: dennis@isi.uu.nl

7.1 Introduction

A block diagram of an MR receiver[1] is shown in Figure 7.1. For high-resolution NMR spectroscopy, the same coil is used for transmitting the RF pulses and receiving the MR signal: as seen in Chapters 1 and 3, the experiments may require RF pulse transmission at many different frequencies. Recent advances have also allowed simultaneous acquisition from different nuclei on high resolution NMR systems.[2] In MRI, typically one coil is used to transmit, whereas a large number of separate coils are used to receive the signal.

The key parameters determining the design of the MR receiver are:

(i) The transmit/receive (T/R) switch must provide very high isolation between the high power input from the RF power amplifier and the low power output connected to the preamplifier and remainder of the receive chain. The switch must also have a very fast response time on the order of tens of microseconds.
(ii) The preamplifier must provide a high degree of amplification while adding as small a noise contribution as possible. The dynamic range of the MR signal can be as high as 60–80 dB and so the entire receiver

New Developments in NMR No. 7
Magnetic Resonance Technology: Hardware and System Component Design
Edited by Andrew G Webb

Published by the Royal Society of Chemistry, www.rsc.org

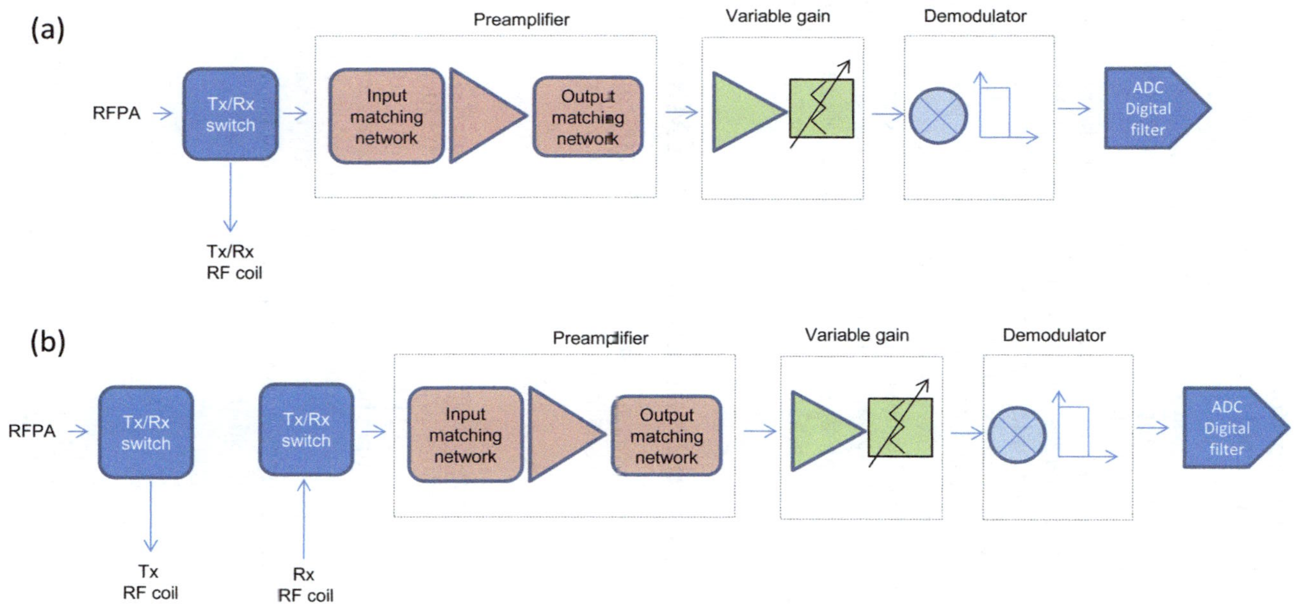

Figure 7.1 Block diagrams of the MR receiver for (a) a single transmit/receive coil, and (b) separate transmit and receive coils. A high power signal is produced from the RF power amplifier (RFPA) and the transmit/receive switch must ideally direct this power exclusively to the RF (transmit) coil without any power passing through the receiver. During signal acquisition, the transmit/receive switch directs the signal from the RF coil to an impedance matched preamplifier. The signal can be further amplified and may be demodulated to match the optimal input range of the analogue-to-digital converter.

chain must be able to cover this large range. The preamplifier must be protected from any signals that pass through the T/R switch. Matching networks must be constructed both for the input and the output of the preamplifier to maximize performance.

(iii) Additional variable gain amplification is provided by the combination of a fixed-gain amplifier and variable attenuator. In many MR experiments, the signal intensity varies significantly during the experiment, and therefore in order to capture the full dynamic range of the signal a time-varying amplification factor can be applied.

(iv) The Larmor frequency for most MR experiments is in the range of tens to hundreds of MHz, whereas the spectral bandwidth of the MR signal is several orders of magnitude lower. In order to minimize the noise in the recorded signal, the bandwidth of the measurement should match that of the MR signal itself. This can be achieved by demodulating the MR signal to a much lower frequency before digitization. Various methods of performing this are discussed later in this chapter.

(v) The signal is digitized by an analogue-to-digital converter (ADC), which should have as high a resolution as possible. Digital filtering, noise shaping and other signal processing methods are typically integrated into modern ADCs.

7.2 Signal Levels and Dynamic Ranges of MR Data

As covered in Chapter 3, for a passive system such as the MR coil plus sample, the root-mean-square Johnson (thermal)[3] noise voltage, N_s, is given by:

$$N_s = \sqrt{4kT\Delta fR} \tag{7.1}$$

where T is the temperature of the RF coil and sample in kelvin, R is resistance, Δf is the bandwidth of the digitizer in Hz (*i.e.* the sampling frequency), and k is the Boltzmann constant. For an MR system with a sampling frequency of 1 MHz, the RF coil impedance-matched to 50 Ω, and a sample temperature of 310 K, the value of N_s is ~1 μV. Note that the noise level of the impedance-matched RF coil is independent of the RF coil geometry or sample size.

The signal, S, detected in an MR experiment based on inductive coupling of the spins with a pickup coil can be expressed by the classic formula of Hoult and Richards:[4]

$$S \propto k_0 \frac{B_1}{I} V_s N \gamma \frac{h^2}{4\pi^2} I(I+1) \frac{\omega_0^2}{3kT\sqrt{2}} \tag{7.2}$$

where V_s is the sample volume, k_0 is a constant that accounts for spatial inhomogeneities in the B_1 field, I is the spin angular momentum quantum number, γ is the gyromagnetic ratio, N is the spin density (number of spins per unit volume), and h is Planck's constant. The factor B_1/I, the magnetic field per unit current, is defined to be the coil sensitivity, and is inversely proportional to the diameter of the coil. In high resolution NMR, the maximum signal might arise from, for example, a protein solution in water using a cryogenic RF probe. In MRI, the corresponding situation might be a three-dimensional proton image of the human body. In both cases, signals in the tens of mV are produced. However, there are many MR measurements for which the signal is actually below the noise level, *e.g.* for a low concentration of a low gyromagnetic ratio nucleus such as ^{13}C. In this case, signal averaging must be used in order to obtain a useable SNR. Assuming that the noise is random, the SNR is proportional to the square root of the number of co-summed signal averages. It is therefore important to note that, in this latter case, the dynamic range of the MR signal may actually be greater than the difference between the highest signal and the thermal noise. The dynamic range of MR signals can easily span 80 dB.

Another consideration is that the dynamic range of the signal in an MR experiment may depend on exactly how the data are acquired. As seen in Chapter 1, in MRI the signal intensity is highest at zero and near-zero values of *k*-space and becomes much smaller at higher *k*-space values, *e.g.* as the strength of the phase encoding gradient increases. In such cases, the receiver gain can be modulated as the inverse of the expected magnitude of the *k*-space signal, as shown in Figure 7.2.[4] This process is known as signal compression since it effectively compresses the dynamic range of the signal, and is well known in other imaging fields, such as ultrasonic imaging. Similar situations may occur in high resolution NMR, where for example the maximum signal varies as a function of t_1-increments in 2D NMR spectroscopy.

As shown in Figure 7.1, signal compression is usually performed using a fixed gain amplifier and a variable attenuator, rather than using a variable gain amplifier. An alternative method of achieving variable gain is to have a number of different amplifiers with different gain factors available, and route the signal to the particular amplifier depending upon the expected signal intensity. Any variable gain applied during the course of data acquisition must be

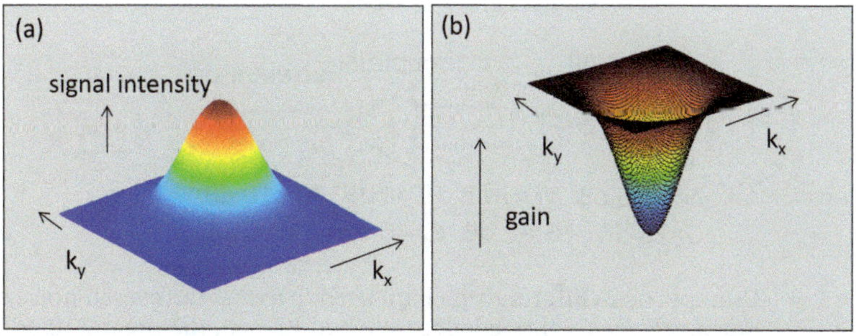

Figure 7.2 Illustration of variable gain applied during MRI data acquisition in order to compress the dynamic range of the MR signal. (a) The signal strength is highest at values of k_x and k_y close to zero, (b) applying a gain factor that is approximately inversely proportional to the signal strength leads to compression of the signal dynamic range.

accompanied by an associated pre-processing of the data before image reconstruction. For example, in the case of variable gain as a function of *k*-space acquisition, the magnitude and phase of the signals can be corrected by a look-up normalization table that characterizes the performance of the variable attenuator. Signal compression performed in this way is also useful in wireless transmission of MR data,[5] covered in the final section of this chapter, since the transmission process often has a limited dynamic range that it can cover.

7.3 Overall Noise Figure of the Receive Chain

Having determined that the receive chain must be able to deal with an input dynamic range of ~80 dB, the next general step is to determine which are the most critical components in defining the final SNR of the digitized MR signal. Each component in the receive chain adds noise, and one can define the reduction in SNR produced by each component in the chain in terms of a noise figure (NF) defined as:

$$\mathrm{NF(dB)} = 20\log\left(\frac{\mathrm{SNR_{in}}}{\mathrm{SNR_{out}}}\right) \tag{7.3}$$

Typical values of NF for each component are shown in Figure 7.3, as well as the gain factors *G*.

The overall NF for the receive chain can be calculated from:

$$\mathrm{NF_{total}} = \mathrm{NF_{preamp}} + \frac{\left(\mathrm{NF_{cable}} - 1\right)}{G_{\mathrm{preamp}}} + \frac{\mathrm{NF_{amp}}}{G_{\mathrm{preamp}}G_{\mathrm{cable}}} + \frac{\mathrm{NF_{attenator}}}{G_{\mathrm{preamp}}G_{\mathrm{cable}}G_{\mathrm{amp}}} \tag{7.4}$$

Using the numbers in Figure 7.3, the overall NF is 0.51 dB, which is very close to the NF of the first-stage amplifier. Indeed, eqn (7.4) shows that the critical component in terms of the overall NF of the receiver is the first stage,

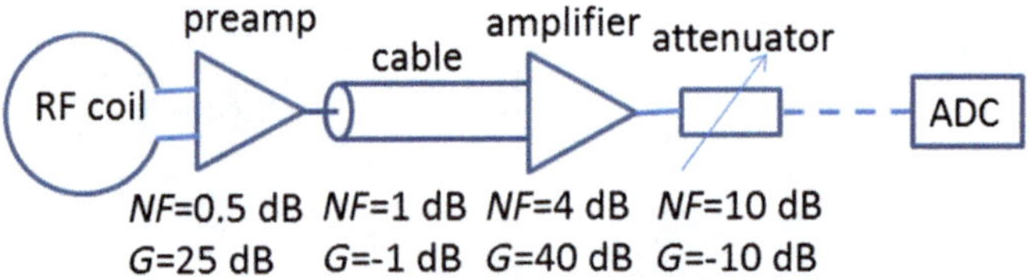

Figure 7.3 Example of an MR receiver chain used to assess the overall noise figure. Starting from the impedance matched RF coil, the preamplifier amplifies the signal by 25 dB, but adds 0.5 dB of noise. The cable attenuates the signal by 1 dB, corresponding to a noise figure of 1 dB since the cable is a passive device. The next stage amplifier has a gain factor of 40 dB with a noise figure of 4 dB, after which the signal is attenuated by a suitable amount (in this example by 10 dB) to reach a level commensurate with the dynamic range of the ADC.

which should have as low an NF and as high a gain as possible. As an example of how important this is, if the RF cable with a 1 dB loss were inserted between the RF coil and preamplifier rather than after the preamplifier, the noise figure of the receive chain would increase by 1 dB. As a result, for maximum SNR it is important to connect the preamplifier as close to the RF coil as possible: this is particularly true for high frequency operation in which cable losses increase significantly compared to lower frequencies.

Having determined the overall performance of the receive chain, the following sections consider the design of specific components of the receive chain in the order, left-to-right, shown in Figure 7.1.

7.4 Design of Transmit/Receive Switches

The role of the T/R switch is to provide a rapidly switchable high isolation between the transmit and receive chains of the RF system, with a very low loss. In transmit mode, the several kW of power from the RFPA must be sent to the RF coil with as low a loss and as high an isolation from the receive chain as possible. In receive mode, the maximum SNR is achieved with as low a loss and as high an isolation from the transmit chain as possible. Typical specifications are an insertion loss between 0.1 and 0.3 dB, and an isolation between the preamplifier and the RF power amplifier of greater than 60 dB.

Although mechanical switches can provide very high isolation, they cannot be switched on the microsecond timescale, which is necessary for MR operation. Therefore, all MR T/R switches are based on electronically switched PIN diodes. As covered in Chapter 3, a PIN diode can be switched from its "on-state" (forward-biased with a positive driving DC voltage above 0.7 volts, acts as a small resistor) to its "off-state" (reverse-biased with a negative driving DC voltage, acts as a very high impedance). There are a number of different designs,[6–8] the basic structure of which is given in Figure 7.4, and consists of a shorted quarter-wavelength transmission line, or its lumped element equivalent circuit.

As shown in Section 3.5, the input impedance, Z_{in}, of a transmission line of length L and characteristic impedance Z_0, which is terminated in an impedance Z_{load}, is given by:

$$Z_{in} = Z_0 \frac{jZ_0 \tan(\beta l) + Z_{load}}{jZ_{load} \tan(\beta l) + Z_0} \tag{7.5}$$

where $\beta = 2\pi/\lambda$. In Figure 7.4(a), during transmission of the RF pulses, diodes D_1 and D_2 are both on. The value of Z_{load} is therefore zero (a short circuit) and so from eqn (7.5) the value of Z_{in} is very high, ideally infinite, preventing any signal from entering the receive chain. During signal reception, both diodes are switched off, Z_{load} is 50 Ω, and the quarter-wavelength cable simply appears as an additional length of transmission line. Since D_1 appears as a very high impedance, all of the MR signal passes to the receiver.

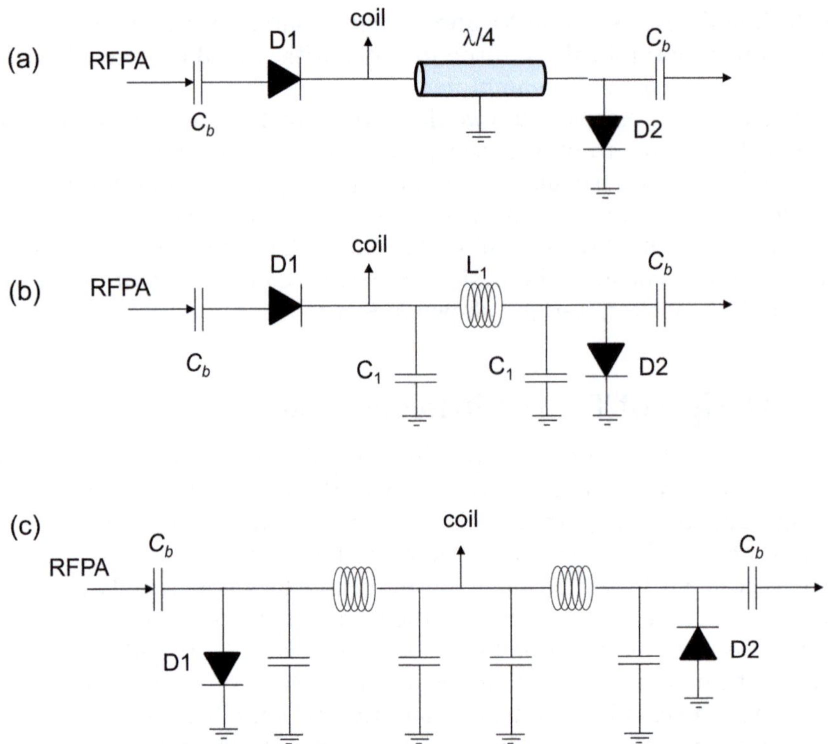

Figure 7.4 Three variations of T/R switch based on a quarter-wavelength transmission line (a) and its lumped element pi-section equivalent (b) and (c). In (c) two pi-sections are used for additional isolation between transmitter and receiver. Capacitors C_b have very high values to provide a DC-block for the diode-driving voltage, and present essentially a short circuit at RF frequencies.

Figure 7.4(b) shows an electrically identical circuit using a pi-network that is equivalent to the quarter-wavelength transmission line. There are many reasons why the lumped element circuit might be preferred, including much smaller physical dimensions (particularly at low frequencies), lower interactions between cables if multiple receive coils are used, and reduced loss which may arise from RF radiation from the cables (particularly at high frequencies). The values of the capacitors, C_1, and inductor, L_1, are given by:

$$L_1 = \frac{50}{\omega}, C_1 = \frac{1}{50\omega} \tag{7.6}$$

In practice (depending on the RF frequency, the particular PIN diode used, and the PIN-diode driving current), a single pi-network can attenuate the RF pulse by approximately 30 dB and has an insertion loss of less than 0.1 dB. Many designs, such as that shown in Figure 7.4(c), therefore contain two or more pi-networks. The circuit in Figure 7.4(c) is also designed with the

Figure 7.5 Photograph of a T/R switch assembled on a PC board.

following criteria in mind: when a PIN diode is turned on (*i.e.* when a positive driving voltage above 0.7 V is applied to the diode), the device does not act as a perfect short (*i.e.* it contains resistive losses), and therefore may increase the noise level entering the receiver chain. When the PIN diode is turned off (*i.e.* when a reverse voltage is applied to the diode), the impedance can be very high and the noise floor remains essentially unchanged. Consequently, an electrical circuit is preferred that passes the RF when the diode is reverse biased, and blocks the RF when the diode is forward biased, as shown in Figure 7.4(c).

At high frequencies one has to be very aware of the connections between different components, and connections to ground. For example, a pin-diode connected to ground *via* a wire, which is meant to behave as a short circuit, may have significant impedance owing to the finite inductance ($Z = j\omega L$) of the connecting wire. In this case, a series capacitance can be added to resonate out the additional inductance and achieve a true short circuit. Similarly, the capacitance in the negatively biased PIN-diode can be on the order of 1–2 pF, which may not be sufficiently high at high frequencies to completely block current flow: a parallel inductor can be added to increase the impedance. A photograph of an assembled T/R switch is shown in Figure 7.5.

There are several other forms of T/R switches that are suitable for MR purposes. One is shown in Figure 7.6, consisting of two series quadrature hybrids. In transmit mode, the diodes are turned on and transmit power is transmitted from the RFPA to the coil and the preamplifier is protected. In receive mode the RFPA is isolated from the coil and the coil receive signal is split between the two ouputs of the first quadrature hybrid. These signals are then recombined at the output of the second quadrature hybrid before entering the preamplifier. This design has been tested using 2 kW input power, giving −0.26 dB transmit loss and 42 dB receive isolation, with a total noise figure of 0.67 dB.[9]

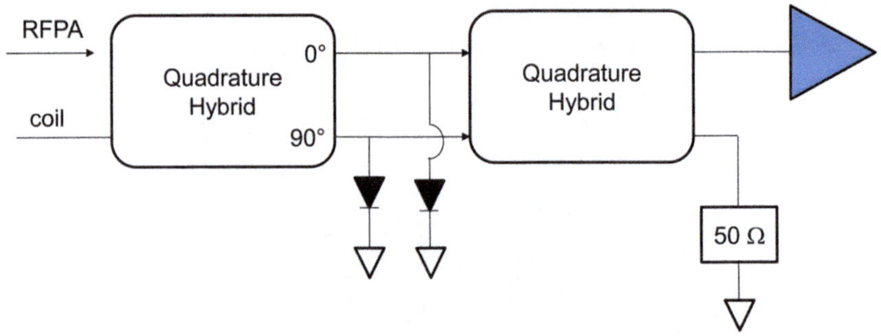

Figure 7.6 Schematic of a high-frequency (300 MHz) T/R switch based on two back-to-back commercial quadrature hybrids. The overall package can be made extremely small.

7.5 Low-Noise Preamplifiers

The first and most critical component of the receiver chain is the preamplifier, which must have a high gain and low noise figure, as described earlier. The preamplifier is ideally incorporated into the coil design to minimize cable losses between the coil and preamplifier, but this means that it must be able to operate in a strong magnetic field. This is commonly performed for clinical MRI systems, but in high resolution NMR and animal imaging systems the preamplifier is usually placed outside the magnet.

The preamplifier most commonly used in MR is based on a gallium arsenide field effect transistor (GaAs-FET), although variations using high-electron-mobility-transistors (HEMTs) are also available. GaAs-FETs have low noise figures (~0.5 dB) and operate linearly over the signal range from the thermal noise level to several tens or hundreds of mV. If the signal level is greater than this, the FET can be damaged or else becomes saturated; this results in reduced gain and an increased noise figure, and therefore the isolation provided by the T/R switch is critical. The input impedance of FETs is very high, several tens of kΩ up to MΩ. The performance of FETs as amplifiers is critically dependent upon the impedance of the input circuitry, as shown in Chapter 6. The gain and noise figure, in particular, vary significantly as a function of the input impedance, and it is not possible to optimize both of these parameters simultaneously. A schematic diagram of a preamplifier assembly, consisting of input and output matching networks, as well as the FET itself, and associated impedances is shown in Figure 7.7. If Z_{in} is matched to 50 Ω this gives the maximum gain, but a higher noise figure than desired. For optimal noise matching the condition is that $Z_s = Z_{opt}$, where Z_{opt} is the input impedance that minimizes the noise figure of the particular FET. The output matching network is used to match the output impedance of the FET to the 50 Ω cable.

In practice, the input matching network is almost always designed to be noise matched, *i.e.* to minimize the noise figure. The required impedance is usually orders of magnitude less than the intrinsic input impedance of the FET. A general circuit for performing noise matching is shown in Figure 7.8.[10,11] In practice, the DC bias voltage V_{DD} is provided by a circuit comprising a voltage regulator, with associated filter capacitors, connected to the drain *via* a high inductance (>1 µH).

As an example, suppose the RF input impedance of a FET is 50 kΩ, while the optimal NF occurs for an input impedance of 1 kΩ. Considering an RF coil that is matched to 50 Ω at a Larmor frequency of 300 MHz, the circuit

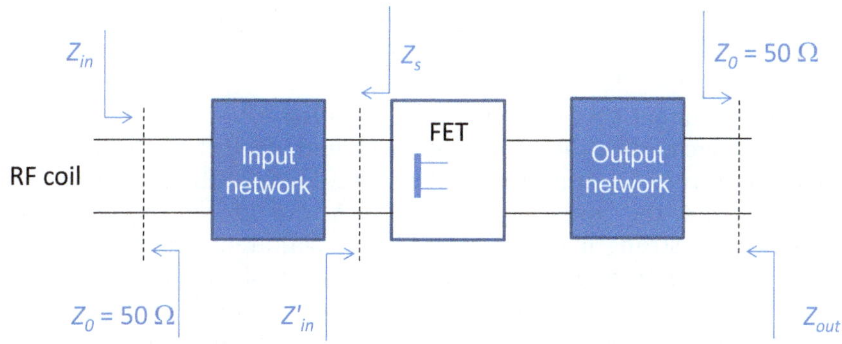

Figure 7.7 Block diagram of a FET-based preamplifier connected on the input side to an RF coil impedance matched to 50 Ω, and on the output side to a 50 Ω cable.

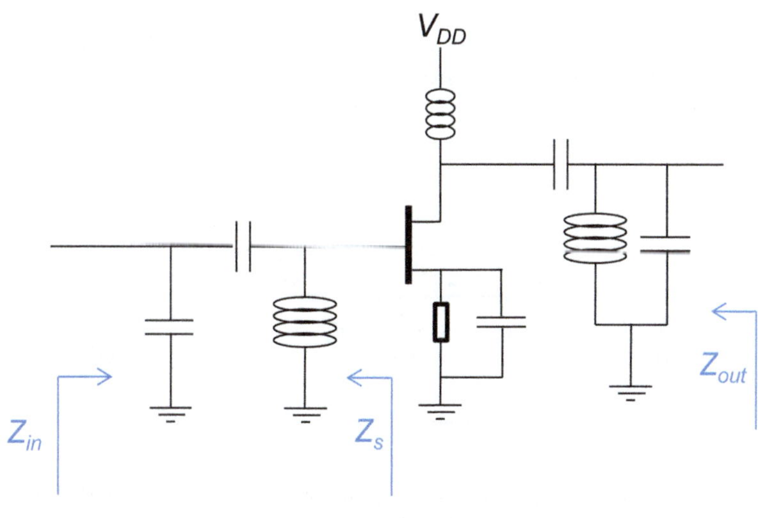

Figure 7.8 Schematics of the circuits used for the input and output matching circuits for the preamplifier, and the DC bias, V_{DD}.

used in Figure 7.9 can be used for noise matching to 1 kΩ. Assuming an input impedance of 50 kΩ at the input of the FET, the noise matching circuitry transforms this into 1 Ω at the input of the matching network (seen by the RF coil).

An important design criterion for the preamplifier is to ensure that it is unconditionally stable, *i.e.* that it satisfies Rollett's stability criterion[12] for a wide range of frequencies, not only those close to the frequency for which it is designed. Detailed derivation of this formula can be found in most textbooks on amplifiers. In terms of the measured *S*-parameters, the value of Rollett's stability constant, *K*, is given by:

$$K = \frac{1 - S_{11}S_{11}^{*} - S_{22}S_{22}^{*} + \left|S_{11}S_{22} - S_{12}S_{21}\right|^{2}}{2\left|S_{12}S_{21}\right|} \tag{7.7}$$

The condition that $K > 1$ corresponds to stability.

As mentioned earlier, the preamplifiers should ideally be placed as close as possible to the coil to minimize the noise figure of the receiver chain. This means that they must operate within a very strong magnetic field. However, when strong magnetic fields (≫1 tesla) are aligned *perpendicular* to the surface of the FET, the Hall effect causes the noise figure of the preamplifier to increase significantly. Therefore care must be taken in the orientation of the preamplifier with respect to the main magnetic field. In addition, when preamplifiers are installed directly at the RF coil port, care must be taken that the output of the preamplifier is not coupled to the RF coil. Any RF coupling should be less than the inverse of the gain of the preamplifier or else oscillations can occur, *i.e.* it can become unstable. If the frequency of the oscillations is within the MR acquisition bandwidth then artifacts will occur in the spectrum or image. If these oscillations occur outside the acquisition bandwidth they can still can cause saturation of the preamplifier, which reduces its gain and increases the noise figure substantially.

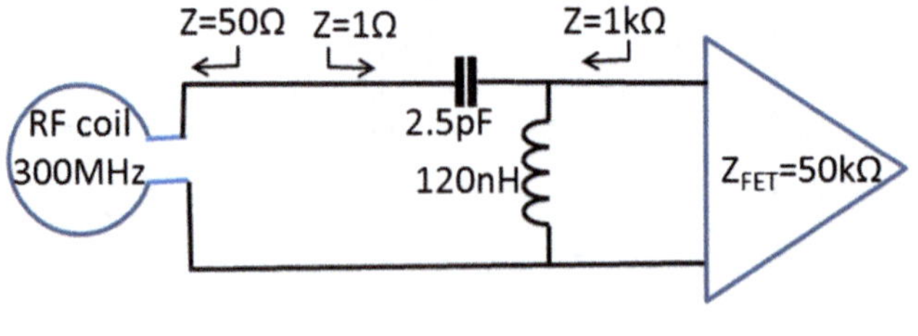

Figure 7.9 Example of a matching circuit for optimal noise figure of the preamplifier. The 50 Ω input impedance of the RF coil at 300 MHz is transformed to the noise matched impedance of 1 kΩ using a capacitor of 2.5 pF and an inductor of 120 nH. The input impedance of the preamplifier (FET) is 50 kΩ, and is transformed into a 1 Ω impedance seen by the RF coil.

7.6 Data Sampling

The Nyquist–Shannon sampling theory states that in order to correctly sample a signal with frequency *f*, the sampling rate must be at least $2/f$, *i.e.* two samples per period must be acquired. So for an MR signal with a bandwidth of 20 kHz, the sampling rate should be at least 40 kHz, *i.e.* the time between samples must be less than or equal to 2.5 ms. If this criterion is not fulfilled, then aliasing of the signal occurs: this is shown in Figure 7.10.

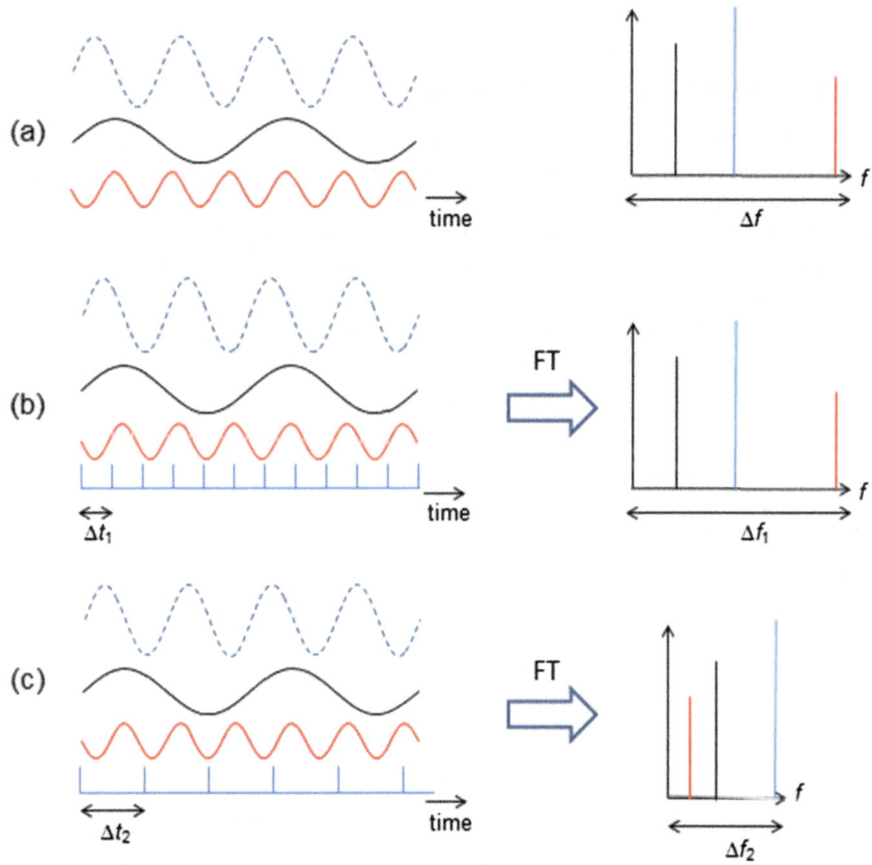

Figure 7.10 Schematic showing the principle of Nyquist sampling and signal aliasing. The figure shows three analogue signals, each with a different frequency and amplitude. The total bandwidth of the signals is given by Δf as shown in (a). In (b) the signals are sampled according to the Nyquist criterion every Δt_1 seconds. If the digitized sampled signal is Fourier transformed the spectrum shows all three components at the correct frequencies. In (c), the sampling time Δt_2 is twice Δt_1, meaning that the highest frequency signal is sampled less than twice per sine-wave. The highest frequency (red) is now aliased back into the spectrum at an incorrect lower frequency.

Typical bandwidths of MR signals are in the tens of kHz for NMR spectroscopy and hundreds of kHz for MRI. The Larmor frequency is, in most cases, several orders of magnitude higher than the required bandwidth. In principle, the MR signal could be sampled directly at the Larmor frequency; however, as will be shown later, the dynamic range and resolution of the analogue-to-digital converter decrease at higher sampling rates, and would be too low to capture the full dynamic range of the MR signal for most situations. There are a number of different approaches that can be used to overcome this limitation. Some of these are discussed in the following sections.

7.6.1 Frequency Demodulation

The most common method of reducing the carrier frequency of the NMR signal is *via* frequency demodulation. This can occur in one or two steps, as described below. In a single-step demodulation, two phase-sensitive detectors (PSDs) are used to produce real and imaginary outputs centred at a lower intermediate frequency, ω_{IF}. Each PSD contains a mixer and low pass filter, as shown in Figure 7.11.

In Figure 7.11, the input MR signal from a linear coil is given by:

$$s_1(t) = A(t)\cos[(\omega_0 + \Delta\omega)t] \tag{7.8}$$

where $\Delta\omega$ represents the different frequency components produced by the frequency-encoding gradient in imaging or chemical shift in spectroscopy. The mixer or demodulator can be modeled most simply as a multiplier, and so the output signal is given by:

$$s_2(t) = A(t)\cos[(\omega_0 + \Delta\omega)t]\cos\omega_{LO}t \tag{7.9}$$

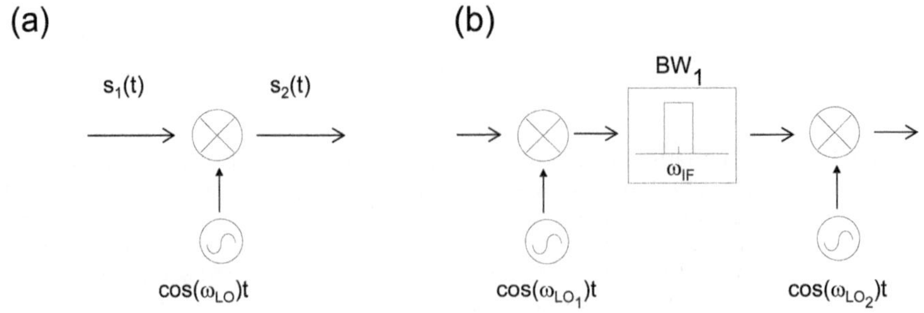

Figure 7.11 (a) Schematic of a mixer. (b) The combination of two mixers with a bandpass filter in-between constitutes the basis of a two-stage demodulator.

Re-arranging the cosine product:

$$s_2(t) = \frac{A(t)}{2}\left\{\cos\left[\left(\omega_0 - \omega_{LO} + \Delta\omega\right)t\right] + \cos\left[\left(\omega_c + \omega_{LO} + \Delta\omega\right)t\right]\right\} \tag{7.10}$$

The low-pass filter removes the right hand term in eqn (7.10), leaving just a term at an intermediate frequency, ω_{IF}, where ω_{IF} is given by $\omega_0 - \omega_{LO}$. Although it is possible to set ω_{LO} to be exactly equal to ω_O, thus giving a signal that has no carrier frequency, it is well-known from receiver design that this is undesirable since the local oscillator frequency lies within the bandwidth of the MR signal, and breakthrough can lead to signal contamination. Typical values for the intermediate frequency are 10–20 MHz. Single-stage demodulation works well for cases in which the Larmor frequency is not too many multiples of the intermediate frequency. However, one can see that for very high frequencies, the higher and lower frequencies in eqn (7.10) would require a very sharp filter in order to achieve good suppression of the higher frequencies. In this case, a two-stage demodulation is often used, as shown in Figure 7.11(b), using a circuit very similar to a double conversion superheterodyne receiver used in many high frequency RF circuits. This type of circuit is also useful in the case that, for example, a 7 tesla human system can be based upon the electronics in a 3 tesla system, simply with one extra stage of signal demodulation.

7.6.2 Direct Detection Using Undersampling

A second method to sample the MR signals is to use direct detection based on undersampling.[13–15] As an example, consider the case of using a sampling frequency of 40 MHz to detect the proton MR signals from a 1.5 T imaging system. The Larmor frequency is 64 MHz and in Figure 7.12 a frequency bandwidth of the MR signal of 1 MHz is assumed, corresponding perhaps to a high bandwidth echo planar imaging scan. According to Nyquist theory, the 40 MHz detector cannot distinguish between signals of 64 MHz and signals of 24 MHz (64–40 MHz), which are superimposed on top of one another. However, if signal amplification is applied to signals at 64 MHz, and a band-pass filter implemented at the same frequency, then any signal or noise components at 24 MHz have a very small contribution to the signal that passes to the next stage of the receiver. In this way undersampling is effectively acting as a single-stage demodulator.

The same principle can be applied for frequency bands even further away from the sampling frequency. In Figure 7.13, a digitizer with a sampling frequency of 38 MHz is used to detect the proton signals on a 7 T MR system that operates at 298 MHz.

Band-pass filters are used to exclude the noise from all frequency bands that can be aliased inside the frequency band of the digitizer. However, the further the carrier frequency is from the sampling frequency, the higher the

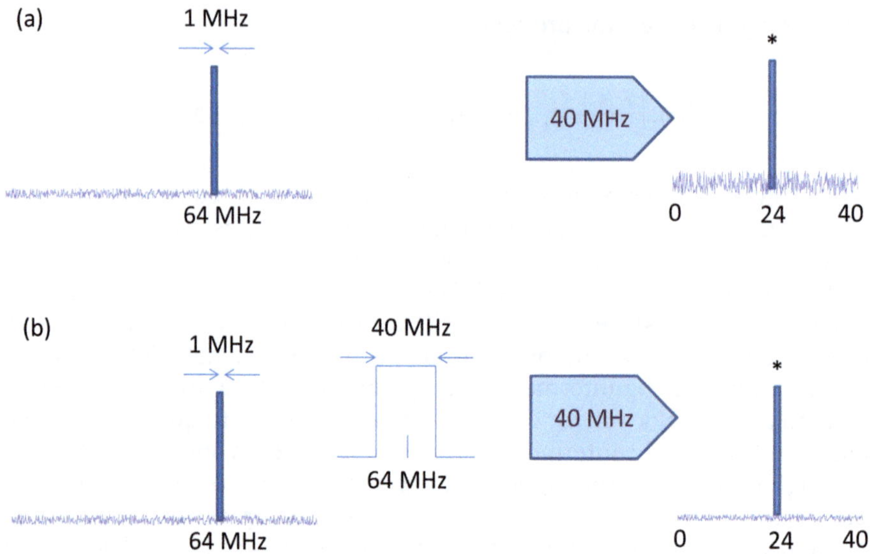

Figure 7.12 Demonstration of the principle of undersampling. The MR signal consists of a 1 MHz bandwidth signal centred at 64 MHz. (a) Using a 40 MHz ADC gives an aliased signal centred at 24 MHz. Any noise or signal components at 24 MHz will be added to the desired signal. (b) In undersampling, the bandpass filter is designed to remove all signal and noise components at 24 MHz.

order of filter required. The example in Figure 7.13 requires a very sharp filter. One filter that has excellent bandpass characteristics is a surface acoustic wave (SAW) filter, whose characteristics are shown in Figure 7.14: for comparison, the properties of a third order conventional lumped element filter are also shown.

7.7 Analogue-to-Digital Converters

The final component in the receive chain is the analogue-to-digital converter (ADC), which not only digitizes the signal but also implements a number of digital signal processing and filtering steps. Important specifications of an ADC include the dynamic range (number of bits), voltage range (maximum-to-minimum), maximum sampling frequency (Hz), and frequency bandwidth (Hz). For an N-bit ADC, the number of different output values is given by 2^N. For example, a 16-bit ADC can output values from 1 to 65 535. The voltage difference between these levels is called the resolution, and its value is given by the voltage range of the ADC divided by the number of levels. This is shown in Table 7.1, which assumes that the maximum voltage of the ADC is 1 volt. The least significant bit (LSB) is defined as one-half the maximum voltage divided by the number of levels.

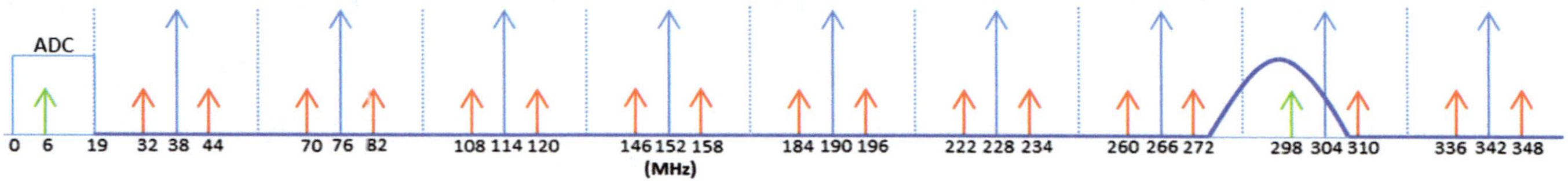

Figure 7.13 Example of using a 38 MHz ADC to detect the MR signals at 298 MHz (7 T). A total of 19 frequency bands (red and green arrows) are displayed, which all fold into the acquisition band of the ADC. To ensure that only the 298 MHz band is acquired without adding noise from the other bands, a band pass filter needs to be applied. The bandwidth of the filter needs to be sufficiently narrow to suppress those sidebands most closely positioned to the 298 MHz band (*i.e.* 272 MHz and 310 MHz).

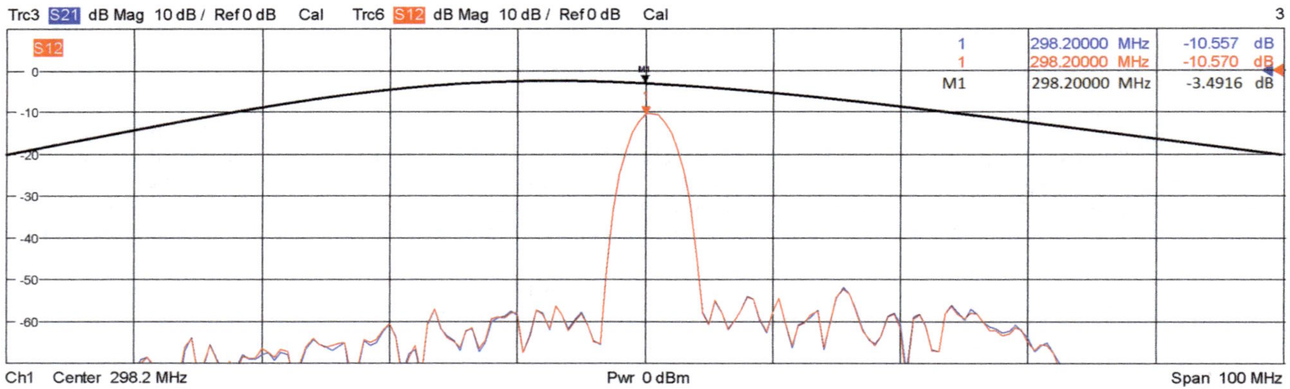

Figure 7.14 A comparison of output *versus* input response of a third order band pass filter (black) and a SAW filter (red and blue) measured with a network analyzer. The third order band pass filter would suppress the closest sidebands in Figure 7.13 by only approximately 5 dB causing a substantial noise amplification. In comparison, the SAW filter suppresses these side bands by more than 60 dB. Note that the SAW filter attenuates the MR signal by 10 dB, which needs to be taken into consideration when optimizing the overall receiver gain and dynamic range of the ADC.

The most significant bit (MSB) is one-half of the maximum voltage. The resolution in dB is calculated as:

$$\text{Resolution(dB)} = 20\log\left(\frac{1}{\text{LSB}}\right) \tag{7.11}$$

Even ADCs with very high resolution cannot reproduce the analogue signal perfectly. The difference between the real analogue input signal and the digitized output is called the quantization error or quantization noise, and this error is smaller the greater the number of bits, as shown in Figure 7.15. It can also been seen that the values of the quantization error lie between 0 and ±1/2 of the ADC resolution. In order to minimize the relative contribution of the quantization errors, therefore, the input voltage to the ADC should be as large as possible, without "saturating", *i.e.*, going above the maximum value of the ADC, which leads to distorted signals.

The root mean square (RMS) quantization noise, $e_{\text{rms}}{}^2$, is given by:

$$e^2_{\text{rms,Nyquist}} = \frac{\Delta^2}{12} \tag{7.12}$$

where Δ is the ADC resolution. In ADCs there are two other measures of error that are typically quoted, the differential non-linearity (DNL) and integrated non-linearity (INL), both of which are shown in Figure 7.16. For an

Table 7.1 Properties of a 1 volt ADC.

Number of bits	Number of levels	Least significant bit 1 volt full scale	Resolution (dB)
8	256	3.91 mV	48.2
10	1024	977 μV	60.2
12	4096	244 μV	72.3
14	16 384	61 μV	84.3
16	65 536	15.3 μV	96.3
20	1 048 576	954 nV	120.4
24	16 777 216	59.5 nV	144.5

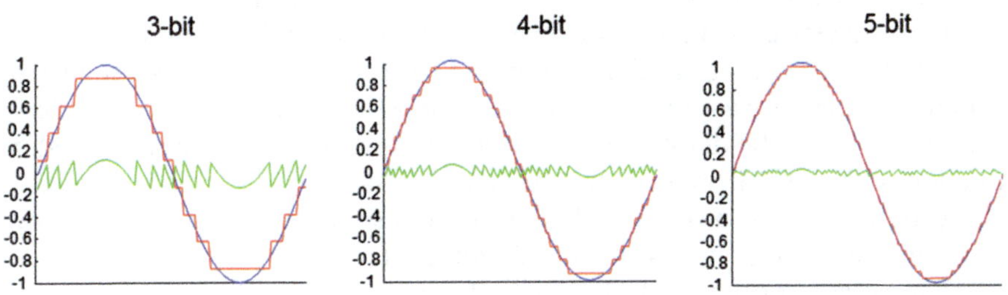

Figure 7.15 The effects of the number of bits on the quantization error of a digitized signal. Blue indicates the analogue signal, red the quantized approximation, and green the quantization error between the two.

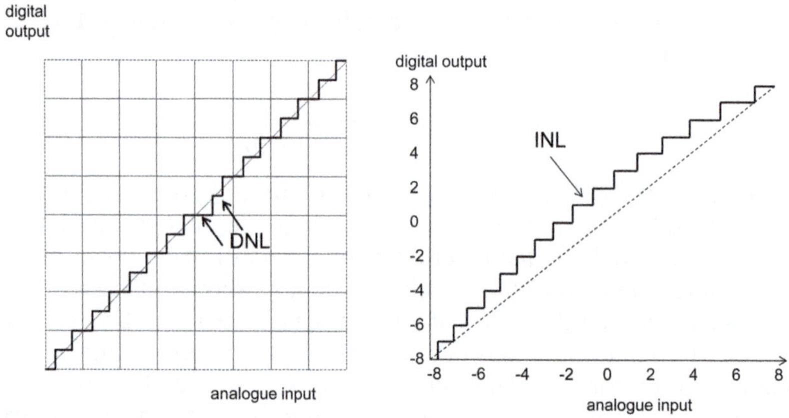

Figure 7.16 (left) The differential non-linearity, and (right) the integrated non-linearity of an ADC.

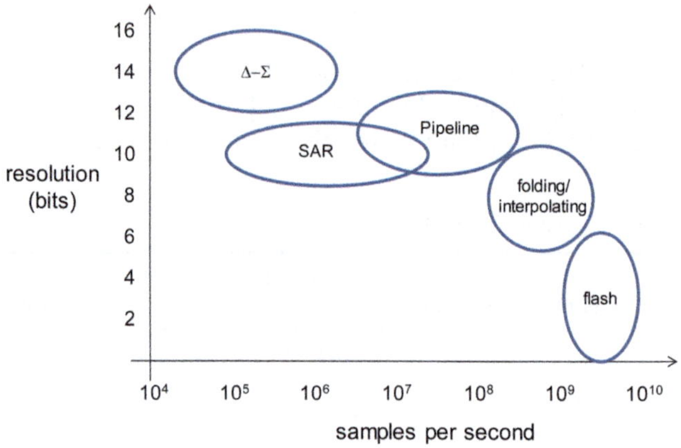

Figure 7.17 Approximate relationship between resolution, samples per second, and the architecture of different types of ADC.

ideal ADC, the output is divided into 2*N* uniform steps, each with a constant width. Any deviation from the ideal step width is called differential non- linearity (DNL) and is measured in number of LSBs. For an ideal ADC, the DNL is 0 LSB. INL is a measure of how closely the ADC output matches its ideal response. INL can be defined as the deviation in LSBs of the actual transfer function of the ADC from the ideal transfer curve.

In general, there is a trade-off between the resolution of the ADC and the maximum sampling rate, as shown in Figure 7.17.

The most commonly used ADC for MR applications is the Δ–Σ. Oversampling, decimation, noise shaping and digital filters are also integrated into a standard Δ–Σ package. In front of the ADC itself is a low-pass filter, as shown in Figure 7.18. Using the example introduced in Section 7.6.2 of using a

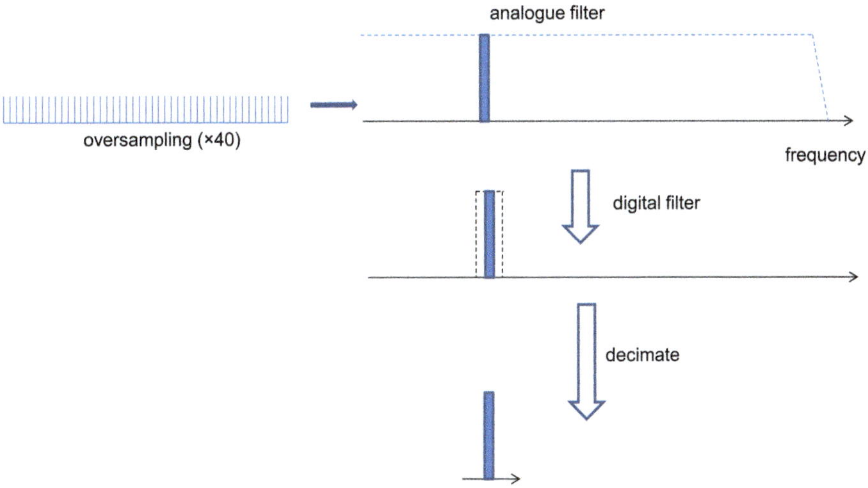

Figure 7.18 Illustration of oversampling, digital filtering and decimation.

40 MHz sampling rate ADC for a 1 MHz bandwidth MR signal centred at 64 MHz, one can see that the requirements for the low-pass filter in front of the ADC are relatively benign. This is one of the major advantages of "oversampling", *i.e.* sampling a larger bandwidth than is occupied by the actual MR signals.

A digital filter is then applied to the sampled data to select only the bandwidth of interest: a digital filter can be designed to have an extremely sharp transition.[16] At this stage, the filtered data has N-times as many data points (where N is the oversampling factor) as would have been acquired without oversampling, and the final step is to "decimate" the data, in which successive data points are averaged together, as shown in Figure 7.18. Since the quantization error in alternate data points is assumed to be random, the quantization noise in the decimated data set is reduced with respect to that of a Nyquist sampled signal by a factor of $\sqrt{M}$, where M is the oversampling factor, *i.e.*

$$e^2_{\text{rms,oversampled}} = \frac{\Delta^2}{12\sqrt{M}} \tag{7.13}$$

In other words, for every factor-of-four in oversampling, the equivalent resolution of the ADC increases by 1-bit.

Digital filters are generally programmed in dedicated digital signal processors (DSPs). These DSPs use weighted averaging of the data input that is shifted over a certain time kernel. The frequency response of the digital filter depends on the weighting factors and time kernel, both of which can be set as a controllable parameter to the DSP. Many different sorts of filter types can be defined, of which Butterworth and Chebychev are most commonly used. Depending on the acceptable level of time response and frequency response,

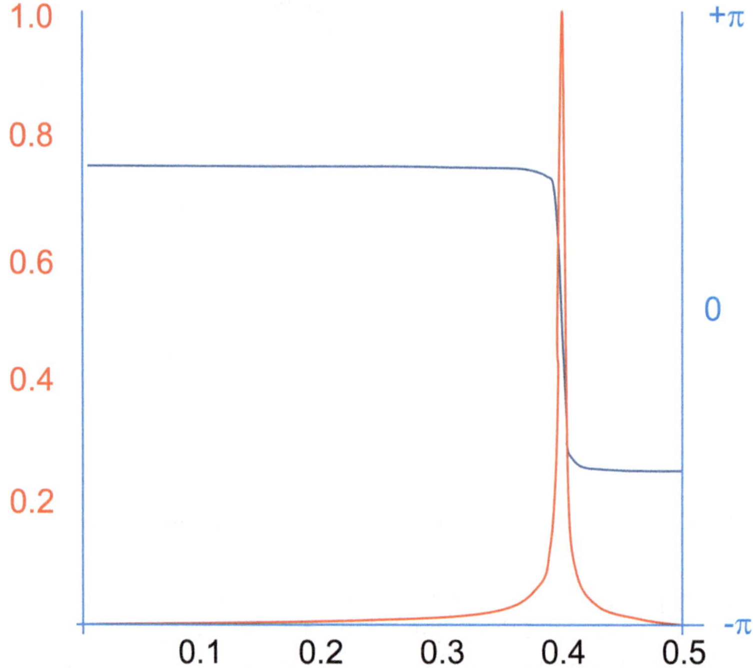

Figure 7.19 Example of the frequency response function of the simple digital filter from eqn (7.14). Both amplitude (left axis) and phase (right axis) are plotted over the relative frequency range compared to the sample frequency. With a 40 MHz sample frequency, the filter selects the 24 MHz band that matches to the aliased frequency of 64 MHz on a 1.5 T MR system.

the order and type of the filter can be chosen. A simple first order Butterworth filter for proton imaging at 1.5 T with a 100 kHz bandwidth and an ADC with 40 MHz sampling frequency can be calculated to be of the form:

$$y[n] = -x[n-2] + x[n] - 0.984y[n-2] - 1.605y[n-1] \tag{7.14}$$

where $x[n]$ is the input to the DSP, $y[n]$ the output, and n the index of the data obtained in time. The frequency response of this filter is shown in Figure 7.19. Note that the DSPs in MR systems in practice use filters with much higher orders than represented in eqn (7.14).

7.8 Optical and Wireless Data Transmission

As mentioned many times, MRI systems are generally equipped with multiple receive coils and multiple receiver channels. With an ever-increasing number of receiver coils and independent receivers, the RF infrastructure of the receiver chain becomes very complicated. Crosstalk between receive channels degrades some of the benefits of parallel acquisition (*i.e.* in terms of acceleration factors in data acquisition and SNR) or can even cause oscillations in the amplifiers in each channel. Crosstalk can be reduced with well-shielded coaxial cables and plugs, but nevertheless the issue of available space within the magnet bore with the patient in place remains. The optimal solution would be to amplify, filter and digitize the signal from each coil as close to that coil as possible, and then to transfer the signal out of the magnet either *via* thin optical fiber[17] or wirelessly. Currently, most commercial systems have fewer receiver channels than the number of receiver coils, which means that the signals from multiple coils must be combined for each receive channel. Direct digitization would also remove this restriction. If digitization is to be performed separately for every coil in an array, then some form of signal lock is needed to tie together each of the separate local clock oscillators that are used for each receiver. The digital system can be used to send signals from a single central system clock to phase-lock each of the local receivers together: this "pilot tone" signal can either be optical or electrical. The digital link can also be used to send signals to tune or detune the coils, again either electrically or optically. The easiest method to convert from electrical to optical signals is to use a directly modulated semiconductor laser diode, in which the light intensity is modulated by the current at its input. Alternative methods for high data rates include incorporating a Mach–Zehnder optical modulator or similar device. The bandwidth of optical data transfer can be very high. MR signals from many receiver elements can be transported over a single fiber optic line particularly when digital filtering or decimation is incorporated prior to data transmission, *i.e.* the ADC is situated very close to the coil itself.

The alternative strategy of wireless transmission has been shown to be feasible using a digital wireless transmission link based on 802.11b.[18] There are 14 standard channels defined for 802.11b from 2.400 to 2.487 GHz. All channels are evenly spaced at 5 MHz intervals and each channel has a bandwidth of 22 MHz. To transmit many MRI signals simultaneously, frequency division multiplexing (FDM) has to be applied, which makes it possible for each transmission channel defined by 802.11b to carry multi-channel MR signals modulated at different frequencies. A differential quadrature phase shift keyed (DQPSK) modulation method has been shown to be capable of data transmission rates of 2 megabytes per second and a bandwidth of 2 MHz. Despite a significant amount of work in the period 2000–2010, this technology has not been widely adopted for commercial MRI systems, with the balance swinging heavily in favour of fibre-optic data transfer.

References

1. D. I. Hoult, *Prog. Nucl. Magn. Reson. Spectrosc.*, 1978, **12**, 41.
2. E. Kupce and L. E. Kay, *J. Biomol. NMR*, 2012, **54**, 1.
3. J. B. Johnson, *Phys. Rev.*, 1928, **32**, 97.
4. D. I. Hoult and R. E. Richards, *J. Magn. Reson.*, 1976, **24**, 71.

5. M. A. Elliott, E. K. Insko, R. L. Greenman and J. S. Leigh, *J. Magn. Reson.*, 1998, **130**, 300.
6. D. I. Hoult, *J. Magn. Reson.*, 1984, **57**, 394.
7. I. J. Lowe and C. E. Tarr, *J. Phys. E: Sci. Instrum.*, 1968, **1**, 320.
8. J. R. Garbow, C. Mcintosh and M. S. Conradi, *Concepts Magn. Reson., Part B*, 2008, **33B**, 252.
9. R. D. Watkins, R. H. Caverly and W. E. Doherty, *Proc. Int. Soc. Magn. Reson. Med.*, 2012, 2686.
10. A. Reykowski, S. M. Wright and J. R. Porter, *Magn. Reson. Med.*, 1995, **33**, 848.
11. X. M. Cao, D. L. Zu, X. N. Zhao, Y. Fan and J. H. Gao, *Sci. China: Technol. Sci.*, 2011, **54**, 1766.
12. J. M. Rollett, *IRE Trans. Circuit Theory*, 1962, **9**, 29.
13. G. Giovannetti, V. Hartwig, V. Viti, G. Gaeta, R. Francesconi, L. Landini and A. Benassi, *Concepts Magn. Reson., Part B*, 2006, **29B**, 107.
14. P. Perez and A. Santos, *Med. Eng. Phys.*, 2004, **26**, 523.
15. P. Perez, A. Santos and J. J. Vaquero, *Magn. Reson. Mater. Phys.*, 2001, **13**, 109.
16. D. Moskau, *Concepts Magn. Reson.*, 2002, **15**, 164.
17. J. Yuan, J. Wei and G. X. Shen, *J. Magn. Reson.*, 2007, **189**, 130.
18. J. Wei, Z. G. Liu, Z. Chal, J. Yuan, J. Y. Lian and G. X. Shen, *J. Magn. Reson.*, 2007, **186**, 358.

CHAPTER 8

Electromagnetic Modelling

CHRISTOPHER M. COLLINS*[a], ANDREW G. WEBB[b], AND JAN PAŠKA[a]

[a]Center for Biomedical Imaging, Department of Radiology, New York University School of Medicine, 660 First Avenue, New York, NY 10016, USA; [b]C.J.Gorter Center for High Field MRI, Department of Radiology, Leiden University Medical Center, Leiden, The Netherlands
*E-mail: c.collins@nyumc.org

8.1 Introduction

As described in the preceding chapters, MRI involves the application of three different magnetic fields (static B_0, radiofrequency B_1 and gradient G) to the object being imaged. These fields interact in fundamentally different ways with materials, including biological tissue. Many of these interactions are necessary for the imaging process, such as the precessional motion described by the Bloch equations, which allows manipulation of magnetization as a function of time and space *via* RF pulses and gradients. Other interactions are less desirable, such as static and RF magnetic field inhomogeneities potentially resulting in image artifacts, and strong electric fields, which can produce tissue heating.

New Developments in NMR No. 7
Magnetic Resonance Technology: Hardware and System Component Design
Edited by Andrew G Webb

Published by the Royal Society of Chemistry, www.rsc.org

In the design process for MR hardware and associated software in terms of imaging protocols, calculations of the interactions between the applied fields and the human body can be performed to optimize data quality and understand image characteristics (including image contrast, SNR, and the origin of image artifacts). In addition, these calculations can be used to predict the behaviour of SNR as a function of B_0, and ensure the safety of the human subjects and imaging staff. Electromagnetic simulations for simple geometries and low magnetic fields can often be performed analytically, or at least semi-analytically, giving valuable insights into the imaging process. As operating frequencies increase, however, the quasi-static approximations become less valid and full numerical simulations become necessary. As an example, simulated RF magnetic fields from a birdcage coil at 4 tesla and 7 tesla loaded with the human head, are shown in Figure 8.1.[1]

This chapter gives an overview of the many purposes, methods, and applications of EM field calculations for MRI, with specific examples and numerous references to works in the literature.

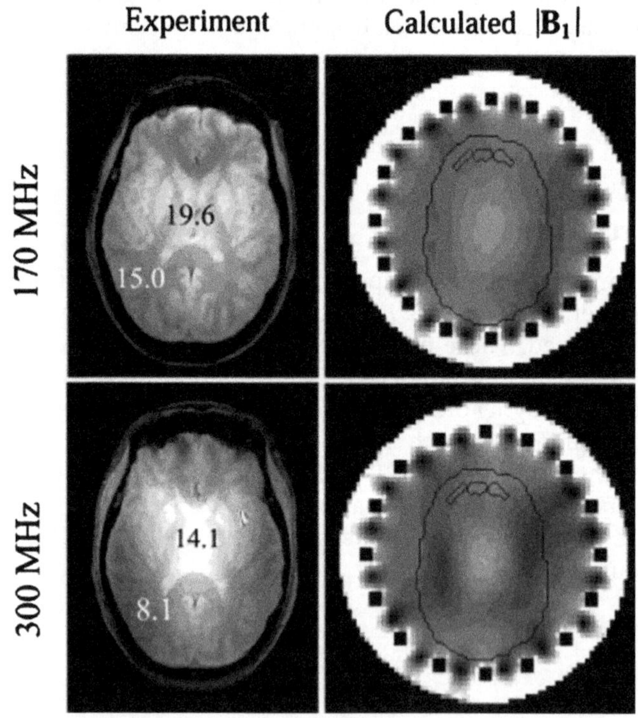

Figure 8.1 Experimental and calculated magnetic fields in the head at two different field strengths, 4 tesla (170 MHz) and 7 tesla (300 MHz). The field is significantly less homogeneous at the higher frequency owing to wavelength effects. Reproduced from ref. 1 with permission from Wiley © 2001 Wiley-Liss, Inc.

8.2 Simulating Electromagnetic Fields for Magnetic Resonance

To first gain some context for the requirements for calculations involving the different fields applied in MRI, consider the Maxwell equations, which can be written in differential or integral form, and contain everything needed for computing fields pertinent to MRI.

	Differential form	Integral form
Gauss' law	$\nabla\cdot\varepsilon\mathbf{E}=\rho_v$	$\oiint_s \varepsilon\mathbf{E}\cdot\mathrm{d}s=\iiint_v \rho_v\,\mathrm{d}v$
Gauss' law for magnetism	$\nabla\cdot\mathbf{B}=0$	$\oiint_s \mathbf{B}\cdot\mathrm{d}s=0$
Faraday's law	$\nabla\times\mathbf{E}=-\frac{\partial}{\partial t}\mathbf{B}$	$\oint_l \mathbf{E}\cdot\mathrm{d}l=-\iint_s \frac{\partial}{\partial t}\mathbf{B}\cdot\mathrm{d}s$
Ampere's law with Maxwell's addition	$\nabla\times\frac{1}{\mu}\mathbf{B}=\left(\sigma\mathbf{E}+\frac{\partial}{\partial t}\varepsilon\mathbf{E}\right)$	$\frac{1}{\mu}\oint_l \mathbf{B}\cdot\mathrm{d}l=\left(\iint_s \sigma\mathbf{E}\cdot\mathrm{d}s+\iint_s \frac{\partial}{\partial t}\varepsilon\mathbf{E}\cdot\mathrm{d}s\right)$

Here bold variables indicate three-dimensional vectors, **B** is the magnetic flux density, **E** the electrical field strength, l a line, s a surface, v a volume, t is time, ρ_v is electric charge density, and μ, σ, and ε are the material properties of magnetic susceptibility, electrical conductivity, and electric permittivity, respectively. μ and ε are often expressed as the products of their values in free space (μ_0 and ε_0) and their relative values in any given material (μ_r and ε_r).

These equations are hopefully somewhat familiar to the reader, even if only from an undergraduate physics course. Nonetheless, written all together they can be somewhat intimidating. Fortunately, in simulating any one of the fields applied in MRI and its interactions with tissue, typically only one sub-group of these equations needs to be considered. For example, in calculating the magnetic fields produced by electrical currents in the static-to-low radiofrequency range for the purpose of designing gradient coils and static magnets, all that is needed is Ampere's law without the time-varying term (Maxwell's addition) since wavelengths are all extremely long in comparison to the dimensions of the human body. Similarly, for calculating the perturbations of the B_0 field caused by the sample, only Gauss's law for magnetism is required.

In practice, different methods are used for simulating B_0, gradient, and RF magnetic (B_1) fields to the necessary degree of accuracy. In the following sections, the most common methods used for simulating these fields in the presence of the human body, or a representation of it, are described. Most attention is devoted to RF fields, which are the vast majority of simulations performed for MRI today.

8.2.1 Static Magnetic (B_0) Fields

Static magnetic fields, including the fields from the shim coils, have been simulated for the purposes of magnet design, predicting the effects of field/tissue interactions on MR images, and understanding the potential biological effects of these fields. The particular method used for simulating the B_0 field depends very much on the specific application.

As covered in Chapter 2, in simulations and calculations of the design of the magnet to produce the B_0 field it is common to ignore the effects of the sample on the field distribution and use only the Biot–Savart law or spherical harmonic expansion approach. The perturbations of the applied field caused by the presence of the sample are typically very small (on the order of a few parts per million) and have a different distribution for every sample or subject being imaged. Thus, while these perturbations can cause notable effects on some MR images, in magnet design it is often not practical or valuable to consider these effects explicitly.

The same is true for the design of conventional shim coils based on spherical harmonics, as covered in detail in Chapter 4. However, it was also shown that to estimate the number of shims (which varies with field strength, sample size and sample geometry) required for B_0 field homogenization, as well as to design new types of shim coils, it is necessary to be able to calculate B_0 field distributions in an arbitrary geometry with arbitrary magnetic susceptibility distribution: for this it is necessary to turn to numerical methods of solution. The fundamental equation is Gauss' law for magnetism with the definition $\mathbf{B} = \mu\mathbf{H}$. One approach to solving this equation for arbitrary objects involves defining a magnetic scalar potential φ such that $\mathbf{H} = -\nabla\varphi$. Gauss' law for magnetism can be written as:

$$\nabla \cdot (\mu_r \nabla \varphi) = 0 \tag{8.1}$$

To solve this with finite difference approximations for an arbitrary spatial distribution of μ_r starting with a reasonable initial guess for the spatial distribution of φ and assuming one eventually converges to a final solution for φ *via* iteration, one can write:

$$\frac{d\varphi}{dT} = \nabla \cdot (\mu_r \nabla \varphi) \tag{8.2}$$

where T is pseudo-time, increasing incrementally by ΔT with iteration number.

With finite difference approximations, a first derivative of some function f with respect to location (x, y, z) or time (t) is approximated as the difference in f over the corresponding difference in the other, *i.e.* the first derivative can be approximated as the rise over the run of a line over a region where the line is approximately straight. It is thus important to ensure that steps in space (Δx, Δy, and Δz) and time (ΔT) are suitably small.

With this knowledge and with some mathematical manipulation eqn (8.2) can be approximated as:[2,3]

$$\begin{aligned}
&\varphi(n+1)\\
&=\varphi+\Delta T\left(\frac{\mu_\mathrm{r}(i+1)-\mu_\mathrm{r}(i-1)}{2\Delta x}\frac{\varphi(i+1)-\varphi(i-1)}{2\Delta x}+\mu_\mathrm{r}\frac{\varphi(i+1)-2\varphi+\varphi(i-1)}{\Delta x^2}\right)\\
&+\Delta T\left(\frac{\mu_\mathrm{r}(j+1)-\mu_\mathrm{r}(j-1)}{2\Delta y}\frac{\varphi(j+1)-\varphi(j-1)}{2\Delta y}+\mu_\mathrm{r}\frac{\varphi(j+1)-2\varphi+\varphi(j-1)}{\Delta y^2}\right)\\
&+\Delta T\left(\frac{\mu_\mathrm{r}(k+1)-\mu_\mathrm{r}(k-1)}{2\Delta z}\frac{\varphi(k+1)-\varphi(k-1)}{2\Delta z}+\mu_\mathrm{r}\frac{\varphi(k+1)-2\varphi+\varphi(k-1)}{\Delta z^2}\right)
\end{aligned} \tag{8.3}$$

where i, j, k, and n are indices in x, y, z, and T, respectively, and where only deviations from (i, j, k, n) are shown in the parenthesis following each occurrence of μ_r and φ. Given an initial guess for φ and a pre-defined arbitrary distribution for μ_r, all values on the right hand side of the equation are known for all interior locations on the grid, and we need merely to step through all locations in space and forward in pseudo-time performing algebraic operations until φ no longer changes with T at each location. Eqn (8.3) can be simplified considerably for efficient computation, but is written here for ease of derivation from eqn (8.2). For MRI with only diamagnetic and paramagnetic materials (including air and all biological tissues), a reasonable first guess for φ corresponds to a perfectly homogeneous **B** field, or a perfect one-dimensional gradient in φ. Accurate results with a reasonable grid size may require additional steps, but it should be possible to see here how, with use of finite difference approximations, an equation involving vector calculus can be approximated as one requiring only algebra that can be solved readily with a computer. For reference, biological tissues tend to be slightly diamagnetic, with μ_r having values between that of a vacuum (exactly 1.0) and water (0.9999909), and air is slightly paramagnetic, with μ_r of 1.0000004. Also, while much faster methods of solving for static magnetic field distributions have been developed with use of Fourier-transform-based methods, as discussed in Chapter 4,[4] for the purposes of this chapter introducing a finite difference method to solve the functional equations directly is more instructive.

Importantly, in MRI we are interested not in the magnetic flux density (**B**) experienced in the bulk material but in that experienced by the nuclei, $\mathbf{B}_\mathrm{nuc}$. Calculation of this quantity from the bulk magnetic susceptibility requires the Lorentz correction.[5]

$$\mathbf{B}_\mathrm{nuc}=\mathbf{B}\left(1-\frac{2}{3}(\mu_\mathrm{r}-1)\right) \tag{8.4}$$

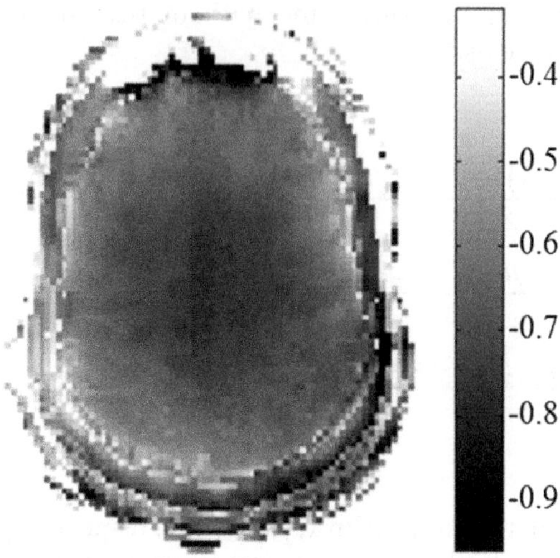

Figure 8.2 Numerically calculated B_0 field magnitudes in ppm deviation from applied field on an axial plane. Deviations occur owing to difference in susceptibility between materials including a variety of tissues and air. Reproduced from C. M. Collins, B. Yang, Q. X. Yang and M. B. Smith, Numerical calculations of the static magnetic field in three-dimensional multi-tissue models of the human head, *Magn. Reson. Imaging*, 2002, **20**, 413–424. Copyright (2002) with permission from Elsevier.[3]

As an example, Figure 8.2 shows a numerically calculated B_0 field in the brain at a static magnetic field of 7 tesla.[3]

8.2.2 Switched Gradient Fields (G_x, G_y, G_z)

As with the design of homogeneous B_0 fields, the presence of the human body has negligible effect on the gradient magnetic fields (which are much weaker than $\mathbf{B}_0$) and the switching frequencies are so low that fields throughout the imaging region are in the quasi-static regime, with the result that evaluation of the fields produced by gradient coil designs can be performed with the Biot–Savart law, as covered in Chapter 5.

As will be discussed later in this chapter, the gradient-induced electrical currents in the subject and the forces/vibrations resulting from gradient coil operation in the B_0 field are very important for the safety and comfort of the patient. Numerical simulations have been used to calculate the electric fields and currents through the human body,[6,7] and to calculate the acoustic vibrations caused by pulsing gradient coils. For calculating electrical currents induced in the subject, one approach is to use finite difference formulations of a simplified set of Maxwell's curl equations that does not include the time-dependent displacement term in the modified Ampere's law. Another approach is to recognize that, as long as wavelengths are always much longer than the dimensions of the problem region so that this displacement contribution is negligible, simulation results calculated at one frequency can be scaled to make predictions at another frequency with the assumption from

Faraday's law that ‖**E**‖/‖**B**‖ will be proportional to the frequency of excitation. This is especially useful when the Finite Difference Time Domain (FDTD) method (described in the next section) is used, because with FDTD the steady state solution is usually achieved more rapidly at a high frequency than at a low one. This approach to determining gradient-induced electrical fields and currents in complicated samples is called "frequency scaling".

8.2.3 Radiofrequency Magnetic (B_1) Fields

As covered in Chapter 1, all magnetic resonance experiments require an RF magnetic (B_1) field to be applied to the sample to excite the nuclei to produce coherent precessing magnetization. The RF field typically has a frequency in the hundreds of MHz range, and from Faraday's law, the rapid change of the RF field with respect to time is associated with significant electric fields. During transmission of the RF pulses these electrical fields can cause tissue heating. During signal reception these electric fields in conductive materials couple to random thermal motion of charges, such as the physical translation of ions, rotations of dipoles, and other effects, resulting in noise. There is a wide range of methods to calculate the RF fields (both magnetic and electric) in MR. Here, a summary of the most common methods is presented with references to MR-relevant simulations in the literature.

8.2.3.1 Analytically Based Methods

As in other fields, much insight in MR can be gained from the use of analytical treatments of important problems, even if they provide a limited representation of reality. Analytical methods of approximating the RF fields in MR have been very useful in exploring, for example, the trends in tissue heating and SNR with increased B_0 field strength and the corresponding increase in B_1 frequency.[8–16]

At B_0 field strengths where the B_1 frequency is low enough that the electrical wavelength in tissue is very large compared to the imaging region, simple quasi-static methods, including the Biot–Savart law, can be used to model magnetic and electric fields. Determining when this criterion is valid simply involves consideration of the relevant electrical wavelength. For example, the wavelength in free space at 300 MHz is almost exactly 1 metre. For MRI applications, tissues within the human body have a range of relative permittivity values, with the highest ones approaching that of water, approximately 78 across the spectrum of frequencies used in MRI of human subjects. As outlined in Chapter 3, the wavelength is inversely proportional to the square root of the relative permittivity, ε_r:

$$\lambda = \frac{v}{f\sqrt{\varepsilon_r}} \tag{8.5}$$

For example, the smallest wavelength we might expect in the human body at 21 MHz (*i.e.* a 0.5 T MRI system) is ~162 cm (in practice the wavelength is longer since most human tissues have a permittivity lower than that of water). As a rule-of-thumb one can say that the quasi-static regime works reasonably well for all MRI up to 1.5 tesla, and for head and extremity imaging up to 3 tesla. It starts to break down for abdominal imaging at 3 tesla or neuroimaging at 7 tesla. In terms of high resolution NMR, the maximum sample sizes are tubes with 20 mm diameter, and therefore even with the highest current operating frequencies ~1 GHz, quasi-static approximations are still very accurate.

Examples of quasi-static methods for RF field calculation include the Biot–Savart law for coil design,[17,18] and estimation of the behaviour of absorbed power with respect to sample size and B_1 frequency up to 3 T.[8–10] The results from the latter investigations led to predictions that SNR should increase roughly linearly with the B_0 field strength (covered in more detail later in this chapter) and that SAR should increase approximately as the square of the B_0 field strength. For power absorbed by the sample as a function of B_0 field strength in the quasi-static regime, Faraday's law dictates that the E-field amplitudes increase proportionally to the B_1 frequency (and thus proportionally to the B_0 field strength). The total power dissipated in the sample, P_{diss}, is

$$P_{\text{diss}} = \int_{v} \sigma \hat{\mathbf{E}}^{H} \hat{\mathbf{E}} \, \mathrm{d}v \tag{8.6}$$

where $\hat{\mathbf{E}}$ is a 3×1 column vector, whose complex components are the phasors of the three orthogonal components of the electric field (E_x, E_y and E_z), and the superscript H denotes the complex conjugate transpose of the vector. Eqn (8.6) leads to the prediction that the total power dissipated increases as the square of the B_0 field strength for a given sample and coil in the quasi-static regime.

More advanced analytically-based methods also find great utility in the quasi-static regime.[15–18] For example, with the use of Dyadic Green's functions, it is possible to find the ultimate SNR and lowest possible SAR as well as their corresponding current patterns, *i.e.* the corresponding RF coil geometry, or one can also compare different RF coil designs in terms of their SNR and SAR in specific locations or regions within spherical or cylindrical samples, which can be treated as very simple models for the human head and torso, respectively.[17,18]

When quasi-static approximations cannot be applied, then sophisticated numerical simulation tools must be employed. From an MRI perspective, this occurs when the presence of the body changes the distributions of the magnetic and electric fields from those that are produced in an empty coil. Since the body contains many complicated structures with different relative permittivity and conductivity values, one of the most important aspects is to have an accurate body model available for the simulations. The three most common simulation methods are detailed in the following sections.

8.2.3.2 Finite Difference Time Domain (FDTD) Method

The FDTD method[21] gained popularity in MRI research and safety assurance when voxel-based high-resolution computer models of the human body first became available several years ago. Most commonly the method involves representation of the imaging subject as well as the RF coil(s) on a rectilinear grid, though grid dimensions can often vary considerably through space. This so-called "adaptive gridding" allows areas containing very fine structures to be gridded more finely than other areas that have only relatively coarse structures.

An understanding of how the method works can be gained by considering Maxwell's curl equations in differential form, but written with the time derivative on the left hand side, *i.e.*

$$\begin{aligned}\frac{\partial \mathbf{B}}{\partial t} &= -\nabla \times \mathbf{E}\\ \frac{\partial \varepsilon \mathbf{E}}{\partial t} &= -\sigma \mathbf{E} + \frac{\nabla \times \mathbf{B}}{\mu}\end{aligned} \tag{8.7}$$

Rewriting these two equations in terms of their orthogonal components in rectilinear equations and applying the curl operators yields:

$$\begin{aligned}&\frac{\partial}{\partial t}\left(B_x\mathbf{i} + B_y\mathbf{j} + B_z\mathbf{k}\right) =\\ &-\left[\left(\frac{\partial E_z}{\partial y} - \frac{\partial E_y}{\partial z}\right)\mathbf{i} + \left(\frac{\partial E_x}{\partial z} - \frac{\partial E_z}{\partial x}\right)\mathbf{j} + \left(\frac{\partial E_y}{\partial x} - \frac{\partial E_x}{\partial y}\right)\mathbf{k}\right]\\ &\frac{\partial}{\partial t}\varepsilon(E_x\mathbf{i} + E_y\mathbf{j} + E_z\mathbf{k}) = -\sigma(E_x\mathbf{i} + E_y\mathbf{j} + E_z\mathbf{k})\\ &+\frac{1}{\mu}\left[\left(\frac{\partial B_z}{\partial y} - \frac{\partial B_y}{\partial z}\right)\mathbf{i} + \left(\frac{\partial B_x}{\partial z} - \frac{\partial B_z}{\partial x}\right)\mathbf{j} + \left(\frac{\partial B_y}{\partial x} - \frac{\partial B_x}{\partial y}\right)\mathbf{k}\right]\end{aligned} \tag{8.8}$$

By separating terms according to the orthogonal components, both parts of eqn (8.8) can now each be separated into three equations, producing:

$$\begin{aligned}\frac{\partial}{\partial t}B_x &= -\left(\frac{\partial E_z}{\partial y} - \frac{\partial E_y}{\partial z}\right)\\ \frac{\partial}{\partial t}B_y &= -\left(\frac{\partial E_x}{\partial z} - \frac{\partial E_z}{\partial x}\right)\\ \frac{\partial}{\partial t}B_z &= -\left(\frac{\partial E_y}{\partial x} - \frac{\partial E_x}{\partial y}\right)\\ \frac{\partial}{\partial t}\varepsilon E_x &= -\sigma E_x + \frac{1}{\mu}\left(\frac{\partial B_z}{\partial y} - \frac{\partial B_y}{\partial z}\right)\\ \frac{\partial}{\partial t}\varepsilon E_y &= -\sigma E_y + \frac{1}{\mu}\left(\frac{\partial B_x}{\partial z} - \frac{\partial B_z}{\partial x}\right)\\ \frac{\partial}{\partial t}\varepsilon E_z &= -\sigma E_z + \frac{1}{\mu}\left(\frac{\partial B_y}{\partial x} - \frac{\partial B_x}{\partial y}\right)\end{aligned} \tag{8.9}$$

These six equations are each composed of scalar components and first derivatives with respect to space and time. Before applying finite difference

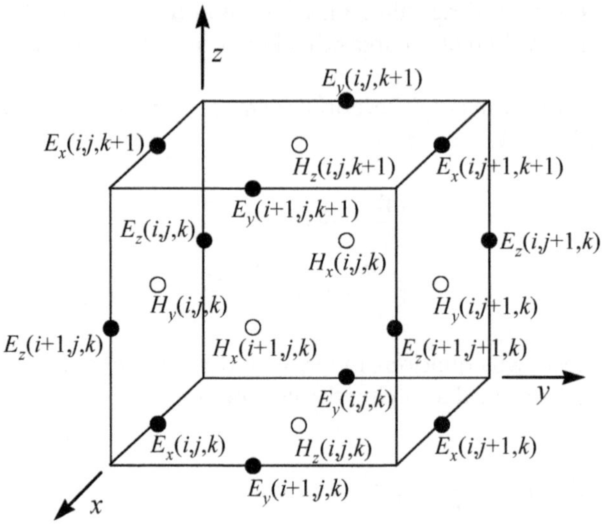

Figure 8.3 Yee cell: strategic staggering of reference locations for electric and magnetic field components to maximize the accuracy for a given grid resolution with the FDTD method.

approximations, it is useful to take advantage of the symmetry of the system by strategically staggering reference locations for electrical and magnetic fields as defined by the Yee cell[19] in order to maximize numerical accuracy for a given grid resolution. The structure of the Yee cell is shown in Figure 8.3.

This method of staggering in space plus staggering reference times allows for central difference approximations to the first derivatives to be taken with one-half the time step-size and grid spacing that would be otherwise possible. Substituting these central difference approximations for the first and fourth of the six separate equations (8.9) gives:

$$\frac{B_x\left(i,j,k,n+\frac{1}{2}\right)-B_x\left(i,j,k,n-\frac{1}{2}\right)}{\Delta_t}=\frac{E_z(i,j+1,k,n)-E_z(i,j,k,n)}{\Delta_y}-\frac{E_y(i,j,k+1,n)-E_y(i,j,k,n)}{\Delta_z} \tag{8.10}$$

and

$$\frac{\varepsilon E_x(i,j,k,n+1)-\varepsilon E_x(i,j,k,n)}{\Delta_t}=-\sigma E_x(i,j,k,n)+\frac{1}{\mu}\left(\frac{B_z\left(i,j,k,n+\frac{1}{2}\right)-B_z\left(i,j-1,k,n+\frac{1}{2}\right)}{\Delta_y}-\frac{B_y\left(i,j,k,n+\frac{1}{2}\right)-B_y\left(i,j,k-1,n+\frac{1}{2}\right)}{\Delta_z}\right) \tag{8.11}$$

By starting with no fields anywhere at time zero and assigning known excitations at specific locations through time, it is possible to rewrite these two and the other four equations so that there is only one unknown—the value at a future time point, *e.g.*:

$$B_x\left(n+\frac{1}{2}\right)=B_x\left(n-\frac{1}{2}\right)+\Delta_t\left(\frac{E_z(j+1)-E_z}{\Delta_y}-\frac{E_y(k+1)-E_y}{\Delta_z}\right) \tag{8.12}$$

and after solving for B_x, B_y, and B_z over all space at time step $n + 1/2$ the *E*-field components at the time step $n + 1$ can be calculated as:

$$\varepsilon E_x(n+1)=\varepsilon E_x + \Delta_t\left[-\sigma E_x+\frac{1}{\mu}\left(\frac{B_z\left(n+\frac{1}{2}\right)-B_z\left(j-1,n+\frac{1}{2}\right)}{\Delta_y}-\frac{B_y\left(n+\frac{1}{2}\right)-B_y\left(k-1,n+\frac{1}{2}\right)}{\Delta_z}\right)\right] \tag{8.13}$$

As before, only departures from (i, j, k, n) are shown here in the parenthesis following the *E* and *B* field components. Using this method, all electromagnetic fields as a function of time and space can be calculated for arbitrary spatial distributions of different materials and arbitrary excitation time courses.

FDTD is an intuitive technique, easy to use and understand, which allows a natural and easy modelling of materials. This is reflected in the fact that most available numerical models of the human body are essentially designed for the FDTD method, as shown in Figure 8.4.

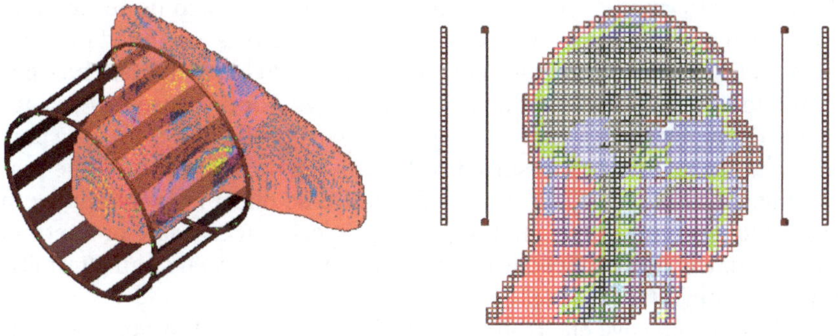

Figure 8.4 (left) Simulation set up for analyzing the EM fields produced by a birdcage coil in the human head. (right) Details of the rectilinear meshing (isotropic in this case) through one slice of the human head. The different colours indicate different tissues, each of which is specified in terms of a relative permittivity and conductivity.

A very fine discretization of the numerical domain is possible; nowadays numerical domains with up to 100 million mesh cells can be solved with graphical processor unit (GPU)-based accelerator technology. The EM fields at multiple frequencies can be solved with only one simulation run, which is useful for the design and evaluation of, for example, double-tuned RF coils and designs of coils that have multiple resonances. With the FDTD method, the most common method of evaluating the frequency response of a structure is to first excite the structure with a brief excitation pulse (often Gauss shaped) and then perform a Fourier Transform of the time-domain response.

On the other hand, FDTD has difficulties in simulating resonant structures with a high quality factor, such as unloaded RF coils. Imperfect boundary conditions can result in more energy remaining in the computational domain than should occur, leading to long computation times, a poor power balance, or even numerical instabilities and erroneous solutions. Geometrical features that do not lie along Cartesian directions cannot be modelled accurately by a rectilinear grid (an effect called "stair-casing"), and a fine discretization in a particular part of a numerical domain leads to a very large number of grid cells throughout, unless advanced sub-gridding techniques are used.

8.2.3.3 Finite Element Method (FEM)

The finite element method (FEM) is a numerical method that approximates the solution to boundary value problems, *i.e.* those that are defined both by a differential equation:

$$L\varphi = f \tag{8.14}$$

and also by specific boundary conditions with respect to the computational domain. Here L is a linear differential operator, φ the unknown function, and f the excitation function. FEM has been used since the 1940s and has a wide range of applications in engineering and mathematics. In FEM the numerical domain is truncated by appropriate boundary conditions, and subdivided into smaller, simple shaped, subdomains, called finite elements. Within each of the subdomains the unknown quantity is expanded in simple basis functions (usually polynomials) with unknown coefficients. The Galerkin method of weighted residuals, or the Ritz method, is used to set up a system of equations. The resulting system matrix, which is a square sparse matrix, has to be inverted to solve for the unknown coefficients.[20] The main advantage of FEM over FDTD is that the staircase effect is overcome by the use of an unstructured tetrahedral (usually) mesh. An example of a tetrahedral mesh of the human head is shown in Figure 8.5, as well as an alternative hexahedral meshing scheme.[21]

In 3D FEM full wave electromagnetic simulations the EM fields are typically solved in the frequency domain. To reduce the number of unknowns, the vector wave equation for the E-field is used to set up the system equation. After solving for the electric field the magnetic field can be obtained

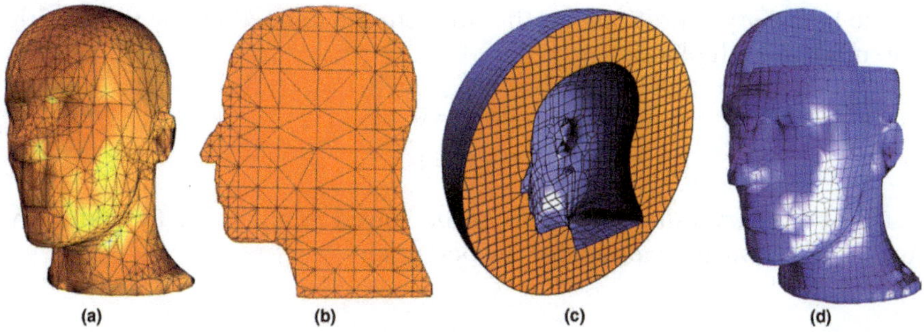

Figure 8.5 Tetrahedral/hexahedral meshes of the human head model: (a and b) show adaptive tetrahedral meshes and (c and d) show the hexahedral meshes. (a) The tetrahedralization of the volume inside the head; (b) a cross-section of (a); (c) the hexahedralization of the volume between the human head and a sphere boundary; (d) the hexahedralization of the volume inside the head. Reproduced from Y. Zhang, C. Bajaj and B.-S. Sohn, 3D finite element meshing from imaging data, *Comput. Meth. Appl. Mech. Eng.*, 2005, **194**, 5083–5106. Copyright (2005) with permission from Elsevier.[21]

by Faraday's law. The basis functions are usually tetrahedral shapes of varying size, which are well-suited to model complex structures and materials, although currently there are not many human models that have been meshed in an appropriate way for FEM analysis. The simulation time for FEM frequency domain solvers is independent of the number of excitations, which is beneficial for coil arrays with a large number of channels. In most EM simulations for MRI the EM behaviour of the coil array is of interest only at the Larmor frequency, and therefore a frequency sweep is not needed. A disadvantage of the FEM is the high memory requirement owing to the required inversion of the large system matrix. Nevertheless, nowadays modern PCs can handle the computational burden of simulating a human body model with a fine enough discretization.

To understand the concept of the finite element method a simple one dimensional example is given to solve for the scalar electric potential φ between the plates of a parallel-plate capacitor. The governing differential equation is the Poisson equation:

$$-\Delta(\varepsilon\Delta\varphi) = \rho \tag{8.15}$$

where ε is the permittivity of the material and ρ is the linear charge density. The capacitor is modeled as two infinite, perfectly conducting parallel planes. The dielectric in between the parallel plates is constant and the material has a varying linear charge density of $\rho = 3x\varepsilon$. The plates are separated by a distance of 1 m, which yields:

$$-\frac{\partial^2\varphi}{\partial x^2} = 3x \quad x \in [0,1] \tag{8.16}$$

The boundary conditions for the problem are defined by the value of the potential on both plates:

$$\begin{aligned}\varphi(0)&=0\\ \varphi(1)&=1\end{aligned} \tag{8.17}$$

This boundary value problem can be solved analytically. By taking the Ansatz of a cubic polynomial for the potential:

$$\varphi(x)=-\frac{1}{2}x^3+\frac{3}{2}x \tag{8.18}$$

This exact solution can be used to validate the numerical approximation using the FEM. One can rewrite the problem in the weak form by multiplying the equation with an arbitrary test-function v and integrating over the whole domain:

$$\int_0^1\left(-\frac{\partial^2\varphi}{\partial x^2}-3x\right)v\,\mathrm{d}x=0 \tag{8.19}$$

Integration by parts yields:

$$\frac{\partial\varphi}{\partial x}v\bigg|_0^1-\int_0^1\frac{\partial\varphi}{\partial x}\frac{\partial v}{\partial x}\mathrm{d}x+\int_0^1 3xv\,\mathrm{d}x=0 \tag{8.20}$$

Dividing the domain into two subdomains of equal size, and defining basis functions v_i for each of the subdomains:

$$v_i=\begin{cases}\dfrac{x-x_{i-1}}{x_i-x_{i-1}} & x_{i-1}\le x\le x_i\\ \dfrac{x-x_{i+1}}{x_i-x_{i+1}} & x_i\le x\le x_{i+1}\\ 0 & \text{otherwise}\end{cases} \tag{8.21}$$

with $x_1 = 0$, $x_2 = 0.5$, and $x_3 = 1$. Now, the unknown potential φ can be approximated by $\hat{\varphi}$ as a superposition of the basis functions in each of the subdomains with unknown coefficients c_i:

$$\hat{\varphi}(x)=\sum_{i=2}^{3}c_iv_i(x) \tag{8.22}$$

The basis function v_1 can be excluded, owing to the boundary condition $\varphi(0) = 0$. Therefore, only the basis functions v_2 and v_3 need to be considered: these are plotted in Figure 8.6.

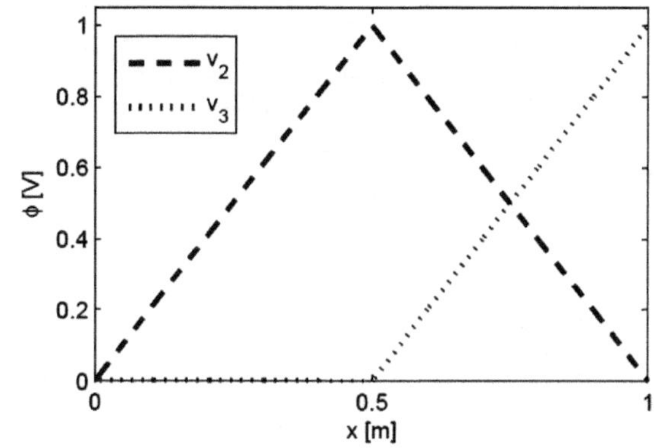

Figure 8.6 Basis functions used in the FEM simulation of the scalar potential for a parallel-plate capacitor.

To solve for the unknown coefficients c_i, eqn (8.21) and (8.22) are inserted into eqn (8.19) to produce the set of equations:

$$\begin{aligned}\int_0^1\left(-\frac{\partial^2\hat{\varphi}}{\partial x^2}-3x\right)v_2\,\mathrm{d}x=-4c_1+2c_2+\frac{3}{4}=0\\ \int_0^1\left(-\frac{\partial^2\hat{\varphi}}{\partial x^2}-3x\right)v_3\,\mathrm{d}x=2c_1-2c_2+\frac{5}{8}=0\end{aligned} \tag{8.23}$$

One can now rewrite this set of equations as a matrix equation:

$$\begin{bmatrix}-4 & 2\\ 2 & 2\end{bmatrix}\begin{bmatrix}c_2\\ c_3\end{bmatrix}=\begin{bmatrix}-\frac{3}{4}\\ -\frac{5}{8}\end{bmatrix} \tag{8.24}$$

Now we can solve for the unknown coefficients by inverting the matrix to obtain $c_2 = 11/16$ and $c_3 = 1$. The exact solution φ and the numerical approximation $\hat{\varphi}$ are plotted in Figure 8.7.

8.2.3.4 *Method of Moments*

The Method of Moments (MoM) has been used since the 1960s to solve electromagnetic problems in the frequency domain in free space. The unknown quantity in the MoM is the surface current present on metallic surfaces. Homogeneous dielectric materials are modelled by the "Surface Equivalence Principle" as electric and magnetic equivalent surface currents enclosing the dielectric. In this way, inhomogeneous dielectric objects can be modelled by defining a magnetic and electric surface current on the dielectric boundaries. The dielectric boundaries and metallic surfaces are discretized in the form of surface triangles. The electric and magnetic surface currents are

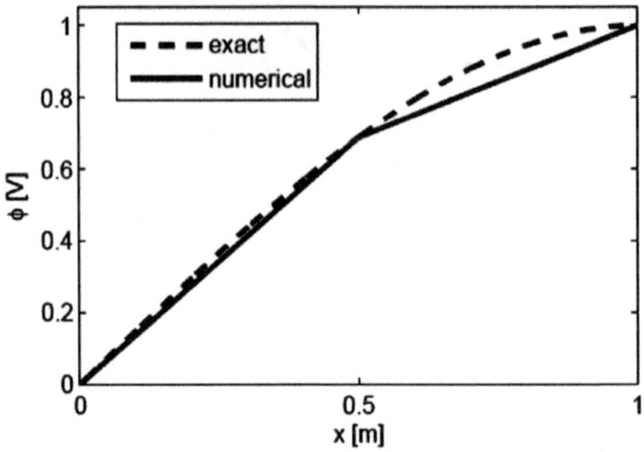

Figure 8.7 Exact solution and numerical FEM-based approximation of the parallel-plate capacitor example, showing the very close agreement between the two approaches.

then developed in basis functions with unknown coefficients within the surface triangles. The Electric Field Integral Equation (EFIE) is written in terms of these discretized surface currents. The equation is then weighted by the basis functions to obtain a set of equations. The resulting system matrix is a full matrix that has to be inverted to solve for the unknown coefficients of the basis functions for the electric and magnetic currents. A more detailed description of the method can be found elsewhere.[22]

MoM is a very powerful simulation tool for the numerical analysis of antenna structures in free space. In MRI it can be used for the simulation of RF coil arrays when unloaded or loaded by simple homogeneous phantoms. The main advantages of MoM over the FDTD and FEM methods are that it is not necessary to discretize the free-space region that surrounds the coil/shield, which reduces the number of unknowns, and since it uses a free-space Green's function it does not require any mesh truncation using perfectly absorbing boundaries, and so does not suffer from truncation errors. However, the MoM is intrinsically not very well-suited for the simulation of large inhomogeneous dielectrics, such as the human body. The complicated geometry and the large surface of dielectric boundaries results in an extremely large system matrix that is not possible to solve on current computers. However, as covered in the next section, for the simulation of the human body, hybrid simulation methods can be used, in which the MoM is used to model the RF coil and FDTD or FEM are used to calculate the EM fields inside the human body model.

8.2.3.5 Hybrid Simulation Methods

As described previously, the FDTD method has the advantage of being able to represent a heterogeneous object in high resolution, but "staircasing effects" resulting from representing objects on a rectilinear grid can significantly reduce the accuracy when modelling conductive surfaces, such as RF coils, with arbitrary orientation. In contrast, other methods, such as FEM or MoM, are capable of modelling conductors with surfaces oriented in arbitrary directions and are thus preferable for accurate modelling of RF coils but require extensive memory and/or computational time if numerous different volume units are required—such as in the representation of the heterogeneous human body in high resolution.

Given the different strengths of the different methods in simulating RF fields for MRI, some research groups have combined methods to take advantage of their respective strengths in different portions of the problem. Often this involves using the FDTD method in the region of the complex human body and a method more suited for modeling conductors, such as MoM or FEM, in the region of the coil.[23,24] A common approach is to first perform a MoM or FEM simulation with a realistic coil and a homogeneous phantom with an external surface similar to the geometry of the body and with electrical properties representing a global average of the tissue permittivity and conductivity. Then the fields from the MoM or FEM simulation, especially at the surface of the body model, serve as the input for the FDTD simulation. An example of such an approach is shown in Figure 8.8.

8.2.3.6 Approaches Using Multiple "Ideal" Current or Voltage Sources

In the methods described above, the RF coil(s) can be excited or driven in a number of possible ways. In analytical methods, coils often have a defined spatially-dependent current distribution with no physical gaps corresponding

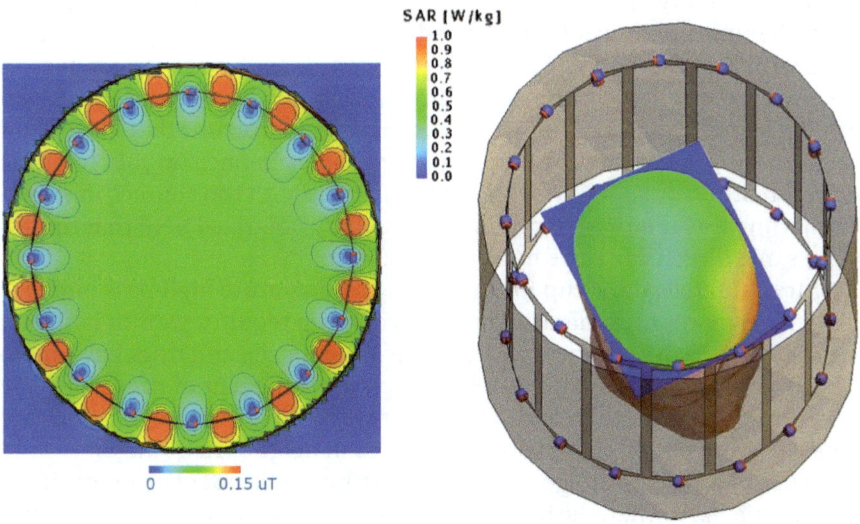

Figure 8.8 An example of a hybrid MoM/FEM approach to simulating EM fields within the human head. On the left, an MoM approach is used to simulate the RF fields from an empty birdcage coil. These act as the inputs for an FEM simulation which contains the human head, shown on the right, with corresponding calculated SAR values.

to the locations of capacitors or the driving voltage sources. With numerical methods, it is possible to explicitly model lumped element components (as will be described further in the next section), and while this may provide an accurate representation of specific RF coils and how their current distribution may be affected by coupling to a lossy sample, or provide guidance in tuning, matching, and decoupling a coil in reality, in other cases adding this level of complexity may not be necessary to obtain the desired results.[8–15,25–30]

In cases where the desired current pattern is known, or where the coil behaviour is well understood, it is possible to perform very efficient and rapid calculations (avoiding an iterative tuning process) with the use of multiple "ideal" voltage or current sources. In the simulated coil a single source is placed across a gap in the conductor, at the location of each capacitor in the actual physical coil. For example, in the case of a circular surface coil, the number of gaps in the conductor is often defined such that no conductor length exceeds about one-tenth of a wavelength in free space. Use of ideal current or voltage distributions requires, of course, first knowing the current distribution to be modelled, placing conductors along the route of those current distributions, and then placing gaps in the conductors as needed to achieve the desired current distribution with a number of sources. Consider the case of a simple loop coil with four gaps. Assuming that the physical coil has equal value capacitors across each gap, each of these capacitors has the same current with the same phase passing through them. The resulting voltage across each capacitor is simply $I/(j\omega C)$, so that these capacitors, as far as the conductive segments of the coil are concerned, can be represented by four voltage sources oriented in the same direction azimuthally, and having the same amplitude and phase. As another example, simulating a quadrature low-pass birdcage coil with N rungs with this method is fairly straightforward, as the voltage sources in all rungs are oriented in the same direction (typically $+z$), and are driven with the same amplitude but with phase equal to the azimuthal location of the given rung, such that the phases of the sources in one rung are in phase with each other, but $\pm 360/N$ degrees out of phase with the sources in neighboring rungs, as required to produce a circularly polarized field that rotates in the same direction as the nuclear precession.[26]

Figure 8.9 shows the setup for simulating a quadrature high-pass birdcage coil in ideal mode 1 resonance. The physical setup requires gaps in both end rings between each set of rungs. The sources in opposite end rings have opposite azimuthal directions and those between the same pair of rungs have the same phase as each other, and (consistent with the azimuthal location) are $\pm 360/N$ degrees out of phase with the sources in neighboring rungs.[30]

The use of an ideal voltage or current distribution is not appropriate in all cases, particularly when the body interacts strongly with the coil, *i.e.* is in very close proximity or very closely coupled, and/or when the coil is very asymmetrically loaded. Nevertheless ideal sources can be used in cases where determining the exact voltage or current distribution is not necessary.[25–30] This approach has been shown to provide very similar results to those acquired experimentally,[28] and also similar to more complicated methods of driving

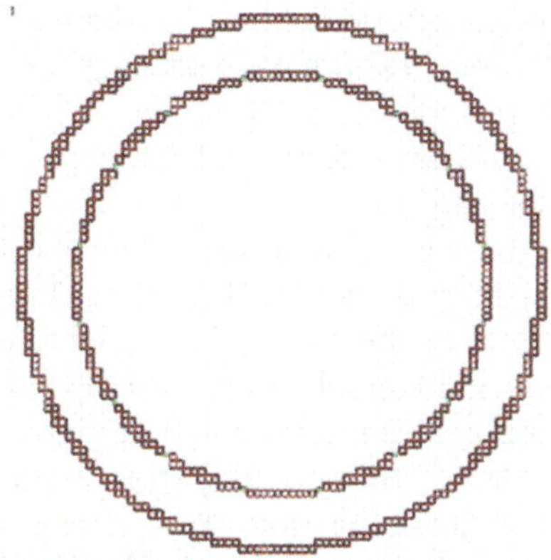

#	Port	Type	(Amp/Phase/Delay)	Dir,(X,Y,Z)	Load/Switch Type	(R,L,C) or Switch Params (Time step,Duration)
1	Y	Current	(1.00/0.00/0)	-X,(77,152,190)	N/A	(50.00,N,N)
2	Y	Current	(1.00/22.50/0)	-X,(54,143,190)	N/A	(50.00,N,N)
3	Y	Current	(1.00/45.00/0)	-Y,(37,125,190)	N/A	(50.00,N,N)
4	Y	Current	(1.00/67.50/0)	-Y,(28,102,190)	N/A	(50.00,N,N)
5	Y	Current	(1.00/90.00/0)	-Y,(28,77,190)	N/A	(50.00,N,N)
6	Y	Current	(1.00/112.50/0)	-Y,(37,54,190)	N/A	(50.00,N,N)
7	Y	Current	(1.00/135.00/0)	X,(54,37,190)	N/A	(50.00,N,N)
8	Y	Current	(1.00/157.50/0)	X,(77,28,190)	N/A	(50.00,N,N)
9	Y	Current	(1.00/180.00/0)	X,(102,28,190)	N/A	(50.00,N,N)
10	Y	Current	(1.00/202.50/0)	X,(125,37,190)	N/A	(50.00,N,N)
11	Y	Current	(1.00/225.00/0)	Y,(143,54,190)	N/A	(50.00,N,N)
12	Y	Current	(1.00/247.50/0)	Y,(152,77,190)	N/A	(50.00,N,N)
13	Y	Current	(1.00/270.00/0)	Y,(152,102,190)	N/A	(50.00,N,N)
14	Y	Current	(1.00/292.50/0)	Y,(143,125,190)	N/A	(50.00,N,N)
15	Y	Current	(1.00/315.00/0)	-X,(125,143,190)	N/A	(50.00,N,N)
16	Y	Current	(1.00/337.50/0)	-X,(102,152,190)	N/A	(50.00,N,N)
17	Y	Current	(1.00/0.00/0)	X,(77,152,93)	N/A	(50.00,N,N)
18	Y	Current	(1.00/22.50/0)	X,(54,143,93)	N/A	(50.00,N,N)
19	Y	Current	(1.00/45.00/0)	Y,(37,125,93)	N/A	(50.00,N,N)
20	Y	Current	(1.00/67.50/0)	Y,(28,102,93)	N/A	(50.00,N,N)
21	Y	Current	(1.00/90.00/0)	Y,(28,77,93)	N/A	(50.00,N,N)
22	Y	Current	(1.00/112.50/0)	Y,(37,54,93)	N/A	(50.00,N,N)
23	Y	Current	(1.00/135.00/0)	-X,(54,37,93)	N/A	(50.00,N,N)
24	Y	Current	(1.00/157.50/0)	-X,(77,28,93)	N/A	(50.00,N,N)
25	Y	Current	(1.00/180.00/0)	-X,(102,28,93)	N/A	(50.00,N,N)
26	Y	Current	(1.00/202.50/0)	-X,(125,37,93)	N/A	(50.00,N,N)
27	Y	Current	(1.00/225.00/0)	-Y,(143,54,93)	N/A	(50.00,N,N)
28	Y	Current	(1.00/247.50/0)	-Y,(152,77,93)	N/A	(50.00,N,N)
29	Y	Current	(1.00/270.00/0)	-Y,(152,102,93)	N/A	(50.00,N,N)
30	Y	Current	(1.00/292.50/0)	-Y,(143,125,93)	N/A	(50.00,N,N)
31	Y	Current	(1.00/315.00/0)	X,(125,143,93)	N/A	(50.00,N,N)
32	Y	Current	(1.00/337.50/0)	X,(102,152,93)	N/A	(50.00,N,N)

Figure 8.9 An example of the model used, and input parameters, for simulating the EM fields produced by a sixteen rung high pass birdcage coil. The copper conductors are displayed in red, with the ideal voltage sources in green. Gaps are introduced into the conductors and the sources placed in these gaps. Each current source is defined by an amplitude and phase, as well as an orientation.

the coil which include tuning with multiple capacitors modelled exactly as for an actual experiment,[29] even for a body-sized birdcage coil at 3 T. Indeed, the appropriate use of ideal voltage or current distributions can be used to compute the values of capacitors required for experimental coils using either analytical[31] or numerical[32] methods.

Exactly how voltage sources or current sources are represented in numerical simulations depends on the particular method being used. For example, in the simplest case a potential difference (voltage) between two points can be represented as an electric field between them where, in the case where the two points are separated by a distance Δx, $|E_x(t)| = V(t)/\Delta x$ where $V(t)$ is the desired voltage. This voltage could also be chosen so as to implement a current source, where the current is calculated by evaluating a line integral of the magnetic fields immediately encircling the source. In most commercial software simulation packages, voltage and current sources are associated with resistances in series or in parallel to present a characteristic impedance so that meaningful modeling of impedance matching circuits can be performed.

In the case that a more complicated coil model is necessary, it is possible with both the FDTD and FEM methods to model lumped element components in the coil and associated circuitry as in the actual physical coils. While, in principle, this could be done by explicitly modeling a capacitor (for example) as a material with a high dielectric constant between two conductive plates[33] or modeling an inductor as a wire wound into a solenoid, these approaches can result in extremely inefficient simulations owing to the high resolution requirements to obtain accurate field distributions in these structures. More practically, it is possible to model lumped-element inductors, capacitors, and resistors with simple equations based on the voltages across, and currents through, the locations associated with these elements.[20,34] Simulating circuit components as in a real coil typically requires tuning the resonant circuit including the coil to the desired frequency, (depending on needs) impedance matching the coil to the source by means of a suitable circuit, and (when necessary) also adding inductive or capacitive circuits to decouple coils from each other.

8.2.3.7 Circuit Co-Simulation

The traditional way to incorporate lumped elements in a 3D full wave EM simulation includes fixing and embedding them in the numerical domain, such that the lumped elements chosen ahead of time determine the calculated EM fields. This method requires no post-processing, but is only practical if the lumped element values are already known. Changing their impedance would then require an additional full wave simulation. In the design process of an RF coil array, for example, the lumped element values are not known in advance, and many iterations would be necessary in order to tune and match the coil array, or achieve a desired field distribution, owing to the complex coupling behaviour within the array.[35] Each iteration involves a full wave 3D EM simulation, and can take several hours of computation time for complex loads such as human models.

An alternative approach allows rapid evaluation of variable lumped elements in combination with any full wave 3D EM-field simulation. The method was first published for MRI by Kozlov *et al.*[36] and by Paska *et al.*,[37] and later by Lemdiasov *et al.*[38] The RF setup is divided into a 3D EM field domain and a network domain containing the lumped elements and ports, which are replaced by sources in the 3D full wave simulation. The corresponding EM fields and the relation between these sources in the form of a scattering matrix are needed. The actual lumped element distribution can then be added in a post-processing step to the results obtained by the 3D simulation. The resulting scattering matrix of the coil array with the actual lumped element distribution can be computed using network theory. The corresponding 3D EM field solution can be obtained simultaneously by superimposing the EM fields in a way such that the boundary conditions at all sources are maintained. The solution for a given set of lumped elements is obtained within seconds, as compared to hours for a 3D EM full wave simulation. In this way, only one full wave 3D simulation is required (for time-domain methods a simulation refers to N simulation runs, with N being the number of sources in the EM domain; for frequency domain methods only one simulation run is required for N sources). The entire process is summarized in Figure 8.10.

The formulation for this approach has been described for fields as a function of lumped elements in terms of either currents and voltages[36] or scattering parameters.[37] Lemdiasov *et al.* derive it in terms of scattering parameters,[38] but without explicit expression of the EM fields as a function of the lumped elements. In the following analysis, the relationship of the EM field domain and the network domain is derived in terms of scattering parameters. The formulae derived in this chapter can be easily implemented in *e.g.* Matlab or Python, to avoid the cumbersome scripting languages of

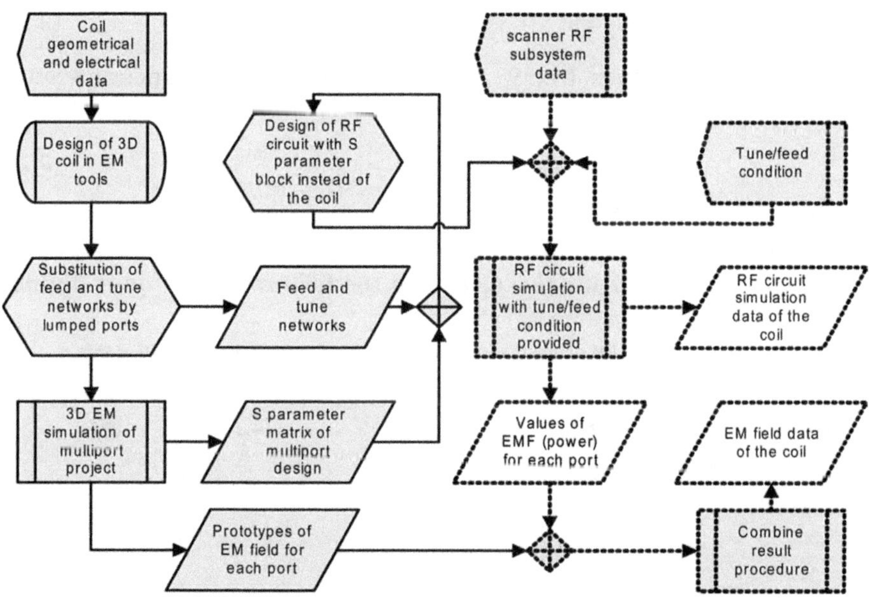

Figure 8.10 A schematic of the setup for circuit co-simulation. The solid line is used for operations and paths that are executed only once for a coil design investigated, the dotted line is used for operations and paths that can be executed repeatedly to obtain data corresponding to different tune and feed conditions. Reproduced from M. Kozlov and R. Turner, Fast MRI coil analysis based on 3-D electromagnetic and RF circuit co-simulation, *J. Magn. Reson.*, 2009, **200**, 147–152. Copyright (2009) with permission from Elsevier.[36]

commercially available 3D full wave solvers, and to implement efficient optimization procedures. The proposed tools are demonstrated in the design process of an eight element transmit stripline RF coil array, similar to that described in Chapter 3, for neuroimaging at 7 T. The lumped element distribution of the array is optimized for operation with eight transmit channels driven in "quadrature mode". Different figures-of-merit at the network level, as well as at the EM-field level, in different regions of interest (whole volume, transverse and sagittal slices only) are investigated.

Lumped elements are replaced in the numerical domain by ports and the values of the lumped elements are added in a postprocessing step. The EM fields and the scattering matrix of the original EM simulation are transformed in a post-processing step by establishing boundary conditions at the ports. The post-processing step is very fast, making it highly suitable for the design of RF coil arrays. The transformation of the original scattering matrix S at the reference plane 0, computed in the EM domain through an RF network represented by the junction scattering matrix S_j can be found by applying the boundary conditions at the reference plane 0 (see Figure 8.11).

The resulting scattering matrix S_1 at the reference plane 1 can be computed as:

$$S_1 = S_{j,11} + S_{j,12} S_0 \left(I - S_{j,22} S S_0\right)^{-1} S_{j,21} \tag{8.25}$$

where I is the identity matrix. $S_{j,kl}$ are block-matrices of the junction scattering matrix:

$$\begin{bmatrix} \mathbf{b}_1 \\ \mathbf{b}_0 \end{bmatrix} = S_j \begin{bmatrix} \mathbf{a}_1 \\ \mathbf{a}_0 \end{bmatrix} \text{with} \quad S_j = \begin{bmatrix} S_{j,11} & S_{j,12} \\ S_{j,21} & S_{j,22} \end{bmatrix} \tag{8.26}$$

with $\mathbf{a}_k$ and $\mathbf{b}_k$ being the forward- and backward-traveling wave vectors at the kth reference plane (k = 0, 1), and the forward direction is defined

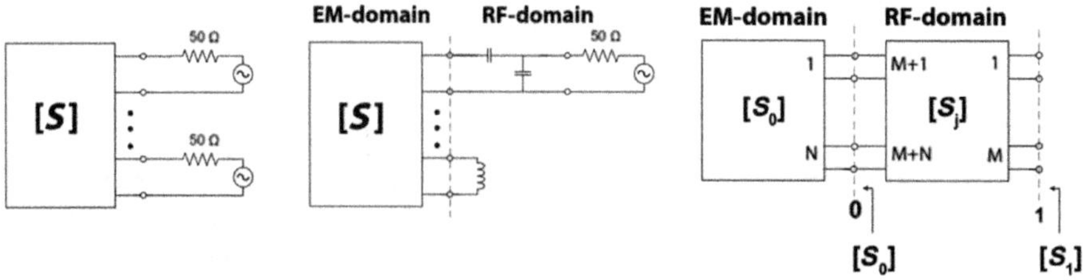

Figure 8.11 After performing simulation(s) required to determine all fields and S-parameters at the desired frequency for a network where all lumped-element circuit components and sources are treated as ports (left), subsequent simulations can be performed rapidly with any desired combination of lumped element circuit components and ports defined in the "RF-Domain" of an RF circuit co-simulator, where these lumped elements and ports can be either positioned across gap locations individually (centre) or by defining a general matrix for more flexible and complete simultaneous matching of all coils (right).

as being towards the junction matrix. The block-matrix $S_{j,11}$ has dimensions $M \times M$ and describes the reflection and coupling between the ports at reference plane 1. The block-matrix $S_{j,22}$ has dimensions $N \times N$ and describes the reflection and coupling between the ports at reference plane 0. The block-matrix $S_{j,12}$ has dimensions $M \times N$ and describes the coupling between ports at reference plane 1 to ports at reference plane 0. Finally, the block-matrix $S_{j,21}$ has dimensions $N \times M$ and describes the coupling between ports at reference plane 0 to ports at reference plane 1. The EM fields can be transformed from reference plane 0 to reference plane 1 according to:

$$\mathbf{E}_1 = \mathbf{E}_0\left(\mathbf{I} - \mathbf{S}_{j,22}\mathbf{S}_0\right)^{-1}\mathbf{S}_{j,21} \tag{8.27}$$

where $\mathbf{E}_0 = [\mathbf{E}_{0,1}...\mathbf{E}_{0,N}]$, and $\mathbf{E}_1 = [\mathbf{E}_{1,1}...\mathbf{E}_{1,M}]$ are electric field vectors at reference planes 0 or 1. The *i*th component of either vector is the electric field that is excited selectively by the forward wave with unit amplitude at the *i*th port, at the respective reference plane. The magnetic fields are transformed in the same way.

Each element of the eight-element array[39,40] consists of a strip, length 250 mm, above a ground plane. The strip is divided into four equal parts with three gaps. Strip capacitors C_{strip} are inserted into the two outer gaps, and a matching circuit followed by an excitation port is inserted into the central gap. The strip is connected at both ends to the groundplane through two equal-valued capacitors C_{end}. At both ends of the ground planes, decoupling capacitors C_{dec} are placed between the elements. The eight elements are covered by an octogonal copper shield. The coil array is loaded by a cylindrical phantom with a diameter of 20 cm and a length of 30 cm, a relative permittivity of 80 and a conductivity of 0.5 S m^{-1}. Each of the lumped elements are replaced by lumped ports in the 3D full wave simulation, resulting in 56 excitation ports. The 3D full wave finite element solver HFSS version 14.0 (Ansys, PA, USA) was used for the simulation, using first-order basis functions. The RF bore was included as a cylinder with the outer shell simulated by a perfect electrical conductor boundary condition and the two faces were terminated with a radiation boundary condition. The simulation setup excluding the RF bore is shown in Figure 8.12.

The optimal lumped element distribution is investigated for the eight channel RF coil array in quadrature drive, *i.e.* with equal phase increments of 45° for adjacent elements. The position of the excitation is chosen in the center of each of the eight stripline elements. The values of the end capacitor C_{end} and the decoupling capacitor C_{dec} were used to optimize the fields and network behaviour of the coil array. To reduce the number of variables the two strip capacitors C_{strip} on each of the stripline elements were short-circuited. It was found that the value of the strip capacitor alone did not have a big influence on the field distribution, only the combined capacitance of the strip capacitor C_{strip} and the end capacitor C_{end}, which allows the simplification of (initially) short-circuiting the strip capacitor.

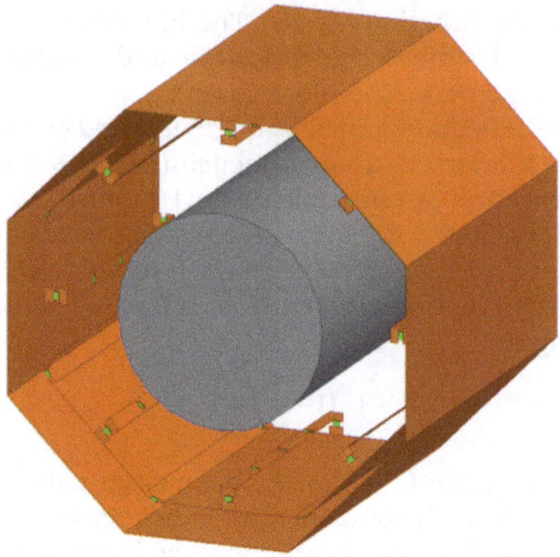

Figure 8.12 Geometry of an 8-channel stripline coil array. Locations of ports for simulation (including all lumped element components and all drive points in the actual coil) are indicated in green.

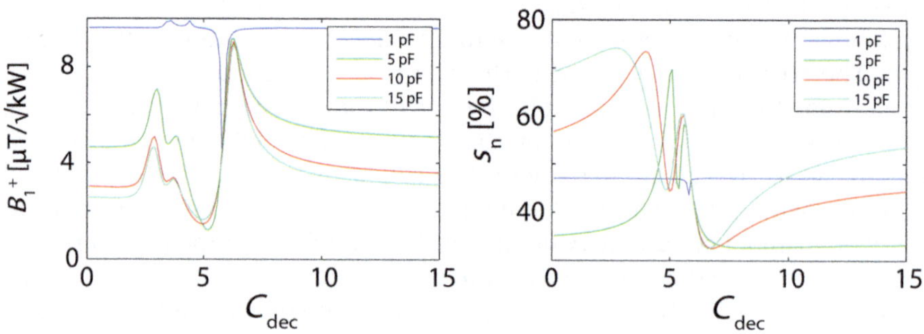

Figure 8.13 Mean B_1^+ and normalized standard deviation s_n in the phantom produced by the coil operated with quadrature drive for an input power of 1 kW with respect to the value of the decoupling capacitor C_{dec} for different end capacitors C_{end} in the entire phantom.

Figure 8.13 shows the mean B_1^+ and normalized standard deviation s_n in the entire phantom for quadrature drive and an input power of 1 kW plotted *versus* the value of C_{dec} for varying end capacitor values. Each channel of the array was matched with a single channel network to 50 Ω, the characteristic impedance of the coaxial cables. The array shows the highest power efficiency for an end capacitor of C_{end} = 1 pF. For this value, the power efficiency as well as the normalized standard deviation remain relatively constant as a function of C_{dec}, except for a value of C_{dec} ≈ 6 pF, where both show a sharp minimum. However, the high degree of power sensitivity of the array to the decoupling capacitor for C_{end} = 1 pF comes at the cost of a high field non-uniformity, as shown in Figure 8.13. A good trade-off is to choose a higher value of C_{end} = 5 pF, which shows a global maximum in power sensitivity for a decoupling capacitor of C_{dec} ≈ 6.2 pF, and a high field uniformity. After selecting this end capacitor value, further optimizations can be performed to determine the other parameters.[41] Optimizations can also be performed for driving each coil individually with a general matching network.[41] Since these optimizations require no new field calculations, they can be completed in seconds. Attempting to perform such optimizations using lumped element circuit components in the field simulation directly would easily require many *months* of simulation!

8.3 The Role of Simulations in Assessing MR Safety and Bioeffects

The strong fields (static, gradient, and high frequency RF) utilized in MRI create an environment in which a number of precautions must be taken to ensure the safety both of subjects and operating personnel. The greatest concerns are best addressed with diligent management of the MRI environment, various aspects of which are discussed at the end of this section. In addition, field simulations play an important role in understanding the interactions and hazards, as well as setting limits on the RF power and gradient waveforms that can be used safely during MRI.

In the following sections we concentrate on how field simulations can be used to understand and evaluate potential bioeffects and ensure safety in the MRI environment. It is important to note that not all bioeffects present safety hazards. For example, the sense of vertigo or of metallic taste that can be experienced while moving in a strong B_0 field, or peripheral nerve stimulation potentially caused by gradient field pulses, should ideally be avoided owing to the discomfort they cause, but in general this discomfort is transitory with no implications for the long-term health of the subject or staff.

Having performed meaningful simulations of the fields in MRI, a good understanding of the processes involved in image formation and the underlying physics are required to interpret these results in a meaningful way with respect to safety and bioeffects. For example, it is possible to find examples in the literature where researchers report the effects of different RF coil designs on SAR for a given input power with no corresponding information on the effects on the B_1 field as a function of the same input power. Of course, in a practical MRI examination, if the B_1 is increased by a factor equal to the square of the SAR value, the measured increase in SAR for one coil compared to the other would not indicate a less safe situation, since maintaining a given B_1 field strength in the ROI would result in exactly the same SAR. This simple example shows that the link between EM simulations and MR physics is absolutely necessary for meaningful interpretation of the data. In the remainder of this section, different methods of interpreting field distributions pertaining to their effects in MRI are presented.

8.3.1 Static Field Effects

Although there are no known harmful effects for humans from static magnetic fields of any strength used for MRI now or in the foreseeable future, motion within these strong fields can cause sensations of dizziness or vertigo or a "metallic taste" on the tongue. These effects become more pronounced at higher magnetic fields, and as a result, a number of qualitative survey studies have been performed recently, particularly with the aim of determining subjects' reactions to 7 tesla human systems compared to lower clinical field strengths.[42,43]

In the case of vertigo, the fluid in the inner ear that is normally pulled downwards by gravity and, by contacting innervated cilia, indicates which way is "down," is electrically conductive, such that the ions in the fluid experience forces other than gravity when moving in a magnetic field. This results in disorientation and vertigo in the subject moving in the magnetic field.[44] In addition, a subject who lies still in a high magnetic field (typically a whole body 7 tesla magnet) experiences persistent nystagmus, *i.e.* motion of the eyes. This occurs owing to the magnetic field interacting with the spontaneous ionic current flowing in labyrinthine endolymph (extracellular fluid with a high concentration of potassium ions, which fills the internal chamber of the labyrinth and bathes the apical surface of the vestibular hair cells), which induces Lorentz forces strong enough to deflect semicircular canal cupulae.[45] These forces are given by:

$$F = LI \times B \tag{8.28}$$

where F is the Lorentz force in newtons and L is the relevant distance in metres. The mechanisms of vertigo and nystagmus have recently been linked by Mian *et al.*[46]

The mechanism causing metallic taste also requires motion in the magnetic field, resulting in electrical currents being induced, *via* Faraday's law, across the surface of the tongue (which is conductive by nature) stimulating the very sensitive neurons that are located near the surface of the tongue.

Using a finite difference-based solution to Faraday's law and the known B_0 field distribution for a self-shielded MRI system, Liu *et al.* calculated the induced electrical currents through the human head and body both for a patient moving into the magnet bore and for a patient voluntarily moving their head in the magnetic field.[47] The method was based on the following considerations. Applying Faraday's law, the electric field generated in the body by the time-varying magnetic field can be expressed as:

$$\mathbf{E} = -\frac{\partial \mathbf{A}}{\partial t} - \nabla \Phi \tag{8.29}$$

where $\mathbf{A}$ is the vector magnetic potential and Φ the scalar electric potential. In conductive samples such as the body, changes in the magnetic field cause a current, with the current density $\mathbf{J}_1$ being given by:

$$\mathbf{J}_1 = -\sigma \frac{\partial \mathbf{A}}{\partial t} \tag{8.30}$$

Differences in tissue conductivity along the current path give rise to a scalar potential Φ, which in turn causes a secondary current, the density $\mathbf{J}_2$ of which is given by:

$$\mathbf{J}_2 = -\sigma\Delta\Phi \tag{8.31}$$

The continuity equation dictates the conservation of current density:

$$\nabla \cdot \mathbf{J} = \nabla \cdot (\sigma \mathbf{E}) = 0 \tag{8.32}$$

This equation can be solved in integral form, and then converted from a volume integral into a closed surface integral:

$$\int_s (\sigma \nabla \Phi) \cdot \mathrm{d}s = \int_s \left(\sigma \frac{\partial \mathbf{A}}{\partial t} \right) \cdot \mathrm{d}s \tag{8.33}$$

The relevant boundary condition is that the *E*-field normal to the surface of the conductive tissue is zero. The scalar potential can now be evaluated by a finite difference method as outlined earlier in this chapter, using a suitably discretized human body model. The electric field components are then derived from the scalar potential. Figures 8.14 and 8.15 show simulated plots of the current densities in transverse and coronal views of the body for a patient bed moving at 0.5 m s^{-1} in a 4 tesla magnet.[47]

There are other known effects of the static magnetic fields on the human body that are not sensed by the subject and have no known adverse effects, but have been studied to ensure safety. The most notable of these has to do with the forces associated with blood flowing through the magnetic field. It is well-known that the electrocardiogram of a subject in a strong magnetic field appears different than it does in the absence of a field. This is not owing to any change in cardiac function, but rather the fact that (as with the fluid in the inner ear) opposite charges experience forces in opposite directions when a conductive fluid moves through a magnetic field. This sets up a polarization within large arteries that can affect the electrical potential sensed at the surface of the body. Investigative calculations regarding potential impediment of blood flow itself owing to these forces have determined that only a 0.2% increase in blood pressure would be necessary to maintain flow for a human subject in a 10 T magnetic field, and thus this effect poses no risk.[48]

8.3.2 Gradient-Induced Peripheral Nerve Stimulation (PNS)

The rapidly switched magnetic field gradients can have a wide range of waveforms (*e.g.* trapezoidal, bipolar, spiral) in MRI. Depending on the thickness of the excitation slice, its orientation, the desired image contrast (determining

(a)
(b)
(c)

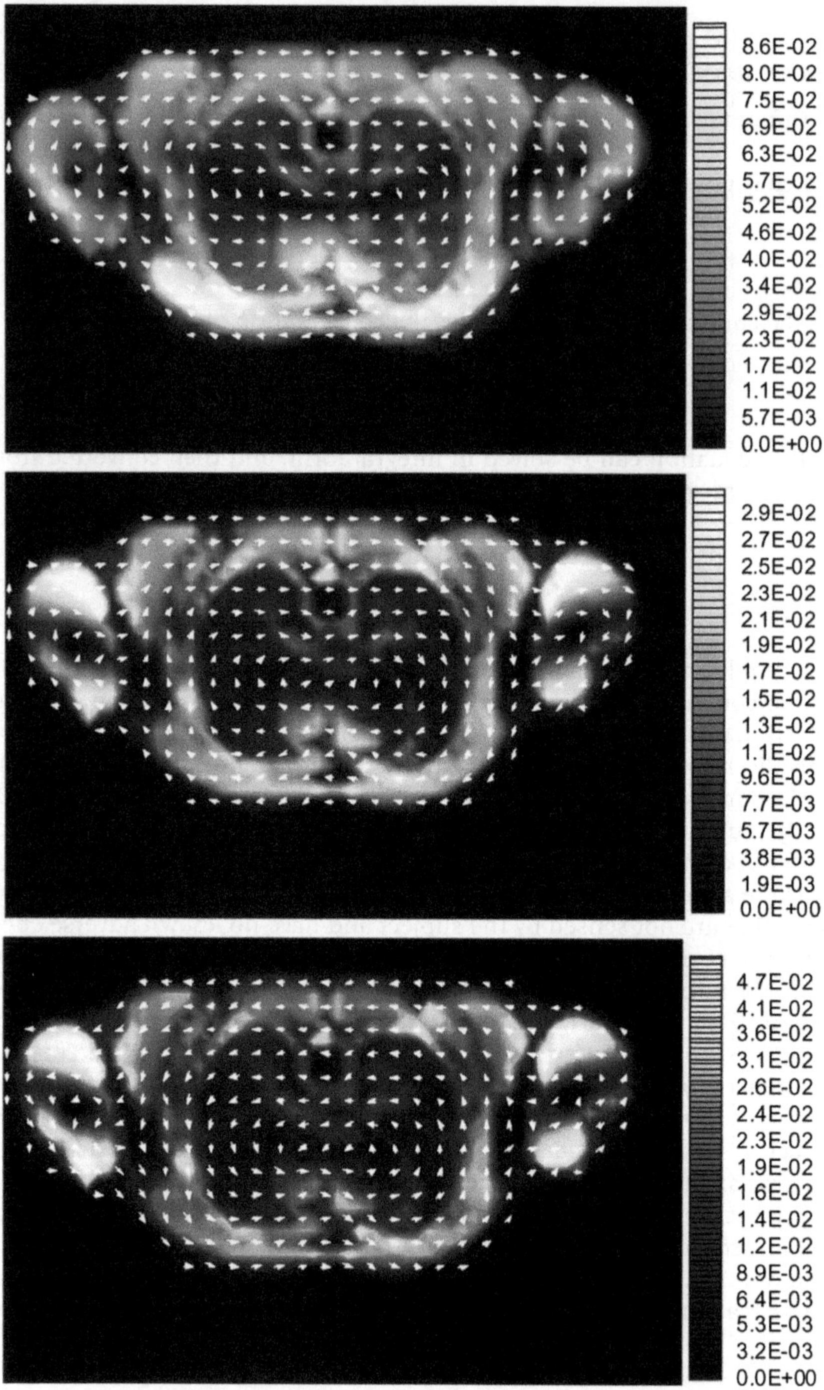

Figure 8.14 The distribution of current densities in a transverse section at the level of the chest for three positions: (a) −2 m; (b) −1.55 m; and (c) −0.8 m with respect to the 4 T magnet center. The patient velocity was 0.5 m s^{-1}. The greyscale shows the magnitude of the induced current densities (A m^{-2}). Reproduced from F. Liu, H. W. Zhao and S. Crozier, Calculations of electric fields induced by body and head motion in high-field MRI, *J. Magn. Reson.*, 2003, **161**, 99–107. Copyright (2003) with permission from Elsevier.[47]

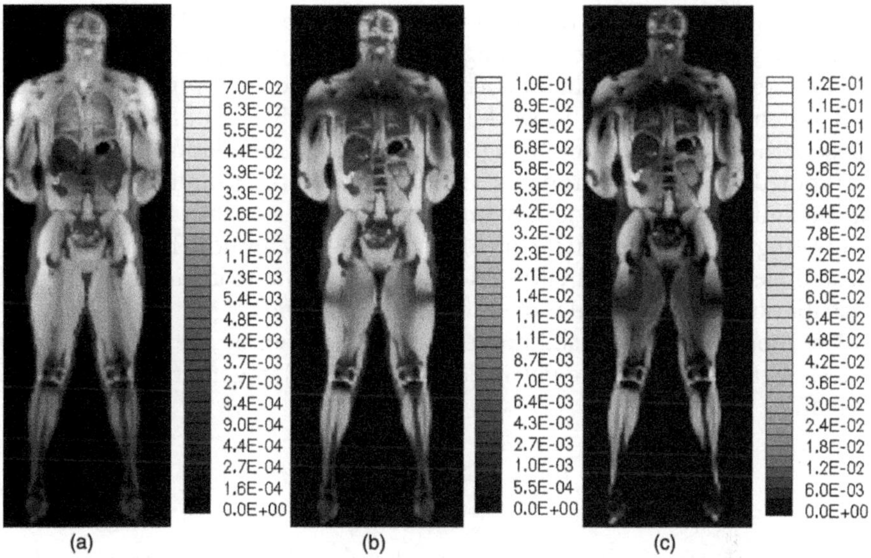

Figure 8.15 The distribution of the current densities in a cross-section at $y = 0.05$ m for three positions: (a) −2 m; (b) −1.55 m; and (c) −0.8 m with respect to the centre of the magnet. Reproduced from F. Liu, H. W. Zhao and S. Crozier, Calculations of electric fields induced by body and head motion in high-field MRI, *J. Magn. Reson.*, 2003, **161**, 99–107. Copyright (2003) with permission from Elsevier.[47]

the TR and TE values), the number of slices or volumes being imaged at once, whether inversion or saturation pulses are to be used, and numerous other considerations, the three primary gradient coils can be put through an effectively infinite variety of pulse shapes and durations. As is clear from Faraday's law, a time-varying magnetic field within the body induces an associated electric field. In the presence of the conductive human body, these electric fields produce electric currents and can result in the stimulation of peripheral nerves, which occurs owing to cellular electrical depolarization. Experiments have shown that, fortunately, it is very difficult to stimulate nerves in deeper-lying tissues that might affect respiratory or cardiac function.[49] As such, PNS is therefore more of an issue of patient discomfort than safety. Nevertheless, it is still important to avoid PNS to ensure as relaxing an experience as possible for the patient, as well as to minimize potential motion of the subject during image acquisition.

Current FDA guidelines do not define any limits on gradient strengths or time-rates of change of gradient strength, and do not "disallow" PNS in general during MRI, but do recommend avoiding painful stimulation.[50] Current IEC guidelines, in contrast, do place a limit on dB/dt, or the maximum rate of change of magnetic fields that can be experienced in the body during MRI, to 20 T s^{-1}.[51] While the early rationale for limiting dB/dt invoked calculations of currents induced in a conductive body by a time-varying homogeneous field, this methodology is clearly flawed in the sense that, by definition, gradient fields are not even remotely homogeneous! Nonetheless, between these calculations and experience in practice, limits have been instituted on commercial systems that seemed to work (*i.e.* do not produce painful stimulation) most of the time. Based on the physiology of nerve stimulation, Chronik and Rutt noted that it would probably be more meaningful to limit a combination of maximum slew rate and maximum gradient strength, rather than concentrating on the individual contributions.[52] A few authors have also investigated electric fields and currents induced by more realistic gradient field distributions in numerical models of the human body.[6,7]

The most commonly used model for PNS is given by the convolution of the slew rate, $\mathrm{SR}_i(t)$, of the particular gradient $G_i(t)$, together with the nerve response:

$$R_i(t) = \frac{1}{\mathrm{SR}_{\mathrm{min}}} \int_0^t \frac{SR_i(\theta)c}{(c+t-\theta)^2} \mathrm{d}\theta \tag{8.34}$$

where $i = (x, y, z)$ is the gradient axis, R is the response, c is the chronaxie time of the gradient coil, and $\mathrm{SR}_{\mathrm{min}}$ is the stimulation slew rate, *i.e.* the threshold value at which an infinitely long ramp with this slew rate leads to a detectable stimulation for 50% of the subjects studied. In terms of gradient coil characteristics the value of $\mathrm{SR}_{\mathrm{min}}$ is given by:

$$\mathrm{SR}_{\mathrm{min}} = \frac{r}{\alpha} \tag{8.35}$$

where α is the effective gradient coil length with rheobase r. The overall PNS threshold in percentage for all the gradient axes combined is given by:

$$P_{\mathrm{thresh}} = 100\sqrt{R_x{}^2 + R_y{}^2 + R_z{}^2} \tag{8.36}$$

The allowed maximum value is $P_{\mathrm{thresh}} = 80\%$ for normal, and $P_{\mathrm{thresh}} = 100\%$ for first- and second-controlled modes. These modes (also used for SAR) are defined as: first level mode of operation of the MR equipment in which one or more outputs reach a value that may cause physiological stress to patients, which needs to be controlled by medical supervision, and second level mode of operation of the MR equipment in which one or more outputs reach a value that may produce significant risk for patients, for which explicit ethical approval is required.

8.3.3 RF-Induced Heating in the Human Body

RF heating of the human body can potentially cause both systemic increases in temperature resulting in discomfort for the subject (and, eventually, thermoregulatory distress), and local heating to the point of causing actual tissue burns. Since RF power is readily monitored during MRI, the most commonly regulated quantity related to RF heating is the specific energy absorption rate, or SAR, defined as the time- and volume-averaged power absorbed in the whole body (SAR_{wb}), head (SAR_{head}), partial body (SAR_{pb}), or 10 g of contiguous tissue (SAR_{10g}), each quantity being divided by the mass (in kg) of the corresponding volume. While the first three of these measures can be estimated with knowledge of the RF input power (derived from the power delivered by the amplifier, the power reflected from the coil, and thereby the power absorbed by the body) and the mass of the exposed region of the body, estimates of local SAR such as SAR_{10g} have been shown to vary greatly depending on the morphology of the subject.[29,53,54] For this reason, numerical methods of calculation with representations of anatomically-accurate human geometries have become increasingly common in the estimation of local SAR. An example is shown in Figure 8.16 for a 3 tesla scan of the brain.

An interesting example of the insights afforded by EM simulations is the very different SAR patterns that are obtained when a subject places their arms in different positions within the magnet, as illustrated in Figure 8.17.[55] The results show that there is a very significant decrease in the torso SAR when the arms are placed above the head.

Since the temperature in the human body depends on numerous factors, including the rate of perfusion of tissue by blood, thermal conduction, respiration, and perspiration, SAR has only a limited direct correspondence to actual tissue temperature.[56–59] For this reason, other quantities have also been proposed for limiting exposure to RF energy in MRI. The current IEC guidelines provide limits on core body temperature and maximum local temperature

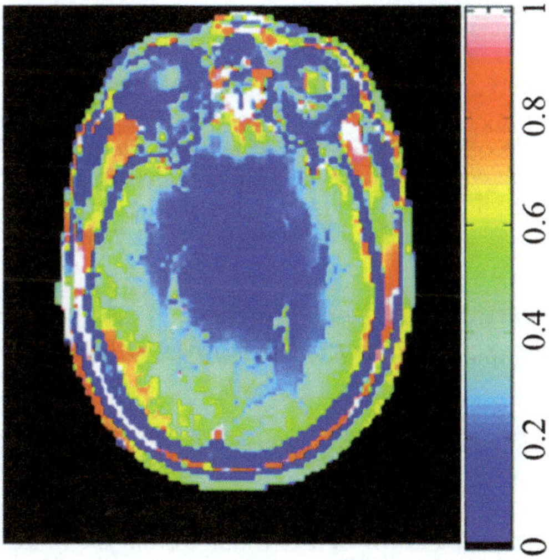

Figure 8.16 SAR distribution in a transverse plane through the human head at 125 MHz. The dependence of SAR on tissue conductivity is apparent, with low SAR in non-conductive tissues such as cortical bone and fat. The effects on SAR of even smaller differences in conductivity, such as those between white and gray matter, are also apparent. Reproduced from ref. 55 by permission. Copyright © 2011 Wiley-Liss, Inc.

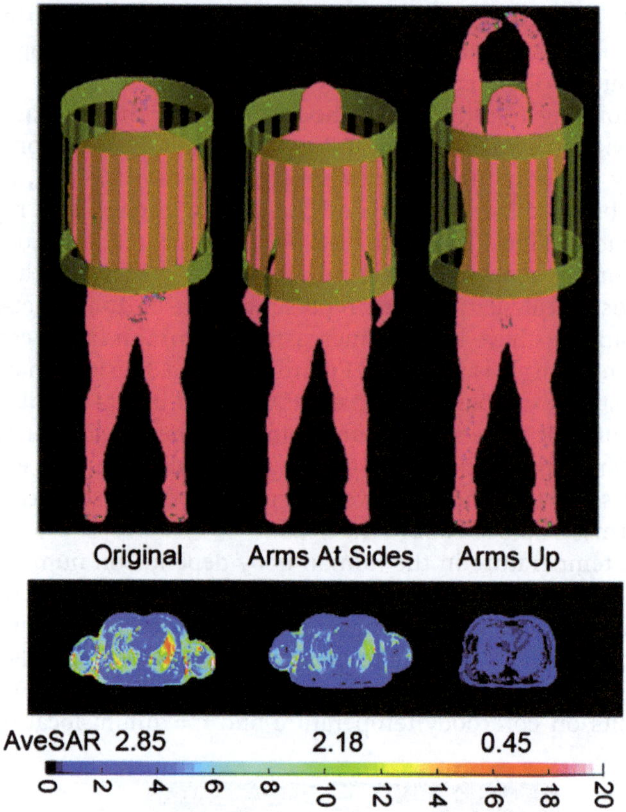

Figure 8.17 Geometry of a human body model in various postures (top) and corresponding SAR distribution (bottom). The colour scale corresponds to a single-cell (5 mm)^3 SAR. Whole-body average SAR is also given below for each case. In all cases the fields are driven to produce a 2 μT B_1^+ at the center of the heart. All SAR values are in W kg^{-1}. Reproduced from ref. 55 by permission. Copyright © 2011 Wiley-Liss, Inc.

depending on the particular region of the body.[59] Perhaps the simplest starting point for temperature calculations is the Pennes Bioheat Equation[56]

$$\rho c \frac{\mathrm{d}T}{\mathrm{d}t} = \nabla \cdot (k \nabla T) + \left[-\rho_{\text{blood}} w c_{\text{blood}} \left(T - T_{\text{core}} \right) \right] + Q_m + \text{SAR}\rho \tag{8.37}$$

where T is temperature, t is time, ρ is mass density, c is heat capacity, Q_m is rate of metabolic heat generation, and w is rate of blood perfusion in tissue. A number of groups have developed methods for simulating temperature throughout the human body during MRI experiments starting with this or other similar relationships that consider SAR, thermal conduction, local perfusion by blood, and convective heat loss to the environment.[56–59] Some implementations also include simple methods for incorporating the effects of respiration and other mechanisms of heat exchange with the environment.[60] Figure 8.18 shows an example of simulated B_1^+, unaveraged SAR, and temperature increase for a human body in a body coil driven at the maximum allowable whole-body average SAR for an extended period of time, with consideration of whole-body temperature increase and heat exchange with the environment through a number of mechanisms.[60] This simulation was performed with a 5 mm isometric grid.

The FDA guidelines for staying within heating limits are currently: 4 W kg^{-1} for whole body averaged over 15 minutes, 3 W kg^{-1} averaged over the head for 10 minutes, and 8 W kg^{-1} in any 1 gram of tissue in the head or torso averaged over 15 minutes, and 12 W kg^{-1} in the extremities averaged over 15 minutes. However, since the likelihood of causing tissue damage depends on the temperature over a period of time and because different tissues have different tolerances for temperature rises of different durations, some authors use the concept of thermal dose to evaluate MR safety.[61–63] The thermal dose is derived by integrating a temperature-weighted function over time. The most common model uses cumulative equivalent minutes at 43 °C (CEM43) in order to calculate accumulated temperatures above 39 °C. The CEM43 is given by:

$$\mathrm{CEM43}(t) = \int_{t_0}^{t_{\mathrm{final}}} R^{(43-T(t))}\,\mathrm{d}t \tag{8.38}$$

where t_0 and t_{final} are the beginning and end of the heating period, and R is a constant equal to 0.5 for $T > 43$ °C, 0.25 for 39 °C $< T <$ 43 °C, and 0 for $T <$ 39 °C. It is certainly possible that future MR safety guidelines will incorporate thermal dose concepts.[61]

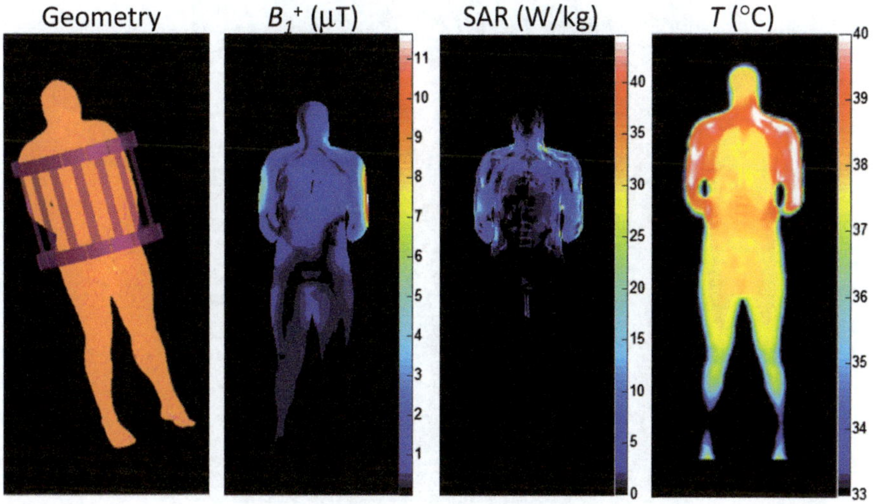

Figure 8.18 Calculated B_1^+, SAR, and temperature distributions for a human body in a commercial 3 T body coil.

8.3.4 Safety of Devices and Implants

Recent years have seen an increasing interest and significant progress in modeling implants in the human body, as well as situations in which conductive materials are in contact with the body during MRI scans.[64–67] In the simplest-to-model cases, these objects or devices are fairly large and are composed of regular shapes such that they can be represented fairly realistically in a numerical model, and simulations have been performed as described in many previous examples in this chapter.[64] In other cases, the implants have fine structures and complex shapes that require more advanced and specialized modeling methods. One example of such a specialized method is the concept of the Huygen's Box to model pacemaker or deep brain stimulator leads.[67] In order to minimize potential heating in MRI, these types of implanted metallic leads have increasingly complex shapes to reduce the amount of RF current that can be induced in them during scanning. In some cases, this can be thought of as making the structure highly inductive in order to impede high-frequency currents. Modeling these very fine structures inside a large volume containing a human body and RF coil(s) can result in extremely large meshes and impracticable memory or time requirements for solution. Applying the Huygen's Box principle,[67] the whole volume is first simulated without the device present, and then a smaller volume containing the device, the "box", is simulated with the fields obtained at the location of the outer surface of the box from the first simulation then used as the input for the second simulation. An example of this approach to model the temperature increase at the end of a pacemaker lead is shown in Figure 8.19.

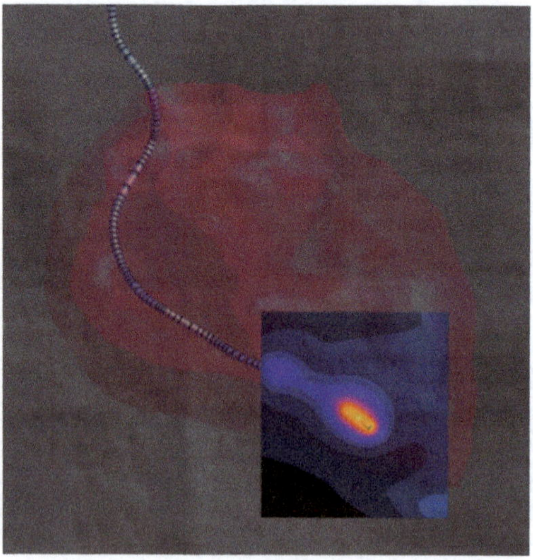

Figure 8.19 The temperature rise at the tip of a cardiac pacemaker lead, simulated with Sim4Life, using the principle of Huygen's Box.

8.3.5 MR Safety in Practice

A wide range of simulations for understanding safety and bioeffects in MRI have been discussed in this section. In some cases, these types of simulations can be used to set limits on RF energy (SAR) or gradient switching rates to ensure the safety and comfort of subjects. In general, it is very important to recognize that no matter what simulations are performed for the design of systems, RF pulses, and imaging sequences concerning safety and bioeffects, it is the everyday diligence of the personnel running the MR system and handling the human subjects that ensures safety in practice.

The most serious injuries in MRI occur when a ferromagnetic object (potentially a device implanted within the patient's body) is introduced into the magnet room, or when conductive objects, such as loose wires or an unconnected RF coil, are placed on the patient inappropriately during an MR examination. For these reasons, MR technologists query human subjects thoroughly about medical history, including any possible implants and/or history working with metal that may have led to small metal filings in the patient's eyes. At most sites, all implants are initially considered "MR Unsafe" and preclude continuation of the examination unless they are found to be "MR Conditional" or "MR Safe." Implants are studied extensively to determine under what circumstances the patient can be imaged safely, and are now even designed with MRI safety in mind. If an implant can be identified as "MR conditional," the study may continue with certain precautions and limitations to avoid any damage to the subject. It is rare for any metallic implants to be considered "MR Safe": indeed, this designation is used to distinguish equipment (oxygen tanks, tools, *etc.*) that can be brought safely into the magnet room, from those that are "MR Unsafe" and must not be brought inside.

As discussed earlier, the switching of the gradient fields in the presence of the B_0 field results in very large forces and strong vibrations resulting in significant acoustic noise. In general, these forces are greater at higher B_0 field strengths, such that great engineering effort is applied to minimize the acoustic noise created in high field systems. Because of the high levels of acoustic noise, patients and anyone else in the magnet room must wear ear protection in the form of earplugs or protective headgear. Figure 8.20 shows the sound pressure levels (SPL) typically produced by MRI scanners.[68]

Under certain circumstances, skin-to-skin contact between different limbs of the patient allows current loops to form with regions of high heating resulting in burns. Contact with the bore of the magnet or other system components can potentially result in high SAR regions, also potentially resulting in injury.[69] For these reasons, MR system manufacturers typically specify that patients be positioned so that no such skin-to-skin contact or skin-to-bore contact occurs, and that nothing can be in contact with the subject during MRI that was not either specifically designed for that purpose, *e.g.* electrocardiogram leads, or necessary for the study, for example contrast agent infusion pumps.

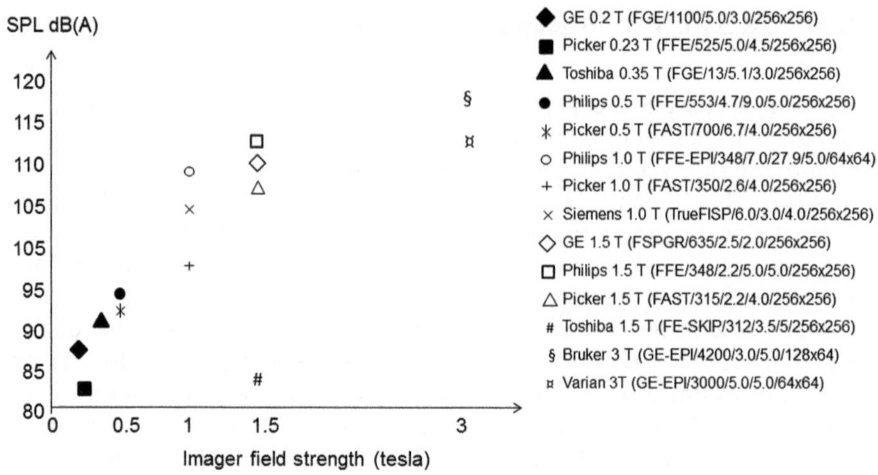

Figure 8.20 SPL (dB) levels *vs.* imager field strength for 15 MRI systems running fast imaging sequences. The legend describes the pulse sequence parameters as follows: sequence type, TR (ms), TE (ms), slice thickness (mm), and matrix size (phase encoding × frequency encoding). All pulse sequences were run in a transverse orientation with a FOV of 250 mm. Reproduced from ref. 68 by permission from Wiley © 2001 Wiley-Liss, Inc.[68]

8.4 Calculating the Effects of Electromagnetic Fields on MR Images

Being able to simulate the interactions between the magnet/gradients/RF coil and the patient, and indeed also between the three hardware elements themselves, is obviously very important in terms of characterizing RF fields, both magnetic and electric, within the patient, as well as enabling safe operation of the MR scanner. However, an important additional step in the entire simulation process is to be able to translate these interactions into being able to predict actual MR data, *i.e.* their effects on the reconstructed MR images. Two examples of this type of extension are shown in the following sections: the first concentrates on quantifying the intrinsic SNR of any type of MR experiment, and the second on integrating EM simulations with Bloch equation simulations of the nuclear spin system to provide a full package to determine the effects of hardware on image appearance.

8.4.1 Calculation of the Intrinsic Signal-to-Noise Ratio (ISNR)

The SNR in MR images depends on a very large number of tissue-dependent factors, such as proton density, T_1-, T_2- and T_2^* relaxation times, magnetic susceptibility, permittivity, conductivity, in relation/addition to sequence-dependent factors such as relaxation delays, the extent and geometry of the k-space coverage, tip angles used and so on. Rather than trying to cover this enormous parameter space, estimating what is called the "intrinsic" SNR (iSNR) allows for a more direct focus on the fields alone.[9] Here the concept of iSNR is introduced and considered with respect to, in particular, the effects of the B_0 field strength.

Random thermal motions of charges and dipoles in the coil and sample induce currents in the RF coil. As covered in Chapter 3, in order to utilize Nyquist's equation for the thermal noise[70] measured at the coil terminals, $V_{\text{noise}} = \sqrt{4kT\Delta\nu R}$ (where $\Delta\nu$ is the MR receiver bandwidth), it is necessary to find effective resistance values for both the coil and sample, R_{coil} and R_{sample}, respectively. Knowing that the dissipated power P is equal to the square of the coil current I times the resistance R ($P = I^2R$), one can state that $R \propto P$, defining P as the dissipated power when $I = 1$ A. This uses an argument based on reciprocity—calculating power lost when the coil is driven in order to indicate the noise voltage induced in the coil by thermal motion at the location of power loss. Treating the coil and sample as two resistors in series:

$$V_{\text{noise}} \propto \sqrt{4kT_{\text{coil}}\Delta\nu P_{\text{coil}} + 4kT_{\text{sample}}\Delta\nu P_{\text{sample}}} \propto \sqrt{P_{\text{coil}} + P_{\text{sample}}} \tag{8.39}$$

assuming that the sample and coil are at approximately the same temperature on an absolute scale. To proceed further requires analysis of P_{coil} and P_{sample} as functions of B_0. This analysis requires different approaches depending upon whether quasi-static approximations can be applied, or whether there are appreciable wavelength effects in the coil and sample.

Alternating electrical current in a "good" conductor tends to flow along the outer surface. The exact current distribution depends on the shape of the conductor, as well as the electromagnetic fields around it,[71–73] but for a round wire one can approximate the cross-sectional area across which the current flows as $2\pi\rho\delta$ where ρ is the wire radius and δ is the skin depth. For a good conductor:

$$\delta = \sqrt{\frac{2}{2\pi\nu\mu_c\sigma_c}} \tag{8.40}$$

where μ_c and σ_c are the magnetic permeability and electrical conductivity of the conductor. The wire resistance R is given by:

$$R \approx \frac{\text{length}}{2\pi\rho\delta} \propto\sim \nu^{\frac{1}{2}} \tag{8.41}$$

For a current of 1 A:

$$P_{\text{coil}} \propto \nu^{\frac{1}{2}} \propto\sim B_0^{\frac{1}{2}} \tag{8.42}$$

Referring to Faraday's law in time-harmonic form:

$$\oint_l \mathbf{E}\cdot d\vec{l} = -j2\pi\nu\int_s \mathbf{B}\cdot d\vec{s} \tag{8.43}$$

For a B-field that is independent of frequency, *i.e.* under quasi-static conditions, then $E \propto \nu$. Since at each location $P = \sigma E^2$, the integrated power dissipated throughout the sample is given by:

$$P_{\text{sample}} \propto\sim \nu^2 \propto\sim B_0^{\ 2} \tag{8.44}$$

Allowing all of the B_0-independent, experiment-specific factors (*e.g.*, coil and sample geometries and temperatures) that determine the exact relationship between P and B_0 to be absorbed into factors a and b, one can now write:[12]

$$\frac{V_{\text{signal}}}{V_{\text{noise}}} \propto \frac{B_0^{\ 2}}{\sqrt{aB_0^{\ 1/2} + bB_0^{\ 2}}} \tag{8.45}$$

As B_0 approaches zero, $aB_0^{\ 1/2}$ (the contribution from coil noise) will become much greater than $bB_0^{\ 2}$ (the contribution from sample noise), indicating that the noise will be dominated by the contribution of the coil. Thus, in the low-field limit we can expect:[74]

$$\frac{V_{\text{signal}}}{V_{\text{noise}}} \propto \frac{B_0^{\ 2}}{\sqrt{aB_0^{\ 1/2}}} \propto B_0^{\ 7/4} \tag{8.46}$$

Assuming well-designed coils, the contribution of sample noise becomes dominant as B_0 is increased above only a fraction of 1 tesla, depending on the coil being used and the anatomy being imaged. If one makes the assumption that wavelength effects do not become important until B_0 is at least 1.5 tesla in human MRI, then for magnetic fields between very low field and 1.5 tesla the SNR is given by:[10]

$$\frac{V_{\text{signal}}}{V_{\text{noise}}} \propto \frac{B_0^{\ 2}}{\sqrt{bB_0^{\ 2}}} \propto B_0 \tag{8.47}$$

In high field MRI, sample noise is still dominant, but the B_1 field distribution varies with frequency, precluding the use of quasi-static assumptions and the approximations for B_1^+, B_1^-, and P_{sample}, which are appropriate in low- and mid-field MRI. This means that at high field the expression for SNR at a single location is more complicated and is given by:[75]

$$\frac{V_{\text{signal}}}{V_{\text{noise}}} \propto \frac{B_0 \omega_0 \sin(\gamma V B_1^+ \tau) B_1^-}{\sqrt{4kT_{\text{sample}} \Delta\nu P_{\text{sample}}}} \propto \frac{B_0^{\ 2} \sin(\gamma V B_1^+ \tau) B_1^-}{\sqrt{P_{\text{sample}}}} \tag{8.48}$$

Since in MRI we are typically interested not in a single location, but in a three-dimensional volume, it is often more appropriate to use an expression such as:[14,30]

$$\frac{V_{\text{signal}}}{V_{\text{noise}}} \propto \frac{B_0^{\ 2} \int_{\text{VOI}} W \sin(\gamma V B_1^+ \tau) B_1^- \, \mathrm{d}\nu}{\sqrt{P_{\text{sample}}}} \tag{8.49}$$

where the integration is performed over the volume-of-interest (VOI) and W is a weighting factor accounting for tissue-specific and sequence-specific factors affecting signal intensity, such as T_1, T_2, proton density, TE, and TR. Further complicating the matter, the signal before Fourier reconstruction also has phase-related terms so that:[14,30]

$$\frac{V_{\text{signal}}}{V_{\text{noise}}} \propto \frac{B_0^2 \int_{\text{VOI}} W \sin\left(\gamma V B_1^+ \tau\right) e^{i\beta_1^+} \left(\hat{B}_1^-\right)^* \mathrm{d}v}{\sqrt{P_{\text{sample}}}} \tag{8.50}$$

where the circumflex indicates that $\hat{B}_1^-$ is complex and β_1^+ is the phase of $\hat{B}_1^+$.

Occasionally, for simplification and discussion purposes, it may be desirable to consider the maximum SNR at one particular location, assuming a tip angle of 90° at that location. In this case, the SNR can be expressed as:

$$\frac{V_{\text{signal}}}{V_{\text{noise}}} \propto B_0^2 \frac{\hat{B}_1^-}{\sqrt{P_{\text{sample}}}} \tag{8.51}$$

The series of expressions above are presented to give some insight into the various considerations in comparing SNR in different situations, such as studying the effects of B_0 or to compare SNR for simulations of different single-channel or quadrature coils.

For multi-channel receive coils it is necessary to consider also the noise correlation between the different channels. This is often calculated considering the electric field distributions from the coils such that:[76]

$$\text{SNR} \propto \sqrt{\hat{B}_1^{-H} \hat{R}^{-1} \hat{B}_1^-} \tag{8.52}$$

where $\hat{B}_1^-$ is a 1D array containing complex values from all N coils at the location of interest, the superscript H indicates the Hermitian, and $\hat{R}^{-1}$ is the $N \times N$ noise resistance matrix with $\hat{R}_{ij} = \int \sigma \hat{E}_i \cdot \hat{E}_j \, \mathrm{d}v$ for coils i and j with the integration taken over the entire sample, when the noise is sample-dominated. In the case where other noise sources become non-negligible (*e.g.* noise due to losses in the coil conductors, in the RF shield, or the shielding wall of the magnet room) the integration has to be expanded accordingly.

Considering that the noise resistance matrix describes power loss by a coil array, an equivalent and faster way of computing the noise resistance matrix is possible if the impedance matrix Z for the coil array is available. By using the relations in the network domain, instead of integrating over the electrical fields in the EM domain, the noise resistance matrix can be computed as the real part of the impedance matrix $\hat{R} = \text{real}\{Z\}$.

8.4.2 Simulating MR Images

In many cases, it would be desirable to extend the simulations of B_0, B_1^+ and B_1^- fields by integrating simulations of the Bloch equations (including gradients) in order to completely simulate how MR images appear when using different pulse sequences, data acquisition parameters, gradient readouts *etc.* Different sequences have very different characteristics with respect to the degree of B_1^+ or B_0 inhomogeneity. As an example, for gradient echo sequences, the image contrast depends on TE, TR, and flip angle, and short TRs combined with low flip angles are often used for speed. For this type of low tip angle sequence, an inhomogeneous B_1^+ results in non-uniform image contrast since the signal intensity is essentially linearly proportional to the local value of B_1^+. However, other types of sequence that use tip angles closer to 90° may be close to being immune to the value of B_1^+, and so the distribution of image contrast would be quite different.

By combining field simulations with the Bloch equations it is possible to simulate a wide array of effects on MR images, including chemical shift displacements (*i.e.* a water/fat shift), signal loss and spatial misregistration owing to B_0 inhomogeneity, and effects of inhomogeneous B_1^+ field or flip angle on the distribution of signal intensity, SNR, and tissue contrast. This can be done by defining magnetization vectors (**M**) throughout space, with the strength of the initial resting state vectors ($\mathbf{M}_0$) being proportional to local proton density and local B_0 field strength; then tracking **M** through time at each location considering local B_1^+, gradient field strengths, chemical shift, and relaxation effects (T_1, T_2, and T_2^*) by using the Bloch equations. The signal strength detected by each coil at each point in time during acquisition can then be calculated by summing the local complex transverse component M_T of the magnetization weighted by the local complex receive coil sensitivity B_1^- at all locations in the sample

$$\text{Signal} = \Delta x \Delta y \Delta z \times \gamma B_0 \sum_{\text{sample}} \left(M_T B_1^{-*} \right) \tag{8.53}$$

Depending on the spatial resolution of the numerical model and that of the simulated image, it may be necessary to use several discrete magnetization vectors, or "isochromats", distributed within each voxel. However, this can be very time- and memory-intensive. In some cases it is possible rather to consider the effect of a continuum of magnetization vectors across each voxel in the model with analytical methods to avoid aliasing effects.[77]

To simulate realistic noise, including noise coupling between coils, it is necessary to generate random numbers for each coil appropriately weighted by the electric field E of each receive coil and the sample conductivity σ throughout space. For example, for an N-coil receive matrix, an $N \times N$ noise matrix $\varphi_{i,j}$ can be calculated as

$$\varphi_{i,j} = \Delta x \Delta y \Delta z \sum_{\text{sample}} \sigma \left(E_i E_j^* \right) \tag{8.54}$$

where i and j both vary from 1 to N. Using this noise matrix and N complex random numbers ζ (having Gaussian distribution with zero mean and unit variance) generated for each coil at each acquisition time, the noise voltage in the ith coil at each point in time during the acquisition can be calculated as:[77]

$$V_{\text{noise},i} = \sqrt{4kT \times BW} \left(\sum_{j=1:N} \varphi_{i,j}^{1/2} \zeta_j \right) \tag{8.55}$$

Models of the head and cylindrical phantoms with different RF coils have been used to simulate results for many different pulse sequences. Figure 8.21 shows some of these effects generated with such an MR simulator.[77,78]

Figure 8.22 shows different brain images simulated either with an array receive coil or a volume transmit/receive coil at different field strengths.

By additionally considering the electric field distributions, it is also possible to simulate images with realistic noise distributions for different acquisition and reconstruction algorithms, including realistic quantitative SNR levels, as shown in Figure 8.23.[77]

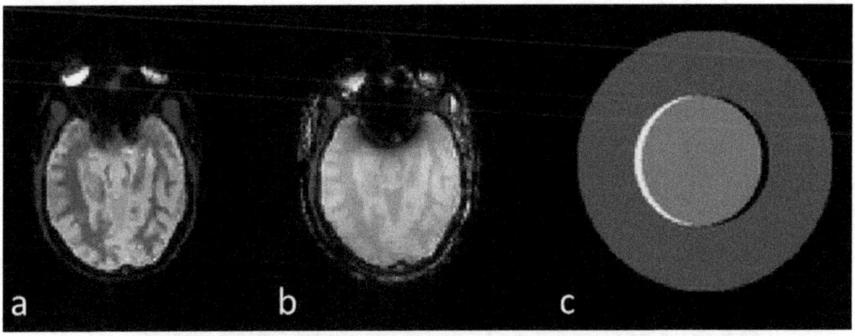

Figure 8.21 Simulated MR images with (a and b) ΔB_0 and (c) chemical shift artifacts. Image distortion and signal loss artifacts occur due to in-plane and through-plane B_0 inhomogeneity with (a) GRE-EPI, (b) long TE GRE, and (c) chemical shift artifact in a simulated phantom consisting of oil surrounded by cerebrospinal fluid. Reproduced from ref. 77 by permission from Wiley © 2013 Wiley Periodicals, Inc.[77]

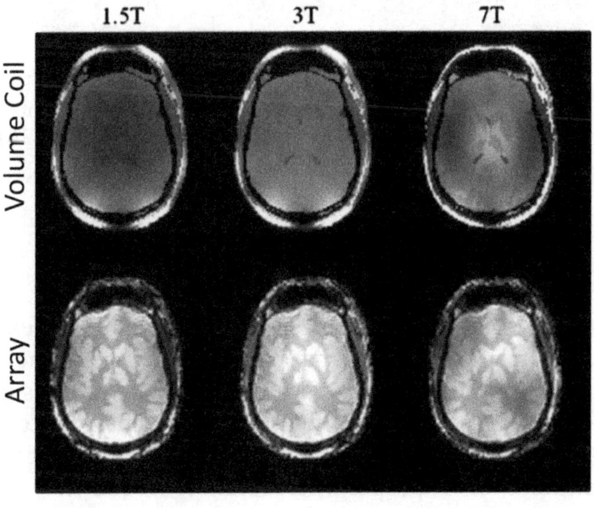

Figure 8.22 Simulated images for a quadrature transmit/receive volume coil (top) and 8-channel transmit-receive coil array (bottom) at different field strengths. B_1 inhomogeneity effects are much less apparent with the array.

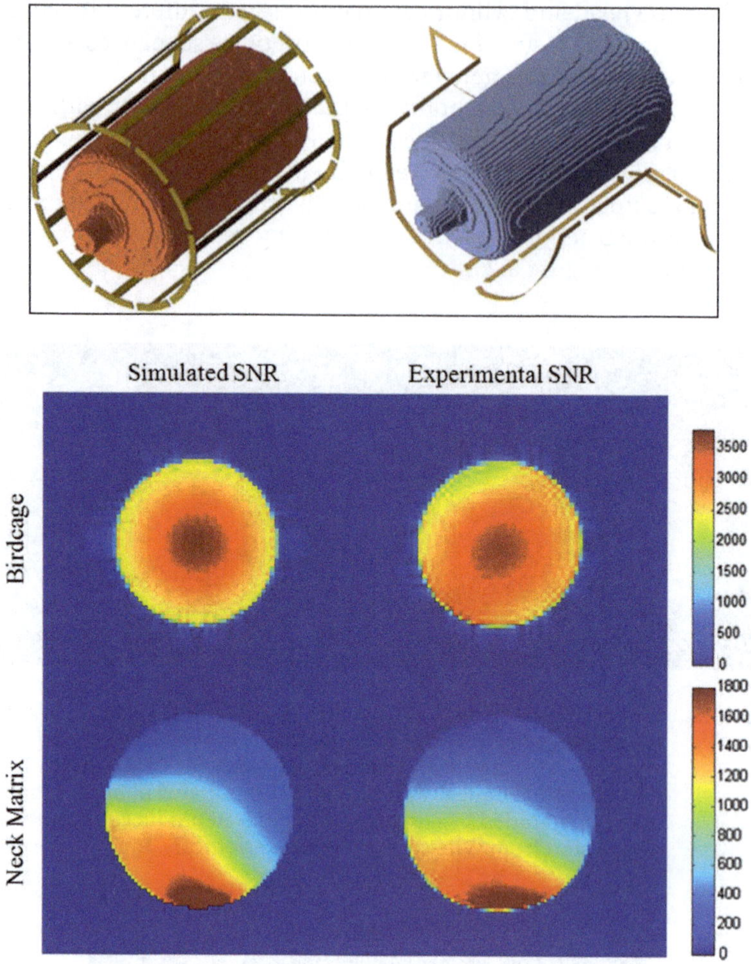

Figure 8.23 Experimental and simulated SNR distribution of a phantom imaged with a birdcage coil and the posterior two elements of a neck matrix. Reproduced from ref. 77 by permission from Wiley © 2013 Wiley Periodicals, Inc.[77]

8.5 Methods for Validating Simulations

In the early stages of gaining experience with EM simulations (especially with complicated commercial software packages), there are numerous ways to make errors in the setup of the simulations and in the processing of results. It is therefore critical to validate simulation results in order to confirm correct modeling of coils, subjects, and samples, as well as correct manipulation of data after simulation for proper interpretation of results.

Depending on the particular purpose of the simulation, the preferred method for validation will vary. In general, validation of simulated quantities can be performed by comparison to analytical solutions,[3] actual MR images,[27] and/or experimentally measured quantities such as the B_1^+ field.[77] For validation of safety-related simulations in particular, which are most commonly performed for RF fields, it is important to compare the simulated results to experimentally measured B_1 fields, electrical fields, and/or temperature increases within phantoms for similar driving conditions using a particular coil or array. The experimental measurements can be performed with probes in a benchtop situation, or utilizing established MR-based field or temperature mapping techniques.[79–85] The easiest experimental MR-based comparison method is probably to map the B_1^+ field, using any number of available techniques, and then to compare this with the simulations. Electric fields are more difficult to measure, but if one obtains good agreement between the B_1^+ simulations and experiments then there is a high degree of probability that the *E*-field simulations are also accurate provided that tissue conductivity values used in the simulation are accurate. One example that shows extremely good agreement between quite complicated B_1^+ patterns is shown in Figure 8.24.

A second example, in which temperature changes measured using the proton reference frequency method are compared with simulations, is shown in Figure 8.25. Again the excellent agreement gives a high degree of confidence in the accuracy of the simulations.

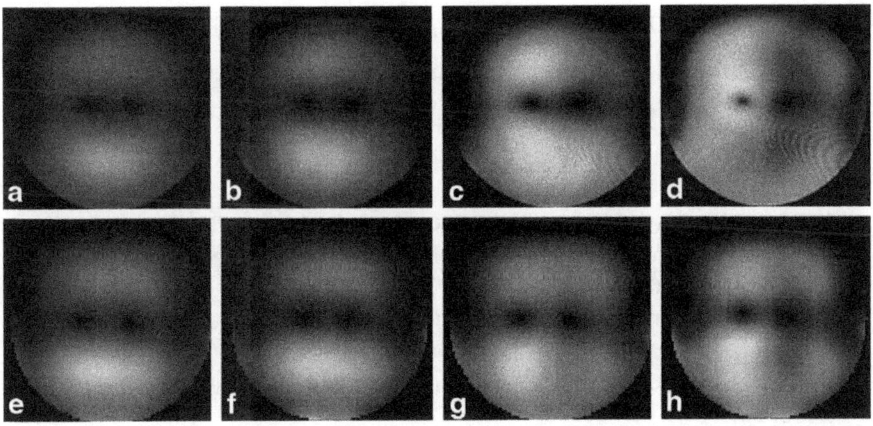

Figure 8.24 Gradient echo axial images (a–d) and corresponding simulated (e–h) images using a quadrature surface coil at 7 tesla with different transmission power levels. There is a 6 dB power increment for each subsequent column of images from left to right. The signal intensity distribution becomes more asymmetric about the vertical centerline of the sample as the transmission power level increases.

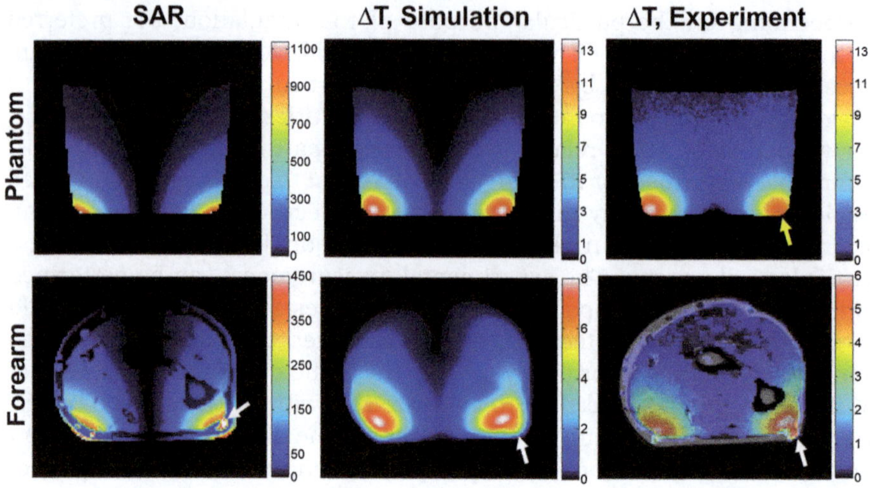

Figure 8.25 Simulated unaveraged SAR map (left column), simulated temperature increase [°C] (center column), and experimentally measured PRF temperature increase [°C] (right column) for an agar-gel phantom (top row), and a human forearm *in vivo* (bottom row). The yellow arrow indicates the region of largest discrepancy between experiment and simulation in phantom. The white arrow indicates a blood vessel that has a high SAR but only a moderate temperature increase in the forearm. The plane shown passes through the center of the coil and is transverse to the long axis of the phantom and forearm. Reproduced from ref. 55 with permission from John Wiley and Sons © 2011 Wiley-Liss, Inc.

Acknowledgements

We are grateful to Dr Giuseppe Carluccio for his assistance in creating Figure 8.18.

References

1. J. T. Vaughan, M. Garwood, C. M. Collins, W. Liu, L. DelaBarre, G. Adriany, P. Andersen, H. Merkle, R. Goebel, M. B. Smith and K. Ugurbil, *Magn. Reson. Med.*, 2001, **46(1)**, 24–30.
2. R. Bhagwandien, M. A. Moerland, C. J. G. Bakker, R. Beersma and J. J. W. Lagenijk, *Magn. Reson. Imaging*, 1994, **12**, 101.
3. C. M. Collins, B. Yang, Q. X. Yang and M. B. Smith, *Magn. Reson. Imaging*, 2002, **20**, 413.
4. J. P. Marques and R. Bowtell, *Concepts Magn. Reson., Part B*, 2005, **25B**, 65.
5. S. C.-K. Chu, Y. Xu, J. A. Balschi and C. S. Springer Jr., *Magn. Reson. Med.*, 1990, **13**, 239.
6. F. Liu, W. Zhao and S. Crozier, *IEEE Trans. Biomed. Eng.*, 2003, **50**, 804.
7. W. Mao, B. A. Chronik, R. E. Feldman, M. B. Smith and C. M. Collins, *Magn. Reson. Med.*, 2006, **55**, 1424.
8. O. Ocali and E. Atalar, *Magn. Reson. Imaging*, 1998, **39**, 462.

9. W. A. Edelstein, G. H. Glover, C. J. Hardy and R. W. Redington, *Magn. Reson. Med.*, 1986, **3**, 604.
10. D. I. Hoult and P. C. Lauterbur, *J. Magn. Reson.*, 1979, **34**, 425.
11. J. R. Keltner, J. W. Carlson, M. S. Roos, T. S. Wong, T. L. Wong and T. F. Budinger, *Magn. Reson. Med*, 1991, **22**, 467.
12. P. Roschmann, *Med. Phys.*, 1987, **14**, 922.
13. J. Tropp, *J. Magn. Reson.*, 2004, **167**, 12.
14. D. I. Hoult, *J. Magn. Reson. Imaging*, 2000, **12**, 46.
15. R. Lattanzi, D. K. Sodickson, A. K. Grant and Y. Zhu, *Magn. Reson. Med.*, 2009, **61**, 315.
16. R. Lattanzi and D. K. Sodickson, *Magn. Reson. Med.*, 2012, **68**, 286.
17. J. Jin, G. Shen and T. Perkins, *Magn. Reson. Med.*, 1994, **32**, 418.
18. T. A. Riauka, N. F. De Zanche, R. Thompson, F. E. Vermeulen, C. E. Capjack and P. S. Allen, *Magn. Reson. Med.*, 1999, **41**, 1180.
19. K. S. Yee, *IEEE Trans. Antennas Propag.*, 1966, **14**, 302.
20. J. Jin, *The finite element method in electromagnetics*, John Wiley & Sons, New York, 1993.
21. Y. Zhang, C. Bajaj and B.-S. Sohn, *Comput. Meth. Appl. Mech. Eng.*, 2005, **194(48–49)**, 5083–5106.
22. J. H. Coggon, *Geophysics*, 1971, **36**, 132.
23. B. K. Liu, F. Liu and S. Crozier, *Magn. Reson. Med.*, 2005, **53**, 125.
24. S. Wang and J. H. Duyn, *Phys. Med. Biol.*, 2008, **53**, 2677.
25. J. W. Hand, J. J. W. Lagendijk, J. V. Hajnal, R. W. Lau and I. R. Young, *J. Magn. Reson. Imaging*, 2000, **12**, 68.
26. J. M. Jin, J. Chen, W. C. Chew, H. Gan, R. L. Magin and P. J. Dimbylow, *Phys. Med. Biol.*, 1996, **41**, 2719.
27. R. W. Singerman, T. J. Denison, H. Wen and R. S. Balaban, *J. Magn. Reson.*, 1997, **125**, 72.
28. C. M. Collins, Q. X. Yang, J. H. Wang, X.-H. Zhu, G. Adriany, S. Michaeli, J. T. Vaughan, X. Zhang, H. Liu, P. Anderson, K. Ugurbil, M. B. Smith and W. Chen, *Magn. Reson. Med.*, 2003, **47**, 1026.
29. W. Liu, C. M. Collins and M. B. Smith, *Appl. Magn. Reson.*, 2005, **29**, 5.
30. C. M. Collins and M. B. Smith, *Magn. Reson. Med.*, 2001, **45**, 684.
31. C. L. Chin, C. M. Collins, S. Li, B. J. Dardzinski and M. B. Smith, *Concepts Magn. Reson., Part B*, 2002, **15**, 156.
32. G. McKinnon and Z. Wang, *Proc. Int. Soc. Magn. Reson. Med.*, Toronto, 2003, p. 2381.
33. Q. X. Yang, H. Maramis, S. Li and M. B. Smith. *Proc Int Soc Magn Reson Med*, San Francisco, 1994; p. 1110.
34. K. S. Kunz and R. Luebbers, *The Finite Difference Time Domain Method for Electromagnetics*, CRC Press, Boca Raton, 1993.
35. K. P. Pruessmann, M. Weiger, M. B. Scheidegger and P. Boesiger, *Magn. Reson. Med.*, 1999, **42(5)**, 952–962.
36. M. Kozlov and R. Turner, *J. Magn. Reson.*, 2009, **200**, 147.
37. J. Paska, J. Froehlich, D. O. Brunner, K. P. Pruessmann and R. Vahldieck, *Proc. Int. Soc. Magn. Reson. Med.*, 2009, p. 3038.

38. R. A. Lemdiasov, A. A. Obi and R. Ludwig, *Concepts Magn. Reson., Part A*, 2011, **38A**, 133.
39. D. O. Brunner, N. Zanche, J. Froehlich, D. Baumann and K. P. Pruessmann, *Proc. ISMRM*, 2007
40. J. Froehlich, D. Baumann, D. O. Brunner, K. P. Pruessmann and R. Vahldieck, *IEEE/MTT-S International Microwave Symposium*, 2007, pp. 2217–2220.
41. J. Paska, Modeling, Design, and Safety of RF-Transmitters for High Field MRI, Doctoral Dissertation, ETH-Zurich, 2014.
42. C. Heilmaier, J. M. Theysohn, S. Maderwald, O. Kraff, M. E. Ladd and S. C. Ladd, *Bioelectromagnetics*, 2011, **32**(**8**), 610–619.
43. M. J. Versluis, W. M. Teeuwisse, H. E. Kan, M. A. van Buchem, A. G. Webb and M. J. van Osch, *J. Magn. Reson. Imaging*, 2013, **38**(**3**), 722–725.
44. P. M. Glover, *et al.*, *Bioelectromagnetics*, 2007, **28**, 349–361.
45. D. C. Roberts, *et al.*, *Curr. Biol.*, 2011, **21**, 1635–1640.
46. O. S. Mian, *et al.*, *Front. Neurol.*, 2015, **6**, 201.
47. F. Liu, H. W. Zhao and S. Crozier, *J. Magn. Reson.*, 2003, **161**, 99.
48. J. R. Keltner, M. S. Roos, P. R. Brakeman and T. F. Budinger, *Magn. Reson. Med.*, 1990, **16**, 139.
49. G. A. Mouchawar, J. A. Nyenhuis, J. D. Bourland and L. A. Geddes, *IEEE Trans. Magn.*, 1993, **29**, 3355.
50. FDA, *Guidance for Industry and FDA Staff – Criteria for Significant Risk Investigations of Magnetic Resonance Diagnostic Devices*, 2003, http://www.fda.gov/cdrh/ode/guidance/793.pdf.
51. IEC, *International standard, medical equipment – part 2: particular requirements for the safety of magnetic resonance equipment for medical diagnosis, 3rd edition, 601-2-33*, International Electrotechnical Commission, Geneva, 2010. p. 32.
52. B. A. Chronik and B. K. Rutt, *Magn. Reson. Med.*, 2001, **45**, 916.
53. P. L. Davis, C. Shang, L. Talagala and A. W. Pasculle, *IEEE Trans. Biomed. Eng.*, 1993, **40**, 1324.
54. C. M. Collins, S. Li and M. B. Smith, *Magn. Reson. Med.*, 1998, **4**, 847.
55. C. M. Collins and Z. Wang, *Magn. Reson. Med.*, 2011, **65**, 1470.
56. H. H. Pennes, *J. Appl. Physiol.*, 1948, **1**, 93.
57. C. M. Collins, W. Liu, J. H. Wang, R. Gruetter, J. T. Vaughan, K. Ugurbil and M. B. Smith, *J. Magn. Reson. Imaging*, 2004, **19**, 650.
58. D. Shrivastava and J. T. Vaughan, *J. Biomech. Eng.*, 2009, **131**, 1.
59. Z. Wang and C. M. Collins, *J. Magn. Reson. Imaging*, 2008, **28**, 1303.
60. G. Carluccio, Y.-S. Ding and C. M. Collins, *Proc. Int. Soc. Mag. Reson. Med.*, 22, 2014, p. 323.
61. IEC, *International standard, medical equipment – part 2: particular requirements for the safety of magnetic resonance equipment for medical diagnosis, 3rd edition, 601-2-33*. International Electrotechnical Commission, Geneva, 2010. pp. 34–35.
62. P. S. Yarmolenko, E. Jung Moon, C. Landon, A. Manzoor, D. W. Hochman, B. Viglianti and M. W. Dewhirst, *Int. J. Hyperthermia*, 2011, **27**, 320.

63. M. Murbach, E. Neufeld, M. Capstick, W. Kainz, D. O. Brunner, T. Samaras, K. P. Pruessmann and N. Kuster, *Magn. Reson. Med.*, 2014, **71**, 421.
64. J. Powell, A. Papadaki, J. Hand, A. Hart and D. McRobbie, *Magn. Reson. Med.*, 2012, **68**, 960.
65. L. M. Angelone, A. Potthast, F. Segonne, S. Iwaki, J. W. Belliveau and G. Bonmassar, *Bioelectromagnetics*, 2004, **25**, 285.
66. S. Feng, R. Qiang, W. Kainz and J. Chen, *IEEE Trans. Microwave Theory Tech.*, 2015, **63**, 305.
67. E. Neufeld, S. Kühn, G. Szekely and N. Kuster, *Phys. Med. Biol.*, 2009, **54**, 4151.
68. D. L. Price, *et al.*, *J. Magn. Reson. Imaging*, 2001, **13**, 288.
69. M. V. Knopp, M. Essig, J. Debus, H. J. Zabel and G. van Kaick, *Radiology*, 1996, **200**, 572.
70. H. Nyquist, *Phys. Rev.*, 1928, **32**, 110.
71. S. Li, Q. X. Yang and M. B. Smith, *Magn. Reson. Imaging*, 1994, **12**, 1079.
72. L. K. Forbes, S. Crozier and D. M. Doddrell, *Meas. Sci. Technol.*, 1996, **7**, 1281.
73. S. Crozier, L. K. Forbes, W. U. Roffman, K. Luescher and D. M. Doddrell, *Concepts Magn. Reson.*, 1997, **9**, 195.
74. D. I. Hoult and R. E. Richards, *J. Magn. Reson.*, 1976, **24**, 71.
75. C. M. Collins and M. B. Smith, *Magn. Reson. Med.*, 2001, **45**, 692.
76. P. B. Roemer, W. A. Edelstein, C. E. Hayes, S. P. Souza and O. M. Mueller, *Magn. Reson. Med.*, 1990, **16**(**2**), 192.
77. Z. Cao, S. Oh, C. T. Sica, J. M. McGarrity, T. Horan, W. Luo and C. M. Collins, *Magn. Reson. Med.*, 2014, **72**, 237.
78. Z. Cao, J. Park, Z.-H. Cho and C. M. Collins, *J. Magn. Reson. Imaging*, 2015, **41**, 1432.
79. Z. Chen, K. Solbach, D. Erni and A. Rennings, *Antennas and Propagation (EuCAP), 2013 7th European Conference on*, IEEE, 2013, pp. 1716–1719.
80. J. Paska, J. Fröhlich, D. O. Brunner, K. P. Pruessmann and R. Vahldieck, *Proc. ESMRMB*, 2009, p. 193.
81. A. Kangarlu, L. Tang and T. S. Ibrahim, *Magn. Reson. Imaging*, 2007, **25**, 1222.
82. ASTM, *Test method for measurement of radio frequency induced heating on or near passive implants during magnetic resonance imaging*, ASTM F2182, 2011.
83. N. N. Graedel, J. R. Polimeni, B. Guerin, B. Gagoski and L. L. Wald, *Magn. Reson. Med.*, 2015, **73**, 442.
84. L. Alon, C. M. Deniz, R. Brown, D. K. Sodickson and Y. Zhu, *Magn. Reson. Med.*, 2013, **69**, 1457.
85. D. O. Brunner and K. P. Pruessmann, *Magn. Reson. Med.*, 2009, **61**, 1480.

Subject Index

absorption-mode spectrum, 17
active approaches, shimming
local active shimming, 198
multi-coil (MC) shimming, 199–202
active shield magnets, 74
adaptive gridding, 339
air-cored resistive magnets, 49–50
Alderman–Grant design, 104, 105
Ampere's law, 84
amplifier architecture
bench measurements, 301–302
selection of, 304–305
testing and comparison of, 301–304
testing using MRI, 303–304
analogue-to-digital converter (ADC), 16, 322–328
apparent diffusion coefficients, measurements of, 23–24
ARF475FL, push–pull device, 298

balanced impedance matching, 99
balanced matching networks, 100
balanced-to-unbalanced (balun) circuits, 91, 99–101
b-factor, 24
Bi-2201, 63
Bi-2212, 63
Bi-2223, 63
Biot–Savart law, 35–36, 84–85, 175, 211–213, 334, 337
examples, 212–213
bipolar junction transistors (BJTs), 265
birdcage coils, 106–107
Bloch equations, 14, 331, 370
Boltzmann's equation, 6
brain, proton image of, 2
B_0 shimming technology, 166–202
building block methods, 213

cable traps, 91, 99–101
calibration error, 283
capacitive impedance matching, 95–96
chemical shielding, 29
chemical shift, 10–12
chemical shift anisotropy (CSA), 29
clinical imaging systems
receive arrays for, 144–145
single-channel and dual-channel transmit coils for, 142–144
coherence order, 22
coil matching network, for LOI amplifier, 297–301
coil workbench measurements, 155–157
correction coils, 18
cross-polarization, 30
cryoprobes, 132–135
cryo-refrigerators, 71–72
current density, 211
methods, 213–219

current source amplifiers, 290–294, 304–305
current source network analysis, 292–294
induced voltage and current, 293
matching networks for, 291–292
MOSFET, 292–294
steeper load-line, 293

data sampling, 319–322
direct detection, undersampling, 321–322
frequency demodulation, 320–321
deuterium lock channel, 26–28
diamagnetic materials, 168
dielectric loss, 91, 102
dielectric resonators, 146–150
HEM_{11} mode resonators, 150
TE_{01} mode resonators, 150
dipolar coupling, 29, 30
dispersion-mode spectrum, 17
double quantum filtered (DQF) COSY, 23
dual-channel transmit systems, 143
dynamic spherical harmonic (SH) shimming, 189–196
of human brain, 193–196
practical considerations for, 190–193
principle of, 189–190

eddy current losses, 102
eddy currents, 191, 193, 210, 215, 217, 218, 219
effective magnetic field (B_{eff}), 11
electrical circuit analysis, 91–104
electric field integral equation (EFIE), 346
electromagnetic modelling, 331–374
validating simulations, 372–374
energy level diagram, 13

Faraday's law, 84, 343, 356
feedback capacitance, 296
ferromagnetic materials, 168
ferromagnetism, 56, 57, 58
field effect transistors (FETs), 266
field monitoring, 26–28
finite difference time domain (FDTD) method, 338–342
finite element method (FEM), 342–345
forward Fourier transformations, 172
Fourier coefficients, 245
Fourier transformation, 16–18
free induction decay (FID), 16
frequency demodulation, 320–321
full-width-half-maximum (FWHM), 17

gallium arsenide field effect transistor (GaAs-FET), 316
Gauss' law for magnetism, 334
Gifford–McMahon cycle, 73
Golay coil, 213
gradient-based shimming, 24–25
gradient coils, 19–26
apparent diffusion coefficients, measurements of, 23–24
gradient-based shimming, 24–25
high-resolution NMR, coherence selection, 22–23
in MRI, 25–26
phase accumulation, 23
transverse magnetization, dephasing, 21–22
gradient coil system, 230, 231–234
gradient control system, 230–231, 241–243
gradient cooling system, 230, 236–241
gradient-echo-based B_0 mapping, 181
gradient efficiency, 220
gradient-induced heating, 73–74

gradient-induced peripheral nerve stimulation, 357–360
gradient power amplifier (GPA), 222, 224, 230, 234–236

harmonic distortion, 283
head gradient coil insert, 255–258
heat load duty cycle, 238
Helmholtz pair, 53
HEM_{11} mode resonators, 150
high and ultra-high field magnets, 76–78
high field human imaging, 145–146
    multi-channel transmit arrays for, 146
high resistance current driver, 278
high resolution NMR
    and hyphenated micro-separations, 127–130
    microprobes for, 127–130
    probes for, 124–127
homogeneous magnetic field, 104–111
    birdcage coils, 106–107
    generation of, 43–44
    loop gap resonators, 109–111
    partial-volume coils, 108–109
    solenoids, 109–111
    TEM resonators, 107–108
human calf muscle, phosphorus spectrum, 2
human heart, electrocardio-gram-triggered proton image, 2
hybrid electromagnetic (HEM) modes, 149
hybrid-pi circuit model, 272

impedance (Z), 92–94
impedance matching circuit, 91
induced magnetization, 170
induced voltage, 16
inductance, 92, 93
inductive impedance matching, 96–97
inductive loss, 90
inhomogeneous line-broadening, 15
intrinsic signal-to-noise ratio (ISNR), 366–369
inverse Fourier transformations, 172
iron-cored resistive magnets, 50
iron-cored superconducting magnets, 50
isochromats, 370

Joule heating, 90

Karplus angle, 13

lactic acid molecule
    ball-and-stick model of, 12
    chemical structure of, 12
Laplace's equation, 40–41, 172
Larmor frequency, 209
least significant bit (LSB), 322
Legendre polynomial functions, 41–42
linearity radius, 221
linear magnetic field gradients, 44–46, 208–209
local active shimming, 198
longitudinal magnetization, 22
loop gap resonators (LGR), 111
Lorentz correction, 335
Lorentz force, 356
Lorentz function, 17
low-noise preamplifiers, 316–318
low output impedance (LOI) amplifier, 294–301, 305
    coil matching network for, 297–301
    matching network for, 295, 296, 298, 299
    MOSFET, 297

MAFIA software package, 246
magic angle spinning (MAS) solid-state NMR, 28–31
    principles and instrumental requirements, 28–31
magnesium di-boride, 61
magnetic dipole, 58

magnetic field generation, 51–60
  basic physics, 51–53
  field homogeneity, 53–55
  magnetic field shimming, 58–60
  magnetic shielding, 56
  system shielding, external interference, 56–58
magnetic field gradient coils. *See* gradient coils
magnetic field gradients, 208–262
  bore-size systems, 225–229
  conventional gradient designs, developments, 221–224
  current density methods, 213–219
  design methods, classification of, 211
  geometric distortion, 209–210
  gradient performance parameters, 219–221
  integrated gradient and RF designs, 225
  linear, 208–209
  non-cylindrical designs, 229–230
  spatial encoding, 209–210
  specific gradient coil designs, 244–262
  spherical harmonics, methods, 219
magnetic field homogenization
  of human brain, 187
  optimizing, 173
magnetic field inhomogeneity, 24, 196, 197
  diameter-of-spherical-volume (DSV), 168
  origins of, 167–172
magnetic field lines, 171
magnetic field mapping, 178–182
  B_0 field mapping, 178–179
  MRI-based B_0 field mapping, 179–182
magnetic fields
  by circular wire loop, 37–38
  generation of, 173, 175
  perturbations, 192
  shimming, 58–60
  spherical harmonic representation of, 39–46
  by straight wire, 36–37
  visual description, 39–40
magnetic resonance imaging (MRI), 1, 27
  electromagnetic fields effects, 366–372
  intrinsic signal-to-noise ratio, 366–369
  principles and instrumental requirements, 31–34
  radiofrequency magnetic (B_1) fields, 337–355
  sequences, 34
  simulating, 369–372
  static magnetic (B_0) fields, 334–336
  switched gradient fields, 336–337
magnetic susceptibility distribution, 169, 170
magnetization, 6, 8, 13, 14
matching networks, 279–282
  basics of, 279–280
  broadband matching, 282
  for current source amplifiers, 291–292
  narrowband matching, 280–281
Mathematica, calculus package, 247
Maxwell coil, 212
Maxwell's equations, 35–36, 84–85
metal–oxide–semiconductor field-effect transistors. *See* MOSFET
Method of Moments (MoM), 345–346
MOSFET, 265, 266
  "active" region, 269
  "cutoff" region, 269
  DC characteristics of, 267–270
  final structure of, 266–267

MOSFET (*continued*)
  forward transconductance (g_{fs}) of, 269
  hybrid-pi model of, 272
  I–V curves, 270
  N-Channel Enhancement mode, 267
  parasitic capacitances of, 270, 271
  RF characteristics, 270–272
  RF power, 266–267
  "saturation" region, 269
  symbolic representation of, 268, 270
  transfer curve of, 267, 269
most significant bit (MSB), 325
MRI-based B_0 field mapping, 179–182
MRI magnets, 48–79
  air-cored resistive magnets, 49–50
  cryostat design, 70–74
  heat transfer, 70–74
  helium-free technology, 78–79
  installation issues, 75–76
  iron-cored resistive magnets, 50
  iron-cored superconducting magnets, 50
  magnetic field generation, 51–60
  permanent magnets, 50
  safety, 74–75
  superconducting cylindrical magnets, 51
  superconductivity, 60–70
  types, 49–51
MR receiver, 308–329
  analogue-to-digital converter, 322–328
  block diagram of, 308, 309
  data sampling, 319–322
  data, signal levels and dynamic ranges of, 310–312
  design, parameters, 308–310
  individual components in, 17
  low-noise preamplifiers, 316–318
  noise figure of, 312–313
  optical data transmission, 328–329
  transmit/receive switches, design, 313–316
  wireless data transmission, 328–329
MR safety and bioeffects, 355–366
  devices and implants, safety, 364
  gradient-induced PNS, 357–360
  RF heating, of human body, 360–363
  safety in practice, 365–366
  static field effects, 356–357
MR signal, receiver coil for, 15–16
MR system
  block diagram of, 4
  hardware components of, 35
multi-coil approach, 53, 55
multi-coil (MC) shimming
  of human brain, 201–202
  principles and considerations in, 199–200
multiple-frequency circuits, 120–124
  multiple-pole circuits, 120–121
  multiple-tuned surface coils, 124
  multiple-tuned volume coils, 122–124
  transformer coupled circuits, 121–122
multiple-pole circuits, 120–121

network analyzer, 155–157
Ni–Fe alloy (permalloy), 196
niobium tin (Nb_3Sn) conductor, 61
niobium titanium alloy, 61
noise voltage, 103
nuclear gyromagnetic ratio, 6
nuclear polarization, 5–7
nuclear spin quantum number, 5
Nyquist–Shannon sampling theory, 319

one-dimensional proton nuclear magnetic resonance (NMR) spectroscopy, 1, 2
optical data transmission, 328–329

paramagnetic material, 168
partial-volume coils, 108–109
passive shimming approach, 196–198
Pennes bioheat equation, 362
performance index (PI), 221
permanent magnets, 50
phase encoding gradient, 32
phase sensitive detectors (PSDs), 16
phase wrapping, 180
Poisson equation, 343
port impedance, 289
preamplifier decoupling, 118–120
precession frequencies, 10–15
    chemical shift, 10–12
    relaxation processes, 13–15
    scalar coupling, 12–13
proton density, 33
pulse width modulation (PWM), 234

quality factor (Q), 101
quasi-static conditions, 85

radiation loss, 102
radiofrequency (RF) amplifiers, 8
    architectures, 301–305
    dynamic range, 284
    efficiency, 284–285
    linearity, 283–284
    for multi-channel transmission, 285–290
    noise gating, 284
    performance considerations, 282–285
    stability, 285
    technical specifications, for RFPAs, 285, 286
    transmit arrays, mutual coupling in, 287–290
radiofrequency coils, 3, 81–82
balanced-to-unbalanced (balun) circuits, 99–101
Biot–Savart law, 84–85
cable traps, 99–101
capacitive impedance matching, 95–96
for clinical imaging systems, 142–145
combined MR/optical histology, 138–142
conservative electric fields, 89–91
detuning circuits, 115–116
dielectric resonators, 146–150
electrical circuit analysis, 91–104
electromagnetic principles for, 82–91
electromagnetic simulations, 91
goals of, 82
high field human imaging, 145–146
homogeneous magnetic field (volume coils), 104–111
impedance (Z), 92–94
inductive impedance matching, 96–97
linear and circular polarization, 87–89
Maxwell's equations, 84–85
for MR microscopy, 138–142
multiple-frequency circuits, 120–124
NMR spectroscopy, 124–135
non-conservative electric fields, 89–91
receive arrays, 116–120
resonant circuits, 94–95
RF coil loading, sample effect, 101–104
small animal imaging, 136–138
surface coils, 111–114
transmission line elements, impedance matching, 98
transmit ($B^+{}_1$) and receive ($B^-{}_1$) magnetic fields, 85–87
travelling wave MRI, 150–155

radiofrequency magnetic (B_1) fields, 337–355
- analytically based methods, 337–338
- circuit co-simulation, 350–355
- finite difference time domain method, 338–342
- finite element method (FEM), 342–345
- hybrid simulation methods, 346–347
- method of moments (MoM), 345–346
- multiple "ideal" current/voltage sources, 347–350

radiofrequency pulses, 7–10
- magnetization effect, 9

rat brain, proton image of, 2
receive arrays, 116–120
- array optimization, 117–118
- preamplifier decoupling in, 118–120

receive-only surface coils, 115–116
receiver electronics, 16–17
reflection coefficient, 93
relaxation processes, 13–15
resonance frequency, 7–8
resonant circuits, 94–95
resonant frequency, 11
RF amplification
- mechanism of, 275–279
- principles of, 266–279

RF power amplifier (RFPA), 264–265
- class A, 272–273
- class AB, 273–274
- class B, 273
- class C, 274
- class G and H, 274
- linear class, 274

RF power MOSFET, 266–267
Rollett's stability constant, 318
root-mean-square Johnson (thermal) noise voltage, 310
root mean square (RMS) quantization noise, 325

saddle coil, 10
safety watchdog, 241–243
scalar coupling, 12–13
shim coil efficiency, calibration, 182–185
shim coils, 3, 18–19, 177
shim degeneracy analysis, 190
signal demodulation, 16–18
signal digitization, 16–18
signal processing, 17–18
signal-to-noise ratio (SNR), 133, 312, 332, 369
single-channel shim coils, 198
six-point approximation, 106
slew rate (SR), 221
solenoids, 109–111
solid-state NMR, probes for, 130–132
specific gradient coil designs, 244–262
- head gradient coil insert, 255–258
- 7 T horizontal bore animal magnet, 244–250
- ultra-strong whole-body gradients, 258–262
- whole-body modular gradient, 250–255

spherical harmonic functions, 173, 174
- mathematical description of, 40–43

spin-active nuclei, 5
standard power amplifiers, 304
static field effects, 356–357
static magnetic (B_0) fields, 334–336
static spherical harmonic shimming
- of human brain, 185–189
- magnetic field mapping, 178–182
- shim coil efficiency, calibration of, 182–185
- theory, 172–178

Stefan–Boltzmann law, 71
Stejskal–Tanner spin-echo experiment, 23, 24

stick-shift movement, 70
stress-induced quenching, 70
stress limits, 67–70
sub-atmospheric operation, superconductor, 72–73
superconducting cylindrical magnets, 51
superconducting magnet, 5–7
superconductivity, 60–70
 quenching, 66, 67
 quench protection, 66, 68
 stress limits, 67–70
 superconducting joints, 65–66
 superconducting magnet, energising, 64
 superconducting materials, 61–63
 superconducting switch, 64–65
surface coils, 111–114
 detuning circuits, 115–116
 quadrature surface coils, 112–114
 transmit/receive surface coils, 111–112
surface equivalence principle, 345
switched gradient fields, 336–337
switching-mode current amplifier, 244
switch-mode amplifiers, 275

TE_{01} mode resonators, 150
temperature supervision, 236–241
transmit-only volume coils, 115–116
transmit/receive switches, 313–316
Transmit SENSE, 286, 290
transmitter coil, 7–10
transverse electromagnetic mode (TEM) resonators, 107–108
transverse magnetization, 15
 dephasing, crusher gradients, 21–22
trap circuits, 124
travelling wave MRI, antennae for, 150–155
triple quantum filtered (TQF) COSY, 23
two-dimensional inverse Fourier transform, 33
two-dimensional proton-nitrogen NMR spectrum, 2
two-loop coil geometry, 38–39
 homogeneous magnetic field, 39
 linear *z*-gradient, 38–39

ultra-strong whole-body gradients, 258–262
undersampling, 321–322

variable field characteristics, 250–255
voltage standing wave ratio (VSWR), 156, 280

waveform diagram, amplifier, 277
whole-body modular gradient, 250–255
wireless data transmission, 328–329

zonal harmonics, 42
zonal magnetic field, 42